CW00920065

Symmetry Principles
in Solid State and Molecular Physics

Melvin Lax

Distinguished Professor of Physics
City College of the City University of New York

and

Former Member of the Technical Staff
Now Consultant
Bell Laboratories
Murray Hill, New Jersey

DOVER PUBLICATIONS, INC.
Mineola, New York

Bibliographical Note

This Dover edition, published in 2001, is an unabridged republication of the original edition published by John Wiley & Sons, Inc., New York, in 1974.

Library of Congress Cataloging-in-Publication Data

Lax, Melvin J.
Symmetry principles in solid state and molecular physics / Melvin Lax.
p. cm.
Originally published: New York : J. Wiley, 1974.
Includes bibliographical references and index.
ISBN-13: 978-0-486-42001-1
ISBN-10: 0-486-42001-9
1. Group theory. 2. Solid-state physics. 3. Molecules. 4. Symmetry (Physics) 5. Lattice theory. I. Title.

QC174.5 .L38 2001
530.4'1—dc21

2001028860

Manufactured in the United States by LSC Communications
42001906 2018
www.doverpublications.com

PREFACE

This text is one of six drafts of lecture notes prepared in connection with courses given at Bell Laboratories (1960,1962,1967), Princeton (1961), Oxford (1961–1962), Trieste (1967), and City College of New York (1972). The first draft proved all the fundamental theorems of group theory and then proceeded to the applications. The second draft was a complete revision in which five key theorems are stated without proof in the first chapter but all subsequent results are derived. Succeeding drafts followed the latter plan.

The second approach has the advantage of leading more directly to the applications and avoids the possibility of spending a long period on the fundamental (and abstract) theorems of group theory. These theorems can be found in many textbooks on group theory, but the way to apply them to solid state and molecular physics is less readily available and constitutes the primary contribution of this book.

I do *not* recommend that the student first attempt to learn the proofs of these fundamental theorems. This textbook is analogous to an automobile driving course. It would not be profitable to learn first the theory of internal combustion engines. Discussion (by the text) of applications of group theory to solids and molecules (driving techniques), followed by the solution of many problems (driving practice), will lead to understanding. At this point, the more mathematical reader may have acquired the motivation to learn the proofs of the fundamental theorems.

Chapter 1 describes the relationship between group theory and quantum (or classical) mechanics, between classes and observables, between representations and states.

The reader should not be deterred by the brevity and abstractness of Chapter 1 from proceeding to Chapter 2, where he will become acquainted with the point groups: the full rotation group, the crystallographic point groups, and their associated double groups (needed for problems with spin).

In Chapter 3 we are in a position, for the first time, to demonstrate the power of group theory by a series of examples including crystal field theory, with and without spin, decomposition of angular momenta, and selection rules. The use of character tables is illustrated.

The first three chapters constitute a distillation of a complete course in group theory (at the point group level) retaining only the most relevant theorems, ideas, and applications. The reader should proceed to the easier chapters which follow, returning to Chapters 1–3 for relevant points as they occur.

Chapter 4 shows how to determine the form and number of constants in the macroscopic tensors of crystals (1) by inspection (the layman's method) where easy, and (2) by character table and invariance methods where needed.

Chapter 5 treats molecular vibrations in detail and works out the vibrations of the ammonia (NH_3) molecule as a full-scale example. This problem requires the diagonalization of a 4 atoms $\times$ 3 dimensions $= 12 \times 12$ matrix. Symmetry vectors and the base vector orthogonality theorem of Chapter 3 are used to demonstrate that no matrix larger than 2×2 need be diagonalized. Such is the power of group theory.

Chapter 6 introduces the reader to the translational properties of crystals and their relation to the point group symmetry of a given crystal. Topics include Bravais lattices, reciprocal lattices, Brillouin zones, and X-ray diffraction.

Chapter 7 discusses electronic energy bands, the reduced and extended zone schemes, and plane wave and $\mathbf{k} \cdot \mathbf{p}$ methods. The dynamics of electron motion and the effective mass description of impurity states are also covered.

Chapter 8 provides the first detailed discussion of space groups (single and double). The nature of the irreducible representations of space groups is discussed. A reduction to much smaller multiplier groups is presented, as well as an algebraic treatment of the latter.

Chapter 9 presents a variety of space group examples, including (1) the form and number of constants in various microscopic tensors, (2) selection rules for photon and phonon transitions, and (3) compatibility relations between states at $\mathbf{k}=0$ and states at small $\mathbf{k}$.

Chapter 10 presents a detailed discussion of time reversal with and without spin. The Frobenius-Schur criterion as to whether time reversal causes additional degeneracy is (*a*) generalized and (*b*) used to derive the associated Herring criterion for space groups. With the help of the generalized mobility theory of Appendix H, the Onsager relations are derived. The modification of selection rules due to time reversal is an original contribution.

Chapter 11 provides a detailed discussion of lattice vibrations: the use of symmetry coordinates, normal coordinates, and the consequences of displacement and rotational invariance, space group symmetry, and time reversal. The connection with elasticity is provided via the method of long waves.

The vibrations of the diamond structure presented in Chapter 12 constitute a full-scale illustration of the methods of applying group theory. Moreover, the presence of screw axes in diamond (resulting in a nonsymmorphic space group) leads to results unexpected on the basis of point group intuition.

In Chapter 13 we return to molecular physics and deal with molecular and valence bond orbitals and the coupling to vibrations via the Jahn-Teller effect. The relationship between many electron wave functions, resonance, and chemical structure is described. In addition, the concept of broken symmetry is related to open shells.

Each chapter has a summary. These summaries, together with the table of contents, provide a detailed survey of the subject matter covered in this book.

After the first draft, the organizational structure did not change but the text was revised and enlarged at the places at which students encountered difficulty. I am indebted not only to these many generations of students but also to E. I. Blount and G. A. Baraff, who taught courses based on these notes at Bell Laboratories, and to B. Segall, who did the same at Case Western Reserve University.

No attempt was made to be complete in the references, which were construed not as a means of establishing priorities but as supplementary reading for students. There was also no effort to add references with each pedagogical revision. Thus the references will appear to be old even if they were not intended to be the *original* references on all subjects. Fortunately the techniques of group theory have grown slowly enough to permit this cavalier treatment even though the fields of application, solid state and molecular theory, have burgeoned.

The encouragement of Bell Laboratories department heads Conyers Herring, Peter Wolff, John Klauder, and Evan Kane, the cheerful services of many secretaries, including Barbara Chippendale, Dot Putyrske, Pauline Potempa, Marian Riley, Dot Luciani, and Joan Lemberger, and the patience of my family helped to make this manuscript possible.

MELVIN LAX

November 1973
New York, New York

CONTENTS

chapter one

RELATION OF GROUP THEORY TO QUANTUM MECHANICS

The aim in a problem in quantum mechanics is to diagonalize the Hamiltonian H. In the presence of degeneracy, the states can be *classified* only by means of quantum numbers in addition to the energy. These are eigenvalues of other operators that commute with H (and with each other). The existence of operators that commute with H reflects *symmetry* properties of the physical system and gives rise to the observed degeneracy. In this chapter we discuss the nature of such symmetry operations, show that they can be used to simplify the solution of a problem, and demonstrate that these operations invariably form a *group*.

Since the elements of such a group do not necessarily commute with one another, they cannot all be simultaneously diagonalized. We show, however, that the elements of a group can be divided into *classes* such that the *sum* of the elements in each class provides us with a set of operators Ω_c, one for each class C, that do form a commuting set. The diagonalization of these "Dirac characters" Ω_c will be shown to be essentially equivalent to the determination of the *irreducible representations* of the group, and the eigenvalues of the Ω_c will be simply related to the *characters* of these representations, usually available in tables. We therefore define irreducible representations and summarize without proof here the key theorems about them that are needed for the applications. Such theorems, available in most textbooks on group theory, will be designated by a superscript$^{\|}$.

1.1. SYMMETRY OPERATIONS

The symmetry operations relevant to a classical (quantum-mechanical) problem are those that leave invariant (commute with) the Hamiltonian of the physical system. Examples of such operations are the following:

1. Rotations and reflections that transform an object—a molecule or crystal—into a new configuration that is not distinguishable from the original.
2. Discrete lattice translations for a crystal.
3. Permutations among identical particles, for example, electrons or nucleons.
4. Time reversal (we shall see that in the absence of spin this means taking the complex conjugate of the Hamiltonian, so that invariance under this operation means the Hamiltonian is real).
5. Infinitesimal displacements and rotations of the system as a whole. These operations impose restrictions on the force constants between atoms in a crystal to ensure free translation and rotation of the system (conservation of linear and angular momentum).
6. Arbitrary rotations in spin space, which are permitted only when no spin-orbit coupling is present.
7. Rotations in isotopic spin space. In nuclear physics, the charge independence of nuclear forces implies that the Hamiltonian is invariant under such operations.

Simplifications Produced by Symmetry

The diagonalization of a Hamiltonian (i.e., the calculation of its eigenfunctions) often requires detailed calculations. These can be greatly facilitated if one restricts oneself to wave functions of the correct symmetry. This can be accomplished by *first* diagonalizing one or more symmetry operators (this is always possible, since the latter commute with the Hamiltonian).

Example 1.1.1

An atom invariant under arbitrary rotations has a Hamiltonian that commutes with arbitrary infinitesimal rotations. In the absence of spin, these rotations can be described in terms of the three independent infinitesimal generators L_x, L_y, L_z, the three components of the angular momentum operator (see Section 2.1 for proof). These operators do not commute with one another and hence cannot *all* be diagonalized simultaneously. However, any one of them, say L_z, commutes with L^2. It is customary then to diagonalize L^2 and L_z. For one electron in a spherical potential the eigenfunctions of these operators are the well-known spherical harmonics.

If we write

$$\mathbf{L}=\mathbf{r}\times\mathbf{p}=\frac{\hbar}{i}(\mathbf{r}\times\nabla)=\hbar\mathbf{l} \qquad (1.1.1)$$

then

$$\mathbf{l}^2 Y_m^l(\theta,\varphi)=l(l+1)\,Y_m^l(\theta,\varphi) \qquad (1.1.2)$$

$$l_z Y_m^l(\theta,\varphi)=mY_m^l(\theta,\varphi) \qquad (1.1.3)$$

where

$$Y_m^l(\theta,\varphi)\propto P_m^l(\cos\theta)\exp(im\varphi)$$

These results are usually obtained in an introductory quantum-mechanics course by the method of separation of variables in spherical coordinates. This procedure succeeds because the Laplacean (i.e., the kinetic energy) separates into a radial and an angular part:

$$\nabla^2=\left(\frac{\partial}{\partial r}+\frac{1}{r}\right)^2+\frac{1}{r^2}(\mathbf{r}\times\nabla)^2 \qquad (1.1.4)$$

The diagonalization of the angular part is essentially the diagonalization of $l^2=-(\mathbf{r}\times\nabla)^2$. The solution is then written in the form

$$\psi_m^l(\mathbf{r})=R_m^l(r)\,Y_m^l(\theta,\varphi) \qquad (1.1.5)$$

The symmetry information has been exhausted in determining the angular part. The radial factor $R_m^l(r)$ depends on the details of the potential and must be computed by variational, numerical, or other techniques.

Example 1.1.2

In this example we use symmetry arguments to simplify the solution of a simple vibrational problem. Consider the arrangement of springs and masses shown in Fig. 1.1.1. We shall not analyze the equations of motion, but instead shall leave such analysis as an exercise.[1] We restrict our

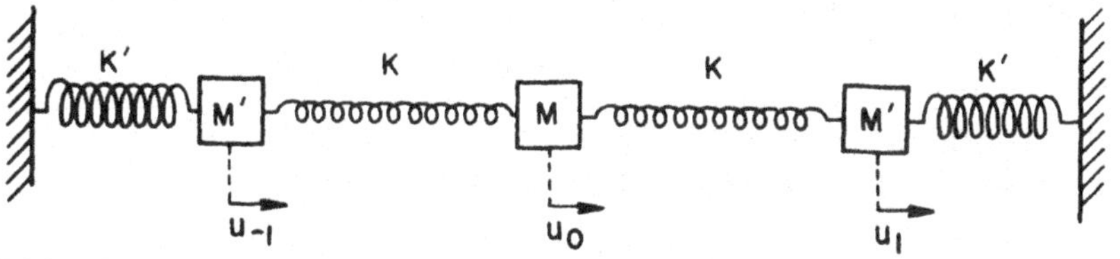

Fig. 1.1.1. Vibrations of a three-mass system that has inversion symmetry about the central atom.

[1] A similar problem with identical symmetry properties is discussed on p. 333 of Goldstein (1950).

discussion to the properties of the solutions that we can determine solely from the inversion symmetry, which is apparent from Fig. 1.1.1. Displacements to the right are regarded as positive. Under the inversion operation I, the (total) *position* vectors of the masses are reversed in sign; this reverses the sign of the *displacements* from equilibrium and measures these displacements from the "inverse atom":

$$\begin{aligned} u_1' &= Iu_1 = -u_{-1} \\ u_0' &= Iu_0 = -u_0 \\ u_{-1}' &= Iu_{-1} = -u_1 \end{aligned} \tag{1.1.6}$$

Since the energy or Hamiltonian is invariant under inversion, (i.e., the physical system is restored to itself by inversion), it follows that the solutions can be chosen as eigenfunctions of the inversion operator. Since $I^2 = 1$, as shown by Eq. 1.1.6, $+1$ and -1 are the possible eigenvalues of I. Thus we diagonalize I by choosing the displacements to be even or odd. For the even "parity" case, that is, the solution even under inversion,

$$I[u_1, u_0, u_{-1}] = [u_1, u_0, u_{-1}]$$

Comparison with Eq. 1.1.6 shows that $u_0 = 0$, $u_1 = -u_{-1} \equiv a$. Thus there exists a solution with only one parameter of the form

$$[u_1, u_0, u_{-1}] = [a, 0, -a]$$

for which the right-hand mass M' is displaced to the right, say, the left-hand mass M' is displaced an equal amount to the left, and the center mass M is at equilibrium. This particular solution is determined by symmetry arguments alone!

Query[2] 1.1.1

Under what circumstances is a solution completely determined by symmetry?

Solutions of the coupled mass problem that are "odd" under I obey

$$I[u_1, u_0, u_{-1}] = -[u_1, u_0, u_{-1}]$$

which, on comparison with Eq. 1.1.6, states that $u_1 = u_{-1}$ with no restriction placed on u_0. The general odd solution then is of the form

$$[u_1, u_0, u_{-1}] = [a, b, a]$$

[2]Queries are designed to stimulate thought. They are usually answered later in the text when the necessary background has been developed.

The dynamical problem has been simplified to the solution of two simultaneous equations in two unknowns. The ratio b/a for the two "odd" modes depends on the mass ratio M'/M and the force constant ratio K'/K, that is, on the solution of a dynamical problem.

These examples illustrate the role of group theory. By first diagonalizing certain operators associated with a group (the Dirac characters of the group), the problem is decomposed into parts each of which involves solutions of only one symmetry type, for example, p waves in an atomic problem, or odd modes in a vibrational problem. (We shall indeed prove in the base vector orthogonality theorem 3.4.3 that the Hamiltonian does not produce interactions between modes of different symmetry.) The interaction between modes ("symmetry coordinates") of the same symmetry is not a matter for group theory but depends on the detailed nature of the Hamiltonian. We show, however, in Theorem 3.4.3 that, for modes of a given type, it is never *necessary* to diagonalize a matrix larger than the number of times that a mode of a given symmetry appears in the space of the problem (e.g., there are two odd modes in the three-dimensional space of Example 1.1.2). The determination of this number is a purely group-theoretical problem referred to as "finding the number of *repetitions* of a given *irreducible representation* in a given, possibly reducible space." This problem, which emphasizes for us the importance of irreducible representations (defined in Section 1.4), is solved and illustrated in Section 3.1.

Symmetry Operations Form a Group

A group G is defined as a set of *distinct* elements $E, A, B, C, \ldots$ with a rule referred to as multiplication for combining an ordered pair of elements so that the following conditions hold:

The product of any two elements A and B, denoted by AB, *exists* as a member of the set and is *unique*. (1.1.7a)

Multiplication is associative:

$$A(BC) = (AB)C \tag{1.1.7b}$$

The group contains an identity element E such that

$$EA = AE = A \tag{1.1.7c}$$

for all A in the set.

Every element A possesses an inverse in the set:

$$A^{-1} \in G \tag{1.1.7d}$$

with A^{-1} defined by $A^{-1}A = AA^{-1} = E$. (Read $\in$ as "exists in" or "is a member of the set.")

A *subgroup* is a subset of elements of a group that themselves obey the group requirements of Eqs. 1.1.7.

Under a symmetry operation A, the Hamiltonian H, which might have the form $H(\mathbf{r},\mathbf{p})$, is transformed into $H' = A^{-1}HA$, which would have the form $H(\mathbf{r}',\mathbf{p}')$, where $\mathbf{r}' = A^{-1}\mathbf{r}A$ and $\mathbf{p}' = A^{-1}\mathbf{p}A$ are the rotated position and momentum operators, as in Eq. 2.5.11.

We can now establish a theorem.

Theorem 1.1.1

The set of symmetry operations that commute with a Hamiltonian H forms a group.

Proof: The symmetry operations of quantum mechanics are automatically associative, and the identity is clearly among them. If A and B commute with H, then $A^{-1}HA = B^{-1}HB = H$.

$$(AB)^{-1}H(AB) = B^{-1}(A^{-1}HA)B = B^{-1}HB = H \tag{1.1.8}$$

Thus a, b, and c are satisfied. If $A^{-1}HA = H$, then $H = AHA^{-1}$, so that A^{-1} is a member of the set.[3]

1.2. ABSTRACT GROUP THEORY

Abstract group theory is a study of those properties of groups that are independent of the nature of the elements. The *order* g of an abstract group G is the number of elements in the group. Each element A of a *finite* group must reduce to the identity when raised to some finite power:

$$A^p = E \tag{1.2.1}$$

and p is the *order* of the *element* A. A group whose elements commute with one another is *Abelian*. A group consisting only of the elements

$$E, A, A^2, A^3, \ldots, A^{g-1} \tag{1.2.2}$$

is *cyclic* as well as Abelian.

The only groups of order 1, 2, 3, 5, and 7 are cyclic.

Query 1.2.1

Are all groups of prime order cyclic?

There are only two groups of order 4. One is cyclic. The other has elements E, A, B, C that obey

$$A^2 = B^2 = E, \qquad AB = BA = C \tag{1.2.3}$$

[3]The existence of A^{-1} is established in Section 10.4, Theorem 10.4.6.

and hence is Abelian. Equation 1.2.3 is typical of the representation of a group in terms of its generating elements. (See Problem 1.2.1.)

An abstract group is completely characterized by its multiplication table. For example, the non-Abelian group of lowest order is represented by the following table:

	E	A	B	C	D	F
E	E	A	B	C	D	F
A	A	E	D	F	B	C
B	B	F	E	D	C	A
C	C	D	F	E	A	B
D	D	C	A	B	F	E
F	F	B	C	A	E	D

(1.2.4)

An entry within the table represents the product of the row-heading entry and the column-heading entry in that order, for example, $AD=B$. Note that in every row or column of the table each element appears once and only once, for $AC \neq AD$ unless $C=D$ (as we can establish by multiplying by A^{-1} and using Eq. 1.1.7d).

We have thus established the *group rearrangement theorem.*

Theorem 1.2.1

If G is a set of elements that form a group, then AG is a rearrangement of this set so that

$$\sum_{R\in G} f(R) = \sum_{R\in G} f(AR) \tag{1.2.5}$$

where f is an arbitrary function and A is a group element.

We may use the multiplication table of a group to express *all elements* in terms of a small number of elements known as *generators*, for example, from multiplication table 1.2.4,

$$B=AD, \qquad C=AD^2, \qquad F=D^2 \tag{1.2.6}$$

The group can then be defined most succinctly by stating the relations obeyed by the generators, for example,

$$A^2=E, \qquad D^3=E, \qquad DA=AD^2 \tag{1.2.7}$$

The last relation permits D to be moved to the right so that any element can be written in the form A^mD^n. The finite order of each generator then causes the group of all elements A^mD^n to be finite.

Two groups are said to be *isomorphic* if they can be arranged to have the same multiplication table. In other words, $G_1=(E,A_1,A_2,\ldots)$ and $G_2=(E',B_1,B_2,\ldots)$ have a one-to-one correspondence $A_i \leftrightarrow B_i$ such that, if $A_iA_j=A_k$, then $B_iB_j=B_k$ and vice versa.

Nonabstract groups that are isomorphic have the same structure and are said to be *realizations* of the same abstract group. But they assign different meanings to the group elements.

Example 1.2.1

The following groups are isomorphic to the non-Abelian group of order 6 with multiplication table 1.2.4:

(*a*) The group C_{3v} of rotations and reflections which leave an equilateral triangle invariant (see Section 2.14 and Fig. 1.2.1).

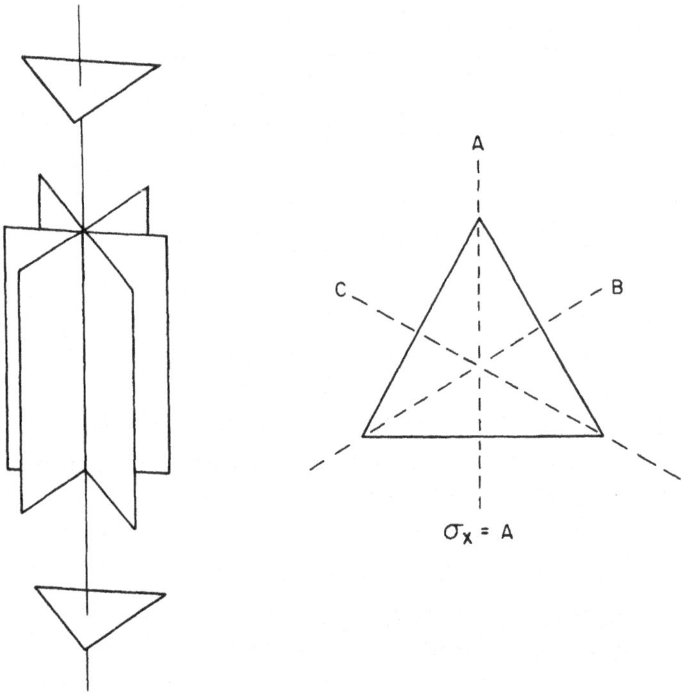

Fig. 1.2.1. The group C_{3v}. This group, the symmetry group of an equilateral triangle, contains one threefold axis (120° rotation indicated by a triangle △) and three reflection planes, A, B, and C. With σ denoting a reflection plane, $A=\sigma_x$ is a plane that takes x into $-x$ in agreement with the matrix representation in Example 1.2.1.

(*b*) The group of permutations of three objects.
(*c*) The six matrices

$$E=\begin{bmatrix} 1 & 0 \\ 0 & 1 \end{bmatrix}, \quad A=\begin{bmatrix} -1 & 0 \\ 0 & +1 \end{bmatrix}, \quad B=\begin{bmatrix} +\frac{1}{2} & \frac{1}{2}\sqrt{3} \\ \frac{1}{2}\sqrt{3} & -\frac{1}{2} \end{bmatrix}$$

$$C=\begin{bmatrix} +\frac{1}{2} & -\frac{1}{2}\sqrt{3} \\ -\frac{1}{2}\sqrt{3} & -\frac{1}{2} \end{bmatrix}, \quad D=\begin{bmatrix} -\frac{1}{2} & -\frac{1}{2}\sqrt{3} \\ +\frac{1}{2}\sqrt{3} & -\frac{1}{2} \end{bmatrix}, \quad F=\begin{bmatrix} -\frac{1}{2} & +\frac{1}{2}\sqrt{3} \\ -\frac{1}{2}\sqrt{3} & -\frac{1}{2} \end{bmatrix}$$

which represent the rotations and reflections of C_{3v} in the x-y plane (see Fig. 1.2.1) in the sense that

$$\begin{bmatrix} x' \\ y' \end{bmatrix} = (R)\begin{bmatrix} x \\ y \end{bmatrix}$$

1.3. COMMUTING OBSERVABLES AND CLASSES

Theorem 1.3.1

If S is an element of group G, and Y is any *object* whose multiplication with group elements R is defined, then

$$A = \sum_{R \in G} RYR^{-1} \tag{1.3.1}$$

commutes with all group elements:

$$[S,A]=0 \tag{1.3.2}$$

Proof:

$$S\sum RYR^{-1} = \sum (SR)Y(SR)^{-1}S$$

By the group rearrangement theorem 1.2.1, the right-hand sum is simply AS. Q.E.D.

Choose Y = a group element T. Two group elements T and U are said to be *conjugate* ($T \approx U$) with respect to G if there exists an element R in G such that

$$RTR^{-1} = U \tag{1.3.3}$$

Since conjugation is reflexive ($T \approx T$), symmetrical (if $T \approx U$, then $U \approx T$), and transitive (if $T \approx U$ and $U \approx W$, then $T \approx W$), *the elements of a group can be divided into classes* such that the elements in each class are the conjugates of one another. See Problems 1.3.1–1.3.3.

Lest the reader think that class is an abstruse concept, we remark that in a group of rotations about a common point all the elements in a given class will involve rotations through the same angle; see Problem 1.3.3. However, two rotations T and U through the same angle about different axes are in the same class if and only if the group contains a rotation R that rotates one axis into the other, for then $U = RTR^{-1}$.

The *Dirac character* Ω_c of a class C is defined as the *sum* of the elements in that class:

$$\Omega_c \equiv \sum_{T \in C} T \tag{1.3.4a}$$

a sum of n_c distinct elements, where $n_c \equiv$ the number of elements in class C.

Class Rearrangement Theorem 1.3.2

A "conjugation" transformation RTR^{-1} of the elements T of a class simply rearranges the class:

$$R\Omega_c R^{-1} = \Omega_c$$

Proof: $R\Omega_c R^{-1}$ contains a sum of n_c distinct terms, all in the same class C, that is, precisely Ω_c. *Thus the Dirac characters Ω_c commute with all elements of the group and with each other.*

If $U \in C$ and $V = SUS^{-1}$ is any other element of C, then

$$\sum_R RVR^{-1} = \sum_R (RS)U(RS)^{-1} = \sum_R RUR^{-1}$$

by the group rearrangement theorem. Thus all elements U in the same class C yield the same result.

$$\sum_R RUR^{-1} = \frac{1}{n_c} \sum_{U \in C} \sum_R RUR^{-1}$$

$$= \frac{1}{n_c} \sum_R R\Omega_c R^{-1} = \frac{1}{n_c} \sum_R \Omega_c = \frac{g}{n_c} \Omega_c$$

Thus we can equally well define the Dirac characters by

$$\Omega_c = \frac{n_c}{g} \sum_{R \in G} RUR^{-1} \tag{1.3.4b}$$

By permitting the addition of group elements as well as multiplication we have entered the domain of *algebra*. The most general element of our group algebra is then a linear combination of the form

$$B = \sum a(T)T \tag{1.3.5}$$

where the $a(T)$ are complex numbers, and the T are elements.

Theorem 1.3.3

The most general element of our group algebra that commutes with each element of the group is a linear combination of the Dirac characters of the group.

Proof: Take RBR^{-1}, replace T by $R^{-1}TR$, and compare the result with B. Equality of the coefficients of T yields $a(R^{-1}TR) = a(T)$. Thus $a(T)$ is a class function (the same for all elements in a class). Q.E.D.

An immediate consequence is the following theorem.

Class Multiplication Theorem 1.3.4

$$\Omega_c\Omega_d = \sum h_{cde}\Omega_e \tag{1.3.6}$$

since $\Omega_c\Omega_d$ also commutes with the group elements.

Example 1.3.1

The group C_{3v} of Example 1.2.1 obeys multiplication table 1.2.4. Its class structure is determined in Problem 1.3.6. With $\Omega_1 = E$, $\Omega_2 = D + F$, $\Omega_3 = A + B + C$, $\Omega_3\Omega_2 = (A + B + C)(D + F) = AD + AF + BD + BF + CD + CF = 2(A + B + C) = 2\Omega_3$, that is, $\Omega_3\Omega_2 = 2\Omega_3$. Similarly, $\Omega_3^2 = 3(\Omega_1 + \Omega_2)$, $\Omega_2^2 = (2\Omega_1 + \Omega_2)$. Since the eigenvalues of a set of commuting operators obey any algebraic equation satisfied by the operators, we can, setting the eigenvalue of $\Omega_1 = 1$, obtain the following possible sets Γ_j of eigenvalues:

	Ω_1	Ω_2	Ω_3
Γ_1	1	2	3
Γ_2	1	2	−3
Γ_3	1	−1	0

(1.3.7)

In summary, then, the Dirac characters constitute a set of commuting observables that can be diagonalized simultaneously with, or, preferably, in advance of the Hamiltonian. The matrices that must be diagonalized to accomplish this depend on the detailed nature of the physical problem (see Section 5.2 for an application of this technique to molecular vibrations). But the eigenvalues are simply a consequence of the group algebra. These eigenvalues are then quantum numbers that characterize the symmetry properties of the state. For the full rotation group, the specification of the Ω's is simply the specification of L^2.

The algebra whose elements are given by Eq. 1.3.5 is a Frobenius algebra and is a special case of a general linear algebra with

$$RS = \sum_{T} a(R,S,T)\,T \tag{1.3.8}$$

where $a(R,S,T)$ are numbers that obey conditions compatible with the requirement that multiplication be associative. We shall be primarily interested in the case

$$RS = \lambda(R,S)\,T \tag{1.3.9}$$

that is, group multiplication with a numerical factor. We shall refer to such an algebra as a "multiplier group." This case is important in the presence of spin, in which case $\lambda = \pm 1$, and we obtain the *double groups*. It is also important for treating representations at the zone boundary in space groups, as discussed in Chapter 8. Equation 1.3.9 with all $\lambda = 1$ defines an ordinary group with which this particular multiplier group is associated. In order to be able to discuss these two groups simultaneously, let us replace the R,S,T notation of 1.3.9 by $O(R), O(S), O(T)$ and reserve R,S,T for ordinary group elements. Then the close correspondence between the two groups can be characterized by rewriting Eq. 1.3.9 in the form

$$O(R)O(S) = \lambda(R,S)\,O(RS) \tag{1.3.10}$$

The above theorems regarding groups generalize immediately to multiplier groups. It can be shown, for example, that

$$\left[\sum_{R\in G} O(R)\,Y[O(R)]^{-1}, O(S)\right] = 0 \tag{1.3.11}$$

so that the Dirac characters can now be written as

$$\Omega_c = \frac{n_c}{g}\sum_{R\in G} O(R)O(T)[O(R)]^{-1} \tag{1.3.12}$$

where n_c is the number of elements in the class of the underlying group, and $[O(R)]^{-1}$ does not necessarily equal $O(R^{-1})$. A detailed discussion of multiplier groups is given in Sections 8.7–8.10.

1.4. REPRESENTATIONS AND IRREDUCIBLE REPRESENTATIONS

Although other kinds of representations exist, we are concerned here only with *matrix representations* of groups. From a mathematical point of view, a matrix representation of a group (or algebra) is any set of matrices $D(R)$ that obey the multiplication table of the group (algebra):

$$D(R)D(S) = D(RS) \tag{1.4.1}$$

or

$$D(R)D(S) = \lambda(R,S)D(RS) \tag{1.4.2}$$

If X is *any* square matrix with the same dimension as $D(R)$, the set of matrices $XD(R)X^{-1}$ obeys the same algebra (1.4.2) as $D(R)$ and is called an *equivalent representation.* By suitable choice of X, the above *similarity* transformation can be used to simplify the representation. If the algebra is Abelian, the $D(R)$ commute and can be simultaneously diagonalized by a suitably chosen X. For non-Abelian algebras, the most that we can hope to do is to introduce one or more zeros in the *same position* in the complete set of matrices $D(R)$. If this is possible, the matrices are said to be reducible; otherwise they are irreducible. In the reducible case, we can arrange to place a set of zeros in the lower left corner:

$$D(R) = \begin{bmatrix} A & 0 \\ 0 & B \end{bmatrix} \quad \text{or} \quad \begin{bmatrix} A & C \\ 0 & B \end{bmatrix} \tag{1.4.3}$$

In Eq. 1.4.3 A, B, C, and 0 are matrices, the last having only zero elements. If the matrices $D(R)$ are unitary, zeros must then appear automatically in the upper right corner as well, and we say that the matrices are fully reducible. Since the representations of a finite group can be made unitary (see Problem 1.5.5), representations of such groups if reducible are fully reducible.

After a sequence of reductions, our matrix representation can therefore be brought to the form

$$D(R) = \left[\begin{array}{c|c|c} D^1(R) & 0 & 0 \\ \hline 0 & D^2(R) & 0 \\ \hline 0 & 0 & D^3(R) \end{array}\right] \tag{1.4.4a}$$

The information contained in representation 1.4.4a can be written more compactly as

$$D(R) = D^1(R) \oplus D^2(R) \oplus D^3(R) \tag{1.4.4b}$$

where the symbol $\oplus$ designates the *direct sum*, which is clearly distinct from the usual sum of matrices. *Each* set of matrices $D^j(R)$, as R runs through the group, constitutes an irreducible representation of the group; the j is a numerical index that names the representation.

Construction of Representations Using Functions as Bases

Suppose that we have a group G of symmetry operators $E, A, B, C, \ldots$ and a function ψ on which these operators can act. [For the sake of visualization $\psi = \psi(\mathbf{r})$ can be a function of the coordinates of one particle, and the symmetry operations can be rotations whose action is described in detail in Section 2.1.] The set of g functions $E\psi, A\psi, B\psi, \ldots$ is *invariant* under the group G in the sense that the action of any $R \in G$ merely rearranges this set of functions. Indeed the space ρ of all functions

$$\varphi = c_1 E\psi + c_2 A\psi + c_3 B\psi + \cdots \tag{1.4.5}$$

(where the c's are complex numbers) is *invariant* under G in the sense that, for any $R \in G$, $R\varphi$ belongs to the same space. Since the functions $E\psi, A\psi, B\psi, \ldots$ may not all be independent, we can always choose a (possibly smaller) set of *independent* functions ψ_μ, $\mu = 1, 2, \ldots, l \leqslant g$, that are *complete* in the sense that all φ in ρ are expressible as a linear combination of the ψ_μ:

$$\varphi = \sum c_\mu \psi_\mu \tag{1.4.6}$$

Thus the ψ_μ are a basis (or set of "basis vectors") in the space ρ and are said to "span" ρ. Since ρ is *invariant* under G, any element S of G takes ψ_ν into another vector in ρ that, by the completeness of the basis vectors ψ_μ, is expressible in the form

$$S\psi_\nu = \sum \psi_\mu D_{\mu\nu}(S) \tag{1.4.7}$$

The coefficients have been written as $D_{\mu\nu}(S)$ rather than $S_{\mu\nu}$ to indicate that they can be regarded as a function of the group element S.

Theorem 1.4.1

Any set of basis vectors ψ_μ that span a space ρ invariant under a group G leads to a set of matrices $D_{\mu\nu}(S)$ via Eq. 1.4.7 that constitute a matrix representation of the group.

Proof: We need only show that the operator RS is represented by the matrix $\mathbf{D}(R)\cdot\mathbf{D}(S)$. We apply R to Eq. 1.4.7; then

$$RS\psi_\nu = \sum (R\psi_\mu) D_{\mu\nu}(S)$$

$$= \sum \psi_\lambda D_{\lambda\mu}(R) D_{\mu\nu}(S)$$

$$= \sum \psi_\lambda D_{\lambda\nu}(RS)$$

so that

$$D_{\lambda\nu}(RS) = \sum D_{\lambda\mu}(R) D_{\mu\nu}(S)$$

as promised. (Note that the first step of this proof assumes that R is a linear operator. See Chapter 10 for a discussion of this point.)

The proof just given uses the completeness of the ψ_μ but not their orthonormality. If the basis vectors are chosen to be *orthonormal*, Eq. 1.4.7 can be inverted to yield

$$D_{\mu\nu}(S) = (\psi_\mu, S\psi_\nu) \qquad (1.4.8)$$

where the scalar product in Eq. 1.4.8 is defined in the usual way as

$$(\varphi,\psi) = \int \varphi^* \psi \, d\tau \qquad (1.4.9)$$

the integral over all arguments τ in ψ or φ of the complex conjugate of the first function times the second.

Example 1.4.1

Construct representations of the group C_{3v}, using various choices $\psi(\mathbf{r})$ as in Eqs. 1.4.5–1.4.7.

Solution: (*a*) Since C_{3v} involves (only) rotations about the z axis or reflections containing this axis, r and z are invariant under the group operations. Thus for any ψ of the form $f(r,z)$ the space of all $R\psi$ consists of the single function $f(r,z)$, and $S\psi = \psi$ leads to the identity representation, that is, $D_{\mu\nu}(S) = \delta_{\mu\nu}$ in Eq. 1.4.7.

(*b*) Let $\psi = x \exp(-r)$, an atomic p state. The invariant factor $\exp(-r)$ can be omitted from the subsequent discussion. The rotations and reflections of C_{3v} take x into some linear combination of x and y. Thus the invariant space ρ of Eq. 1.4.5 is spanned by the two basis vectors $\psi_1 = x$, $\psi_2 = y$. Under the reflection $\sigma_x \equiv A$ (see Example 1.2.1 and Fig. 1.2.1) a point $\mathbf{r} = (x,y)$ in the x-y plane is taken into

$$\mathbf{r}' = (x',y') = (-x,y) \quad \text{or} \quad \begin{bmatrix} x' \\ y' \end{bmatrix} = \begin{bmatrix} -1 & 0 \\ 0 & 1 \end{bmatrix} \begin{bmatrix} x \\ y \end{bmatrix}$$

or

$$\mathbf{r}' = \mathbf{A}\mathbf{r} \quad \text{with } A = \begin{bmatrix} -1 & 0 \\ 0 & 1 \end{bmatrix}$$

Under a counterclockwise 120° rotation $C_3 \equiv D$ we obtain (cf. Eqs. 2.5.9 and 2.5.16)

$$\mathbf{r}' \equiv \mathbf{D}\mathbf{r} \quad \text{with } \mathbf{D} = \begin{bmatrix} -\frac{1}{2} & -\frac{1}{2}\sqrt{3} \\ \frac{1}{2}\sqrt{3} & -\frac{1}{2} \end{bmatrix}$$

We can verify easily that the matrices just found for A and D obey the defining relations 1.2.7 of the group generators, in particular, the nontrivial relation $DA = AD^2$. Thus the remaining matrices computed from the generators via Eq. 1.2.6 will, of necessity, yield a representation of the group. The resulting two-dimensional representation is that given in Example 1.2.1. Since no two elements of the group are represented by the same matrix, this representation is called *faithful.*

Strictly speaking, we have obtained the above matrix representation through tha action of rotations on the position operator $\mathbf{r}$ as in Eq. 2.5.16. A consistent definition in terms of wave functions is provided by Eqs. 2.5.15 and 2.5.17:

$$S\psi_\nu(\mathbf{r}) = \psi_\nu(\mathbf{rS}) = \Sigma\psi_\mu(\mathbf{r})D_{\mu\nu}(S) \tag{1.4.10}$$

where $\mathbf{S}$ is the 2×2 or 3×3 defining representation of the rotation. It appears as a postmultiplicative factor in order that

$$RS\psi_\nu(\mathbf{r}) = R\psi_\nu(\mathbf{rS}) = \psi_\nu((\mathbf{rR})\mathbf{S}) = \psi_\nu(\mathbf{rRS}) \tag{1.4.11}$$

that is, so that the product of two operators will be represented by the product of the corresponding matrices in the same order. A complete discussion of rotation operators is given in Chapter 2.

(*c*) Let $\psi = x^2$, an atomic d state with the factor exp $(-r)$ understood. The rotations and reflections of C_{3v} do not remove a vector from the x-y plane. All $R\psi$ are homogeneous polynomials of degree 2 and are therefore expressible in terms of $\psi_1 = x^2$, $\psi_2 = y^2$, $\psi_3 = 2xy$. Applying the transformations

$$\mathbf{r}' = \mathbf{rA} \qquad \text{and} \qquad \mathbf{r}' = \mathbf{rD}$$

where $\mathbf{A}$ and $\mathbf{D}$ are the 2×2 matrices found in part b of this example, we obtain three-dimensional representations in the form

$$\left[(x')^2,(y')^2,2x'y'\right] = [x^2,y^2,2xy][\mathbf{D}(R)]$$

where the representatives $\mathbf{D}(A)$ and $\mathbf{D}(D)$ of the reflection plane $A=\sigma_x$ and the positive (counterclockwise) 120° rotation $C_3=D$ are given by

$$\mathbf{D}(A)=\begin{bmatrix} 1 & 0 & 0 \\ 0 & 1 & 0 \\ 0 & 0 & -1 \end{bmatrix}, \qquad \mathbf{D}(D)=\frac{1}{4}\begin{bmatrix} 1 & 3 & 2\sqrt{3} \\ 3 & 1 & -2\sqrt{3} \\ -\sqrt{3} & \sqrt{3} & -2 \end{bmatrix}$$

These matrices obey $DA=AD^2$, and relations 1.2.6 can be used to generate a complete three-dimensional representation of the group. Alternatively $\mathbf{D}(R)$ can be generated using $\psi_\nu(\mathbf{rR})$ with the same result. This three-dimensional representation is not unitary because the base vectors x^2, y^2 are not orthogonal to one another. The representation is also reducible because C_{3v} has no three-dimensional irreducible representations (see Eq. 1.5.9, Example 1.5.1, or Table 1.5.1).

Although $x^2 \sim \cos^2\varphi$ and $y^2 \sim \sin^2\varphi$ are not orthogonal to one another, $x^2+y^2 \sim 1$ and $x^2-y^2 \sim \cos 2\varphi$ are. Moreover, x^2+y^2 is invariant under the operators of C_{3v}, that is, it generates the identity representation of C_{3v}. We can use $2xy \sim \sin 2\varphi$ and $x^2-y^2 \sim \cos 2\varphi$ as a pair of orthogonal (and identically normalized) base vectors to span the part of the space ρ that is orthogonal to $x^2+y^2 \sim 1$. Our representation in the form

$$\left[(x')^2+(y')^2, 2x'y', (x')^2-(y')^2\right]=[x^2+y^2, 2xy, x^2-y^2][\mathbf{D}(R)]$$

with the generators represented by

$$\mathbf{D}(A)=\left[\begin{array}{c|cc} 1 & 0 & 0 \\ \hline 0 & -1 & 0 \\ 0 & 0 & 1 \end{array}\right], \qquad \mathbf{D}(D)=\left[\begin{array}{c|cc} 1 & 0 & 0 \\ \hline 0 & -\frac{1}{2} & -\frac{1}{2}\sqrt{3} \\ 0 & \frac{1}{2}\sqrt{3} & -\frac{1}{2} \end{array}\right]$$

is now unitary and is a direct sum of the identity representation and a 2×2 representation *identical* to that generated by $[x,y]$. Thus we can write

$$[2xy, x^2-y^2] \sim [x,y] \tag{1.4.12}$$

where $\sim$ means "transforms as," that is, yields the identical representation. In Example 1.5.1 we show that this 2×2 representation is indeed irreducible.

Our decomposition of the reducible space spanned by x^2, y^2, xy into its irreducible components used the fastest procedure—inspection. More systematic methods are available (when necessary) in Chapter 3.

Theorem 1.4.2

Symmetry implies degeneracy, and the size of the irreducible representation governs the *minimum degeneracy* that can occur.

Proof: Let ψ^i be an eigenfunction of the Hamiltonian H, with eigenvalue E^i, that is, $H\psi^i = E^i\psi^i$. If R is a symmetry operator that commutes with H, we may apply R to the Schrödinger equation to obtain

$$H(R\psi^i) = E^i(R\psi^i) \tag{1.4.13}$$

so that $R\psi^i$ is also a solution (possibly not independent of the first) with energy E^i. If we have a group or algebra of elements $E, A, B, C, \ldots$ that commute with H, the functions in the set $E\psi^i, A\psi^i, B\psi^i, C\psi^i, \ldots$ are all degenerate with (have the same energy as) ψ^i and, indeed, they span a space ρ of all functions whose energy is required to be E^i for symmetry reasons.

Let us denote by ψ^i_μ a complete set of *independent* functions that span ρ. [We always use a Latin superscript i, j, etc., to give the name of a set of functions (or the representation they generate) and a Greek subscript μ, ν, λ, etc., to designate the *partners* or members of the set; for example i may refer to the p states of an atom, and $\mu = 1, 2, 3$ can label three independent p states, such as $xf(r), yf(r), zf(r)$, which transform as x, y, z, respectively.]

The representation $D^i(R)$ generated by the ψ^i_μ:

$$R\psi^i_\nu = \sum \psi^i_\mu D^i_{\mu\nu}(R) \tag{1.4.14}$$

is almost always irreducible. If it were reducible, this would mean that the ψ^i_λ could be split into two sets that did not mix under the operations of the group. Since states in one set cannot be obtained from states in the other by an operation R as in Eq. 1.4.10, there is no symmetry reason why these states must be degenerate. Nevertheless, such additional degeneracy could arise because of some neglected symmetry element (e.g., time reversal). There are also *accidental degeneracies*, which arise for special choices of the Hamiltonian and are removed if the Hamiltonian is modified without changing its symmetry (Herring, **1937b**).

Except, then, for accidental degeneracies, the degeneracy produced by symmetry has the dimension of the irreducible representation. Conversely, *Abelian groups,* whose elements can be diagonalized and whose representations are therefore one dimensional, *give rise to no degeneracy* (see Theorem 6.2.1).

It is important to note that the nature and number of a group of irreducible representations depend on the Hamiltonian H only in the sense that H specifies which group is relevant.

1.5. RELATION BETWEEN REPRESENTATIONS, CHARACTERS, AND STATES

Theorem 1.5.1

The construction of the irreducible representations of a group automatically diagonalizes the Dirac characters. Indeed it makes them a constant within each irreducible representation. Conversely, the diagonalization of the Dirac characters facilitates but does not complete the construction of the irreducible representations.

Since by the class rearrangement Theorem 1.3.2, the Dirac characters automatically commute with all the elements of the group, the first part of Theorem 1.5.1 is an immediate consequence of the following lemma.

Schur's Lemma 1.5.2$^{\|}$ [4]

A matrix that commutes with every member of a set of irreducible matrices must be a constant, that is, a number times the unit matrix.

We have already provided a proof of this lemma (with mild restrictions) by showing that H, which commutes with the symmetry operators R, is diagonal within any irreducible representation: see Theorem 1.4.2.

In the solution of a physical problem we search for eigenvectors in a space that is originally large or possibly infinite. The diagonalization of the Dirac characters simplifies the problem by confining our search to a subspace appropriate to multiples of a single irreducible representation. This is illustrated by the separation in Example 1.1.2 into the spaces appropriate to even and odd representation. If an irreducible representation of dimension l_i is repeated a_i times, it is essential to choose the several sets of basis vectors so that they transform in the same way. Theorem 3.4.3 will then restrict dynamical interactions to *corresponding partners* of the same irreducible representation. The solution of the dynamical problem will require only the diagonalization of an $a_i \times a_i$ matrix rather than an $a_i l_i \times a_i l_i$ matrix. (If a four-dimensional representation is repeated twice, we need solve a 2×2 and not an 8×8 problem.) This additional simplification can usually be achieved by diagonalizing some group operator in addition to the Dirac characters. For the rotation group, diagonalization of L^2 corresponds to diagonalizing the Dirac characters and can be used to restrict us to p waves. Diagonalization of L_z could restrict us further to solutions of the form $zg(r)$ [or $(x \pm iy)g(r)$, respectively]. These points are discussed further in Sections 3.2 and 3.4, and a detailed application is given in Chapter 5.

[4]The superscript $\|$ designates a theorem presented without proof. Proofs of these general theorems are readily available in standard group theory textbooks.

Characters of Representations

The *character* $\chi(R)$ (not the Dirac character) of a representation is defined as the *trace* of the representation:

$$\chi(R)=\sum_{\mu} D_{\mu\mu}(R)=\chi(C) \tag{1.5.1}$$

Since the trace of a matrix is invariant under similarity transformations, and the elements of a class are related by such transformations, $\chi(R)$ is a class function and can be written as $\chi(C)$. The character of irreducible representation i will be denoted as $\chi^i(R)=\chi^i(C)$. With the help of Lemma 1.5.2 we can use the constancy of Ω_c to write:

$$\text{Trace}\,\Omega_c^i=\sum_{R\in C}\chi^i(R)$$

or

$$l_i\Omega_c^i=n_c\chi^i(C)$$

$$\chi^i(C)=\frac{l_i\Omega_c^i}{n_c} \tag{1.5.2}$$

where l_i is the dimension of the representation, n_c is the number of elements in the class, and Ω_c^i is an eigenvalue of Ω_c that can be obtained by solving the algebra of the Dirac characters; see Example 1.3.1. The dimension l_i is not yet known but will be available shortly in Eq. 1.5.8.

We now state without proof the following Theorem.

Matrix Orthogonality Theorem 1.5.3||

$$\frac{1}{g}\sum_{R\in G} D_{\mu\nu}^i(R)\left[D^j(R)\right]_{\nu'\mu'}^{-1}=\frac{\delta_{ij}\delta_{\mu\mu'}\delta_{\nu\nu'}}{l_i} \tag{1.5.3}$$

Here i and j refer to two irreducible representations of a group *or multiplier group* that are either distinct or *identical. If* we restrict ourselves to ordinary groups (not multiplier groups), we can write

$$\left[D^j(R)\right]^{-1}=D^j(R^{-1}) \tag{1.5.4}$$

But we shall *not* restrict ourselves to ordinary groups. *If* the representation is chosen to be *unitary*, we can write

$$\left[D^j(R)\right]_{\nu'\mu'}^{-1}=D_{\mu'\nu'}^j(R)^* \tag{1.5.5}$$

If we use Eq. 1.5.5, set $\mu=\nu$ and $\mu'=\nu'$, and sum over $\mu\mu'$ in 1.5.3, we obtain

$$\frac{1}{g}\sum_{R\in G}\chi^i(R)\chi^j(R)^* = \delta_{ij} \tag{1.5.6}$$

which is valid for multiplier as well as ordinary groups. The assumption that the representation is unitary is only a minor restriction on the validity of Eq. 1.5.6, since we know from the unitarity theorem (Weyl, **1946**) that every representation of a finite or compact group is surely *equivalent* to a unitary one, and relations between traces are not affected by an equivalence (i.e., similarity) transformation.

Equation 1.5.6 can be rewritten as a Theorem.

First Character Orthogonality Theorem 1.5.4

$$\sum_{c} n_c\chi^i(C)\chi^j(C)^* = g\delta_{ij} \tag{1.5.7}$$

for ordinary and multiplier groups.

The use of Eq. 1.5.2 in 1.5.7 with $j=i$ now yields

$$l_i^2\sum_{c}\left[\frac{|\Omega_c^i|^2}{n_c}\right] = g \tag{1.5.8}$$

which determines l_i. With the help of Eq. 1.5.2 we can now convert Eq. 1.3.7 into a conventional character table Table 1.5.1. The line across the top lists first the group and then the separate classes in the group. For example, $2C_3$ denotes the fact that there are two threefold rotations in the class C_3. Similarly $3\sigma_v$ implies that there are three mirror planes, that is to say, three elements in this class. Of course, E is the identity element and forms a class in itself. Since $\chi^i(E)$ is the *trace of the unit matrix*, it is evidently equal to *the dimension of the representation.* Thus one may read off the dimensions of the representations from the character tables.

Table 1.5.1. Character Table for Group C_{3v}

	C_{3v}			E	$2C_3$	$3\sigma_v$
x^2+y^2, z^2	z	A_1	Γ_1	1	1	1
	R_z	A_2	Γ_2	1	1	-1
$x^2-y^2, 2xy$ (xz, yz)	(x,y) (R_x, R_y)	E	Γ_3	2	-1	0

Actually the fact that this group has two one-dimensional representations and one two-dimensional representation could have been ascertained directly from two basic theorems of representation theory. The first of these (proof suggested in Problem 3.1.2) is as follows.

Burnside's Theorem 1.5.5

$$\sum_i (l_i)^2 = g = \text{order of the group} \tag{1.5.9}$$

This can be used as a measure of the completeness of a set of irreducible representations. For our case, $g=6$ and this equation has only two possible solutions: (a) $l_1=l_2=1$, $l_3=2$, or (b) $l_1=l_2=l_3=l_4=l_5=l_6=1$. Solution ($b$) is ruled out by a second fundamental theorem.

Theorem 1.5.6||

The number of irreducible representations equals the number of classes.

Example 1.5.1

Discuss the character table and representations of group C_{3v} without using the class multiplication theorem.

Solution: The two-dimensional representation found in Example 1.4.1c has characters $\chi(E)=2$, $\chi(A)=0$, $\chi(D)=-1$, so that

$$\frac{1}{g}\sum_R |\chi(R)|^2 = \tfrac{1}{6}\left\{[\chi(E)]^2+3[\chi(A)]^2+2[\chi(D)]^2\right\}=1$$

and Eq. 3.1.6 implies that this representation is therefore irreducible. We can now construct a tentative character table:

	E	(D,F)	(A,B,C)
Γ_1	1	1	1
Γ_2	1	d	a
Γ_3	2	-1	0

The first line is the *identity* representation possessed by every group. The last line is the irreducible representation just found. The middle line reflects the fact that the remaining irreducible representation must be one dimensional because of Theorems 1.5.5 and 1.5.6. The parameters a and d are required by the first character orthogonality theorem 1.5.4 to obey $1+2d+3a=0$, $2-2d=0$ or $d=1$, $a=-1$.

The Second Character Orthogonality Theorem[||] **1.5.7**

$$\sum_i \chi^i(C)\chi^i(C')^* = \frac{g}{n_c}\delta(C,C') \qquad (1.5.10)$$

yields the relations $1+d-2=0$, $1+a=0$ with the same solution. (Perhaps the simplest procedure for one-dimensional representations is to look for numerical solutions of the generator relationships, in this case $DA = AD^2$, $D^3 = A^2 = 1$.)

The second representation Γ_2 is the *determinantal* representation; element R is replaced by determinant R. But, by definition, all proper rotations obey $\det R = +1$, and all improper rotations obey $\det R = -1$. Such a representation can be written immediately for any group of proper and improper rotations (see Section 2.10).

The representations of C_{3v} can also be generated from the representations of its subgroup C_3 by a procedure known as the method of induced representations (see Section 3.8).

Theorem 1.5.8

Equivalent representations have the same characters.

Proof: Characters are invariant under similarity transformations.

Theorem 1.5.9

Irreducible representations with the same characters are equivalent.

Proof: If the representations were inequivalent, the characters would be orthogonal, not identical.

Comment: This theorem is readily extendable to reducible representations.

The labels of the representations are purely conventional. The Γ labeling of Bethe (**1929**) and Bouckaert, Smoluchowski, and Wigner (**1936**) is popular in solid-state physics. In molecular physics an alphabetical scheme due to Mulliken (**1933**) is more conventional; A and B are used for one-dimensional representations, and E, F, G, and H for two-, three-, four-, and five-dimensional representations. The first two columns in Table 1.5.1 list basis functions for the representations such as those found in Example 1.4.1 or Section 3.8. A complete set of character tables for all point groups is given in Appendix A.

Use of Representations to Label States

The terminology $s,p,d,\ldots$ of spectroscopy implies that there are electron wave functions that are products of a radial wave function and a spherical harmonic of order $l=0,1,2,\ldots$. The terminology S,P, or D refers to a configuration of a many-electron atom, whose wave function is not necessarily factorable into a radial and an angular part but which transforms under proper rotations in the same way as s,p,d. The S,P,D labeling can also be replaced by Γ_0, Γ_1, and Γ_2 for $l=0,1,2$, respectively. In either case, when a state is labeled by an irreducible representation, this conveys information concerning how the state *transforms* under the symmetry operations of the group, rather than about the wave function itself. This information is often sufficient, however, to answer important qualitative questions. For example, the wave function of an atom in a P atomic state transforms as (x,y,z). If the atom is placed in an electrostatic potential with the symmetry of the point group C_{3v}, and with symmetry axis in the z direction, we expect the threefold degeneracy to be split into a two fold degeneracy plus a nondegenerate state:

$$P \rightarrow A_1 + E$$

that is,

$$(x,y,z) \rightarrow z + (x,y) \tag{1.5.11}$$

Let us now restate this conclusion in a physical and a mathematical way: The state ψ_z (which transforms as z) need no longer be degenerate with ψ_x, since there is no operation R in C_{3v} that can, by Eq. 1.4.10, take ψ_x into ψ_z. More generally, there is no R such that

$$\psi_z = R(a\psi_x + b\psi_y)$$

Thus the reason for the degeneracy derived in Theorem 1.4.2 between ψ_z and either of ψ_x, ψ_y has been eliminated.

The corresponding mathematical statement is that under the operations of C_{3v} (the 120° rotations C_3 about the z axis, and the mirror planes containing the z axis) a vector in the x-y plane cannot be taken out of the plane. Thus, with respect to C_{3v}, $(x$-y-$z)$ space is no longer irreducible because it possesses the x-y plane as an invariant subspace. It is left as an exercise for the reader to show that the latter space is irreducible under C_{3v} but reducible under group C_3 (cf. Section 2.8) into the $(x+iy)$ and $(x-iy)$ subspaces. In this language, with respect to a specified group, a space is *irreducible* if it possesses no *invariant subspace*.

We found it easy in Eqs. 1.5.11 to guess how the degeneracies are lifted. For more complicated states or groups, we discuss in Section 3.1 a systematic procedure for decomposing a space into its irreducible components.

The important point to remember is that the label E or Γ_3 for a state in group C_{3v} conveys information similar to that imparted by the label P for a free atomic state.

1.6. ▲[5] CONTINUOUS GROUPS

The Commutator Algebra

For the purposes of this book the only continuous group with which we are concerned is the rotation group. In Chapter 2 we shall show that the irreducible representations of the rotation group can be constructed readily from well-known representations of the angular momentum operators. The purpose of the present section is simply to remark that the elementary quantum-mechanical procedures for dealing with angular momentum operators are in fact just what Lie's[6] (1893) rigorous theory of continuous groups suggests.

Lie groups are groups whose elements $R=R(a)$ are continuous differentiable functions of a finite number of parameters $a=a_1,a_2,\ldots,a_n$, for example, rotation angles. The parameters can be chosen so that the identity operator is $R(0)$. Group multiplication is defined by

$$R(c)=R(a)R(b) \tag{1.6.1}$$

with $c=\varphi(a,b)$ or $c_i=\varphi_i(a_1\cdots a_n;\, b_1\cdots b_n)$ a continuous differentiable function. The group then has n infinitesimal generators, given by

$$I_j=\left.\frac{\partial R(a_1\cdots a_n)}{\partial a_j}\right|_{a\to 0} \tag{1.6.2}$$

or

$$R(a_1\cdots a_n)\approx 1+\Sigma a_i I_i \tag{1.6.3}$$

for sufficiently small a_i.

We must now exploit two theorems.

Theorem 1.6.1

The finite elements of a continuous group can be built up by successive applications of infinitesimal elements.

Proof:

$$R(a)=R\,[\varphi(a,b)]R(b^{-1}) \tag{1.6.4}$$

where b^{-1} stands for parameters appropriate to the transformation inverse to $R(b)$. We differentiate with respect to a_i, and then let $b^{-1}\to a$, that is, $\varphi\to 0$, to obtain:

$$\frac{\partial R(a)}{\partial a_i}=\frac{\partial R\,[\varphi(a,b)]}{\partial \varphi_j(a,b)}\,\frac{\partial \varphi_j(a,b)}{\partial a_i}R(b^{-1}) \tag{1.6.5}$$

[5]Material delimited by a shaded triangle (▲) should be omitted or skimmed in a first reading. (Read for the ideas and the theorem statements. Omit the proofs.)

[6]See also Lyubarskii (1960) and Hamermesh (1962).

$$\frac{\partial R(a)}{\partial a_i} = I_j M_{ij}(a) R(a) \tag{1.6.6}$$

where

$$M_{ij}(a) = \left.\frac{\partial \varphi_j(a,b)}{\partial a_i}\right|_{b \to a^{-1}} \tag{1.6.7}$$

and we have employed the summation convention (repeated indices are understood to be summed over). Thus

$$R(a+\Delta a) = [1 + \Delta \mathbf{a} \cdot \mathbf{M}(a) \cdot \mathbf{I}] R(a) \tag{1.6.8}$$

By the successive use of this equation it is possible to progress from $a=0$ to any finite a with whatever accuracy is desired by using a small enough Δa and a large enough number of steps:

Corollary 1.6.2

The representations of all the elements of a continuous group are uniquely determined by the representations of its generators.

Furthermore, the representatives of the generators are determined by their commutator algebra Eq. 1.6.9, which is closed in accord with the following theorem.

Theorem 1.6.3

The commutator of every pair of generators is a linear combination of the finite set of generators

$$[I_i, I_j] = \Sigma c_{ijk} I_k \tag{1.6.9}$$

In a sense, this theorem is obvious, since the commutator is a measure of the difference found when the operators are applied in two different orders: To the second order of accuracy, one can write

$$R = \exp(r) \approx 1 + r + \tfrac{1}{2}r^2; \qquad R^{-1} \approx 1 - r + \tfrac{1}{2}r^2$$

$$S \approx 1 + s + \tfrac{1}{2}s^2; \qquad S^{-1} \approx 1 - s + \tfrac{1}{2}s^2$$

$$R^{-1}S^{-1}RS - 1 \approx [r,s] \tag{1.6.10}$$

But this difference is also an infinitesimal operator and must be expressible in terms of the infinitesimal generators.

Formal Proof:

$$\frac{\partial R}{\partial a_i} = M_{ik}(a) I_k R(a)$$

$$\frac{\partial^2 R}{\partial a_i \partial a_j} = \frac{\partial M_{ik}}{\partial a_j} I_k R + M_{ik} I_k \frac{\partial R}{\partial a_j}$$

$$= \frac{\partial M_{ik}}{\partial a_j} I_k R + M_{ik} I_k M_{jl} I_l R \qquad (1.6.11)$$

where repeated indices are summed over. Comparison between the different orders of operation in Eqs. 1.6.10 is now made by the requirement

$$\frac{\partial^2 R}{\partial a_i \partial a_j} = \frac{\partial^2 R}{\partial a_j \partial a_i} \qquad (1.6.12)$$

This leads immediately to the requirement:

$$M_{ik} M_{jl} [I_k, I_l] = \left(\frac{\partial M_{jk}}{\partial a_i} - \frac{\partial M_{ik}}{\partial a_j} \right) I_k \qquad (1.6.13)$$

for all a. We set $a = 0$, at which point $M_{ik} = \delta_{ik}$, to obtain the desired equation 1.6.9 with

$$c_{ijk} = \left[\frac{\partial}{\partial a_i} (M_{jk}) - \frac{\partial}{\partial a_j} (M_{ik}) \right]_{a \to 0} \qquad (1.6.14)$$

as a set of constants antisymmetric in i and j.

Thus the construction of representations of an *infinite, continuous*, n-parameter *group* is reduced to finding the representations of a *commutator algebra,* Eq. 1.6.9, *containing n elements.*

The Hurwitz Invariant Integral

Sums over group elements must, for continuous groups, be replaced by integrals over the group parameters $a = a_1, a_2, \ldots, a_n$:

$$\Sigma f(R) \equiv \int f(R)\, dR = \int f[R(a)] \rho(a)\, da \qquad (1.6.15)$$

$$da = da_1, da_2, \ldots, da_n$$

where $\rho(a)$, the density of group elements in the interval da, must be chosen so that the group rearrangement theorem remains valid:

$$\int f[R(a)]\rho(a)\,da = \int f[R(a)R(b)]\rho(a)\,da \tag{1.6.16}$$

If we write

$$R(c) = R(a)R(b), \qquad c = \varphi(a,b) \tag{1.6.17}$$

this is equivalent to the requirement

$$\rho(a)\,da = \rho(c)\,dc \tag{1.6.18}$$

or

$$\frac{\rho(a)}{\rho(c)} = \frac{J(c)}{J(a)} = \frac{J[\varphi(a,b)]}{J(a)} \tag{1.6.19}$$

where $J(c)/J(a)$ is the Jacobian of the transformation from a to c. If we let $a \to 0$ (the identity), then $c = \varphi(a,b) \to b$ and

$$\frac{\rho(0)}{\rho(b)} = \det\left[\frac{\partial \varphi_i(a,b)}{\partial a_j}\right]_{a\to 0}. \tag{1.6.20}$$

whereas, if we let $b \to a^{-1}$, and $c \to 0$, we find

$$\frac{\rho(a)}{\rho(0)} = \det\left[\frac{\partial \varphi_i(a,b)}{\partial a_j}\right]_{b\to a^{-1}} = \det M_{ij}(a) \tag{1.6.21}$$

where M_{ij} is defined by Eq. 1.6.7. Thus the weight factor $\rho(a)$ is determined from the group multiplication functions $\varphi_i(a,b)$.

Even if $f(R)$ is a bounded function of R, not all sums will converge when replaced by integrals. But the key theorems on the unitarity and orthonormality of representations are really averages rather than sums:

$$\frac{1}{g}\sum_R f(R) \to \frac{\int f(R)\,dR}{\int dR} = \frac{\int f[R(a)]\rho(a)\,da}{\int \rho(a)\,da} \tag{1.6.22}$$

and the results are meaningful for bounded $f(R)$. Thus the character orthonormality theorem can be written as

$$\langle \chi^i(R)\chi^j(R)^* \rangle = \frac{\int \chi^i[R(a)]\chi^j[R(a)]^*\rho(a)\,da}{\int \rho(a)\,da} = \delta_{ij} \tag{1.6.23}$$ ▲

1.7. SUMMARY

The solution of a physical problem is simplified by diagonalizing a complete set of commuting observables. In this way the physical space of the problem is decomposed into parts that are not coupled by the Hamiltonian, thus reducing a large problem to a set of small ones. Symmetry operations when present form a group. The Dirac characters Ω_c of a group (or algebra) form a set of commuting observables that should be diagonalized in the physical space of the problem. The importance of group representation theory is that the diagonalization of the Ω's is closely related to finding the irreducible representations (and their characters) of the group. Indeed, the eigenvalues of the Ω_c can be determined from the group character table, using Eqs. 1.5.2 and 1.5.8. Conversely, if the character table is unavailable, the algebra of the Ω's can be applied to determine their eigenvalues and hence the character table.

We show by example (1) how the action of symmetry operators on functions can be used to generate reducible and irreducible matrix representations and (2) how the character orthogonality theorems facilitate the completion of a character table.

The generators of a continuous group obey a commutator algebra whose irreducible representations define those of the entire group.

problems

1.1.1. Write out the potential and kinetic energies for the coupled mass problem of Fig. 1.1.1. Show that these energies are indeed invariant under the inversion operation of Example 1.1.2. Write out the equations of motion, and find the normal modes and natural frequencies for this problem.

1.2.1. Show that the following groups are isomorphic to the noncyclic Abelian group of order 4:

(1) the four matrices

$$\begin{pmatrix} 1 & 0 \\ 0 & 1 \end{pmatrix}, \quad \begin{pmatrix} 1 & 0 \\ 0 & -1 \end{pmatrix}, \quad \begin{pmatrix} -1 & 0 \\ 0 & 1 \end{pmatrix}, \quad \begin{pmatrix} -1 & 0 \\ 0 & -1 \end{pmatrix};$$

(2) the four functions of z

$$f_1(z)=z, \qquad f_2(z)=-z, \qquad f_3(z)=z^{-1}, \qquad f_4(z)=-z^{-1},$$

where multiplication is defined by

$$f_i f_j(z) \equiv f_i\,[\,f_j(z)\,];$$

(3) the four numbers 1, 3, 5, and 7, where multiplication is defined as ordinary multiplication modulo 8;

(4) the point groups D_2, C_{2v}, and C_{2h}, described in Chapter 2.

1.2.2. Consider a group with generators $A_1, A_2, \ldots$. Show that all elements can be expressed in the form $A_1^u A_2^v A_3^w \cdots$ and that the set of such products is finite provided that

$$A_1^p = E, \qquad A_2^q = E, \qquad A_3^r = E, \qquad \text{etc.}$$

and we are given a set of relations of the form

$$A_j A_i = A_1^u A_2^v A_3^w \cdots$$

for all $j > i$, where $u \geqslant 0$, $v \geqslant 0$, $w \geqslant 0$, etc.

1.3.1. Show that the reciprocal of the elements of a class form a class, possibly not distinct from the original.

1.3.2. Show that no element can be a member of two classes.

1.3.3. Show that two elements in the same class must be of the same order.

1.3.4. A permutation on five objects can be represented by an element such as

$$\begin{pmatrix} 1 & 2 & 3 & 4 & 5 \\ 2 & 4 & 5 & 1 & 3 \end{pmatrix} = (1\ \ 2\ \ 4)(3\ \ 5)$$

in which the upper symbol is carried into the lower symbol. On the right-hand side, the same element is represented as a product of cycles

$$1 \to 2 \to 4 \to 1 \qquad \text{and} \qquad 3 \to 5 \to 3$$

(*a*) Show that any permutation of n objects

$$\begin{pmatrix} 1 & 2 & 3 & \cdots & n \\ a_1 & a_2 & a_3 & \cdots & a_n \end{pmatrix}$$

can be decomposed into cycles.

(*b*) Show that the permutations on n objects form a group called S_n (the symmetric group of degree n) of order $n!$

(*c*) Show that elements with the same decomposition into cycles are in the same class.

1.3.5. Prove Cayley's theorem: Every group G of order n is a subgroup of the symmetric group of degree n. *Hint:* If the group elements $E = A_1, A_2, \ldots, A_n$ are multiplied on the left by A_j, the result is a permutation of these symbols.

1.3.6. Use the multiplication table of Eq. 1.2.4 to show that the elements of the group C_{3v} obey $A^2 = B^2 = C^2 = E$, $D^3 = F^3 = E$. Problem 1.3.3 suggests that A, B, and C can be in the same class and D and F can be in another class. Using the multiplication table verify that this is indeed true.

1.4.1. Let M_{ij} be an $n \times n$ matrix possessing a complete set of eigenvectors S_{jp}, $p = 1, 2, \ldots, n$, in the sense that

$$\sum_j M_{ij} S_{jp} = m_p S_{ip}$$

where m_p is the pth eigenvalue. Show that this equation can be rewritten in matrix form as

$$MS = Sm \qquad \text{or} \qquad S^{-1}MS = m$$

where $m_{p'p} = m_p \delta(p', p)$ is a diagonal matrix, that is, the problem of diagonalizing M by a similarity transformation is identical to that of finding a complete set of eigenvectors.

1.4.2. Show that, if Eq. 1.4.7 were replaced by

$$S\psi_\nu = \Sigma D_{\nu\mu}(S)\psi_\mu$$

with the matrix as a prefactor rather than a postfactor, products of operators would still be represented by products of matrices—but in reverse order.

1.4.3. Show that a Hermitian operator is represented by a Hermitian matrix in any base system. Show that a unitary operator is represented by a unitary matrix if and only if the system of basis vectors is an orthonormal system.

1.4.4. Show that, if $D_{\mu\nu}(R)$ is a representation of a group, so is the complex conjugate set of matrices $D_{\mu\nu}(R)^*$.

1.5.1. Generate the representation in the space (x,y) of the cyclic group $\mathcal{C}_3$ whose elements are $C_3, C_3^2, C_3^3 = E$ ($C_3 = 120°$ rotation). Show by direct use of Schur's lemma that this representation is reducible. What are the irreducible subspaces of (x,y)? What are the irreducible representations of the above group?

1.5.2. Show that the two-dimensional representation of C_{3v} in Example 1.2.1 obeys

$$D + F = \begin{pmatrix} -1 & 0 \\ 0 & -1 \end{pmatrix}; \qquad A + B + C = \begin{pmatrix} 0 & 0 \\ 0 & 0 \end{pmatrix}$$

in accord with the conditions $\Omega_2 = -1$, $\Omega_3 = 0$ of Eq. 1.3.7 for the irreducible representation Γ_3.

1.5.3. Show that if two irreducible representations are equivalent

$$D^j(R)' = S^{-1}D^j(R)S$$

the similarity transformation matrix S is unique except for a numerical factor. *Hint:* Assume two matrices S and T. Prove that ST^{-1} obeys Schur's lemma.

1.5.4. If the two irreducible representations of Problem 1.5.3 are unitary, show that by removal of a numerical factor (which cancels out of similarity transformations) S can be made unitary.

1.5.5. Show that $\Sigma_R D(R)^\dagger D(R)$ commutes with all elements of a finite group (or a multiplier group). Use Schur's lemma to prove that all representations of finite groups are equivalent to unitary representations (Wigner, **1959**).

1.5.6. (*a*) Determine the class structure of the group C_{4v}. Show that there are five classes.

(*b*) Evaluate the algebra of the Dirac characters for C_{4v}. Construct a table analogous to 1.3.7.

(*c*) Apply the operators of C_{4v} to 1 or z, R_z, x, x^2, xy, x^3, etc. See whether you can generate the five irreducible representations of this group. (See Example 1.4.1 and Theorem 1.5.6.)

(*d*) Show, using Burnside's theorem, Eq. 1.5.9, that C_{4v} has four one-dimensional representations and one two-dimensional representation. Find the one-dimensional representations, using the algebra of the group generators.

1.5.7. Show that a unitary operator R with character $\chi^i(R) = l_i\lambda$, where $|\lambda| = 1$ in representation i of dimension l_i (not necessarily irreducible), is in fact represented by λ times the unit matrix of dimension l_i. *Hint*: First show that each diagonal element of the representation cannot have a magnitude exceeding unity.

1.6.1. Show that every *one-parameter* Lie group can, by a parameter transformation $t = \int_0^a M(a)\,da$, be rewritten in the Abelian form

$$R(t)R(t') = R(t+t')$$

1.6.2. With the definitions

$$g_{ab} = c_{aij}c_{bji}, \qquad g^{\mu a}g_{a\nu} = \delta_{\mu\nu}$$

where c_{aij} is defined by Eq. 1.6.9, show that the Casimir (**1931**) operator

$$C = g^{cd}I_c I_d$$

commutes with every generator I_j of the group, and is the generalization of the operator L^2 for the rotation group $R(3)$.

1.6.3. Consider the group of unitary unimodular matrices (these have the Cayley form $\begin{bmatrix} a & b \\ -b^* & a^* \end{bmatrix}$ of Eq. 2.6.12, where $|a|^2+|b|^2=1$). Choose a suitable set of three independent parameters. Determine the infinitesimal generators of the group and their Lie algebra. Discuss the representations of this algebra.

chapter two

POINT GROUPS

Our aim in this chapter is to give a sufficient account of the nature and nomenclature of point groups to permit us to proceed quickly to applications. Although the representations of such groups can be efficiently determined by the method of induced representations[||], this information is summarized for our use in the character tables of Appendix A. For present purposes, we merely need to know that such information can always be obtained by "brute-force" use of the class multiplication theorem 1.3.4.

A *point group* is defined as a group of rotations and/or reflections about one fixed point. *Any group that restores some finite body to itself leaves the center of mass fixed and hence is a point group.* The space group of an infinite crystal is not restricted in this way and admits operations of the form

$$\mathbf{r}' = (\boldsymbol{\alpha}|\mathbf{a})\mathbf{r} \equiv \boldsymbol{\alpha}\cdot\mathbf{r} + \mathbf{a}$$

that combine rotations $\boldsymbol{\alpha}$ with translations $\mathbf{a}$. (See Chapter 8.) The rotational parts $\boldsymbol{\alpha}$ must separately form a group (cf. Section 8.2) known as the point group of the space group. Such point groups are called *crystallographic point* groups and will be useful later in building up space groups. Moreover, the point groups are subgroups of the full rotation group $R(3)$ in three dimensions. This relationship is important in determining the crystal field splittings of atomic energy levels. For this purpose, we need the characters of the full rotation group. Many properties of the point group elements are, however, valid for arbitrary rotations, and it is best to establish the algebra of rotation operators within the scheme of the full rotation group. This procedure will lead to the introduction of spin or doubled-valued representations in a natural way. The need for double point groups will then be an obvious consequence.

In Sections 2.1–2.6 we consider the subgroup $R^+(3)$ of $R(3)$ containing all *proper* rotations (i.e., rotations that preserve the right-handedness of our coordinate system). We find operators to represent finite rotations and show the action of such rotations on operators and on functions. In accord with Section 1.6 we display the structure of $R^+(3)$ by determining the infinitesimal generators of the group, the commutator algebra of these generators, and the irreducible representations that follow from this algebra.

2.1. GENERATORS OF THE PROPER ROTATION GROUP $R^+(3)$

Let us consider a physical system whose only relevant variable is its angle φ in the x-y plane. In quantum-mechanical terms this system is described by a wave function $\psi(\varphi)$, and the mean ("expectation value of ") the angle is given by

$$\langle\varphi\rangle = \int \psi(\varphi)^* \varphi \psi(\varphi)\, d\varphi \tag{2.1.1}$$

Consider the positive (counterclockwise) rotation of this system by an angle α about the z axis. (See Fig. 2.1.1.) The new measured value $\langle\varphi\rangle'$ is given by

$$\begin{aligned}\langle\varphi\rangle' &= \langle\varphi\rangle + \alpha \\ &= \int \psi(\varphi)^* (\varphi + \alpha) \psi(\varphi)\, d\varphi\end{aligned} \tag{2.1.2}$$

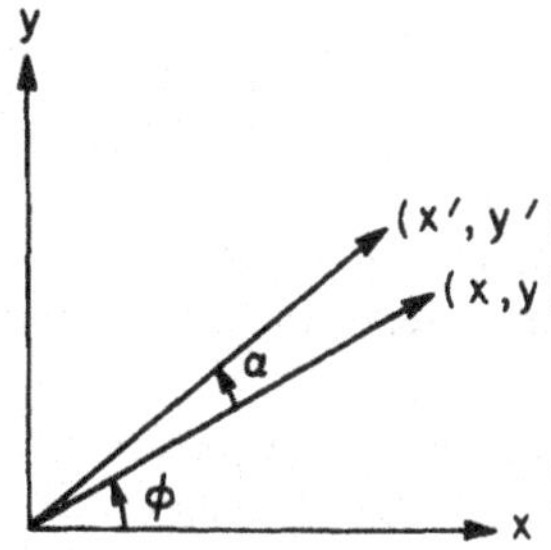

Fig. 2.1.1. Positive rotation α about the z axis.

$$= \int \psi(\varphi-\alpha)^* \varphi \psi(\varphi-\alpha)\, d\varphi \tag{2.1.3}$$

In form 2.1.3 the transformation is described as an action on the *wave function* leading to a new wave function,

$$\psi'(\varphi) \equiv \psi(\varphi-\alpha) \tag{2.1.4}$$

In form 2.1.2 the transformation is described as an action on the *operator,*

$$\varphi' = \varphi + \alpha \tag{2.1.5}$$

Equation 2.1.4 can be written in operator form:

$$\psi'(\varphi) \equiv O_R \psi(\varphi) = \exp\left(-\alpha \frac{\partial}{\partial\varphi}\right)\psi(\varphi) = \psi(\varphi-\alpha) \tag{2.1.6}$$

since the expansion in powers of α is simply the Taylor expansion of $\psi(\varphi-\alpha)$. Let us now evaluate

$$\begin{aligned} O_R^{-1}\varphi O_R \psi(\varphi) &= \exp\left(\alpha \frac{\partial}{\partial\varphi}\right)\varphi \exp\left(-\alpha \frac{\partial}{\partial\varphi}\right)\psi(\varphi) \\ &= \exp\left(\alpha \frac{\partial}{\partial\varphi}\right)[\varphi\psi(\varphi-\alpha)] \\ &= (\varphi+\alpha)\psi(\varphi) \end{aligned} \tag{2.1.7}$$

Thus we see that the appropriate definition of the transformed operator is

$$\varphi' = O_R^{-1}\varphi O_R \tag{2.1.8}$$

Equating 2.1.2 and 2.1.3 in the form

$$\int \psi^* O_R^{-1}\varphi O_R \psi\, d\varphi = \int (O_R\psi)^* \varphi O_R \psi\, d\varphi = \int \psi^* (O_R)^\dagger \varphi O_R \psi\, d\varphi$$

where the last step is an immediate consequence of the definition of the adjoint operator $O_R^\dagger$, we can conclude that

$$O_R^\dagger = O_R^{-1} \tag{2.1.9}$$

that is, the rotation operator O_R is unitary. This is indeed a property of all symmetry operators in quantum mechanics that do not reverse the time. (See Section 10.4.)

Since the properties of a continuous group have been shown in Section 1.6 to be uniquely determined by the *infinitesimal generators* of that group, we examine the operators representing infinitesimal rotations.

To lowest order in α, we can write in Cartesian coordinates

$$O_R\psi(\varphi)=\psi(\varphi-\alpha)\approx\psi(x+\alpha y,y-\alpha x)\approx\left[1+\alpha\left(y\frac{\partial}{\partial x}-x\frac{\partial}{\partial y}\right)\right]\psi(x,y) \tag{2.1.10}$$

Thus the infinitesimal generator is defined by

$$O_R\approx 1+\alpha I_z \tag{2.1.11}$$

$$I_z=-\frac{\partial}{\partial\varphi}=y\frac{\partial}{\partial x}-x\frac{\partial}{\partial y} \tag{2.1.12}$$

In terms of the angular momentum operator

$$\mathbf{L}=\mathbf{r}\times\mathbf{p}=\mathbf{r}\times\frac{\hbar\nabla}{i}\qquad\text{or}\qquad \mathbf{l}=-i(\mathbf{r}\times\nabla)=\frac{\mathbf{L}}{\hbar} \tag{2.1.13}$$

we have

$$I_z=-il_z \tag{2.1.14}$$

and the finite operator of Eq. 2.1.6 is given by

$$O_R=\exp(-i\alpha l_z) \tag{2.1.15}$$

Of course, L_z, the canonical conjugate variable to φ, is expected, in classical mechanics, to generate changes in φ (Goldstein, **1950**, p. 261).

There is nothing special about the direction z. We simply understand l_z to be the component of $\mathbf{l}$ along the axis of rotation. If the latter is described by a unit vector $\mathbf{n}$, we must write

$$O_R(\alpha,\mathbf{n})=\exp(-i\alpha\mathbf{l}\cdot\mathbf{n}) \tag{2.1.16}$$

for a counterclockwise rotation of α about direction $\mathbf{n}$.

2.2. THE COMMUTATOR ALGEBRA OF $R^+(3)$

The commutator algebra of the infinitesimal generators of the proper rotation group in three dimensions, $R^+(3)$, is simply the well known set of commutation rules for the angular momentum operators:

$$[l_x,l_y]=il_z,\quad [l_y,l_z]=il_x,\quad [l_z,l_x]=il_y \tag{2.2.1}$$

In deriving expression 2.1.14 for the infinitesimal generators, we tacitly assumed that we were dealing with a one-body system without spin, since we introduced a wave function $\psi(x,y)$. If we had used $\psi=\psi(\mathbf{r}_1,\mathbf{r}_2,\ldots,\mathbf{r}_n)$, a wave function for many spinless particles we would have obtained $\mathbf{l}=\mathbf{l}_1+\mathbf{l}_2+\ldots+\mathbf{l}_n$, where $\mathbf{l}_s=\mathbf{r}_s\times(\nabla_s/i)$. Direct calculation would again give Eq. 2.2.1. Thus it is not the specific form of the operators $\mathbf{l}$ that is important, but only the structure of the commutator algebra.

Since the representations and characters of a continuous group are uniquely determined by the structure of its commutator algebra, we abandon the specific meanings of the operators l_x,l_y,l_z and retain only the algebra in the form

$$[J_i,J_j]=ie_{ijk}J_k \tag{2.2.2}$$

where e_{ijk} is the Levi-Civita tensor density:

$$\begin{aligned} e_{ijk}&=0 && \text{if any two indices are equal} \\ e_{ijk}&=+1 && \text{if } (ijk)=\text{even permutation of } (123) \\ e_{ijk}&=-1 && \text{if } (ijk)=\text{odd permutation of } (123) \end{aligned} \tag{2.2.3}$$

In each application J will have a specific meaning (the total angular momentum of the system, including spin). Therefore we are not restricted to the purely orbital solutions found in Example 1.1.1 or to representations whose parity is $(-1)^l$. (Parity is the eigenvalue of the inversion operator.)

2.3. IRREDUCIBLE REPRESENTATIONS OF $R^+(3)$

The first step in finding the irreducible representations of a continuous group is to find the irreducible representations of the commutator algebra—in this case Eq. 2.2.2. Theorem 1.5.1 is a guide in accomplishing this: To find the representations of a group, diagonalize the Dirac characters of the group, since they commute with all group elements. For $R^+(3)$, J^2 plays the role of the Dirac character because it commutes with all elements J_x,J_y,J_z of the commutator algebra. (See Problem 1.6.2 for the generalization of J^2 to other continuous groups.) Although the eigenvalue of J^2 labels the representation, we shall also diagonalize J_z in order to label the different partners of the representation.

The algebraic procedure for diagonalizing J^2 and J_z simultaneously, using $J_\pm=J_x\pm iJ_y$ as raising and lowering operators, is readily available in quantum-mechanics textbooks (Feenberg and Pake, **1953**; Rose, **1957**; Edmonds, **1957**; Dirac, **1958**; Schiff, **1968**) and hence will not be repeated here. The key results are as follows:

$$\begin{aligned} J^2\Psi^j_m &= j(j+1)\Psi^j_m \\ J_z\Psi^j_m &= m\Psi^j_m \\ J_+\Psi^j_m &= [j(j+1)-m(m+1)]^{1/2}\Psi^j_{m+1} \\ J_-\Psi^j_m &= [j(j+1)-m(m-1)]^{1/2}\Psi^j_{m-1} \end{aligned} \tag{2.3.1}$$

The coefficients in the last two equations have been chosen real. This choice fixes the relative phases of the Ψ^j_m for given j. For $j=l=$an integer, this is the choice of relative phases of the spherical harmonics $Y_{lm}(\theta,\varphi)$ made by Condon and Shortley (**1967**). The latter also make $Y_{lm}(0,0)$ real (see, e.g., Eq. 2.5.5, thus fixing the phase completely and making $Y^*_{lm}=(-1)^m Y_{l,-m}$.

The operator J_+ acts on Ψ^j_{-j} and after $2j$ actions raises m by integral steps to Ψ^j_j. Thus $2j$ is required to be an integer. The case $j=$an integer simply repeats results available from Example 1.1.1, the orbital case. But the half-integer results (i.e., half-odd integer) are new and will be commented on shortly.

2.4. CHARACTERS OF THE IRREDUCIBLE REPRESENTATIONS OF $R^+(3)$

Since

$$\exp(i\alpha J_z)\Psi^j_m = \exp(i\alpha m)\Psi^j_m \tag{2.4.1}$$

the rotation operator $\exp(-i\alpha J_z)$ is diagonal in the representation based on Ψ^j_m, and its trace or character is given by

$$\chi^j[\exp(-i\alpha J_z)] = \sum_{m=-j}^{j} \exp(-i\alpha m) = \frac{\sin\alpha(j+\frac{1}{2})}{\sin\frac{1}{2}\alpha} \tag{2.4.2}$$

We can again argue that the choice of a particular direction as z axis was arbitrary, and the same conclusion applies to an axis parallel to any unit vector $\mathbf{n}$:

$$\chi^j(\alpha) \equiv \chi^j[\exp(-i\alpha\mathbf{J}\cdot\mathbf{n})] = \frac{\sin\alpha(j+\frac{1}{2})}{\sin\frac{1}{2}\alpha} \tag{2.4.3}$$

This argument is closely related to a statement proved later (see Corollary 2.7.3) that all rotations of a given angle α, regardless of the orientation of the axis $\mathbf{n}$, are in the same class.

We note that

$$\chi^1(\alpha) = 1 + 2\cos\alpha \tag{2.4.4}$$

and

$$\chi^j(\alpha + 2\pi) = (-1)^{2j}\chi^j(\alpha) \tag{2.4.5}$$

Note the double valuedness in the character of the half-integer representations.

▲The Hurwitz invariant volume element for the rotation group using $\alpha\mathbf{n} = (\alpha_1, \alpha_2, \alpha_3)$ as parameters is found by means of Eq. 1.6.21 to be

$$\rho(\alpha_1\alpha_2\alpha_3)\,d\alpha_1\,d\alpha_2\,d\alpha_3 = \rho(\alpha)\alpha^2\,d\alpha\,d\Omega \tag{2.4.6}$$

where

$$\rho(\alpha) = 2\rho(0)\frac{1-\cos\alpha}{\alpha^2} \tag{2.4.7}$$

(For details see Wigner, **1959**, p. 152.)

The irreducibility of the representations found in this way can be tested by setting $j = i$ in the character orthogonality theorem, Eq. 1.6.23. This equation is satisfied and the representations are then shown to be irreducible. It is also easy to demonstrate that the representations are complete by showing that no even function of α can be orthogonal to all $\chi^j(\alpha)$, that is, to all $\chi^j(\alpha) - \chi^{j-1}(\alpha) = 2\cos j\alpha$ (for j integral and half-integral) over the interval $0 \leqslant \alpha \leqslant \pi$. ▲

2.5 THE THREE-DIMENSIONAL REPRESENTATION $j = 1$ OF $R^+(3)$

Equations 2.3.1 yield for $j = 1$ the representations

$$J_z = \begin{bmatrix} 1 & 0 & 0 \\ 0 & 0 & 0 \\ 0 & 0 & -1 \end{bmatrix}, \quad J_+ = \begin{bmatrix} 0 & \sqrt{2} & 0 \\ 0 & 0 & \sqrt{2} \\ 0 & 0 & 0 \end{bmatrix}, \quad J_- = \begin{bmatrix} 0 & 0 & 0 \\ \sqrt{2} & 0 & 0 \\ 0 & \sqrt{2} & 0 \end{bmatrix} \tag{2.5.1}$$

The eigenvalues of J_z are ± 1 and 0. Since any matrix must obey its own characteristic equation (Mirsky, **1955**, p. 206), we have

$$J_z^3 = J_z \tag{2.5.2}$$

Thus any polynomial in J_z can be reduced to one of degree 2 or less. In particular, we find that

$$\exp(i\alpha J_z) = 1 + i\alpha J_z + \left[\frac{(i\alpha)^2}{2!}\right]J_z^2 + \left[\frac{(i\alpha)^3}{3!}\right]J_z$$

$$+\left[\frac{(i\alpha)^4}{4!}\right]J_z^2 + \cdots$$

$$= 1 + iJ_z \sin\alpha - J_z^2(1-\cos\alpha) \tag{2.5.3}$$

In the $j=1$ representation, we have

$$\exp(i\alpha J_z) = \begin{bmatrix} \exp(i\alpha) & 0 & 0 \\ 0 & 1 & 0 \\ 0 & 0 & \exp(-i\alpha) \end{bmatrix} \tag{2.5.4}$$

The triply degenerate $j=1$ representation, like the p states in spherical coordinates, has basis functions that transform not, as (x,y,z) but, because of the recursion relations 2.3.1, as[1]

$$\Psi_1^1 \sim -\frac{1}{\sqrt{2}}(x+iy), \qquad \Psi_0^1 \sim z, \qquad \Psi_{-1}^1 \sim \frac{1}{\sqrt{2}}(x-iy) \tag{2.5.5}$$

The representation in an (x,y,z) basis can be obtained by transformation or, more directly, by replacing J_z by l_z. Equation 2.1.8 shows that the action on any operator is described by

$$\mathbf{r}' = O_R^{-1}\mathbf{r}O_R = \exp(i\alpha l_z)\mathbf{r}\exp(-i\alpha l_z) \tag{2.5.6}$$

$$= [\exp(i\alpha l_z)]\mathbf{r} \tag{2.5.7}$$

since the only effect of $\exp(-i\alpha l_z)$ is to cancel the action of $\exp(i\alpha l_z)$ on any wave function that could appear on the right. Using Eqs. 2.5.3, 2.1.12, and

[1]Note the minus sign in Ψ_1^1. The $\sim$ means "transforms as." An additional common phase factor is still allowed. Fano and Racah (1959) add a phase factor i^j to the Condon-Shortley Ψ_m^j so that $(\Psi_m^j)^* = (-1)^{j-m}\Psi_{-m}^j$.

2.1.14, we obtain

$$\mathbf{r}' = \left[1 + \sin\alpha\left(x\frac{\partial}{\partial y} - y\frac{\partial}{\partial x} \right) + (1-\cos\alpha)\left(x\frac{\partial}{\partial y} - y\frac{\partial}{\partial x} \right)^2 \right]\mathbf{r} \qquad (2.5.8)$$

or

$$x' = x\cos\alpha - y\sin\alpha, \qquad y' = x\sin\alpha + y\cos\alpha, \qquad z' = z \qquad (2.5.9)$$

Equations 2.5.9 simply confirm our original transformation in three space, shown in Fig. 2.1.1. The trace of the matrix in 2.5.9 and 2.5.4 is $1+2\cos\alpha$, as claimed in Eq. 2.4.4.

The above procedure can be immediately generalized to a rotation of angle α about axis $\mathbf{n}$:

$$\begin{aligned} \mathbf{r}' &= \exp(i\alpha\mathbf{J}\cdot\mathbf{n})\mathbf{r}\exp(-i\alpha\mathbf{J}\cdot\mathbf{n}) = \exp(\alpha I)\mathbf{r}\exp(-\alpha I) \\ &= [1 + I\sin\alpha + (1-\cos\alpha)I^2]\mathbf{r} \end{aligned} \qquad (2.5.10)$$

where (using $\mathbf{E}$ to represent the identity dyadic)

$$I = i\mathbf{l}\cdot\mathbf{n} = \mathbf{n}\cdot\mathbf{r}\times\nabla = \mathbf{n}\times\mathbf{r}\cdot\nabla$$

$$\nabla\mathbf{r} = \mathbf{E}, \qquad I\mathbf{r} = \mathbf{n}\times\mathbf{r}\cdot\mathbf{E} = \mathbf{n}\times\mathbf{r}$$

$$I^2\mathbf{r} = I(\mathbf{n}\times\mathbf{r}) = \mathbf{n}\times(I\mathbf{r}) = \mathbf{n}\times(\mathbf{n}\times\mathbf{r}) = \mathbf{n}(\mathbf{n}\cdot\mathbf{r}) - \mathbf{r}$$

so that finally

$$\mathbf{r}' = \exp(i\alpha\mathbf{J}\cdot\mathbf{n})\mathbf{r} = \mathbf{r}\cos\alpha + \mathbf{n}(\mathbf{n}\cdot\mathbf{r})(1-\cos\alpha) + \mathbf{n}\times\mathbf{r}\sin\alpha \qquad (2.5.11)$$

a result that the reader can verify by elementary methods.

By definition, a vector transforms in the same way as $\mathbf{r}$, so that

$$\exp(i\alpha\mathbf{J}\cdot\mathbf{n})\mathbf{A}\exp(-i\alpha\mathbf{J}\cdot\mathbf{n}) = \mathbf{A}\cos\alpha + \mathbf{n}(\mathbf{n}\cdot\mathbf{A})(1-\cos\alpha) + (\mathbf{n}\times\mathbf{A})\sin\alpha \qquad (2.5.12)$$

This equation determines the commutation rules of an arbitrary vector $\mathbf{A}$ with total angular momentum $\mathbf{J}$. A comparison of the terms linear in α for infinitesimal α yields

$$i[\mathbf{n}\cdot\mathbf{J}, A_j] = (\mathbf{n}\times\mathbf{A})_j$$

or

$$in_i[J_i, A_j] = e_{jik}n_iA_k$$

or

$$[J_i, A_j] = ie_{ijk}A_k \qquad (2.5.13)$$

Conversely, if the commutation rules 2.5.13 are accepted, the finite result of Eq. 2.5.12 follows from solution of the differential equation

$$\frac{d\mathbf{A}(\alpha)}{d\alpha} = \mathbf{n} \times \mathbf{A}(\alpha) \tag{2.5.14}$$

where $\mathbf{A}(\alpha)$ is the left-hand side of Eq. 2.5.12.

To avoid 2^n possible ambiguities of sign, we consistently employ the following conventions in this book:

1. A right-handed coordinate system is used.
2. The body, not the coordinate system, is rotated.
3. Positive rotations are counterclockwise (e.g., a small positive rotation about the $+z$ axis moves a vector in the x direction toward the y axis, as in Eqs. 2.5.9 and Fig. 2.1.1.
4. The action of an operator on a state vector produces a matrix as a *postmultiplicative* factor. Thus with $O_R \equiv \exp(-i\alpha\mathbf{l}\cdot\mathbf{n})$

$$O_R \Psi_\nu^j \equiv R\Psi_\nu^j = \sum \Psi_\mu^j D_{\mu\nu}^j(R) \tag{2.5.15}$$

$$\mathbf{r}' = O_R^{-1}\mathbf{r}O_R = \mathbf{r}\cdot\mathbf{D}(R^{-1}) = \mathbf{D}(R)\cdot\mathbf{r} \equiv \mathbf{R}\cdot\mathbf{r} \tag{2.5.16}$$

$$O_R\Psi(\mathbf{r}) = \Psi(O_R\mathbf{r}O_R^{-1}) = \Psi(\mathbf{rR}) = \Psi(\mathbf{R}^{-1}\mathbf{r}) \tag{2.5.17}$$

where

$$O_R \equiv \exp(-i\alpha\mathbf{J}\cdot\mathbf{n}) \tag{2.5.18}$$

Thus, when acting on wave functions, the "opposing action" with O_R represented by $\mathbf{R}^{-1}$ as in Eqs. 2.1.6 and 2.5.17 occurs only when we insist on writing $\mathbf{R}$ as a premultiplication factor. We have used $\mathbf{R}$ as a brief notation for the 3×3 matrix representation $\mathbf{D}(R)$ in the x–y–z base system. This representation is defined for a rotation α about $\mathbf{n}$ by a comparison of Eqs. 2.5.16 and 2.5.11. Note also that $\mathbf{rR} = \tilde{\mathbf{R}}\mathbf{r} = \mathbf{R}^{-1}\mathbf{r}$, since $\mathbf{R}$ is a real, orthogonal matrix (defined by Eq. 2.9.1).

2.6. THE SPIN REPRESENTATION $j=\frac{1}{2}$ OF $R^+(3)$

The 2×2 representation $j=\frac{1}{2}$ will be studied in detail because (1) it is in some respects the simplest representation of the rotation group: the spin algebra provides the simplest method of multiplying together arbitrary rotations; and (2) it is a double-valued representation that contains within it the "origin" of all the other double-valued representations.

To avoid repeated use of fractions, it is conventional in this two-dimensional representation $j=\frac{1}{2}$ to define the Pauli spin operators by

$$\boldsymbol{\sigma}=2\mathbf{J} \tag{2.6.1}$$

Equations 2.3.1 then yield the matrix representations:

$$\begin{aligned} \tfrac{1}{2}(\sigma_x+i\sigma_y)=J_+&=\begin{bmatrix}0&1\\0&0\end{bmatrix}\\ \tfrac{1}{2}(\sigma_x-i\sigma_y)=J_-&=\begin{bmatrix}0&0\\1&0\end{bmatrix}\end{aligned} \tag{2.6.2}$$

so that

$$\sigma_x=\begin{bmatrix}0&1\\1&0\end{bmatrix},\quad \sigma_y=\begin{bmatrix}0&-i\\i&0\end{bmatrix},\quad \sigma_z=\begin{bmatrix}1&0\\0&-1\end{bmatrix} \tag{2.6.3}$$

Since the eigenvalues of $\sigma_z=\pm 1$, we have $\sigma_z^2=1$, and generally:

$$(\boldsymbol{\sigma}\cdot\mathbf{n})^2=\sigma_x^2=\sigma_y^2=\sigma_z^2=1 \tag{2.6.4}$$

Our original commutation rules in this notation

$$[\sigma_x,\sigma_y]=2i\sigma_z \tag{2.6.5}$$

can be simplified by using

$$0=[\sigma_y^2,\sigma_z]=2i(\sigma_y\sigma_x+\sigma_x\sigma_y)$$

to

$$\sigma_x\sigma_y=i\sigma_z=-\sigma_y\sigma_x \tag{2.6.6}$$

Theorem 2.6.1

Any function of σ_x, σ_y, and σ_z can be reduced to a linear function of these matrices and the unit matrix.

Proof: For a polynomial function, the above algebraic relations can be used to reduce all powers of $\sigma_x,\sigma_y,\sigma_z$ to linear functions and constants. For an arbitrary function, we note that the two-dimensional unit matrix and $\sigma_x,\sigma_y,\sigma_z$ constitute *four independent* matrices that span all possible 2×2 matrices.

In view of Theorem 2.6.1 and Eq. 2.6.4, we can immediately specialize the rotation operator $R(\varphi,\mathbf{n})=\exp(-i\varphi\mathbf{J}\cdot\mathbf{n})$ to

$$R(\varphi,\mathbf{n})=\exp(-i\tfrac{1}{2}\varphi\boldsymbol{\sigma}\cdot\mathbf{n})=\cos\tfrac{1}{2}\varphi-i\boldsymbol{\sigma}\cdot\mathbf{n}\sin\tfrac{1}{2}\varphi \tag{2.6.7}$$

for a rotation of φ about axis $\mathbf{n}$ in this two-dimensional spin representation. If $\mathbf{n}$ has components (l,m,n), the particular "Pauli" representation 2.6.3 leads to

$$\boldsymbol{\sigma}\cdot\mathbf{n}=\begin{bmatrix} n & l-im \\ l+im & -n \end{bmatrix} \tag{2.6.8}$$

and hence to a simple 2×2 matrix for $R(\varphi,\mathbf{n})$. However, it is the algebra of Eqs. 2.6.4 and 2.6.6 that is important, not the particular representation chosen. All significant results can be derived most simply without reference to the explicit representation. For example, we find that

$$R(\varphi,\mathbf{n})R(-\varphi,\mathbf{n})=E, \quad \text{the identity} \tag{2.6.9}$$

and the Hermiticity of $\boldsymbol{\sigma}$ permits us to take the adjoint of Eq. 2.6.7:

$$R(\varphi,\mathbf{n})^{\dagger}=R(-\varphi,\mathbf{n})=[R(\varphi,\mathbf{n})]^{-1} \tag{2.6.10}$$

so that $R(\varphi,\mathbf{n})$ is a *unitary* operator or representation. Since the eigenvalues of $\boldsymbol{\sigma}\cdot\mathbf{n}$ are $+1$ and -1, those of $R(\varphi,\mathbf{n})$ are $\exp(\pm i\frac{1}{2}\varphi)$ and the *determinant* is

$$\det R(\varphi,\mathbf{n})=\exp(i\tfrac{1}{2}\varphi)\exp(-i\tfrac{1}{2}\varphi)=1 \tag{2.6.11}$$

as is most easily seen in the representation in which $\boldsymbol{\sigma}\cdot\mathbf{n}=+1$ and -1. Thus the $R(\varphi,\mathbf{n})$ is *isomorphic* to the group $U(2)$ of unitary, unimodular (determinant $=1$) matrices in two space. The latter are often represented in the form

$$U=\begin{bmatrix} a & b \\ -b^* & a^* \end{bmatrix} \tag{2.6.12}$$

where the complex parameters a,b (the Cayley–Klein parameters) obey the constraint

$$|a|^2+|b|^2=1 \tag{2.6.13}$$

leaving three real independent parameters. The parameters $\varphi,\mathbf{n}$ constitute such a set of three real independent parameters, chosen so as to display the parallelism between the group $U(2)$ and the proper rotation group $R^+(3)$. The parallelism is not quite an isomorphism, since

$$R(\varphi+2\pi,\mathbf{n}) \equiv \bar{R}(\varphi,\mathbf{n}) = -R(\varphi,\mathbf{n}) \qquad (2.6.14)$$

so that the two rotations φ and $\varphi+2\pi$, which are not regarded as distinct in $R^+(3)$, are represented by distinct elements in $U(2)$. Such a two-to-one, or, more generally, many-to-one correspondence between the elements in two groups is referred to as a *homomorphism*.

It is clear that if three group elements in $R^+(3)$ obey

$$RS = T \qquad (2.6.15)$$

the corresponding two-dimensional representation obeys

$$RS = \pm T \qquad (2.6.16)$$

We can say that the two-dimensional "spin" representation is not an ordinary ("vector") representation of the group $R^+(3)$. Rather it is a multiplier ("projective") representation of the group or a representation of the "multiplier group" (see Eq. 1.3.9). The reason why we permit such double-valued representations is that

$$R(2\pi,\mathbf{n})\psi = -\psi \qquad (2.6.17)$$

and $R(2\pi,\mathbf{n})\varphi = -\varphi$ does not interfere with the single valuedness of observables $(\varphi, A\psi)$.

Alternatively, we can say that the spin representation is a single-valued representation of the group $U(2)$, or of the *double group* $R^+(3)'$ formed by adding the element

$$\bar{E} = R(2\pi,\mathbf{n}) \qquad (2.6.18)$$

to $R^+(3)$. Thus the double group $R^+(3)'$ is a *covering group*[2] for $R^+(3)$.

Note that the formation of a double group from a single group involves not only the addition of an element $\bar{E}$, but also a modification of the multiplication table of the group. For example,

$$\left[R\left(\frac{2\pi}{p}\right)\right]^p = R(2\pi) = \bar{E} \neq E \qquad (2.6.19)$$

Fortunately, the multiplication table

$$R(\varphi,\mathbf{n}) = R(\varphi_1,\mathbf{n}_1)\cdot R(\varphi_2,\mathbf{n}_2) \qquad (2.6.20)$$

[2]A group whose single-valued (ordinary) representations include all the possible multiplier representations of a group H is referred to as a covering group for H.

can be written immediately, using Eq. 2.6.7 and the relation

$$(\boldsymbol{\sigma}\cdot\mathbf{n}_1)(\boldsymbol{\sigma}\cdot\mathbf{n}_2)=\mathbf{n}_1\cdot\mathbf{n}_2+i\boldsymbol{\sigma}\cdot(\mathbf{n}_1\times\mathbf{n}_2) \qquad (2.6.21)$$

with the result

$$\cos\tfrac{1}{2}\varphi=\cos\tfrac{1}{2}\varphi_1\cos\tfrac{1}{2}\varphi_2-(\mathbf{n}_1\cdot\mathbf{n}_2)\sin\tfrac{1}{2}\varphi_1\sin\tfrac{1}{2}\varphi_2 \qquad (2.6.22)$$

$$\mathbf{n}\sin\tfrac{1}{2}\varphi=(\mathbf{n}_1\times\mathbf{n}_2)\sin\tfrac{1}{2}\varphi_1\sin\tfrac{1}{2}\varphi_2+\mathbf{n}_1\sin\tfrac{1}{2}\varphi_1\cos\tfrac{1}{2}\varphi_2+\mathbf{n}_2\sin\tfrac{1}{2}\varphi_2\cos\tfrac{1}{2}\varphi_1 \qquad (2.6.23)$$

These equations distinguish between φ and $\varphi+2\pi$, that is, they tell whether RS is T or $\bar{T}$. The multiplication table of $R^+(3)$ is recovered, however, by simply ignoring this distinction (and is much more difficult to obtain by any direct method). Alternatively, it is possible to work always with the double group $R^+(3)'$.The representations in which $\bar{E}=+1$ will yield all the single-valued representations of $R^+(3)$. Those in which $\bar{E}=-1$ yield all the double-valued representations. [Since $\bar{E}$ commutes with all elements of $R^+(3)'$, it must be diagonal in each irreducible representation. The fact that $(\bar{E})^2=E=1$ guarantees that the only possibilities are $\bar{E}=\pm 1$.]

2.7. CLASS STRUCTURE OF POINT GROUPS

The class structure of the full rotation group, as well as of the point groups, can be elucidated immediately, using the $j=\frac{1}{2}$ representation, Eq. 2.6.7, as the simplest means of carrying out the group multiplications. We shall use the relation

$$U^{-1}f(A,B)U=f(U^{-1}AU,U^{-1}BU) \qquad (2.7.1)$$

(prove this by considering $A+B, AB, AB^{-1}$, etc.) in the form

$$U^{-1}\exp(-\tfrac{1}{2}i\psi\boldsymbol{\sigma}\cdot\mathbf{n})U=\exp(-\tfrac{1}{2}i\psi\boldsymbol{\sigma}'\cdot\mathbf{n}) \qquad (2.7.2)$$

where $\exp(-\frac{1}{2}i\psi\boldsymbol{\sigma}\cdot\mathbf{n})\equiv R(\psi,\mathbf{n})$ is a rotation of ψ about axis $\mathbf{n}$ and

$$\boldsymbol{\sigma}'=U^{-1}\boldsymbol{\sigma}U \qquad (2.7.3)$$

If we choose

$$U=\exp(-\tfrac{1}{2}i\theta\boldsymbol{\sigma}\cdot\mathbf{u})\equiv R(\theta,\mathbf{u}) \qquad (2.7.4)$$

Eqs. 2.6.7 and 2.6.21 yield

$$\boldsymbol{\sigma}'=\boldsymbol{\sigma}\cos\theta+\mathbf{u}(\mathbf{u}\cdot\boldsymbol{\sigma})(1-\cos\theta)+(\mathbf{u}\times\boldsymbol{\sigma})\sin\theta \qquad (2.7.5)$$

precisely the result Eq. 2.5.12 that must hold for an arbitrary vector. If we define

$\mathbf{n}'$ by

$$\boldsymbol{\sigma}'\cdot\mathbf{n}=\boldsymbol{\sigma}\cdot\mathbf{n}'$$

then

$$\mathbf{n}'=\mathbf{n}\cos\theta+\mathbf{u}(\mathbf{u}\cdot\mathbf{n})(1-\cos\theta)-(\mathbf{u}\times\mathbf{n})\sin\theta \tag{2.7.6}$$

Comparison with Eq. 2.5.11 shows that vector $\mathbf{n}$ has been rotated through an angle $-\theta$ about $\mathbf{u}$ to become $\mathbf{n}'$.

Theorem 2.7.1

$R(\varphi,\mathbf{n})$ and $R(\varphi',\mathbf{n}')$ are in the same class, provided that $\varphi'=\varphi$ and the group contains an element $R(\theta,\mathbf{u})$ such that $R(-\theta,\mathbf{u})$ carries $\mathbf{n}$ into $\mathbf{n}'$.

Corollary 2.7.2

$R(\varphi,\mathbf{n})$ and $R(-\varphi,\mathbf{n})$ are in the same class (the axis $\mathbf{n}$ is "bilateral"), provided that the group contains an element $R(\theta,\mathbf{n})$ which reverses the axis $\mathbf{n}$.

Proof: Use Theorem 2.7.1 and $R(\varphi,-\mathbf{n})=R(-\varphi,\mathbf{n})$. Note that such a reversing element, using Eq. 2.7.6, must have $\theta=\pi$ and $\mathbf{u}\cdot\mathbf{n}=0$, that is, there must be a twofold rotation axis normal to $\mathbf{n}$.

Corollary 2.7.3

For the proper rotation group $R^+(3)$, all $R(\varphi,\mathbf{n})$ for fixed φ and arbitrary $\mathbf{n}$ are in the same class, including $R(-\varphi,\mathbf{n})=R(\varphi,-\mathbf{n})$, that is, all elements with the same $|\varphi|$ are in the same class, and conversely.

Theorem 2.7.4

Theorems 2.7.1–2.7.3, by the nature of their derivation are applicable to double as well as single groups. By the converse of Corollary 2.7.3, $R(\varphi,\mathbf{n})$ and $\bar{R}(\varphi,\mathbf{n})=R(\varphi+2\pi,\mathbf{n})$ *can* be in the same class *only* if

$$|\varphi|=|\varphi+2\pi| \tag{2.7.7}$$

that is, $\varphi=-\pi$, $\varphi+2\pi=\pi$. But the equivalence of $R(-\pi,\mathbf{n})$ with $R(\pi,\mathbf{n})$ is the statement that the axis $\mathbf{n}$ is bilateral:

Theorem 2.7.5

In a double point group, only twofold axes *can* be in the same class as their "barred elements." They *will* be in the same class if and only if the group contains a twofold element at right angles to the original axis to make the latter bilateral.

2.8. THE PROPER POINT GROUPS

The proper point groups are most simply defined in terms of their generators, which must be proper rotations of the form

$$n \equiv C_n = R\left(\frac{2\pi}{n}, \mathbf{n}\right)$$

that is, an n-fold rotation axis. Here C_n is the Schoenflies notation, and n is the briefer International notation. (See Appendix B for a comparison of these systems.) A group containing C_n will also contain $C_n^p = R(2\pi p/n, \mathbf{n})$. In order of increasing complexity the proper point groups are as follows.

1. Groups with Only One Axis: C_n or n. These are labeled by their generating element and are cyclic groups.

2. Groups with Twofold Axes but Only One n-Fold Axis, $n > 2$: $D_n \equiv n2$ for n odd, $D_n \equiv n22$ for n even. The only way that we can add a twofold axis without creating a new n-fold axis is at right angles to the principal axis. The action of C_n creates n twofold axes. Figure 2.8.1 shows D_3 and D_4. Twofold axes are indicated by the symbol (), threefold axes by a triangle △, fourfold axes by a square □. The notation $n2$ already describes the generators of the group. Perhaps $n22$ is used when n is even because then alternating twofold axes fall into two nonequivalent classes, whereas for n odd all twofold elements are in the same class. Corollary 2.7.2 indicates that the twofold axes make the n-fold axis bilateral. Then C_n^p and C_n^{-p} are in the same class.

3. Groups with More Than One Principal Axis, $n > 2$. If two axes $n > 2$ and $n' > 2$ are present, the action of $C_{n'}$ will create at least one other axis n. Two axes n, by their mutual action, will create a whole set of axes n. Hamermesh (**1962**)

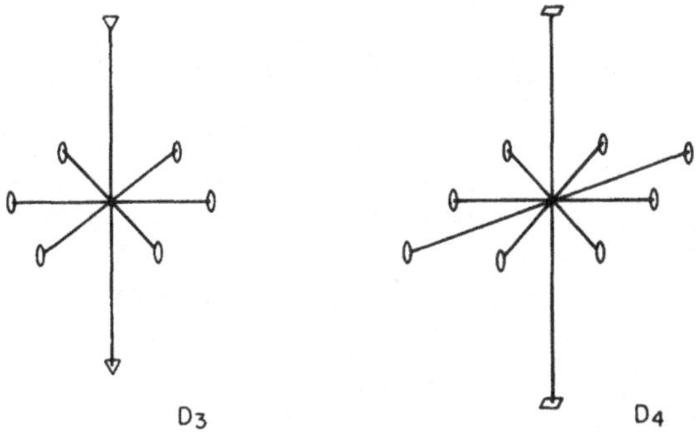

Fig. 2.8.1. Twofold (), threefold △ and fourfold □ symmmetry operations are shown for symmetry groups D_3 and D_4.

has shown that, if the directions of these axes are plotted as points on a sphere, the points are the vertices of a regular polyhedron. And the number of such regular polyhedra is known to be *finite*. Thus the number of point groups is finite.

Another argument can be based on Eqs. 2.6.22 and 2.6.23. The possible elements of the proper group are severely restricted by the requirement that the product of any two must also be a rotation of a rational fraction of 2π; see Heine (**1960**). The only groups that arise are the tetrahedral group T, the octahedral group O, and the icosahedral group Y.

The Tetrahedral Group T or 23. This is the group of proper rotations that restores the tetrahedron to itself. If we consider the tetrahedron to be embedded in a cube so that the four vertices of the tetrahedron coincide with four alternate vertices of the eight cube vertices (see Fig. 2.8.2), the group contains twofold rotations about axes perpendicular to each of the three pairs of cube faces (the group D_2) plus threefold rotations about the cube diagonals. The group is generated by a twofold and a threefold axis, not at right angles—hence the notation 23 rather than 32, which is D_3.

The Octahedral Group 0 or 432. This is the group of *proper* rotations that leave a cube invariant. It may be obtained from T by adding a fourfold axis perpendicular to any cube face.

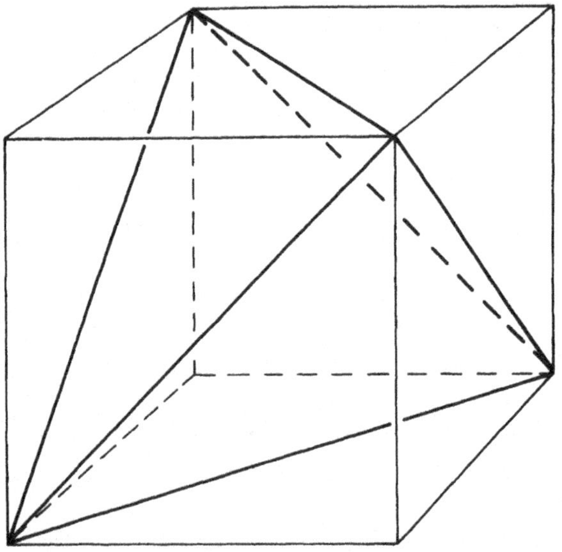

Fig. 2.8.2. Tetrahedral symmetry.

The classes and characters of the above groups are listed in Appendix A.

The Icosahedral Group Y. The icosahedral group, the symmetry group of the icosahedron and dodecahedron, has fivefold axes. Although a few molecules have this symmetry, we shall show presently that no crystal can have a fivefold axis. We shall not discuss the icosahedral group further than to remark that it is isomorphic to the group of even permutations on five objects of order $5!/2=60$.

2.9. NATURE OF IMPROPER ROTATIONS IN A FINITE GROUP

Our discussion up to this point has been restricted to proper rotations. We now consider the enlargement of a proper group by the addition of improper rotations, for example, reflections since many physical systems possess reflection symmetry. In this section we deduce the nature and multiplicative properties of all possible improper elements. The suceeding sections will relate the structure and representations of a group containing improper elements, called briefly an *improper group*, to the structure and representations of the proper group from which it is derived.

We are concerned with rotation groups $R(n)$ in an n-dimensional space that *can* be defined by real orthogonal $n \times n$ matrices R obeying

$$\tilde{R}R = E \tag{2.9.1}$$

where $\tilde{R}$ is the transpose of R. Elements are proper or improper according as $\det R = +1$ or -1, respectively.

Theorem 2.9.1

Every *improper* orthogonal matrix R obeys

$$\det(E+R) = 0 \tag{2.9.2}$$

Proof: The product of $\tilde{R}$ with $E+R$ is $\tilde{R}(E+R)=\tilde{R}+E$; hence $\det\tilde{R}\times \det(E+R)=\det(E+\tilde{R})$. But $\det\tilde{R}=-1$; therefore $-\det(E+R)=\det(E+\tilde{R})= \det(E+R)=0$.

Corollary 2.9.2

Every improper orthogonal matrix R has an eigenvalue -1, and an associated eigenvector $\mathbf{x}$ that lies along a *reversal axis* (i.e., an axis reversed by R), that is, $R\mathbf{x}=-\mathbf{x}$.

Proof: Since $\det(E+R)=0$, one eigenvalue of $E+R$ must be 0. Therefore one eigenvalue of R is -1.

In an odd number of dimensions, $-R$ is improper if R is proper.

Corollary 2.9.3

Every proper orthogonal matrix R in an odd number of dimensions has an eigenvalue $+1$, and an axis left invariant by R: $R\mathbf{x}=\mathbf{x}$.

This corollary has already been assumed in describing the elements of $R^+(3)$ by $R(\varphi,\mathbf{n})$ with an axis $\mathbf{n}$. The full rotation group $R(3)$ can be obtained from the proper rotation group $R^+(3)$ by adding the elements $R(\varphi,\mathbf{n})I$, where I is the inversion $I\mathbf{r}=-\mathbf{r}$. Therefore the improper *generating* elements of any *finite* point group in three space must have the form:

$$\bar{n}\equiv C_n I = R\left(\frac{2\pi}{n}\right)I \tag{2.9.3}$$

In an even number of dimensions this procedure fails, since I is then a proper element. Whether the number of dimensions is even or odd, the presence of a reversal axis ensures that the improper rotations can be generated using elements of the form[3]

$$\tilde{n}\equiv S_n = C_n\sigma_h = \sigma_h C_n \tag{2.9.4}$$

That is, a rotation of $2\pi/n$ followed (or preceded) by a reflection in a plane perpendicular to the rotation axis (rotary-reflection axis). (For the double group case, we find it convenient to follow the conventions of Appendix B and define $S_n = C_n\sigma_h\bar{E}$.)

For n odd,

$$(S_n)^n = \sigma_h \qquad (2.9.5)^{*\,4}$$

so that both σ_h and C_n are separately members of the group, whereas for n even this need not be the case.

The inversion element $I\mathbf{r}=-\mathbf{r}$ in this notation is

$$I\equiv S_2 = C_2\sigma_h = \sigma_h C_2 \tag{2.9.6}$$

The product of any two of I, C_2, σ_h yields the third. Since I is a constant matrix, it commutes with all elements of the group. We now state some elementary theorems.

[3] In Eq. 2.9.4 σ_h stands for a horizontal reflection plane, that is, one perpendicular to the axis; σ_v would stand for a vertical plane containing the axis. See Appendix B.

[4] Relations marked with an asterisk in this section may be modified by a factor $\bar{E}$ for double group operations. See Problems 2.16.3–2.16.6.

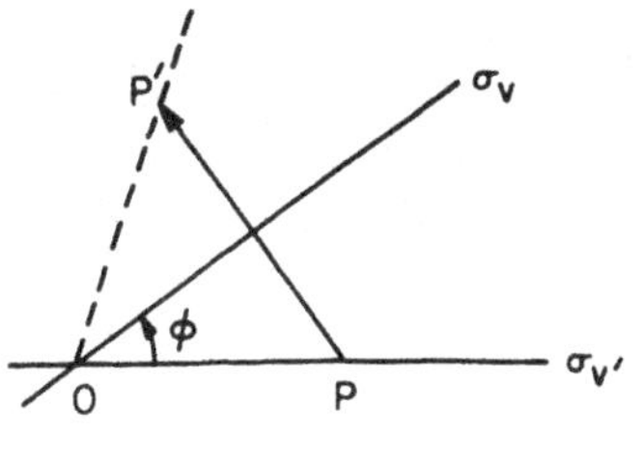

Fig. 2.9.1. Product of two mirror planes σ_v and $\sigma_{v'}$ yield a rotation $C(2\varphi)$.

Theorem 2.9.4*

The product of two reflection planes $\sigma_v \sigma_{v'}$ making an angle φ with respect to one another results in a rotation $C(2\varphi)$ of an angle 2φ about the intersection line as axis, in the direction from the plane of $\sigma_{v'}$ to that of σ_v. (See Fig. 2.9.1.)

Proof: The product contains an even number of reflections and is therefore a proper rotation. The intersection of the planes is invariant under the reflections and must be the axis of the rotation. A point p in plane $\sigma_{v'}$ is left invariant by the first reflection and traverses an angle 2φ on the second reflection. Thus

$$\sigma_v \sigma_{v'} = C(2\varphi), \qquad \sigma_{v'} = \sigma_v C(2\varphi) \tag{2.9.7}^*$$

Theorem 2.9.5 *

Two (intersecting) twofold axes making an angle φ with respect to one another generate $C(2\varphi)$ about an axis pp' perpendicular to the plane of the first two. (See Fig. 2.9.2.)

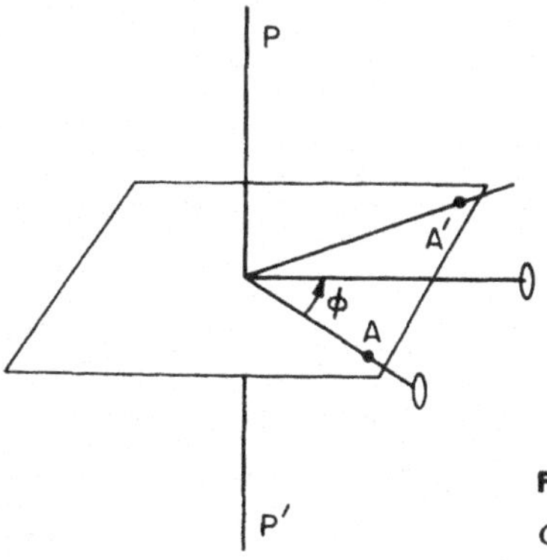

Fig. 2.9.2. Product of two twofold axes yields a rotation $C(2\varphi)$.

Proof: The first 180° rotation takes p into p', and the second takes p' back into p. Thus pp' is the axis of the resulting proper rotation. A point A on the first axis is unaffected by the first rotation and is taken to A' at an angle 2φ by the second. Q.E.D.

Theorem 2.9.6

The following operations commute:

1. Two rotations about the same axis.
2. Reflections in two mutually perpendicular planes:* $\sigma_v \sigma_{v'} = C(\pi)$ about their intersection.
3. Rotations about two perpendicular twofold axes:*

$$C_2 C_2' = C_2' C_2 = C_2''$$

where C_2'' is about the third perpendicular axis.

4. A rotation C_n and a reflection σ_h in a plane perpendicular to the axis.
5. Inversion and a rotation axis or a reflection plane that passes through the inversion point.
6. A twofold axis and a reflection plane σ_v containing the axis* (with the product a reflection plane $\sigma_{v'}$, also containing the axis, and at right angles to the first). See Theorem 2.9.4.

2.10. RELATION BETWEEN IMPROPER AND PROPER GROUPS

Let G be a group containing improper rotations M and proper rotations H:

$$G = H + M \tag{2.10.1}$$

where H contains at least the identity element. Since the product of two proper elements must be proper, H must be a subgroup of G. Since the product of any two improper elements is proper, and the product of a proper and an improper element is improper, we can conclude that H is an *invariant or normal subgroup*; this means that, for any $X \in G$, the set of elements H obeys

$$XHX^{-1} = H \tag{2.10.2}$$

If we multiply G by any improper element M_1, the group is rearranged by interchanging the proper and improper elements:

$$M_1(H + M) = M + H \tag{2.10.3}$$

thus requiring them to be equal in number. (See also Problem 2.3.1.) An immediate consequence of Eq. 2.10.3 is that we can write

$$G = H + M_1 H \tag{2.10.4}$$

(This is a special case of a coset expansion discussed in Section 8.5), that is, an improper group G can be obtained from a proper group H by adding any *one* of the improper elements M_1 as a generator. In order that no new proper element be created by addition of M_1, we must have the conditions that

$$(M_1)^2 \in H \tag{2.10.5}$$

and that

$$M_1 H_n M_1^{-1} \in H \tag{2.10.6}$$

where H_n is each generator of H. For example, we can add a reflection plane σ_h normal to a rotation axis **n**, or a plane σ_v containing the axis, but not a reflection plane at an arbitrary angle to the axis.

Since the inversion commutes with all rotations, we can generate from each proper group a correspondingly improper group by means of

$$G = H + IH \tag{2.10.7}$$

This procedure yields all groups that contain the inversion element. If an improper group $G = H + M_1 H$ does not contain the inversion element, then

$$G_p = H + (IM_1)H \tag{2.10.8}$$

is a proper group *isomorphic* to the original improper group, with the *same set of irreducible representations and characters and a corresponding set of classes.*

A simple way to construct the improper groups that do not contain the inversion is, then, to examine each proper group G_p to see whether it possesses a subgroup H of order h with *index* 2 (i.e., $g/h = 2$). The elements $H + I(G_p - H)$ constitute the improper group G. *If* the character table of the proper group G_p is available, every subgroup of index 2 can be determined quickly by looking at the representations whose elements (hence characters) all have the value ± 1. The positive elements belong to H, and the negative to $G_p - H$, in accord with Eq. 2.10.4. These are just the determinantal representations of the "improper groups" since R is represented by $\det R$.

2.11 REPRESENTATIONS OF GROUPS CONTAINING THE INVERSION

Since I commutes with all elements of the group, it will be in a class by itself, taking the eigenvalue $+1$ and -1. Thus, if $D^j(R)$ is an irreducible representation of H, then $j+$ and $j-$ obeying

$$D^{j\pm}(R) = D^j(R), \qquad D^{j\pm}(IR) = \pm D^j(R) \tag{2.11.1}$$

for proper R will be two irreducible representations of $G = H + IH$. If C is a class in H, C and IC will be distinct classes in G. The resulting character table will look like this:

$$\begin{array}{c|cc} & C & IC \\ \hline \Gamma_{j+} & \chi^j(C) & \chi^j(C) \\ \Gamma_{j-} & \chi^j(C) & -\chi^j(C) \end{array} \tag{2.11.2}$$

Thus the character table of a group containing the inversion can be written immediately from the corresponding proper group. In particular, the representations Γ_j of the proper rotation group $R^+(3)$ become the representations Γ_j^+ and Γ_j^- of the full rotation-reflection group $R(3)$. This seems to contradict the well-known assertion in elementary quantum-mechanics textbooks that the states of angular momentum l have parity $(-1)^l$. The latter statement is true for one-electron wave functions, but two-electron states can have even parity and odd angular momentum, and vice versa. For example, $(\mathbf{r}_1 \times \mathbf{r}_2)f(|\mathbf{r}_1 - \mathbf{r}_2|)$ constitutes a set of three even P states. Odd S states are exceptional in that they require at least three-body wave functions, for example, $\mathbf{r}_1 \cdot \mathbf{r}_2 \times \mathbf{r}_3$.

2.12. PRODUCT GROUPS

The particularly simple results of Eq. 2.11.2 are a consequence of the fact that G can be written as a direct product of the group H with the group $\mathcal{C}_i = (E, I)$.

Definition: PRODUCT GROUPS (outer Kronecker product)

The direct product $H \times K$ of two groups

$$H = (E \equiv A_1, A_2, \ldots, A_h), \qquad K = (E \equiv B_1, B_2, \ldots, B_k)$$

is a group whose elements are denoted as $A_i \times B_j$ with multiplication defined by

$$(A_i \times B_j)(A_m \times B_n) = (A_i A_m \times B_j B_n) \tag{2.12.1}$$

so that the group properties are automatically fulfilled.

As defined, there is no connection between the H and K operators. In a realization of the groups, the A and B operators may act on entirely different spaces. For example, the symmetry of a molecule can be made up of operators that act on the nuclear coordinates and those that act on the electronic coordinates. The ordinary operator product relationship

$$A_i B_j A_m B_n = A_i A_m B_j B_n \tag{2.12.2}$$

is then valid because the A's commute with the B's.

An example of operators that act on the same variables yet commute with each other is one in which H represents a group of rotations acting on some particles and K represents the permutations among the particles. An even simpler example is the present one: any group of proper rotations, and the group consisting of inversion (and the identity). The order of a product group $H \times K$ is hk, where h and k are the orders of H and K, respectively. (This assumes that H and K have only the identity element in common.) If C is a class of H, and D a class of K, then $C \times D$ is a class of $H \times K$ and all the classes of the latter are formed in this manner.

2.13. REPRESENTATIONS OF AN OUTER PRODUCT GROUP

If $D^i_{\mu\mu'}(A)$ is a representation of group H, and $D^j_{\nu\nu'}(B)$ is a representation of group K, then

$$D^{i\times j}_{\mu\nu;\mu'\nu'}(A\times B)\equiv D^i_{\mu\mu'}(A)D^j_{\nu\nu'}(B) \tag{2.13.1}$$

will be a representation of the outer product group $H \times K$, since the rule for multiplication of ordered pairs, Eq. 2.12.1, simply states that the A and B matrices should be multiplied separately. The indices $(\mu\nu)$ run over a set of $l_i l_j$ values. Thus

$$l_{i\times j}=l_i l_j \tag{2.13.2}$$

The characters may be obtained by setting $(\mu\nu)\equiv(\mu'\nu')$ and summing:

$$\chi^{i\times j}(A\times B)=\chi^i(A)\chi^j(B) \tag{2.13.3}$$

We will now prove that, if i is irreducible in H and j is irreducible in K, then $i\times j$ is irreducible in $H \times K$, using the character normalization relation 1.5.6 (see also Eq. 3.1.6) as a criterion:

$$\frac{1}{hk}\sum_{A\times B}|\chi^{i\times j}(A\times B)|^2=\frac{1}{hk}\sum_{A,B}|\chi^i(A)|^2|\chi^j(B)|^2$$

$$=\left[\frac{1}{h}\sum_A|\chi^i(A)|^2\right]\left[\frac{1}{k}\sum_B|\chi^j(B)|^2\right]=1 \tag{2.13.4}$$

The completeness of the set of representations can be tested by Burnside's theorem (Theorem 1.5.5):

$$\sum_{i\times j}(l_{i\times j})^2=\sum_{i,j}(l_i)^2(l_j)^2=hk=g \tag{2.13.5}$$

Thus the representations that we have obtained by multiplying a proper group by $\mathcal{C}_i \equiv (E, I)$ are irreducible and exhaust the set of irreducible representations of the larger group.

Our results concerning improper groups can be summarized as follows:

Theorem 2.13.1

The improper groups that do not contain the inversion are isomorphic to and have the same representation as the proper group formed by multiplying each improper element by the inversion. The improper groups containing the inversion can be written as a direct product of a proper group with the group $\mathcal{C}_i$ and are represented by the product of the representations of these two groups.

This theorem applies to double as well as single groups.

2.14. ENUMERATION OF THE IMPROPER POINT GROUPS

In accord with Eq. 2.10.4 the improper groups can be constructed by adding to a proper group H a single improper element, say M, as generator. Since M^2 is proper, we may have $M^2 = E$; in other words, $M = I$ or σ or, more generally,

$$M^2 = C_n \qquad \text{or} \qquad M = S_{2n}$$

Note: If $M^2 = (C_n)^p$, it is always possible to find another M in the group such that $M^2 = C_n$. (Prove this.)

Clearly each proper point group can lead only to a finite number of new groups containing improper elements, since the new element added, I, σ, or S_{2n}, must be such as to create no new proper rotations. A reflection plane, for example, must either be perpendicular to a single principal axis or contain it. The following new groups are obtained in this way.[5]

The Rotary Reflection Groups: $\mathcal{S}_{2n} \equiv (\overline{2n})$, *n even;* $\mathcal{S}_{2n} \equiv \bar{n}$, *n odd.* This is a cyclic group of $2n$ elements. Since $(S_{2n})^n = (\sigma_h C_{2n})^n = \sigma_h^n C_2$ and $\sigma_h C_2 = I$, we note that, for n odd, $\mathcal{S}_{2n} \equiv \mathcal{C}_n \times \mathcal{C}_i$, where $\mathcal{C}_i$ contains only E and I, and in particular $\mathcal{S}_2 \equiv \mathcal{C}_i$.

The $\mathcal{C}_{nh}$ Groups: n/m, $m =$ Mirror Plane σ_h (but $\mathcal{C}_{3h} = \bar{6}$). Add a mirror plane σ_h perpendicular to the axis of C_n. The group is Abelian with elements $(C_n)^p$ and $\sigma_h (C_n)^p$, $p = 1, 2, \ldots, n$ and (for $p = n$) contains σ_h itself. For n even, $\sigma_h (C_n)^{n/2} = \sigma_h C_2 = I$ and $C_{nh} = C_n \times C_i$. The simplest group $\mathcal{C}_{1h}$, also known as $\mathcal{C}_s$, contains E and σ_h. The group $\mathcal{C}_{3h}$ is designated by its generator, $\bar{6}$.

The $\mathcal{C}_{nv}$ Groups: nm for n odd, and nmm for n even. Add one vertical plane σ_v. Then C_n creates $n-1$ other planes that intersect in the axis and make an angle

[5] The relation between International and Schoenflies notation is summarized in Appendix B.

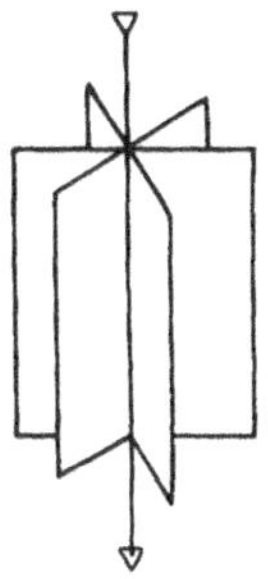

Fig. 2.14.1. Symmetry operations of the group C_{3v}.

π/n. (See Fig. 2.14.1.) The group $\mathcal{C}_{nv}$ is isomorphic to D_n and is obtained from it by replacing the twofold generator U_2 at right angles to C_n by $IU_2=\sigma_v$.

The D_{nh} Groups: $\frac{n}{m}\frac{2}{m}\frac{2}{m}$ *or, briefly,* $D_{2h}=mmm$, $D_{3h}=\bar{6}\,m2$, $D_{4h}=4/mmm$, $D_{6h}=6/mmm$.

$$D_{nh}=D_n\times\mathcal{C}_s \tag{2.14.1}$$

where $\mathcal{C}_S=(E,\sigma_h)$, and σ_h commutes with the elements of D_n. (See Fig. 2.14.2.) For n even, $(C_n)^{n/2}=C_2$ and $C_2\sigma_h=I$, so that $D_{nh}=D_n\times\mathcal{C}_1$.

The D_{nd} Groups: $D_{3d}=\bar{3}\frac{2}{m}$; $D_{2d}=\bar{4}2m$. To the elements of D_n: $(C_n)^k$, $k=0,\ldots,n-1$, and n twofolds U_2 normal to C_n, add a vertical reflection plane σ_d (d for diagonal) midway between a pair of twofold axes; see Fig. 2.14.3. Thus σ_d makes an angle $\pi/2n$ with the nearest twofold U_2. For this pair $U_2\sigma_d=S_{2n}$.

For n odd, one U_2 is perpendicular to one σ_d, so that inversion is in the group and $D_{2p+1,d}=D_{2p+1}\times\mathcal{C}_i$.

The Group T_d: $\bar{4}3m$. We can achieve the full symmetry of the tetrahedron by adding to T a reflection plane that passes through one edge of the tetrahedron

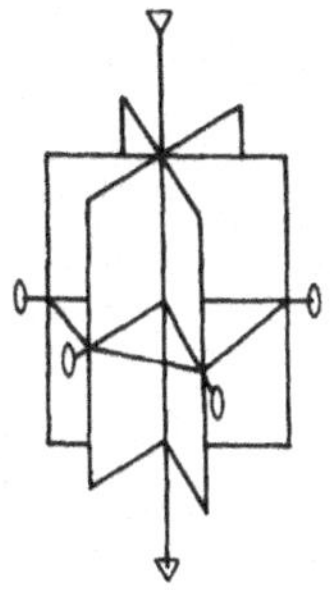

Fig. 2.14.2. Symmetry operations of the group D_{3h}.

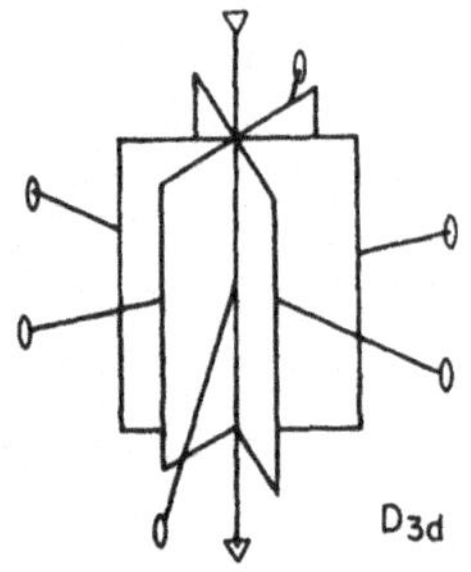

Fig. 2.14.3. Symmetry operations of the group D_{3d}.

and bisects the opposite edge. This plane contains one twofold axis and two threefold axes (see Fig. 2.14.4), thus rendering the latter bilateral.

The reflection planes (as in D_{2nd}) convert the twofold axes to fourfold rotary reflection axes S_4.

The Group T_h: $\frac{2}{m}\bar{3}$ *or, briefly*, $m3$. This group is obtained from T by adding an inversion center: $T_h = T \times \mathcal{C}_i$.

The Group O_h: $\frac{4}{m}\bar{3}\frac{2}{m}$ *or* $m3m$.

$$O_h = O \times \mathcal{C}_i$$

has the *full* symmetry of the cube.

The Group Y_h.

$$Y_h = Y \times \mathcal{C}_i$$

completes the list of possible improper point groups.

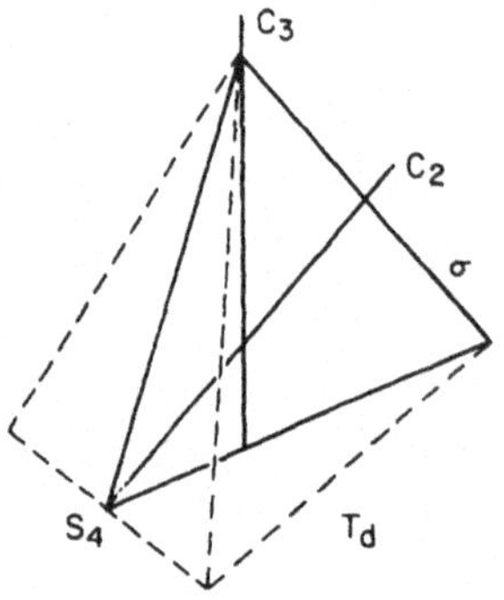

Fig. 2.14.4. Symmetry operations of the group T_d.

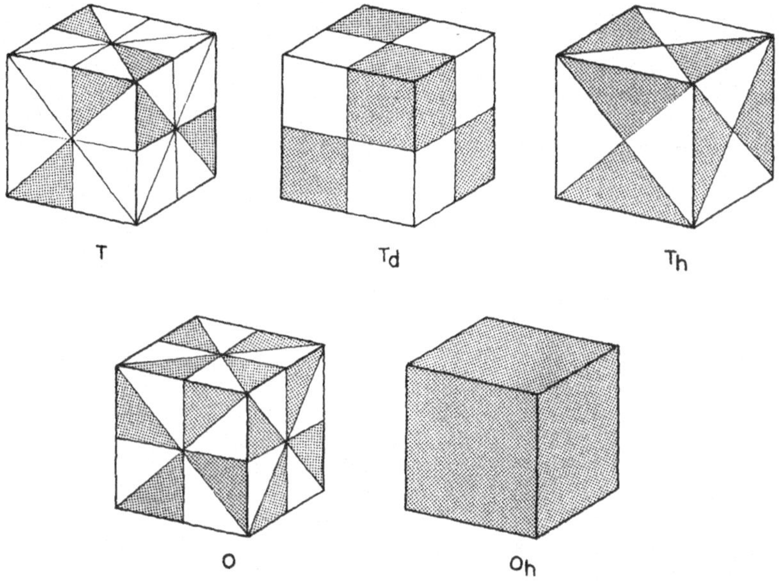

Fig. 2.14.5. The above cubes, colored black and white, are restored to themselves by the operations of one of the appropriately labeled cubic groups.

Figure 2.14.5 illustrates the differences between the five "cubic" groups T, T_d, T_h, O, O_h. They are referred to as cubic because, as we shall see in Chapter 6, they occur only in crystals with cubic Bravais lattices.

2.15. CRYSTALLOGRAPHIC POINT GROUPS

Point groups that leave some space lattice invariant are referred to as crystallographic point groups.

In an appropriate set of axes (primitive translation vectors; see Section 6.1) the lattice points are represented by integers. A rotation that transforms any set of integers into a new set of integers must, in this base system, be represented by matrices whose elements are integers. Their *characters*, which are the same in any base system, must therefore be represented by integers. Hence a rotation through angle φ must obey

$$1+2\cos\varphi = \text{an integer} \tag{2.15.1}$$

The only possible integers are -1, 0, 1, 2, and 3. The corresponding rotation angles are $180°, 120°, 90°, 60°$, and $0°$, that is, C_2, C_3, C_4, C_6, and C_1 (the identity). Thus fivefold axes and axes of order greater than 6 are excluded. The same conditions apply to rotary-reflection axes.

Table 2.15.1. Relation of Improper to Proper Groups

Proper Point Group G	Improper Group Without Inversion Isomorphic to G	$G\times\mathcal{C}_i$	Proper Group Isomorphic to $G\times\mathcal{C}_i$
$\mathcal{C}_1$		$\mathcal{C}_1\times\mathcal{C}_i=\mathcal{C}_i$	$\mathcal{C}_2$
$\mathcal{C}_2$	$\mathcal{S}_1\equiv\mathcal{C}_s$	$\mathcal{C}_2\times\mathcal{C}_i=\mathcal{C}_{2h}$	D_2
$\mathcal{C}_3$		$\mathcal{C}_3\times\mathcal{C}_i=\mathcal{S}_6$	$\mathcal{C}_6$
$\mathcal{C}_4$	$\mathcal{S}_4$	$\mathcal{C}_4\times\mathcal{C}_i=\mathcal{C}_{4h}$	
$\mathcal{C}_6$	$\mathcal{C}_{3h}\equiv\mathcal{C}_3\times\mathcal{C}_s$	$\mathcal{C}_6\times\mathcal{C}_i=\mathcal{C}_{6h}$	
D_2	$\mathcal{C}_{2v}$	$D_2\times\mathcal{C}_i=D_{2h}$	
D_3	$\mathcal{C}_{3v}$	$D_3\times\mathcal{C}_i=D_{3d}$	D_6
D_4	$\mathcal{C}_{4v}, D_{2d}$	$D_4\times\mathcal{C}_i=D_{4h}$	
D_6	$\mathcal{C}_{6v}, D_{3h}\equiv D_3\times\mathcal{C}_s$	$D_6\times\mathcal{C}_i=D_{6h}$	
T		$T\times\mathcal{C}_i=T_h$	
O	T_d	$O\times\mathcal{C}_i=O_h$	
$\mathcal{C}_n$		$\mathcal{C}_n\times\mathcal{C}_h=\mathcal{C}_{nh}$	
D_n	$\mathcal{C}_{nv}$	$D_n\times\mathcal{C}_h=D_{nh}$	
$\mathcal{C}_{2n}$		$\mathcal{C}_{2n}\times\mathcal{C}_i=\mathcal{C}_{2nh}$	
D_{2n}	$D_{nd}, \mathcal{C}_{2nv}$	$D_{2n}\times\mathcal{C}_i=D_{2nh}$	
$\mathcal{C}_{2n+1}$		$\mathcal{C}_{2n+1}\times\mathcal{C}_i=\mathcal{S}_{4n+2}$	
D_{2n+1}	$\mathcal{C}_{2n+1,v}$	$D_{2n+1}\times\mathcal{C}_i=D_{2n+1,d}$	
$\mathcal{C}_{4n}$	$\mathcal{S}_{4n}$		

We are thus restricted to 11 possible proper point groups, $\mathcal{C}_1$, $\mathcal{C}_2$, $\mathcal{C}_3$, $\mathcal{C}_4$, $\mathcal{C}_6$, D_2, D_3, D_4, D_6, T, and O, and a total of 32 distinct point groups (only 18 of which are nonisomorphic). Each improper point group either is the direct product of a point group with $\mathcal{C}_i$ or is isomorphic to the proper point group formed by replacing each improper element by the inversion I times itself. See Section 2.10. These relationships are shown in Table 2.15.1.

2.16. DOUBLE POINT GROUPS[6]

If any set of proper rotation elements $R(\varphi,\mathbf{n})$, regarded either as abstract elements or as three-dimensional rotations, obeys the multiplication table of a point group G, the corresponding set of 2×2 matrices

$$R(\varphi,\mathbf{n})=\exp\left(-\tfrac{1}{2}i\varphi\boldsymbol{\sigma}\cdot\mathbf{n}\right) \qquad (2.16.1)$$

[6] Bethe (**1929**) and Opechowski (**1940**).

obeys the same multiplication table except for the appearance of some minus signs. These 2×2 matrices and their multiplication rules then define the double group G' that corresponds to G:

Theorem 2.16.1

There is a double group G' that corresponds to each *proper* group G.

Since angular momentum $\mathbf{r}\times\mathbf{p}$ is invariant under inversion, we may conclude that $\mathbf{J}$ and $\boldsymbol{\sigma}$ commute with I. Thus the *inversion operator* continues to *commute* with *all elements*, even in a *double group*. If a point group G contains the inversion, then

$$G = H\times \mathcal{C}_i$$

and the corresponding double groups obey the same relation,

$$G' = H'\times \mathcal{C}_i \qquad (2.16.2)$$

Similarly, a group G that does not contain the inversion is isomorphic to a corresponding proper group,

$$G = H + M_1 H \sim H + (IM_1)H$$

in accord with Eq. 2.10.4, and the same relation holds for the corresponding double groups:

$$G' = H' + M_1 H' \sim H' + (IM_1)H' \qquad (2.16.3)$$

Thus the class structure, representations, and character tables of the improper double groups can be obtained immediately from those of the proper double groups by means of the direct product relationship 2.16.2 or the isomorphism equation 2.16.3.

The class structure of the proper double groups is defined by the following theorem (Opechowski, **1940**).

Theorem 2.16.2

If C is a class of n-fold rotations in a single group, then C and $\overline{C}$ are distinct classes in the double group except that, for $n=2$, C and $\overline{C}$ can be joined to form a single class $C+\overline{C}$ if any twofold axis in C is bilateral.

Proof: If $XRX^{-1} = S$ in the single group G, then

$$XRX^{-1} = \pm S \text{ in } G' \qquad (2.16.4)$$

Thus, if C of G has c elements, there is a set of c elements in G' conjugate to R. There is also a set of c elements conjugate to $\overline{R}$. We shall call these sets C and $\overline{C}$. For $n\neq 2$, by Theorem 2.7.4, R and $\overline{R}$ cannot be in the same class. Hence C and

$\bar{C}$ are distinct classes. For $n=2$, they can be united if any twofold element is made bilateral by some element of G.

The theorem has now been established. The only possible point of ambiguity is this: Which elements are in C, and which are in $\bar{C}$? This question is answered by Theorem 2.7.1 in the following form: Two elements in the same class are rotations through the same angle in the same sense about equivalent directed axes. One possible point of confusion occurs in the group D_3. Elements C_3 and C_3^2 are listed as in the same class in the single group. Are we to take

$$C=C_3, C_3^2, \qquad \bar{C}=\bar{E}C_3, \bar{E}C_3^2 \tag{2.16.5}$$

or

$$C=C_3, C_3^{-1}, \qquad \bar{C}=\bar{E}C_3, \bar{E}C_3^{-1}$$

The latter choice is the correct one, by Corollary 2.7.2. It is always $R(-\varphi,\mathbf{n}) =[R(\varphi,\mathbf{n})]^{-1}$ that is in the same class as $R(\varphi,\mathbf{n})$, and not $R(2\pi-\varphi,\mathbf{n})$.

2.17. SUMMARY

The group $R^+(3)$ of all proper rotations in three space (whose defining representation is a set of 3×3 matrices stated implicitly in Eq. 2.5.11) is found to have double as well as single-valued irreducible representations. This double valuedness arises because the operation $\bar{E}=R(2\pi,\mathbf{n})$ [where $R(\varphi,\mathbf{n})$ is a rotation of angle φ about axis $\mathbf{n}$] can take the value $+1$ or -1 when acting on a wave function describing a physical system with total spin integral or half-integral, respectively. Double groups G' are then defined by introducing an additional generator $\bar{E}$ into the original group G and redefining the multiplication table to distinguish between $R(\varphi,\mathbf{n})$ and $R(\varphi+2\pi,\mathbf{n})$. The two-dimensional ($j=\frac{1}{2}$) representation of $R^+(3)'$, namely,

$$R(\varphi,\mathbf{n})=\cos\tfrac{1}{2}\varphi - i\boldsymbol{\sigma}\cdot\mathbf{n}\sin\tfrac{1}{2}\varphi$$

distinguishes between $R(\varphi,\mathbf{n})$ and $R(\varphi+2\pi,\mathbf{n})=\bar{E}R(\varphi,\mathbf{n})$ and, moreover, provides the simplest calculational scheme (Eqs. 2.6.20–2.6.24) for multiplying both single and double group operations. With $R(\varphi,\mathbf{n})$ as the defining representation, the class structures of single and double point groups of proper rotations are elucidated.

A point group G containing improper rotations is shown either to be a direct product between a proper point group H and the group $\mathcal{C}_i$ or to be isomorphic to the proper point group G_p formed from G by replacing each improper element M by IM. The characters and representations of single (and double) improper point groups are thus immediately expressed in terms of the corresponding proper point groups H and G_p.

problems

2.3.1. Show that, if a group contains an element S such that $S^2=(C_n)^p$ (where p and n have no common factors), it also contains an element R such that $R^2=C_n$.

2.5.1. Let $l_z=-i(\mathbf{r}\times\nabla)_z=-i(x\frac{\partial}{\partial y}-y\frac{\partial}{\partial x})$

(*a*) Show that

$$l_z[x,y,z]=[iy,-ix,0]$$
$$=[x,y,z]\mathbf{D}(l_z)$$

where

$$\mathbf{D}(l_z)=\begin{bmatrix}0 & -i & 0\\ i & 0 & 0\\ 0 & 0 & 0\end{bmatrix}$$

(*b*) Similarly show that

$$\mathbf{D}(l_x)=\begin{bmatrix}0 & 0 & 0\\ 0 & 0 & -i\\ 0 & i & 0\end{bmatrix},\qquad \mathbf{D}(l_y)=\begin{bmatrix}0 & 0 & i\\ 0 & 0 & 0\\ -i & 0 & 0\end{bmatrix}$$

(*c*) Show that the matrices $\mathbf{D}(l_x)$, $\mathbf{D}(l_y)$, $\mathbf{D}(l_z)$ obey the commutation rules 2.2.2 of the J operators, and constitute the representation appropriate to $j=1$ in a base system isomorphic to $[x,y,z]$.

2.5.2. Show that

$$\exp(-i\alpha\mathbf{l}\cdot\mathbf{n})[x,y,z]=[x,y,z]\cdot\mathbf{R}$$

where $\mathbf{R}=\mathbf{D}^p(R)$, the polar vector representation, is given by

$$\mathbf{R}=\exp[-i\alpha\mathbf{n}\cdot\mathbf{D}(\mathbf{l})]$$

where $\mathbf{D}(\mathbf{l})=\mathbf{i}\mathbf{D}(l_x)+\mathbf{j}\mathbf{D}(l_y)+\mathbf{k}\mathbf{D}(l_z)$ is determined by Problem 2.5.1. Verify that this result is consistent with Eq. 2.5.11, that is, that

$$\exp(-i\alpha\mathbf{l}\cdot\mathbf{n})\mathbf{r}=\mathbf{r}\cos\alpha+\mathbf{r}\cdot\mathbf{n}\mathbf{n}(1-\cos\alpha)+\mathbf{r}\times\mathbf{n}\sin\alpha$$

or

$$\mathbf{R}=\begin{bmatrix} \cos\alpha+l^2(1-\cos\alpha) & lm(1-\cos\alpha)-n\sin\alpha & ln(1-\cos\alpha)+m\sin\alpha \\ ml(1-\cos\alpha)+n\sin\alpha & \cos\alpha+m^2(1-\cos\alpha) & mn(1-\cos\alpha)-l\sin\alpha \\ nl(1-\cos\alpha)-m\sin\alpha & nm(1-\cos\alpha)+l\sin\alpha & \cos\alpha+n^2(1-\cos\alpha) \end{bmatrix}$$

where $[l,m,n]\equiv\mathbf{n}$.

2.5.3. Show that requirement 2.5.17,

$$O_R\psi(\mathbf{r})=\psi(\mathbf{R}^{-1}\cdot\mathbf{r})=\psi(\mathbf{rR})$$

obeys the group property

$$O_RO_S=O_{RS}$$

whereas the opposite choice

$$O_R\psi(\mathbf{r})=\psi(\mathbf{R}\cdot\mathbf{r})$$

does not.

2.6.1. Multiply $\boldsymbol{\delta}_{3xyz}$ and $\boldsymbol{\delta}_{3x\bar{y}\bar{z}}$ together as single group operations and as double group operations. (See Appendix B for operator notation.)

2.6.2. If ξ and η transform as the fundamental spin states $\psi^{1/2}_{1/2}$ and $\psi^{1/2}_{-1/2}$, respectively, then, according to Eq. B6,

$$R(\alpha,\mathbf{n})[\xi,\eta]=[\xi,\eta]\exp\left(-\tfrac{1}{2}i\alpha\boldsymbol{\sigma}\cdot\mathbf{n}\right)$$

Show that $\xi_1\eta_2-\xi_2\eta_1$ is invariant under such a spinor transformation.

2.6.3. Show that the set of functions

$$|j\mu\rangle=\frac{\xi^{j+\mu}\eta^{j-\mu}}{[(j+\mu)!(j-\mu)!]^{1/2}}$$

constitutes a set of basis functions for the irreducible representations of the special unitary group $SU(2)$, whose defining elements are $\exp(-\tfrac{1}{2}i\alpha\boldsymbol{\sigma}\cdot\mathbf{n})$.

2.6.4. Construct the matrix elements appropriate to the $j=3/2$ representation of the angular momentum operators J_x, J_y, J_z. Construct the corresponding matrices for the finite rotation $\exp(-i\alpha\mathbf{J}\cdot\mathbf{n})$. (Compare with Problem 2.5.2.)

2.7.1. Show that $\boldsymbol{\sigma}_v$ makes an axis bilateral but $\boldsymbol{\sigma}_h$ does not.

2.9.1. Show that $(S_3)^3=\bar{E}\boldsymbol{\sigma}_h$, whereas $(S_5)^5=\boldsymbol{\sigma}_h$.

2.10.1. Show that any subgroup of index 2 must be an invariant subgroup.

2.10.2. Prove the subgroup class theorem: The elements of G appear in an invariant subgroup H in complete classes or not at all. Conversely, if several classes of G from a subgroup of G, the subgroup is invariant. *Note*: The elements of a class of G that appear in H need not be in the same class in H.

2.10.3. Show that H and K are invariant subgroups of $H \times K$.

2.15.1. Consider a proper point group, represented by its defining three-dimensional representation. Let S represent the sum of the elements of the group. Show that

$$R_1 S\mathbf{t} = R_2 S\mathbf{t} = S\mathbf{t}$$

where $\mathbf{t}$ is an arbitrary vector and R_1, R_2 are group elements. Show that, if the group contains more than one axis, $S=0$. By taking the character of this relation prove that

$$s_2 = 3 + s_4 + 3s_6$$

where s_n is the number of n-fold rotation axes in the group (Heine, **1960**, Eq. 16.7).

2.16.1. Show that in a double group an n-fold axis is made bilateral by a reflection plane containing the axis. Show that a mirror plane is made bilateral by a two fold axis in the plane or a mirror plane perpendicular to the first plane.

2.16.2. (*a*) Show that, if A and $\bar{A}$ are in the same class, $\chi(A)=0$ in all double-valued representations.
(*b*) Show that in this case no one-dimensional (double-valued) representations exist.
(*c*) Show that if, for all classes, C and $\bar{C}$ are distinct, a one-dimensional representation must exist.

2.16.3. Show that for double group operations Eq. 2.9.7 must be replaced by

$$\sigma_v \sigma_{v'} = \bar{E}C(2\varphi)$$

where φ is the angle between the two reflection planes.

2.16.4. Show that for rotations about two perpendicular axes the double group operations obey

$$C_{2x}C_{2y} = -C_{2y}C_{2x} = C_{2z}; \ (C_{2z})^{-1} = -C_{2z}$$

where the corresponding results in Theorem 2.9.6 for the single group have the minus signs above replaced by plus signs.

2.16.5. Show that $\mathcal{C}_6 = \mathcal{C}_3 \times \mathcal{C}_2$ is a single group, but that this direct product relationship is not valid for double groups: $\mathcal{C}_6' \neq \mathcal{C}_3' \times \mathcal{C}_2'$.

2.16.6. Show that the definition $S_n = \sigma_h C_n \bar{E}$ of Eq. 2.9.4 and the following text is compatible with the definitions

$$S_4^{-1} = IC_4, \qquad S_6^{-1} = IC_3, \qquad \text{and} \qquad S_3^{-1} = IC_6$$

of Appendix B.

chapter three

POINT GROUP EXAMPLES

In this chapter we illustrate the methods of group theory by discussing a number of examples that require a knowledge only of the point groups and the representation theorems of Chapter 1. Each example is chosen to illustrate a widely used technique: finding the irreducible components of a space, determining compatibility relations, applying selection rules, and so forth. Each section, then, starts by discussing a technique and ends by illustrating it. Simple examples are chosen so as not to require an involved discussion before the technique can be applied. The answers are often intuitively obvious. We use sledge hammers to crack peanuts here because the problems that arise in the application of space groups will be tougher nuts requiring heavy tools.

3.1. ELECTRIC AND MAGNETIC DIPOLES: IRREDUCIBLE COMPONENTS OF A REDUCIBLE SPACE

In Section 3.4 we shall be concerned with the determination of selection rules for matrix elements of the form

$$(\psi_\mu^i, V\psi_\nu^j) = \int (\psi_\mu^i)^* V\psi_\nu^j \, d\tau \tag{3.1.1}$$

In order to determine these selection rules, however, it is desirable to decompose V into parts that transform according to some irreducible representation

$$V = \sum_{m\lambda} V_{\lambda}^{m} \tag{3.1.2}$$

We shall illustrate the procedure for the cases

$$V = \mathbf{p} = \sum e_i \mathbf{r}_i = \text{the electric moment or polarization vector}$$

and

$$V = \mathbf{m} = \frac{e\hbar}{2mc} \sum (\mathbf{l}_i + 2\mathbf{s}_i) = \text{the magnetic moment vector} \tag{3.1.3}$$

The components of $\mathbf{p}$ (or $\mathbf{m}$) span a three-dimensional invariant space. We wish to find out whether, with respect to a given group, this space is irreducible, and if not what its irreducible components are. These questions can be answered by computing the characters $\chi(R)$ in the space spanned by $\mathbf{p}$ (or $\mathbf{m}$) of the operations R of the group, and attempting to expand these in terms of the irreducible representations of the group:

$$\chi(R) = \sum k_j \chi^j(R) \tag{3.1.4}$$

Since a reduced representation of the block form 1.4.4 has a trace that is the sum of the traces of the individual blocks, each number k_j will be an integer representing the number of times that the irreducible representation appears in the original space.

Equation 3.1.4 is valid for any reducible representation even if the reduction has not yet been performed, since the latter step means a similarity transformation that does not modify traces.

The integers k_j can usually be read off by inspection, with the help of a character table. In any case, the character orthogonality theorem, Eq. 1.5.6, permits us to solve for k_j in the form

$$\begin{aligned} k_j &= \frac{1}{g} \sum_{R \in G} \chi(R) \chi^j(R)^* \\ &= \frac{1}{g} \sum_{C} n(C) \chi(C) \chi^j(C)^* \end{aligned} \tag{3.1.5}$$

On the other hand, the relation

$$\frac{1}{g} \sum_{R} |\chi(R)|^2 = \sum_{j} k_j^2 \tag{3.1.6}$$

can be used without a character table to tell us whether or not a *representation is irreducible*, since for the irreducible representation the sum must be unity.

With j, the "angular momentum," defined in Eq. 2.3.1, the representations of the full rotation-reflection group are denoted as Γ_j^+ or Γ_j^- according to whether they are even (+) or odd (−) with respect to inversion. The close relationship to the terminology of atomic spectroscopy permits these states also to be labeled as $\Gamma_0^\pm = S^\pm, \Gamma_1^\pm = P^\pm, \Gamma_2^\pm = D^\pm, \Gamma_3^\pm = F^\pm$, and so on. Without computation, we know then that

$$\mathbf{p} \sim \Gamma_1^- = P^-, \qquad \mathbf{m} \sim \Gamma_1^+ = P^+ \tag{3.1.7}$$

since **p** and **m** both transform as vectors (x,y,z), that is, as first-order spherical harmonics, or P atomic states, but **p** and **m** are odd and even, respectively, under inversion. (Vectors such as **m**, even under inversion, are referred to as axial vectors or *pseudovectors*.)

Suppose, however, that we want to compute electric dipole or magnetic dipole radiative transitions for an atom in a solid at a site having symmetry O. Then we need the characters in the space spanned by **p** or **m** for operations of the cubic group O. These characters are among those displayed in Table 3.1.1.

Table 3.1.1. Characters of a Vector p and Pseudovector m for All Possible Finite Point Group Elements

(Characters of R^{-1} are identical to those of R.)

	1	2	3	4	6	$\bar{1}$	$\bar{2}$	$\bar{3}$	$\bar{4}$	$\bar{6}$
	E	C_2	C_3	C_4	C_6	I	σ	S_6	S_4	S_3
$\mathbf{m}, \Gamma_1^+$	3	−1	0	1	2	3	−1	0	1	2
$\mathbf{p}, \Gamma_1^-$	3	−1	0	1	2	−3	1	0	−1	−2

The characters in Table 3.1.1 can be produced directly by applying an arbitrary rotation to **p** or **m** and taking the trace of the resulting 3×3 matrix. Alternatively, we can use the known characters for the representations $\Gamma_1^\pm$ of the rotation-reflection group (Eqs. 2.4.4 and 2.11.1):

$$\begin{aligned} \chi^{1\pm}[R(\varphi)] &= 1 + 2\cos\varphi = 1 + 2\cos\frac{2\pi}{n} \\ \chi^{1\pm}[IR(\varphi)] &= \pm(1 + 2\cos\varphi) = \pm\left(1 + 2\cos\frac{2\pi}{n}\right) \end{aligned} \tag{3.1.8}$$

for $\varphi = 2\pi/n$.

We next present Table 3.1.2 for the point group O. In this character table, the twofold rotations about the x-y direction or equivalent directions are denoted as δ_{2xy} or C_2, whereas those about the x (or equivalent) direction are designated as

Table 3.1.2. Character Table for the Point Group O

Last line shows characters of a general vector.

	E	$3C_4^2$	$6C_4$	$6C_2$	$8C_3$
	ϵ	δ_{2x}	δ_{4x}	δ_{2xy}	δ_{3xyz}
Γ_1	1	1	1	1	1
Γ_2	1	1	−1	−1	1
Γ_{12}	2	2	0	0	−1
Γ_{15}	3	−1	1	−1	0
Γ_{25}	3	−1	−1	1	0
m, **p**	3	−1	1	−1	0

δ_{2x} or $C_4^2=(\delta_{4x})^2$. With respect to the representations generated by **m** and **p** there is no distinction between different twofold axes. If the characters of $\Gamma_1^{\pm}$ for the elements of O are transcribed into the last line of Table 3.1.2, we see that this line is identical[1] to the one for Γ_{15}. Thus we find that our original vector space is irreducible with respect to O, a fact that could also have been checked using Eq. 3.1.6 if the elements of O were known without recourse to the character table of O. This conclusion is not surprising—an (x,y,z) space cannot be broken up into parts invariant under the cubic operations. We shall see later with the help of the *polarization theorem* (see Problem 4.5.3) that a consequence of the irreducibility of **p** or **m** is that selection rules are, for the group O, independent of the polarization of the radiation.

Since both **p** and **m** transform according to Γ_{15}, the selection rules for electric and magnetic dipole radiation are identical for an atom at a site with the symmetry of point group O. In fact, vectors (like **p**) and pseudo vectors (like **m**) can be distinguished only if the group contains some improper elements, since they differ only under inversion. The group $O_h=O\times C_i$ has representations Γ_{15}^{+}

Table 3.1.3. Characters for Irreducible Representations of C_{3v} and of a Magnetic and an Electric Vector m and p

	E	$2C_3$	$3\sigma_v$	Basis Functions
Γ_1 A_1	1	1	1	z
Γ_2 A_2	1	1	−1	R_z
Γ_3 E	2	−1	0	$(x,y),(-R_y,R_x)$
m	3	0	−1	(R_x,R_y,R_z)
p	3	0	+1	(x,y,z)

[1]For the nomenclature of these representations see Section 12.2.

and Γ_{15}^-, which are appropriate for **m** and **p**, respectively, and result in differing selection rules.

Let us now consider an atom at a site of point symmetry C_{3v}. Table 3.1.3 lists the characters of C_{3v}, together with those for **m** and **p** obtained from Table 3.1.1. Inspection of the character table (or use of Eq. 3.1.5) immediately leads to the results

$$\mathbf{m} \to A_2 + E = \Gamma_2 + \Gamma_3 \tag{3.1.9}$$

$$\mathbf{p} \to A_1 + E = \Gamma_1 + \Gamma_3 \tag{3.1.10}$$

These results are also immediately evident from the basis functions shown.

Thus Eq. 3.1.10 describes the decomposition of (x,y,z) space into z and (x,y). We shall not be surprised, therefore, to learn in Section 3.4 that light polarized along the z axis and perpendicular to that axis obeys different selection rules under the group C_{3v}.

3.2. CRYSTAL FIELD THEORY WITHOUT SPIN: COMPATIBILITY RELATIONS

Crystal field theory consists in a study of the modifications of atomic energy states in regard to energy and symmetry by interaction with a crystalline environment. Crystal field theory is typical of a large class of quantum-mechanical problems in which one starts with a Hamiltonian H_0 that commutes with the elements of a large group G. A perturbation λV is then added of lower symmetry than H_0, so that $H_0 + \lambda V$ commutes only with the elements of a subgroup H of G. *Compatibility relations* represent the conditions of the possible symmetries of the perturbed eigenfunctions that arise from a particular unperturbed representation.

We first note that *any* representation $D^i(R)$ of G provides a representation of H simply by selecting only those $D^i(R)$ for which $R \in H$—hence the term *subduced* representation. If $D^i(R)$ is *irreducible* in G, the subduced representations may be *reducible* in H, since only a smaller number of matrices need then be reduced.

If the representations and characters of G are denoted as Γ_i and $\chi^i(R)$, whereas those of H are called Δ_j and $\varphi^j(R)$, then, using the methods of Section 3.1, we can *decompose* Γ_i into its irreducible components and write

$$\chi^i(R) = \sum k_{ij} \varphi^j(R), \qquad R \in H \tag{3.2.1}$$

Equation 3.2.1 with the k_{ij} understood to be integers constitutes the *First Frobenius theorem*.

By the orthonormality theorem, Eqs. 1.5.6 and 1.5.7, we can write

$$k_{ij}=\frac{1}{h}\sum_{R\in H}\chi^i(R)\varphi^j(R)^* \tag{3.2.2}$$

$$=\frac{1}{h}\sum_{C\in H}n_H(C)\chi^i(C)\varphi^j(C)^* \tag{3.2.3}$$

where $n_H(C)$ is the number of elements in class C of group H. *Thus the compatibility relations* 3.2.1 *are in fact a property of the two groups G and H* and are not dependent on any other information concerning the Hamiltonian $H_0+\lambda V$.

Equation 3.2.1 corresponds to a relation

$$\Gamma_i=\sum k_{ij}\Delta_j \tag{3.2.4}$$

or

$$D^i(R)\approx\sum k_{ij}\Delta^j(R) \tag{3.2.5}$$

between the representations. The $\approx$ symbol in Eq. 3.2.5 means equivalence with respect to a similarity transformation. The sum Σ is really a *direct sum*; see Eq. 1.4.4. The integer k_{ij} is the number of times that a matrix representation $\Delta^j(R)$ (or its equivalent) appears in the decomposition of $D^i(R)$.

Let us illustrate the above remark by considering a free atom with states belonging to irreducible representation D^+ or D^-, that is, $\Gamma_2^{\pm}$ of the full rotation-reflection group $R(3)$, located at a site with point symmetry C_{3v}. To reduce these representations with respect to the group C_{3v}, we use the general formulas, based on Eqs. 2.4.3 and 2.11.1,

$$\begin{aligned}\chi^{j\pm}[R(\varphi)]&=\frac{\sin\frac{1}{2}(2j+1)\varphi}{\sin\frac{1}{2}\varphi}\equiv\chi^j[R(\varphi)]\\ \chi^{j\pm}[IR(\varphi)]&=\pm\chi^j[R(\varphi)]\end{aligned} \tag{3.2.6}$$

to evaluate the characters of Γ_2^+ and Γ_2^- for the elements E, C_3, and σ of C_{3v}. The results are shown in the last two lines of Table 3.2.1. (See also Table 3.6.1 as a source of characters.)

Let us emphasize the procedure in the general case. Pick one element R in each class of the subgroup H. Look up, in the table for G, the character $\chi^i(R)$ of the class that contains the element R (this element is surely included in G). This yields the characters of the representation $\chi^i(R)$ subduced on H. Comparison of these characters with the character table for H yields by inspection or by the character orthonormality theorem the decomposition of $\chi^i(R)$ into the irreducible representations $\varphi^j(R)$ of H.

Table 3.2.1. Character Table for Group C_{3v} of Its Irreducible Representations and the Representation D^+ and D^-

	E	$2C_3$	$3\sigma_v$	Basis Functions
Γ_1 A_1	1	1	1	x^2+y^2, z^2
Γ_2 A_2	1	1	-1	
Γ_3 E	2	-1	0	$(2xy, x^2-y^2), (xz, yz)$
D^+	5	-1	1	
D^-	5	-1	-1	

For the present example, inspection of Table 3.2.1 or use of Eq. 3.2.3 yields the compatibility relations

$$D^+ = A_1 + 2E \equiv \Gamma_1 + 2\Gamma_3$$
$$D^- = A_2 + 2E \equiv \Gamma_2 + 2\Gamma_3 \tag{3.2.7}$$

The first of these relations is also evident from the basis functions listed for A_1 and E in view of the fact that the five D^+ states can be thought of as xy, yz, zx, and any two independent combinations of x^2, y^2, z^2 other than $(x^2+y^2+z^2)$, which has S symmetry.

Since there is no symmetry reason for different representations, or even different repetitions of the same representation, to have the same energy, we see in Fig. 3.2.1 that the fivefold degeneracy of the D^+ state is split into two twofold states and one nondegenerate state. The new eigenstates are found in Example 3.4.2.

If ψ_μ^i are the eigenfunctions of H_0 belonging to representations i of G, the reducibility of this representation with respect to H implies that it is possible to find a set of functions φ_m that span the same space as $\{\psi_\mu^i\}$:

$$\varphi_m = \sum \psi_\mu^i A_{\mu m} \tag{3.2.8}$$

and break up into sets $m = (j\nu)$, $\nu = 1, \ldots, l_j$, that do not mix with one another under the operations of H and therefore constitute bases for irreducible representations of the latter group: $R\varphi_\nu^j = \Sigma_\lambda \varphi_\lambda^j \Delta_{\lambda\nu}^j(R)$. Then the representation generated by the $\varphi_m = \varphi_\alpha^j$ (if the latter are chosen orthonormal),

$$\Delta_{\alpha\beta}^j(R) = \sum_{\lambda,\mu} A_{\lambda j\alpha}^* D_{\lambda\mu}^i(R) A_{\mu j\beta} \tag{3.2.9}$$

is a completely reduced matrix in block form containing representation $\Delta^j(R)$ k_{ij} times, where $A_{\mu j\beta}$ is the matrix defined in Eq. 3.2.8 for $m = j\beta$.

Let us now suppose that the perturbation is sufficiently weak that we can ignore any mixing between wave functions of the degenerate set ψ_μ^i and func-

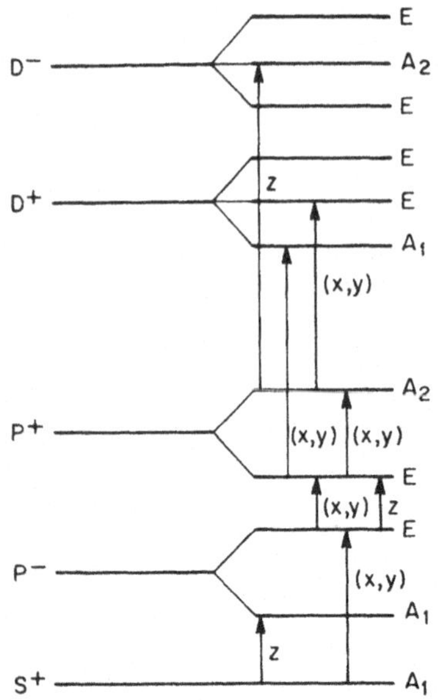

Fig. 3.2.1. Splitting of atomic states in a field of symmetry C_{3v}. Each type of allowed electric dipole transition is shown once by an arrow. Arrows marked z are polarized along the threefold axis. Those marked (x,y) involve radiation polarized transverse to the threefold axis.

tions of some *other* set with the same character. Then, in the lowest order of degenerate perturbation theory, the correct eigenfunctions will be linear combinations of the original set ψ_μ^i. *If a particular representation j is not repeated* ($k_{ij}=1$), the linear combinations $\varphi_m = \varphi_\nu^j$ chosen in Eq. 3.2.9 to reduce the representation will in fact be the correct eigenfunctions, that is, *the form of the eigenfunctions is determined by the symmetry rather than the dynamics of the problem.* If, however, representation j is repeated, several sets of basis functions φ_ν^{ja}, $a=1,2,\ldots$, each of which transforms according to Δ^j, can be formed. Each set may be referred to as a set of "symmetry vectors or orbitals." These representations have the same symmetry and are therefore coupled by $H_0+\lambda V$. *The correct eigenfunctions are then linear combinations of symmetry orbitals of the same type* (i.e., with the same j and ν) with the combination coefficients depending on the actual matrix elements of $H_0+\lambda V$, that is, on the dynamics.

The proof of these remarks follows immediately from the basevector orthogonality theorem 3.4.3 (see Eq. 3.4.10), to be established shortly. These remarks are generally true regardless of the strength of the perturbation. Weakness of the perturbation merely makes it possible to ignore symmetry coordinates of the same type in the more distant levels.

3.3. PRODUCT REPRESENTATIONS AND DECOMPOSITION OF ANGULAR MOMENTUM

Outer Product Representations

Let us consider two independent physical systems described by variables **r** and **r**′. Thus the Hamiltonian can be decomposed into two parts:

$$H = H_1(\mathbf{r}) + H_2(\mathbf{r}') \tag{3.3.1}$$

where H_1 acts only on the variable **r**, and H_2 only on the variable **r**′. Let G_1 be a group of operations that act only on **r** but commute with H_1, and G_2 be a group that acts on **r**′ and commutes with H_2. Then G_2 automatically commutes with H_1, and G_1 with H_2. Thus the group appropriate to H is the direct product group ("outer Kronecker product") of Section 2.12:

$$G^{\text{out}} = G_1 \times G_2 \tag{3.3.2}$$

Let φ_μ^i be a basis for an irreducible representation of G_1, and ψ_ν^j of G_2. We wish to show that

$$\Psi_{\mu\nu}^{i\times j}(\mathbf{r},\mathbf{r}') = \varphi_\mu^i(\mathbf{r})\psi_\nu^j(\mathbf{r}') \tag{3.3.3}$$

is a basis for an irreducible representation of the outer product group. Let $A \in G_1$ and $B \in G_2$; then

$$\begin{aligned}(A\times B)\Psi_{\mu\nu}^{i\times j}(\mathbf{r},\mathbf{r}') &= A\varphi_\mu^i(\mathbf{r})B\psi_\nu^j(\mathbf{r}')\\ &= \sum \varphi_{\mu'}^i(\mathbf{r})D_{\mu'\mu}^i(A)\sum \psi_{\nu'}^j(\mathbf{r}')D_{\nu'\nu}^j(B)\\ &= \sum \Psi_{\mu'\nu'}^{i\times j}(\mathbf{r},\mathbf{r}')D_{\mu'\nu',\mu\nu}^{i\times j}(A\times B)\end{aligned} \tag{3.3.4}$$

where

$$D_{\mu'\nu',\mu\nu}^{i\times j}(A\times B) \equiv D_{\mu'\mu}^i(A)D_{\nu'\nu}^j(B) \tag{3.3.5}$$

We showed in Section 2.13 that Eq. 3.3.5 is an irreducible representation of the outer product group.

Inner Product Group

If, however, our two previously independent systems are coupled:

$$H = H_1(\mathbf{r}) + H_2(\mathbf{r}') + H_{12}(\mathbf{r},\mathbf{r}') \tag{3.3.6}$$

then operations that act separately on $\mathbf{r}$ and $\mathbf{r}'$ will no longer commute with H. For example, if H_{12} involves $\mathbf{r}\cdot\mathbf{r}'$ only, *simultaneous identical rotations* on $\mathbf{r}$ and $\mathbf{r}'$ will leave H invariant. We are therefore restricted to operations of the form

$$R\psi(\mathbf{r},\mathbf{r}') = \psi(\mathbf{rR},\mathbf{r}'\mathbf{R}) \tag{3.3.7}$$

In this case, there is only one group G^{in}. If G_1 has elements $A_1, A_2, \ldots,$ and G_2 has corresponding elements $B_1, B_2, \ldots,$ such that B_k performs the identical action on $\mathbf{r}'$ that A_k performs on $\mathbf{r}$, the elements of G^{in} are the "inner products"

$$R_k = A_k B_k \tag{3.3.8}$$

whereas previously they were the much larger set of elements $A_k B_l$ (the outer products). In this case, the new inner product group G^{in} is a subgroup of the outer product group G^{out}, and the product representation generated by $\Psi^{i\times j}_{\mu\nu} = \varphi^i_\mu(\mathbf{r})\psi^j_\nu(\mathbf{r}')$, although irreducible on G^{out}, is no longer irreducible on G^{in}. Since A_k and B_k are identical in G^{in}, we can call them both R; the representation and character of the product are simply

$$\begin{aligned} D^{i\times j}_{\mu\nu,\mu'\nu'}(R) &= D^i_{\mu\mu'}(R)\, D^j_{\nu\nu'}(R) \\ \chi^{i\times j}(R) &= \chi^i(R)\chi^j(R) \end{aligned} \tag{3.3.9}$$

and this product representation can be decomposed in terms of the complete set of irreducible representations $\chi^m(R)$ of G^{in} in the usual way by using the character orthogonality relations or by inspection.

Although we have written G_1, G_2, and G^{in} to keep our meaning clear, the three groups are now isomorphic and differ only in whether the elements act on $\mathbf{r}, \mathbf{r}'$, or both. The process that we have just described is referred to as forming a product of two representations of a *single* group and decomposing the product in terms of irreducible representations of the *same* group.

Decomposition of Angular Momentum

The best known application of the preceding theory is the coupling together of two systems with angular momenta j_1 and j_2, respectively. The new irreducible representations are eigenfunctions of $J^2 = (\mathbf{J}_1 + \mathbf{J}_2)^2 = j(j+1)$ and are customarily chosen to be eigenfunctions of $J_z = J_{1z} + J_{2z} = m$. The product representation then decomposes in the form

$$\chi^{j_1}(\alpha)\chi^{j_2}(\alpha) = \sum a_j \chi^j(\alpha) \tag{3.3.10}$$

where

$$\chi^j(\alpha) = \sum_{m=-j}^{+j} \exp(im\alpha) \tag{3.3.11}$$

applies to all terms present. A comparison of the number of times that a particular exp $(im\alpha)$ appears on the right- and on the left-hand sides of Eq. 3.3.10 is used by Wigner (**1959**, p. 186)[2] to show that

$$\begin{aligned} a_j &= 1 \quad \text{if} \quad |j_1 - j_2| \leqslant j \leqslant j_1 + j_2 \\ a_j &= 0 \quad \text{otherwise} \end{aligned} \tag{3.3.12}$$

An independent proof is suggested in Problem 3.3.4. This decomposition theorem can thus be written as

$$\Gamma_{j_1} \times \Gamma_{j_2} = \Gamma_{|j_1 - j_2|} + \Gamma_{|j_1 - j_2| + 1} + \cdots + \Gamma_{j_1 + j_2 - 1} + \Gamma_{j_1 + j_2} \tag{3.3.13}$$

Tensors

As a simple application of Eq. 3.3.13 we note that a vector $\mathbf{r}$ transforms according to an irreducible representation $\Gamma_1^-(j=1$, odd parity) of the full rotation group. A tensor of the second rank transforms as the product of the displacements of two particles $\mathbf{r}$ and $\mathbf{r}'$:

$$T_{ij} \sim r_i r'_j \tag{3.3.14}$$

that is, T_{ij} transforms under rotations according to the direct product $\Gamma_1^- \times \Gamma_1^-$. This representation is, however, reducible:

$$\Gamma_1^- \times \Gamma_1^- = \Gamma_0^+ + \Gamma_1^+ + \Gamma_2^+ \tag{3.3.15}$$

that is, we can decompose T into irreducible parts:

$$\mathsf{T} = \mathsf{T}^{(0)} + \mathsf{T}^{(1)} + \mathsf{T}^{(2)} \tag{3.3.16}$$

which transform as S^+, P^+, and D^+ states, respectively. In this case, the parts can be written down by inspection:

$$T_{11} + T_{22} + T_{33} \sim \mathbf{r} \cdot \mathbf{r}' \sim S^+ \tag{3.3.17}$$

In other words, the trace of the tensor behaves as a scalar,

$$(T_{23} - T_{32}, T_{31} - T_{13}, T_{12} - T_{21}) \sim \mathbf{r} \times \mathbf{r}' \sim P^+ \tag{3.3.18}$$

that is, the antisymmetric part $T_{[ij]}$ of the tensor transforms as a cross product between vectors (i.e., as a *pseudovector*). What remains is then the traceless part of the *symmetric* part $T_{(ij)}$ of the tensor, with five independent components.

[2]See also Schiff (**1968**, p. 213).

These five components can be chosen to parallel the five (complex) spherical harmonics of order 2.

An alternative choice groups the five into a set of three and a set of two real components:

$$(T_{(23)}, T_{(31)}, T_{(12)}; T_{11}-T_{22}; 2T_{33}-T_{11}-T_{22}) \sim (\Gamma_{25}^{+}, \Gamma_{12}^{+}) \sim D^{+} \quad (3.3.19)$$

that also display the decomposition of a D^{+} state in a crystal of O_h symmetry:

$$D^{+} = \Gamma_{25}^{+} + \Gamma_{12}^{+} \equiv F_2 + E \quad (3.3.20)$$

Higher-order tensors can be formed in a similar way:

$$T_{ijk} \sim r_i r_j' r_k'' \quad (3.3.21)$$

and yield higher-order product representations that can be decomposed into irreducible parts by successive use of the method, illustrated above, of composition of angular momentum.

Alternative procedures are based on successive decompositions of a tensor into parts that are "constant" (i.e., δ_{ij}) in any pair of indices and parts that are traceless. This decomposition is then followed by a further decomposition, based on taking certain permutations of the indices, that exploits a close relationship between the full linear group and the symmetric (permutation) group. The reader is referred to Fano and Racah (**1959**), Weyl (**1946**, Chapter VB), and Hamermesh (**1962**) for more detailed discussion of these points.

Product Representations in the Same Space

Let us consider the representation generated by the product functions $\Psi_{\mu\nu}^{i\times j}(\mathbf{r}) \equiv \varphi_\mu^i(\mathbf{r})\psi_\nu^j(\mathbf{r})$. Since $\mathbf{r}'=\mathbf{r}$, the same operation R acts on both factors, thus yielding a representation of the *inner* product group. The procedure of Eq. 3.3.5 with $A=B=R$ yields the representation

$$D_{\mu'\nu',\mu\nu}^{i\times j}(R) = D_{\mu'\mu}^{i}(R) D_{\nu'\nu}^{j}(R) \quad (3.3.22)$$

A word of caution is needed, however, if the basis vectors are *identical*, $\varphi_\mu^i(\mathbf{r}) = \psi_\mu^j(\mathbf{r})$, or if their ratio $\varphi_\mu^i(\mathbf{r})/\psi_\mu^j(\mathbf{r})$ belongs to the identity representation.[3] In this case, the number of independent product functions is reduced, since

$$\Psi_{\mu\nu}^{j\times j} = \psi_\mu^j(\mathbf{r})\psi_\nu^j(\mathbf{r}) = \psi_\nu^j(\mathbf{r})\psi_\mu^j(\mathbf{r}) = \Psi_{\nu\mu}^{j\times j} \quad (3.3.23)$$

[3]This ratio $\varphi_\mu^i(\mathbf{r})/\psi_\mu^j(\mathbf{r})$ is not guaranteed to belong to the identity representation if $i=j$. For example, $[x,y,z]$ and $[x^3,y^3,z^3]$ are both bases for Γ_{15} for the group O_h. The ratios x^2,y^2,z^2 certainly do not belong to Γ_1.

(although the two sides would be distinct if one variable were labeled $\mathbf{r}'$). Thus basis functions 3.3.23 yield the part of the product representation that is symmetric in the indices μ and ν, the "symmetrized product representation." Indeed, we shall show that the $j\times j$ space *can be split in an invariant way into a part symmetric and a part antisymmetric in the partner indices*. Since

$$R\Psi_{\mu\nu}^{j\times j}=\sum\Psi_{\mu'\nu'}^{j\times j}D_{\mu'\mu}^{j}(R)\,D_{\nu'\nu}^{j}(R) \tag{3.3.24}$$

then

$$R\Psi_{\nu\mu}^{j\times j}=\sum\Psi_{\mu'\nu'}^{j\times j}D_{\mu'\nu}^{j}(R)\,D_{\nu'\mu}^{j}(R) \tag{3.3.25}$$

If we define the symmetric and antisymmetric parts by

$$\Psi_{\mu\nu}^{\pm}=\tfrac{1}{2}\left(\Psi_{\mu\nu}^{j\times j}\pm\Psi_{\nu\mu}^{j\times j}\right) \tag{3.3.26}$$

then

$$R\Psi_{\mu\nu}^{\pm}=\sum\Psi_{\mu'\nu'}^{j\times j}\tfrac{1}{2}\left[D_{\mu'\mu}^{j}(R)\,D_{\nu'\nu}^{j}(R)\pm D_{\mu'\nu}^{j}(R)\,D_{\nu'\mu}^{j}(R)\right] \tag{3.3.27}$$

But the expression in brackets is symmetric (antisymmetric) in $\mu'\nu'$ for the upper (lower) sign. Thus only the part of $\Psi_{\mu'\nu'}^{j\times j}$ symmetric (antisymmetric) in its indices survives the sum, and we obtain

$$R\Psi_{\mu\nu}^{\pm}=\sum\Psi_{\mu'\nu'}^{\pm}D_{\mu'\nu',\mu\nu}^{j\times j\pm}(R) \tag{3.3.28}$$

$$D_{\mu'\nu',\mu\nu}^{(j\times j)\pm}(R)=\tfrac{1}{2}\left[D_{\mu'\mu}^{j}(R)\,D_{\nu'\nu}^{j}(R)\pm D_{\mu'\nu}^{j}(R)\,D_{\nu'\mu}^{j}(R)\right] \tag{3.3.29}$$

which establishes that the symmetrical and antisymmetrical parts do not mix under a transformation. The characters of the symmetrical and antisymmetrical parts can be obtained by setting $(\mu'\nu')=(\mu\nu)$ and taking the trace:

$$\chi^{(j\times j)\pm}(R)=\tfrac{1}{2}\left[\chi^{j}(R)\right]^{2}\pm\tfrac{1}{2}\chi^{j}(R^{2}) \tag{3.3.30}$$

As an elementary example, let us consider the symmetric tensor $\mathbf{T}\sim\mathbf{rr}$ rather than $\mathbf{rr}'$. Then the antisymmetric vector part $\mathbf{r}\times\mathbf{r}'$ vanishes, that is, we have the decomposition

$$(\Gamma_1^-\times\Gamma_1^-)_{\rm sym}=\Gamma_0^+ +\Gamma_2^+ \tag{3.3.31}$$

This is akin to the statement that it is impossible to construct an even p state using a wave function in one variable. This result could have been deduced from the character formula 3.3.30. The symmetric product is given by

$$[\chi^1(\alpha)\times\chi^1(\alpha)]_{\text{sym}} = \tfrac{1}{2}[\chi^1(\alpha)]^2 + \tfrac{1}{2}\chi^1(2\alpha)$$

$$= \tfrac{1}{2}(1+2\cos\alpha)^2 + \tfrac{1}{2}(1+2\cos 2\alpha)$$

$$= 1 + (1+2\cos\alpha+2\cos 2\alpha)$$

or

$$\chi^{(1\times 1)}(\alpha) = \chi^0(\alpha) + \chi^2(\alpha) \tag{3.3.32}$$

where we use the notation $(j\times j)$ to denote a symmetric product and $[j\times j]$ indicates an antisymmetric product.

A natural generalization of Eq. 3.3.31 is

$$\Gamma_{(l\times l)} = (\Gamma_l\times\Gamma_l)_{\text{sym}} = \Gamma_{2l} + \Gamma_{(2l-2)} + \cdots + \Gamma_2 + \Gamma_0 \tag{3.3.33}$$

$$\Gamma_{[l\times l]} = (\Gamma_l\times\Gamma_l)_{\text{asym}} = \Gamma_{2l-1} + \Gamma_{2l-3} + \cdots + \Gamma_1 \tag{3.3.34}$$

for $l=$ integer. The terms omitted in Eq. 3.3.33 are the "even parity states of odd l," as expected. The character formula 3.3.30 and the decomposition formulas 3.3.33 and 3.3.34 will be particularly useful in Chapter 4 in determining the form and number of independent constants in tensors of various symmetries.

3.4. SELECTION RULES

Selection rules are concerned with the vanishing of matrix elements (i.e., integrals). Let us see how we can use the symmetry operations R of a group G to investigate the possible vanishing of an integral of the form $\int F(\mathbf{r})\,d\mathbf{r}$. By a change of variables $\mathbf{r}\to\mathbf{rR}$ we obtain

$$\int F(\mathbf{r})\,d\mathbf{r} = \int F(\mathbf{rR})\,d(\mathbf{rR}) = \int F(\mathbf{rR})\,d\mathbf{r} \tag{3.4.1}$$

using $|\det\mathbf{R}|=1$. We average over all group operations to obtain

$$\int F(\mathbf{r})\,d\mathbf{r} = \int \bar{F}(\mathbf{r})\,d\mathbf{r} \tag{3.4.2}$$

where

$$\bar{F}(\mathbf{r}) = \frac{1}{g}\sum_R F(\mathbf{rR})$$

A simple example has inversion, I, as the only nontrivial element. We then arrive at the familiar result

$$\int F(\mathbf{r})\,d\mathbf{r} = \tfrac{1}{2}\int [F(\mathbf{r}) + F(-\mathbf{r})]\,d\mathbf{r} \tag{3.4.3}$$

Observe that Eq. 3.4.3 vanishes unless $F(\mathbf{r})$ contains an even part.

The more general statement can be expressed as follows: The integral $\int F(\mathbf{r})\,d\mathbf{r}$ vanishes when $\bar{F}(\mathbf{r})$ vanishes. But $\bar{F}(\mathbf{r})$ belongs to the identity representation of the group, since

$$S\bar{F}(r) = \bar{F}(\mathbf{rS}) = \frac{1}{g}\sum_R F(\mathbf{rSR}) = \frac{1}{g}\sum_T F(\mathbf{rT}) = \bar{F}(\mathbf{r}) \tag{3.4.4}$$

by the group rearrangement theorem.

Theorem 3.4.1

Let $\chi(R)$ be the character of the invariant space spanned by the functions $F(\mathbf{rS})$ (which may not all be independent; see Section 1.4). Then $\int F(\mathbf{r})d\mathbf{r}$ vanishes if

$$k_1 \equiv \frac{1}{g}\sum_R \chi(R) \doteq 0 \tag{3.4.5}$$

Proof: By definition $\bar{F}(\mathbf{r})$ is expressible in this space. It will vanish, however, if the space does not contain the identity representation. Moreover, by Eq. 3.1.5 k_1 is the number of times that the identity representation, whose characters are

$$\chi^{\text{identity}}(R) = 1 \tag{3.4.6}$$

is contained in the original space.

Theorem 3.4.2

The matrix element of any operator H *invariant* under the elements of a group G *vanishes* if taken between functions belonging to two *different* irreducible representations:

$$(\varphi_\nu^i, H\psi_{\nu'}^j) \equiv \int (\varphi_\nu^i)^* H\psi_{\nu'}^j\,d\tau \propto \delta_{ij} \tag{3.4.7}$$

where τ denotes all the variables contained in the wave functions.

Proof: Since H belongs to the identity representation, it can be ignored in the symmetry arguments which follow.

The functions $(\varphi_\nu^i)^*\psi_{\nu'}^j$, generate a representation whose character is $\chi^i(R)^*$

$\times\chi^j(R)$. The identity representation will be contained if

$$k_1 = \frac{1}{g}\sum_R \chi^i(R)^*\chi^j(R) \neq 0$$

but by the character orthonormality theorem, Eq. 1.5.6, we have $k_1 = \delta_{ij}$.

Query 3.4.1

The statement that the $(\varphi_\nu^i)^*$ generate the complex conjugate representation $D^i(R)^*$ seems to depend on the assumption that

$$R[\varphi(\mathbf{r})]^* = [R\varphi(\mathbf{r})]^*$$

Is this always true?

Answer: For rotations described by a real 3×3 matrix

$$R\varphi^*(\mathbf{r}) = \varphi^*(\mathbf{rR}) = \varphi^*(\mathbf{rR}^*) = [\varphi(\mathbf{rR})]^* = [R\varphi(\mathbf{r})]^*$$

The statement is true. For rotations defined by Eq. 3.5.15 appropriate to the spin case, however, it is *not* true. Moreover, it need not be true for a general unitary operation. Theorem 3.4.2 remains valid, however, because it relates to scalar products,

$$(\Phi_\nu^i, \Psi_{\nu'}^j) = (R\Phi_\nu^i, R\Psi_{\nu'}^j)$$

A factor $D_{\mu\nu}^i(R)^*$ then appears on the right-hand side because one *first* acts with R and *then* takes the complex conjugate in removing $D_{\mu\nu}^i(R)$ from the left-hand ("bra") factor in the scalar product. Thus we could write

$$R(\varphi_\nu^i)^\dagger = \sum (\varphi_\mu^i)^\dagger D_{\mu\nu}(R)^* \tag{3.4.8}$$

where $(\varphi_\nu^i)^\dagger$ is $(\varphi_\nu^i)^*$, but the $\dagger$ serves to remind us that the φ^* came from the bra position[4] and the R operation is understood to act *first*.

To demonstrate that this viewpoint is indeed correct, we shall work exclusively with the scalar product notation to provide a rigorous proof of a more general theorem.

Base Vector Orthogonality Theorem 3.4.3

$$(\Phi_\nu^i, \Psi_{\nu'}^j) = \delta_{ij}\delta_{\nu\nu'}A \tag{3.4.9}$$

where

$$A = \frac{1}{l_i}\sum_\mu (\Phi_\mu^i, \Psi_\mu^i) \equiv (\Phi^i, \Psi^i)$$

[4]In the spin case the $\dagger$ also converts a column spinor into a row spinor.

or, more generally,

$$(\Phi^i_\nu, H\Psi^j_{\nu'}) = \delta_{ij}\delta_{\nu\nu'}\langle H\rangle^i \tag{3.4.10}$$

where

$$\langle H\rangle^i = \frac{1}{l_i}\sum_\mu (\Phi^i_\mu, H\Psi^i_\mu)$$

in which H is any operator that commutes with all the elements of G, and the *representations* generated by Φ^i_ν and Ψ^j_ν are assumed to be distinct $(i \neq j)$ or *identical*. Thus overlap integrals exist only between *corresponding components* of the *same* representation.

Proof: Since Eq. 3.4.9 is a special case, with $H=1$, of Eq. 3.4.10, we prove the latter. The unitarity of the elements R permits us to write

$$(\Phi^i_\nu, H\Psi^j_{\nu'}) = (R\Phi^i_\nu, RH\Psi^j_{\nu'})$$

Commuting through the R and averaging over the group gives

$$(\Phi^i_\nu, H\Psi^j_{\nu'}) = \frac{1}{g}\sum_R (R\Phi^i_\nu, HR\Psi^j_{\nu'})$$

$$= \sum_{\mu\mu'} (\Phi^i_\mu, H\Psi^j_{\mu'})\frac{1}{g}\sum_R D^i_{\mu\nu}(R)^* D^j_{\mu'\nu'}(R)$$

For *unitary* representations, the matrix orthogonality theorem 1.5.3 can be written, by means of Eq. 1.5.5, as

$$\frac{1}{g}\sum_R D^i_{\mu\nu}(R)^* D^j_{\mu'\nu'}(R) = \frac{\delta_{ij}\delta_{\mu\mu'}\delta_{\nu\nu'}}{l_i} \tag{3.4.11}$$

from which our desired equation 3.4.10 follows immediately. Since all representations of a finite group are equivalent to unitary representations (see Problem 1.5.5), it follows from Eq. 1.5.3 that

$$\frac{1}{g}\sum_R D^i_{\mu\nu}(R)^* D^j_{\mu'\nu'}(R) \propto \delta_{ij} \tag{3.4.12}$$

even if the representations are *not* unitary. Equation 3.4.12 then confirms Eq. 3.4.7 even for the nonunitary case. The $\delta_{\nu\nu'}$ orthogonality for $i=j$, however, requires the representation to be unitary. But unitary operators are represented by unitary matrices (only) in orthonormal base systems. Our theorem thus says

$$(\Phi^i_\nu, H\Psi^i_{\nu'}) \propto \delta_{\nu\nu'} \qquad \text{if } (\Phi^i_\nu, \Phi^i_{\nu'}) = \delta_{\nu\nu'}, (\Psi^i_\nu, \Psi^i_{\nu'}) = \delta_{\nu\nu'} \tag{3.4.13}$$

and if Φ^i_ν and Ψ^i_ν generate the identical representation—reasonable requirements.

We illustrate these ideas by means of an example.

Example 3.4.1

Consider the coupling of two sets of p states:

$$\varphi_1(\mathbf{r}) = xf(r), \qquad \varphi_2(\mathbf{r}) = yf(r), \qquad \varphi_3(\mathbf{r}) = zf(r)$$

$$\psi_1(\mathbf{r}) = xg(r), \qquad \psi_2(\mathbf{r}) = yg(r), \qquad \psi_3(\mathbf{r}) = zg(r)$$

(eigenstates of Hamiltonian H_0) under a spherically symmetric perturbation $V(r)$.

Solution: The factor δ_{ij} in Eq. 3.4.10 tells us that $V(r)$ introduces no coupling to s, d, f states, etc. The $\delta_{\nu\nu'}$ factor tells us that integrals of the form

$$\int xf(r)^* V(r) yg(r)\, d\mathbf{r} = 0$$

From the fact that $\langle H \rangle$ is independent of μ we know that

$$\int x^2 u(r)\, d\mathbf{r} = \int y^2 u(r)\, d\mathbf{r} = \int z^2 u(r)\, d\mathbf{r}$$

where

$$u(r) = V(r) f(r)^* g(r) \qquad \text{or} \qquad V(r)|f(r)|^2 \qquad \text{or} \qquad V(r)|g(r)|^2$$

Thus, if we set ψ equal to an arbitrary linear combination of these six states and seek a solution by minimizing $(\psi, H\psi)/(\psi,\psi)$, where $H = H_0 + V(r)$, the 6×6 matrix that we must diagonalize reduces to the form

$$\begin{bmatrix} A & B & & & & \\ B^* & C & & & & \\ & & A & B & & \\ & & B^* & C & & \\ & & & & A & B \\ & & & & B^* & C \end{bmatrix}$$

where

$$A = \int x^2 V(r) |f|^2 d\mathbf{r}, \qquad C = \int x^2 V(r) |g|^2 d\mathbf{r}$$

$$B = \int x^2 V(r) f^* g\, d\mathbf{r}$$

Thus only one 2×2 problem need actually be solved. In other words, if $xh(r)$ is a solution where $h(r) = af(r) + bg(r)$, then $yh(r)$ and $zh(r)$ are also solutions with the same a and b.

From the above example we see that a perturbation with the full symmetry of a group mixes only *corresponding* partners of identically transforming representations—and it mixes them with the same coefficients. No matter how big the dimension l_i of the representation is, if it is repeated n_i times, the $n_i l_i \times n_i l_i$ matrix factors into l_i identical $n_i \times n_i$ matrices, and only one of the latter need be diagonalized.

Let Φ_μ^{ia} for $a = 1,2,\ldots,n_i$ be a set of identically transforming symmetry vectors, that is, basis vectors for irreducible representation i. If we seek solutions of $H\Psi = E\Psi$ in the form

$$\Psi = \sum_{i,a,\mu} q_\mu^{ia} \Phi_\mu^{ia} \tag{3.4.14}$$

the expectation value of the Hamiltonian by Eq. 3.4.10 must take the form

$$(\Psi, H\Psi) = \sum_{i,a,b} A_{ab}^i \sum_\mu (q_\mu^{ia})^* q_\mu^{ib}; \qquad A_{ab}^i \equiv (\Phi_\mu^{ia}, H\Phi_\mu^{ib}) \tag{3.4.15}$$

with no coupling between i and j or between μ and ν, and with the same coefficient A_{ab}^i for each μ. For each i, then, we need only diagonalize the $n_i \times n_i$ matrix A_{ab}^i. If no repetition occurs, $n_i = 1$ and Φ_μ^{i1}, the only symmetry orbital available, is already the eigenvector of H, as claimed in our discussion at the end of Section 3.2.

Example 3.4.2

Consider Eq. 3.2.7, the crystal field splitting $D^+ \to A_1 + 2E$ of a D^+ atomic state in a C_{3v} crystalline field. What will be the nature of the zero-order eigenstates in the crystal field?

Solution: State D^+ contains five partners that in the usual Ψ_m^j notation transform roughly as exp $(\pm i2\varphi) \sim (x \pm iy)^2$ for $m = \pm 2$, exp $(\pm i\varphi) \sim (x \pm iy)z$ for $m = \pm 1$, and $3z^2 - r^2$ for $m = 0$. The $3z^2 - r^2$ partner already transforms as A_1, that is, the identity representation of C_{3v}, and is uncoupled to the other states. Hence we already have one solution. By taking linear combinations of the original functions we can construct two identically transforming sets of symmetry orbitals for E:

$$[2xy, x^2 - y^2] \sim [xz, yz]$$

(cf. Eq. 1.4.12 and Section 3.8); thus we can assume solutions of the form

$$a2xy + bxz \qquad \text{and} \qquad a(x^2 - y^2) + byz$$

with the same a and b. The ratio of b to a is determined by the details of the Hamiltonian. The 2×2 problem that we must solve will lead to two solutions for b/a corresponding to the two nondegenerate sets of E states in Fig. 3.4.1.

We are now in a position to state a most important theorem concerning selection rules.

Theorem 3.4.4

The complete set of matrix elements

$$V_{\mu\nu\lambda} = (\varphi_\mu^i, V_\lambda^m \psi_\nu^j) \tag{3.4.16}$$

for all $\mu\nu\lambda$ vanishes if

$$c = \frac{1}{g} \sum_{R \in G} \chi^i(R)^* \chi^m(R) \chi^j(R) \tag{3.4.17}$$

vanishes. Conversely, if $c \neq 0$ there is no symmetry reason for all $V_{\mu\nu\lambda}$ to vanish.

Proof: Here $(\varphi_\mu^i)^* V_\lambda^m \psi_\nu^j$ generates a representation whose character is $\chi^i(R)^* \times \chi^m(R)\chi^j(R)$, and c represents the number of times that the identity representation is contained in this space. By Theorem 3.4.1, if $c=0$, all $V_{\mu\nu\lambda}=0$ since the identity representation is not contained in the space.

This procedure treats representations i, m, and j on an equal footing. In a less symmetrical way, we can note that $V_\lambda^m \psi_\nu^j$ generates $\Gamma^{m\times j} = \Gamma^m \times \Gamma^j$; by Theorem 3.4.2 only the part of $\Gamma^m \times \Gamma^j$ that contains Γ^i will contribute, and this is again given by c. (A proof by direct calculation that all $V_{\mu\nu\lambda}=0$ when $c=0$ is given in Problem 3.4.11.)

Dipole Radiation Selection Rules

Let us now consider one of the simplest possible examples: electric dipole absorption of light by an atom in a crystalline field of symmetry C_{3v}. In this case the perturbation is $\mathbf{p}\cdot\mathbf{A} \propto \mathbf{p}\cdot\mathbf{e}$, where $\mathbf{p}$ is the electron momentum (a polar vector), and $\mathbf{A}$ and $\mathbf{e}$ are the vector potential and the polarization vector of the radiation field, respectively. A possible energy level scheme consistent with the compatibility relations discussed in Section 3.2 is shown in Fig. 3.2.1.

Selection rules for transitions between states i and j can be obtained by taking the triple product of characters $(\chi^i)^*\chi^j\chi^P$, where χ^P is the character of a polar vector, shown in the last line of Table 3.1.3. However, we saw in Eq. 3.1.10 that such a vector decomposes into $A_1 + E$, that is, a part that transforms as z (i.e., light polarized along the threefold axis) and as (x,y) (i.e., light polarized transverse to the threefold axis). Thus we can obtain different selection rules for these two types of radiation. Since $A_1 =$ the identity, $\chi^{A_1}\chi^j = \chi^j$, and our selection rule is $i=j$, that is, z radiation can only connect states of the same

Table 3.4.1. Character Table for C_{3v}, Showing Product Representations Needed for Radiation Transverse to Axis

	E	$2C_3$	$3\sigma_v$	Basis Functions
Γ_1 A_1	1	1	1	z
Γ_2 A_2	1	1	-1	R_z
Γ_3 E	2	-1	0	(x,y)
$A_2 \times E$	2	-1	0	
$E \times E$	4	1	0	

symmetry. For the transverse (E) radiation, Table 3.4.1 shows that

$$A_1 \times E = E, \qquad A_2 \times E = E, \qquad E \times E = A_1 + A_2 + E$$

so that *starting* in states of types A_1, A_2, and E, transverse (E) radiation can be used to reach, respectively, E, E, and A_1, A_2, or E states. The transition A_1 to A_2 is forbidden for any polarization. See Fig. 3.2.1.

3.5. SPIN AND SPIN-ORBIT COUPLING

Experimental Support for the Spin Hypothesis

The frequent occurrence of "spin doublets" (i.e., pairs of closely spaced lines) in the optical spectra of atoms and the Zeeman splitting of such lines provided evidence (1) that two electronic states can exist where only one would ordinarily be expected, and (2) that electrons have a magnetic moment $\mu = -(e\hbar/2mc)$. These experimental facts induced Uhlenbeck and Goudsmit (**1925**, **1926**) and Pauli (**1927**) to endow electrons with an extra intrinsic degree of freedom that could take two values, thus providing a doubling of the number of states. The magnetic moment was also associated with an intrinsic angular momentum or "spin," just as an orbital angular momentum gives rise to a magnetic moment. The economy of the theory is that it is the electron spin itself that gives rise to the two internal states. The elegance of the theory of intrinsic angular momentum is that it follows directly from our general analysis of angular momentum operators in Sections 2.3–2.6 as soon as *double-valued representations* are admitted.

In our language, the spin doublets are a consequence of the composition of orbital angular momentum l with spin $\frac{1}{2}$:

$$\Gamma_l \times \Gamma_{1/2} = \Gamma_{l+1/2} + \Gamma_{l-1/2} \tag{3.5.1}$$

The Zeeman spectra of atoms can be explained by giving them a magnetic moment:

$$\boldsymbol{\mu} = -\frac{e}{2mc}(\mathbf{L} + 2\mathbf{S}) \tag{3.5.2}$$

$$\mathbf{L} = \hbar \sum \mathbf{l}_j \tag{3.5.3}$$

$$\mathbf{S} = \tfrac{1}{2}\hbar \sum \boldsymbol{\sigma}_j \tag{3.5.4}$$

(where $\mathbf{l}_j$ and $\frac{1}{2}\boldsymbol{\sigma}_j$ are the orbital and spin angular momenta of each electron), by adding to the usual one-electron Hamiltonian

$$H_0 = \tfrac{1}{2}m^{-1}\left(\mathbf{p} - \frac{e}{c}\mathbf{A}\right)^2 + V(\mathbf{r}) \tag{3.5.5}$$

(where $e = -|e|$ is the electronic charge) the spin-dependent terms

$$H_{\text{spin}} = -\frac{e\hbar}{2mc}\boldsymbol{\sigma}\cdot\mathbf{H} + \frac{\hbar}{4m^2c^2}\left[\nabla V(\mathbf{r}) \times \left(\mathbf{p} - e\frac{\mathbf{A}}{c}\right)\cdot\boldsymbol{\sigma}\right] \tag{3.5.6}$$

Here $\mathbf{H}$ is an external magnetic field, $\mathbf{A}$ is the corresponding vector potential, and $\boldsymbol{\sigma}$ is a vector whose components are the Pauli spin matrices of Eq. 2.6.3. Thus the Hamiltonian is a 2×2 matrix (whose elements are operators), and it must act on a two-component Schrödinger function

$$\psi(\mathbf{r}) = \begin{bmatrix} \varphi_1(\mathbf{r}) \\ \varphi_2(\mathbf{r}) \end{bmatrix}$$

The probability of finding the electron at $\mathbf{r}$ with spin up or spin down is given by $|\varphi_1(\mathbf{r})|^2$ or $|\varphi_2(\mathbf{r})|^2$, respectively. The term linear in $\mathbf{A}$ in Eq. 3.5.5 gives rise to the expected orbital magnetic moment contribution, $-(e/2mc)\mathbf{L}\cdot\mathbf{H}$. The first term in H_{spin} is the energy associated with the spin magnetic moment. The second term in H_{spin} is referred to as the *spin-orbit coupling term*, since for spherically symmetric $V(\mathbf{r})$, as in atoms,

$$\begin{aligned} \nabla V(r) \times \mathbf{p}\cdot\boldsymbol{\sigma} &= \frac{1}{r}\frac{\partial V}{\partial r}\mathbf{r}\times\mathbf{p}\cdot\boldsymbol{\sigma} \\ &= \frac{1}{r}\frac{\partial V}{\partial r}\mathbf{L}\cdot\boldsymbol{\sigma} \end{aligned} \tag{3.5.7}$$

Since $\boldsymbol{\sigma}\cdot\mathbf{L}=(2/\hbar)\mathbf{S}\cdot\mathbf{L}$ is left invariant only by simultaneous rotations of space and spin, it provides the coupling between spin and orbital angular momentum that produces the composition of the two as in Eq. 3.5.1.

Both the magnetic moment and the spin-orbit coupling terms are relativistic in origin and are natural consequences of the Dirac equation for the electron. Foldy and Wouthuysen (**1950**) have shown how to make a consistent nonrelativistic expansion in powers of $1/c$, starting from the Dirac equation. To order $1/c^2$ they found H_{spin} of Eq. 3.5.6 and the expected kinetic energy correction terms $-p^4/8m^3c^2$, plus an additional non-spin-dependent term $(e/8m^2c^2)\nabla^2 V$, whose effects were first discussed by Darwin (**1928**). See Blount (**1962b**) and Messiah (**1962**, Eq. XX.202).

Transformation Properties of Spin Functions

Under a rotation α about axis $\mathbf{n}$ the vectors $\mathbf{r}$ and $\boldsymbol{\sigma}$ must transform into the

$$\mathbf{r}'\equiv O_R^{-1}\mathbf{r}O_R=\mathbf{r}\cos\alpha+\mathbf{n}(\mathbf{n}\cdot\mathbf{r})(1-\cos\alpha)+(\mathbf{n}\times\mathbf{r})\sin\alpha \tag{3.5.8}$$

$$\boldsymbol{\sigma}'\equiv O_R^{-1}\boldsymbol{\sigma}O_R=\boldsymbol{\sigma}\cos\alpha+\mathbf{n}(\mathbf{n}\cdot\boldsymbol{\sigma})(1-\cos\alpha)+(\mathbf{n}\times\boldsymbol{\sigma})\sin\alpha \tag{3.5.9}$$

of Eqs. 2.5.11 and 2.7.5 with $U\equiv O_R$. If space and spin are to be rotated simultaneously, we must then choose

$$O_R=\exp(-i\alpha\mathbf{l}\cdot\mathbf{n})\exp(-i\tfrac{1}{2}\alpha\boldsymbol{\sigma}\cdot\mathbf{n}) \tag{3.5.10}$$

In this case the infinitesimal generator of the rotation group is

$$\mathbf{J}=\mathbf{l}+\tfrac{1}{2}\boldsymbol{\sigma} \tag{3.5.11}$$

If there were many spin-$\frac{1}{2}$ particles, the generator of the simultaneous rotations of all spaces and spins would be

$$\mathbf{J}=\sum_i(\mathbf{l}_i+\tfrac{1}{2}\boldsymbol{\sigma}_i) \tag{3.5.12}$$

Let us consider one particle of spin $\frac{1}{2}$, described by the two-component wave function

$$\psi(\mathbf{r})=\begin{bmatrix}\varphi_1(\mathbf{r})\\ \varphi_2(\mathbf{r})\end{bmatrix} \tag{3.5.13}$$

The action of the rotation operator O_R on this state is given by

$$\psi(\mathbf{r})' \equiv \exp(-i\alpha\mathbf{l}\cdot\mathbf{n})\exp(-i\tfrac{1}{2}\alpha\boldsymbol{\sigma}\cdot\mathbf{n})\begin{bmatrix}\varphi_1(\mathbf{r})\\ \varphi_2(\mathbf{r})\end{bmatrix} \tag{3.5.14}$$

$$=\exp(-i\tfrac{1}{2}\alpha\boldsymbol{\sigma}\cdot\mathbf{n})\begin{bmatrix}\varphi_1(\mathbf{rR})\\ \varphi_2(\mathbf{rR})\end{bmatrix} \tag{3.5.15}$$

where $\mathbf{rR} = O_R\mathbf{r}O_R^{-1} = \mathbf{r}\cos\alpha + (\mathbf{r}\cdot\mathbf{n})\mathbf{n}(1-\cos\alpha) + (\mathbf{r}\times\mathbf{n})\sin\alpha$ can be obtained by replacing α by $-\alpha$ in Eq. 3.5.8 to replace $(O_R)^{-1}$ by O_R.

To describe the action of improper rotations we decompose them into RI, a proper rotation times inversion. Inversion commutes with angular momentum vectors such as $\mathbf{r}\times\mathbf{p}$. In general the assumption

$$\mathbf{J}' \equiv I^{-1}\mathbf{J}I = \pm\mathbf{J} \tag{3.5.16}$$

is consistent with commutation rules 2.2.2 only if $\mathbf{J}'=\mathbf{J}$. Thus I commutes with all angular momentum operators and all rotation operators $\exp(i\alpha\mathbf{J}\cdot\mathbf{n})$. By Schur's lemma, Eq. 1.5.2, it must be a constant in each irreducible representation. Normalization requires this constant to have a magnitude of unity. Without modifying any results of physical interest (e.g., matrix elements), we can choose this irrelevant phase factor to be unity. Thus we can take

$$O_I\psi = \begin{bmatrix}I\varphi_1(\mathbf{r})\\ I\varphi_2(\mathbf{r})\end{bmatrix} = \begin{bmatrix}\varphi_1(-\mathbf{r})\\ \varphi_2(-\mathbf{r})\end{bmatrix} \tag{3.5.17}$$

and

$$O_{IR}\psi = \exp(-i\tfrac{1}{2}\alpha\boldsymbol{\sigma}\cdot\mathbf{n})\begin{bmatrix}\varphi_1(-\mathbf{rR})\\ \varphi_2(-\mathbf{rR})\end{bmatrix} \tag{3.5.18}$$

For a crystal, we cannot use arbitrary rotations $R(\alpha,\mathbf{n})$, but only those that restore the crystal to itself, that is, those for which

$$O_R^{-1}V(\mathbf{r})O_R \equiv V(\mathbf{Rr}) = V(\mathbf{r})$$

The spin-orbit term then automatically has the desired invariance:

$$O_R^{-1}\nabla V(\mathbf{r})\times\mathbf{p}\cdot\boldsymbol{\sigma}O_R=[\nabla' V(\mathbf{r}')\times\mathbf{p}']\cdot\boldsymbol{\sigma}'$$
$$=\mathbf{R}[\nabla V(\mathbf{Rr})\times\mathbf{p}]\cdot\mathbf{R}\boldsymbol{\sigma}$$
$$=\nabla V(\mathbf{Rr})\times\mathbf{p}\cdot\boldsymbol{\sigma}=\nabla V(\mathbf{r})\times\mathbf{p}\cdot\boldsymbol{\sigma}$$

because $\boldsymbol{\sigma}$ and $\nabla V\times\mathbf{p}$ both transform as vectors under the proper rotation R. Since $\nabla V\times\mathbf{p}$ and $\boldsymbol{\sigma}$ are in fact pseudovectors (i.e., even) under inversion, the invariance of $V(\mathbf{r})$ with respect to an improper rotation also implies the corresponding invariance of the spin-orbit term.

Thus the point or space group required by $V(\mathbf{r})$ is also to be used in the presence of spin-orbit coupling. But the presence of spin-orbit coupling forces us to use the corresponding double group, since O_R for α and O_R for $\alpha+2\pi$ differ by a factor $\exp(i\pi\boldsymbol{\sigma}\cdot\mathbf{n})=-1$.

3.6. CRYSTAL FIELD THEORY WITH SPIN

As an example we discuss the splitting of atomic energy levels of a Co^{2+} ion inserted into a site of octahedral (O) symmetry in MgO by crystal fields and spin-orbit coupling. The free ion belongs to a ^4F state, that is, $S=3/2$. Let us consider first the case in which crystalline field effects dominate and spin-orbit effects are a minor perturbation to be treated later. Crystal field effects cause the representation $F=\Gamma_3$ of the full rotation group to decompose with respect to O:

$$F\equiv\Gamma_3=\Delta_2+\Delta_{15}+\Delta_{25} \tag{3.6.1}$$

The characters of Γ_3 for the rotations in O can be obtained from Eq. 2.4.3 or Table 3.6.1. These are inserted in row 9 of Table 3.6.2 for comparison with the characters of O to yield the decomposition of Eq. 3.6.1.

The total spin $S=3/2$ can still rotate freely, that is, belongs to the representation $\Gamma_{3/2}$ of the rotation group $R(3)$. Thus a complete specification of the decomposition is

$$\Gamma_{3/2}\times\Gamma_3=\Gamma_{3/2}\times(\Delta_2+\Delta_{15}+\Delta_{25}) \tag{3.6.2}$$

Here $\Gamma_{3/2}\times\Delta_{15}$ is a (an outer) product representation of the outer product of the group of all spin rotations times the group O of space rotations. Thus $\Gamma_{3/2}\times\Delta_{15}$ can be abbreviated as $^4\Delta_{15}$ (quartet Δ_{15}) to remind us of the fourfold spin degeneracy, and Eq. 3.6.2 can be written as

$$^4F={}^4\Delta_2+{}^4\Delta_{15}+{}^4\Delta_{25} \tag{3.6.3}$$

Table 3.6.1. Characters of Rotations in the Full Rotation Group

Γ	E	C_2	C_3	C_4	C_6
Γ_0	1	1	1	1	1
Γ_1	3	-1	0	1	2
Γ_2	5	1	-1	-1	1
Γ_3	7	-1	1	-1	-1
Γ_4	9	1	0	1	-2
Γ_5	11	-1	-1	1	-1
Γ_6	13	1	1	-1	1
Γ_7	15	-1	0	-1	2
$\Gamma_{1/2}$	2	0	1	$\sqrt{2}$	$\sqrt{3}$
$\Gamma_{3/2}$	4	0	-1	0	$\sqrt{3}$
$\Gamma_{5/2}$	6	0	0	$-\sqrt{2}$	0
$\Gamma_{7/2}$	8	0	1	0	$-\sqrt{3}$
$\Gamma_{9/2}$	10	0	-1	$\sqrt{2}$	$-\sqrt{3}$
$\Gamma_{11/2}$	12	0	0	0	0
$\Gamma_{13/2}$	14	0	1	$-\sqrt{2}$	$\sqrt{3}$
$\Gamma_{15/2}$	16	0	-1	0	$\sqrt{3}$

Comments

$$\Gamma_j^{\pm}(R^{-1})=\Gamma_j^{\pm}(R), \qquad \Gamma_j^{\pm}(IR)=\pm\Gamma_j(R)$$

$$IC_3=S_6^{-1}, \qquad IC_4=S_4, \qquad IC_6=S_3^{-1}$$

indicating that the quartet spin degeneracy is unaffected by the crystalline field. (See the left-hand part of Fig. 3.6.1.)

Equation 3.6.3 was derived using only the single group representations of O, since F is a single-valued representation of $R(3)$. We have, in short, tacitly assumed the following theorem.

Theorem 3.6.1

Single-valued ("ordinary") and double-valued ("spin") representations are not compatible with one another.

Proof: This conclusion seems obvious if the character relation 3.2.1:

$$\chi^i(R)=\sum k_{ij}\varphi^j(R)$$

is applied for both R and $\bar{R}$. Alternatively, we can write

$$k_{ij}=\frac{1}{2g}\sum_{R\in G}[\chi^i(R)\varphi^j(R)^*+\chi^i(\bar{R})\varphi^j(\bar{R})^*] \tag{3.6.4}$$

Table 3.6.2. Characters for Operations of the Group O

	E	$3C_4^2$	$6C_2$	$8C_3$	$6C_4$	Type of Representation
Δ_1	1	1	1	1	1	
Δ_2	1	1	-1	1	-1	
Δ_{12}	2	2	0	-1	0	Single
Δ_{15}	3	-1	-1	0	1	
Δ_{25}	3	-1	1	0	-1	
Δ_6	2	0	0	1	$\sqrt{2}$	
Δ_7	2	0	0	1	$-\sqrt{2}$	Double
Δ_8	4	0	0	-1	0	
Γ_3	7	-1	-1	1	-1	Rotation group
$\Delta_{15}\times\Gamma_{3/2}$	12	0	0	0	0	Products
$\Delta_{25}\times\Gamma_{3/2}$	12	0	0	0	0	
$\Delta_2\times\Gamma_{3/2}$	4	0	0	-1	0	
$\Gamma_{3/2}$	4	0	0	-1	0	
$\Gamma_{5/2}$	6	0	0	0	$\sqrt{2}$	Rotation group
$\Gamma_{7/2}$	8	0	0	1	0	
$\Gamma_{9/2}$	10	0	0	-1	$-\sqrt{2}$	

$$\chi^i(\bar{R})=\pm\chi^i(R), \qquad \varphi^j(\bar{R})=\pm\varphi^j(R) \tag{3.6.5}$$

where the plus signs refer to single-valued representations for which a rotation of 2π is equivalent to the identity operation, and the minus signs to the double-valued (spin) representations for which such a rotation yields a factor of -1. Thus k_{ij} vanishes if i and j are representations of opposite type. Q.E.D.

When they are of the *same* type, we can, however, use the abbreviated formula

$$k_{ij}=\frac{1}{g}\sum_{R\in G}\chi^i(R)\varphi^j(R)^* \tag{3.6.6}$$

This is the formula that we would have obtained directly if we had used Eq. 1.5.6 as a character orthonormality relation appropriate to *multiplier groups*; in other words, if we apply Theorem 3.6.1, it is not necessary to use a complete double group character table. The only information contained in the characters of the barred elements is already included in Theorem 3.6.1.

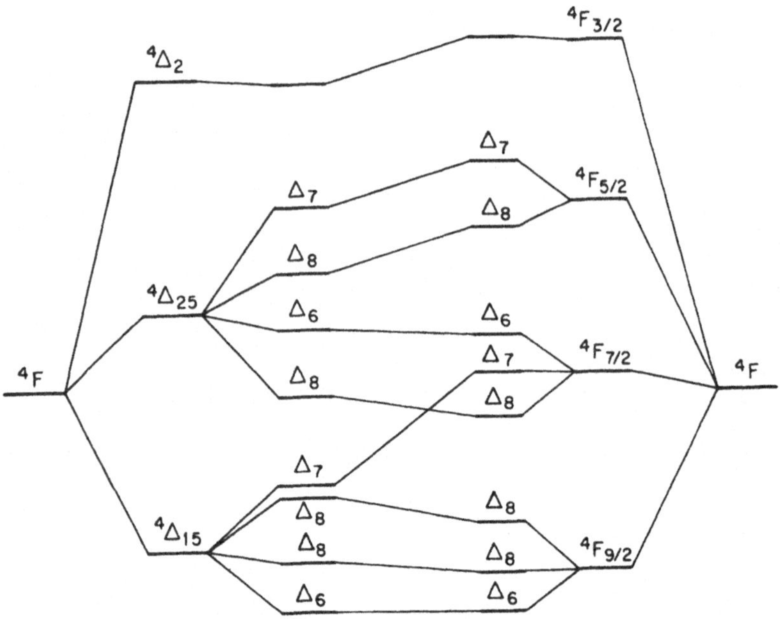

Fig. 3.6.1. Splitting of a ^4F state by a crystalline field of octahedral (O) symmetry and by spin-orbit coupling. The left-hand side of the diagram shows the case in which crystal field splitting predominates; the right-hand side, the case in which spin-orbit coupling predominates. The intermediate region is sketched in, so that crossings between equivalent representations do not occur.

Let us now consider the further splitting of the $^4\Delta_{15}=\Gamma_{3/2}\times\Delta_{15}$ level by spin-orbit coupling. This perturbation, as shown in Eq. 3.3.6 and the following text reduces the symmetry to that of simultaneous space-spin rotations belonging to O. With respect to this *inner* product group the representation is reducible, and we obtain

$$^4\Delta_{15}\equiv\Gamma_{3/2}\times\Delta_{15}=\Delta_6+\Delta_7+2\Delta_8 \tag{3.6.7}$$

In deriving this result, we have simplified the arithmetic by an obvious theorem.

Theorem 3.6.2

The product of two ordinary or two spin representations is an ordinary (single-valued) representation. The product of one ordinary and one spin representation is a spin (double-valued) representation.

With this theorem, decomposition 3.6.7 can be carried out using only the information shown in Table 3.6.2, that is, the characters of only the unbarred elements. In a similar way, we learn that

$$^4\Delta_{25} = \Gamma_{3/2} \times \Delta_{25} = \Delta_6 + \Delta_7 + 2\Delta_8$$

$$^4\Delta_2 = \Gamma_{3/2} \times \Delta_2 = \Delta_8 \tag{3.6.8}$$

A diagram showing the possible splitting is given in Fig. 3.6.1. (Read this diagram from the left when spin-orbit coupling is weaker than crystalline field effects.)

Let us now consider the opposite situation, in which the crystalline field is regarded as small compared to the spin-orbit splitting. In this case, we first form the composition of angular momenta **L** and **S**, using Eq. 3.3.13:

$$\Gamma_3 \times \Gamma_{3/2} = \Gamma_{3/2} + \Gamma_{5/2} + \Gamma_{7/2} + \Gamma_{9/2} \tag{3.6.9}$$

and then decompose each of these representations in terms of appropriate spin representations of O with the help of Table 3.6.2:

$$\begin{gathered} \Gamma_{1/2} = \Delta_6, \qquad \Gamma_{3/2} = \Delta_8, \qquad \Gamma_{5/2} = \Delta_7 + \Delta_8 \\ \Gamma_{7/2} = \Delta_6 + \Delta_7 + \Delta_8, \qquad \Gamma_{9/2} = \Delta_6 + 2\Delta_8 \end{gathered} \tag{3.6.10}$$

Similar results for the decomposition of all Γ_j into all the point groups are summarized in Appendix C. Figure 3.6.1 illustrates how the results for large spin-orbit coupling are related to those for small spin-orbit coupling. The lines connecting the two cases are subject to the rule that two inequivalent representations do not interact. Hence their lines can cross. But two equivalent representations can interact and thus will not, in general, cross.[5]

[5]When two levels approach each other, we can neglect all others and find the eigenvalues from the roots of a secular determinant of the form

$$\begin{vmatrix} E - E_1, & V_{12} \\ V_{12}^*, & E - E_2 \end{vmatrix} = 0$$

or

$$E = \tfrac{1}{2}(E_1 + E_2) \pm \tfrac{1}{2}\left[(E_1 - E_2)^2 + |V_{12}|^2\right]^{1/2} \tag{3.6.11}$$

Thus a crossing occurs only if V_{12} vanishes at the same point at which $E_1 = E_2$. As a perturbation parameter λ is varied, $E_1 - E_2$ will be an approximately linear function $A(\lambda - \lambda_0)$ vanishing at λ_0, and V_{12} can be given its value at λ_0 so that the energy gap varies roughly as

$$\Delta E \approx \left[A^2(\lambda - \lambda_0)^2 + |V|^2\right]^{1/2}$$

and the individual levels are the nearly crossing hyperbolas shown in Fig. 3.6.2.

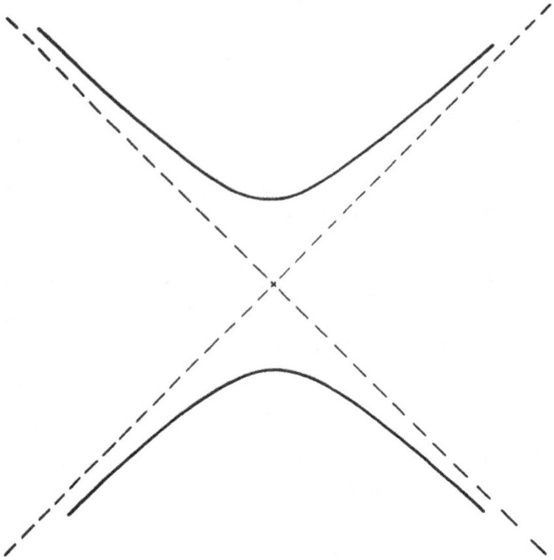

Fig. 3.6.2. Behavior of two neighboring energy levels near a crossing as a function of the perturbation parameter.

A more detailed consideration of the symmetries of the individual perturbation matrix elements, using group-theoretical[6] or the sometimes more approximate spin Hamiltonian[7] methods, can serve to determine some of the ratios of the spacings shown in Fig. 3.6.1. The actual magnitudes of the spacings and the order of the levels are dynamical rather than group-theoretical properties. The order of levels has been guided by Hund's rule that the state of smallest J (i.e., $J=3/2$) is lowest, except when the shell is more than half filled, as is the case here for Co^{2+}.

It may be noted that all of the levels Δ_6, Δ_7, Δ_8 are at least twofold degenerate. This illustrates a theorem due to Kramers that for an odd number of electrons a degeneracy exists because of time reversal that cannot be split by any crystalline field; see Theorem 10.4.7. Furthermore, the level Δ_6 is referred to as a "Kramers doublet" because its two components transform under rotations just as an isolated spin $\frac{1}{2}$ particle would. This is verified by the compatibility relation

$$\Gamma_{1/2}=\Delta_6 \tag{3.6.12}$$

[6]Pryce (1950), Koster (1958), Griffith (1961).

[7]Bleaney and Stevens (1953), Stevens (1963), Bowers and Owen (1955), Nierenberg ((1957), Abragam and Pryce (1951), Judd (1963).

in a field of O symmetry. It is therefore possible to introduce an "effective" spin operator $\mathbf{S}$ whose action on the two Δ_6 states is just that for a free spin $\frac{1}{2}$ particle.

The Δ_6 level is in fact split by hyperfine interactions with the nuclear magnetic moment and the external magnetic fields. Paramagnetic resonance experiments are performed on the lowest Δ_6 state by observing transitions between levels split by the magnetic field. The observed level scheme can be fitted by a "*spin Hamiltonian*" of the form

$$H_{\text{eff}} = g\beta \mathbf{H}\cdot\mathbf{S} + A\mathbf{I}\cdot\mathbf{S} - g_n\beta_n\mathbf{H}\cdot\mathbf{I} \tag{3.6.13}$$

where g, A, and g_n are empirical constants. Here $\mathbf{S}$ is an effective spin operator (not the true spin on the electron), but H_{eff} is written in such a way as to yield the correct matrix elements for the true Hamiltonian within an appropriate subset of states.

A rigorous method of *equivalent Hamiltonians* for *initially degenerate* states was introduced by Van Vleck (**1932**) and Pryce (**1950**); for almost degenerate states a Brillouin-Wigner approach was presented by Lax (**1950**) and Löwdin (**1951**). A unitary transformation, which provides a Hermitian equivalent Hamiltonian (in Problem 11.10.7), was used to derive Eq. 7.5.7, which appears in the 1963 draft of this book.

3.7. PROJECTION OPERATORS

If we multiply the representation relation

$$R\psi_{\nu'}^{j} = \sum \psi_{\mu'}^{j} D_{\mu'\nu'}^{j}(R) \tag{3.7.1}$$

by $[D^i(R)]_{\nu\mu}^{-1}$ and sum on R, the matrix orthogonality theorem 1.5.3 yields

$$P_{\mu\nu}^{i}\psi_{\nu'}^{j} = \psi_{\mu}^{i}\delta_{ij}\delta_{\nu\nu'} \tag{3.7.2}$$

where

$$P_{\mu\nu}^{i} = \frac{l_i}{g}\sum_{R}[D^i(R)]_{\nu\mu}^{-1}R \tag{3.7.3}$$

$$P_{\mu\nu}^{i} = \frac{l_i}{g}\sum_{R} D_{\mu\nu}^{i}(R)^{*}R \tag{3.7.4}$$

and the second form (3.7.4) is restricted to unitary representations. If

$$\psi = \sum_{j\nu'} C_{\nu'}^{j}\psi_{\nu'}^{j} \tag{3.7.5}$$

is an arbitrary function, then

$$P^i_{\mu\mu}\psi = C^i_\mu \psi^i_\mu \tag{3.7.6}$$

selects the component of ψ along the direction of the "unit vector" ψ^i_μ and hence is referred to as a *projection* operator, whereas

$$P^i_{\mu\nu}\psi = C^i_\nu \psi^i_\mu \tag{3.7.7}$$

is a *transfer* operator, since it selects ψ^i_ν and then converts it to ψ^i_μ. In general, a transfer operator P_{nm} can be defined by the equations

$$P_{nm}\psi_s = \delta_{ms}\psi_n \tag{3.7.8}$$

$$P_{nm}P_{lk} = \delta_{ml}P_{nk} \tag{3.7.9}$$

with Eq. 3.7.9 stating that a transfer from k to l and then from m to n is, if $l=m$, a transfer from k to n. With $P_n = P_{nn}$, 3.7.9 specializes to an equation

$$P_n P_k = \delta_{kn} P_n \tag{3.7.10}$$

characteristic of projection operators. Our operator $P^i_{\mu\nu}$ is a special case of P_{nm} with $n=(i\mu)$, $m=(i\nu)$. A projection operator is called *primitive* if it projects into a one-dimensional subspace. An operator such as

$$P^i = \sum_\mu P^i_{\mu\mu} = \frac{l_i}{g}\sum \chi^i(R)^* R \tag{3.7.11}$$

which obeys

$$P^i P^j = \delta_{ij} P^i \tag{3.7.12}$$

projects into a subspace that transforms according to representation i and is surely nonprimitive if $l_i > 1$. Indeed, $P^i_{\mu\mu}$ can be nonprimitive if ψ is in a space that contains the representation i more than once.[8]

Equation 3.7.7 tells us that, if we start with any ψ such that $C^i_\nu \neq 0$, then, for fixed ν, the set of functions

$$\varphi^i_\mu = P^i_{\mu\nu}\psi \tag{3.7.13}$$

transforms as a basis for the representation $D^i_{\mu\mu'}(R)$. [Although the functions $P^i_{\mu\mu}\psi$ span the same space, the factor C^i_μ varies with μ, causing the representation generated by $P^i_{\mu\mu}\psi$ to differ from $D^i_{\mu\mu'}(R)$.]

Equation 3.7.13 tells us that *if* we have a matrix representation we can find

[8]In this case, the right-hand side of Eq. 3.7.6 is to be regarded as an abbreviation for $\Sigma_a C^{ia}_\mu \psi^{ia}_\mu$, where a is the repetition index for representation i.

basis functions for that representation in any space that is not orthogonal to the representation. This seems backwards, since a knowledge of a representation usually requires knowledge of at least *one* set of basis functions. Actually, the situation is not as bad as it sounds. Often one set of basis functions can be obtained by inspection. For example, we found that the polar vector representation Γ_p, spanned by (x,y,z), decomposed into $E=(x,y)$ and $A_1=z$ for the group C_{3v}, thus giving us a set of basis functions for E. On the other hand, in the treatment of the ammonia molecule vibration problem discussed in Section 5.2 we want the basis functions in the twelve-dimensional vector space of the three-dimensional motion of four particles that transform as E, and these are harder to obtain by inspection.

If only the character table is known, it is, of course, always possible to calculate $P^i\psi$ using Eq. 3.7.11, which will produce some function in the irreducible space i. Group operators can then be used to create enough independent functions to span the space. These must then be orthonormalized to yield a unitary representation. Basis functions for the point groups can be obtained in this way by setting $\psi=$an arbitrary polynomial of the first, second, third, etc., degree.

If *one* set of orthonormal basis functions ψ_μ^i is already available, a second (identically transforming) set of basis functions φ_μ^i can be created by Eq. 3.7.13 or, alternatively, *without explicitly writing out the matrices* $\mathbf{D}^i(R)$, by

$$\frac{l_i}{g}\sum_R R\left[(\psi_\nu^i)^\dagger\psi\right]=\sum_\mu(\psi_\mu^i)^\dagger\varphi_\mu^i \tag{3.7.14}$$

where R acts on both factors in the brackets, and the basis functions φ_μ^i of Eq. 3.7.13 are simply read off as the coefficients of the $(\psi_\mu^i)^\dagger$. The simplicity of this procedure is demonstrated in the next section.

3.8. CRYSTAL HARMONICS

In the Wigner-Seitz method of constructing electronic energy bands in solids, one starts with the approximation of treating the potential in each cell as spherically symmetrical. This suggests seeking solutions of the form $f_{nl}(r)$ $Y_{lm}(\phi,\varphi)$, where the Y_{lm} are spherical harmonics. However, the Wigner-Seitz unit cell (cf. Section 6.1) over which boundary conditions must be satisfied has the symmetry of the point group. Thus one would like angular functions that are irreducible representations of the point group, and not, like Y_{lm}, of the full rotational group, for example, cubic harmonics rather than spherical harmonics (Von der Lage and Bethe, **1947**).

Since $r^lY_{lm}(\phi,\varphi)$ is a polynomial of degree l, the problem of finding the *crystal harmonics* is *equivalent to that of finding the polynomial basis functions for the representations of the point groups.* We illustrate the procedure for the point group C_{3v}. We saw in Section 3.1 that the polar vector representation $\Gamma_p=\Gamma_1^-=(x,y,z)$ decomposed into $E=(x,y)$ and $A_1=z$ under the group C_{3v}. We now seek another set of basis functions for E by the "brute-force" use of the projection procedure 3.7.14 among the quadratic polynomials. In order that C_1^i of Eq. 3.7.7 should not vanish, we must choose ψ to be odd in x just as $\psi_1^i=X$ is. Let us choose $\psi=2xy$ and use Eq. 3.7.14:

$$\frac{l_i}{g}\sum_R R[X2xy]=\tfrac{2}{6}(E+\sigma+C_3^{-1}\sigma+C_3\sigma+C_3+C_3^{-1})X2xy \tag{3.8.1}$$

where we have used the group multiplication table 1.2.4 or 1.2.6 to express all the elements in terms of the generators $A=\sigma=$ a reflection plane that reverses the sign of x and $D=C_3=$ a 120° counterclockwise rotation. Since Xxy is even under σ, Eq. 3.8.1 can be written as

$$\begin{aligned}\tfrac{4}{3}(E+C_3+C_3^{-1})Xxy&=\tfrac{4}{3}Xxy\\ &+\tfrac{4}{3}\left(-\tfrac{1}{2}X-\tfrac{1}{2}\sqrt{3}\,Y\right)\left(-\tfrac{1}{2}x-\tfrac{1}{2}\sqrt{3}\,y\right)\left(\tfrac{1}{2}\sqrt{3}\,x-\tfrac{1}{2}y\right)\\ &+\tfrac{4}{3}\left(-\tfrac{1}{2}X+\tfrac{1}{2}\sqrt{3}\,Y\right)\left(-\tfrac{1}{2}x+\tfrac{1}{2}\sqrt{3}\,y\right)\left(-\tfrac{1}{2}\sqrt{3}\,x-\tfrac{1}{2}y\right)\\ &=X(2xy)+Y(x^2-y^2)\end{aligned} \tag{3.8.2}$$

from which the pair of basis functions $[2xy,x^2-y^2]$ that transform precisely as $[x,y]$ can be read off immediately.

The procedure just described can be tedious for a group containing a large number of elements. It is sometimes easier to work backwards: to find out what is contained in a space rather than to project into it. For example, we showed in Section 3.2 that the space of the representation $D^+=\Gamma_2^+$ contains five functions, xz,yz,xy, and, say, x^2 and y^2, and contains representations A_1+2E. Alternatively, the symmetric product

$$(E\times E)_{\text{sym}}=A_1+E \tag{3.8.3}$$

with $E=(x,y)$ consists of the three functions x^2,y^2,xy. It is clear that $[xz,yz]$ $\sim[x,y]$ and x^2+y^2 is A_1. This leaves x^2-y^2 and xy to span E. Furthermore, xy is the odd partner. Thus we could choose $[2xy,a(x^2-y^2)]$ as the basis and fix a by the requirement that this pair transform as $[x,y]$. Thus, by comparison with Examples 1.2.1, and 1.4.1, we impose

$$C_3(2xy)=-\tfrac{1}{2}(2xy)-\tfrac{1}{2}\sqrt{3}\,a(x^2-y^2) \tag{3.8.4}$$

which is satisfied only if $a=1$.

A third procedure for groups containing three and six fold elements is to start with the cyclic subgroup and to construct its one-dimensional representations by diagonalizing C_3 (or C_6). Since $(x \pm iy)^m \sim \exp(\pm im\varphi)$,

$$C_3(x+iy)=\epsilon(x+iy); \qquad C_3(x-iy)=\epsilon^2(x-iy)$$
$$C_3(x-iy)^2=\epsilon(x-iy)^2; \qquad C_3(x+iy)^2=\epsilon^2(x+iy)^2 \tag{3.8.5}$$

where, in accord with Eq. A2, $\epsilon=\exp(-2\pi i/3)$. Comparing the results for the linear and quadratic functions, we see that

$$(x+iy)\sim(x-iy)^2, \qquad (x-iy)\sim(x+iy)^2 \tag{3.8.6}$$

where $\sim$ means "transform alike." If we add a mirror plane σ to get C_{3v}, this mixes the $x+iy$ and $x-iy$ representations:

$$\sigma(x+iy)=-x+iy=-(x-iy), \quad \text{that is, } \sigma\psi_1=-\psi_2$$

whereas

$$\sigma(x-iy)^2=(-x-iy)^2=-\left[-(x+iy)^2\right], \quad \text{that is, } \sigma\varphi_1=-\varphi_2$$

where $[\psi_1,\psi_2]$ and $[\varphi_1,\varphi_2]$ are a pair of identical transforming functions so that

$$[(x+iy),(x-iy)]\sim a\left[(x-iy)^2,-(x+iy)^2\right] \tag{3.8.7}$$

and by taking corresponding linear combinations:

$$[x,y]\sim -ia\left[2xy,(x^2-y^2)\right] \tag{3.8.8}$$

These types of procedures[9] have yielded the sets of corresponding basis functions shown in the character tables of Appendix A.

For representations of dimension l_j, we do not list more than l_j sets of crystal harmonics, in view of the following discussion.

[9]A more powerful systematic procedure is illustrated in Chapter 5 for constructing the symmetry coordinates (i.e., basis vectors for irreducible representations) for the vibrations of NH_3 with less effort than is required by direct use of the projection procedure 3.7.13.

▲ Number of Independent Sets of Crystal Harmonics[10]

Theorem 3.8.1

Let Γ_j be an irreducible representation (other than the identity) of a point group. Let $\psi_\mu^{ja}(\mathbf{r})$, $\mu = 1,2,\ldots,l_j$, and fixed a be a set of basis functions for Γ_j. Let $a = 1,2,\ldots$ run over independent sets of such basis functions. *Then precisely l_j such sets are independent modulo the identity representation.* In other words, an arbitrary set of basis functions $\psi_\mu^j(\mathbf{r})$ can be expanded in terms of a set of l_j such sets:

$$\psi_\mu^j(\mathbf{r}) = \sum_{a=1}^{l_j} F^{ja}(\mathbf{r})\psi_\mu^{ja}(\mathbf{r}) \tag{3.8.9}$$

where the $F^{ja}(\mathbf{r})$ are invariant under all group operations.

Proof: By using spherical harmonics of all orders, we can certainly construct an infinite number of sets of a given representation that are independent in the sense of not being linearly related. We choose l_j such sets. The determinant $\det\psi_\mu^a(\mathbf{r})$ does not vanish identically (the index j was suppressed for clarity), and hence does not vanish in any finite neighborhood of any point $\mathbf{r}$. Thus "almost everywhere" we can regard $\psi_\mu^{ja}(\mathbf{r})$, for fixed $\mathbf{r}$ and a, as a vector of l_j components on μ. But there can be no more than l_j independent vectors in a space of dimension l_j. Thus any additional vector $\psi_\mu^j(\mathbf{r})$ can be expanded in terms of this fundamental set with coefficients $F^{ja}(\mathbf{r})$ that depend on $\mathbf{r}$ as in Eq. 3.8.9. (Conversely, at least l_j such sets are needed for completeness.)

The action of R on Eq. 3.8.9 yields

$$\sum \psi_\lambda^j(\mathbf{r}) D_{\lambda\mu}^j(R) = \sum_{a=1}^{l_j} [RF^{ja}(\mathbf{r})] \sum \psi_\lambda^{ja}(\mathbf{r}) D_{\lambda\mu}^j(R)$$

We multiply by $[D^j(R)]_{\mu\nu}^{-1}$ on the right and sum on μ:

$$\psi_\nu^j(\mathbf{r}) = \sum_{a=1}^{l_j} [RF^{ja}(\mathbf{r})]\psi_\nu^{ja}(\mathbf{r})$$

But for (almost) any fixed $\mathbf{r}$, this expansion of a vector into components is unique. Therefore on comparison with Eq. 3.8.9

$$RF^{ja}(\mathbf{r}) = F^{ja}(\mathbf{r})$$

which completes our proof.

[10]The statement of Theorem 3.8.1 and its reasonableness are established by Hopfield (1960). The proof presented here evolved from discussions with D. R. Fredkin.

Although there are an infinite number of possible invariant functions $F(\mathbf{r})$, the usefulness of Theorem 3.8.1 is enhanced by the fact that *any* invariant polynomial can be expressed as a function of a *finite number of basic invariants*.[11] These basic invariants can in fact be written down quickly by the methods developed in Section 4.5 and illustrated in Section 4.7. ▲

3.9. SUMMARY

In Section 3.1 we indicate character techniques for decomposing a space into irreducible subspaces. The perturbation of a set of degenerate levels by a crystal field splits an originally irreducible representation of the full rotation group into its irreducible components with respect to the crystal point group.

Perhaps the most important theorem of this chapter relating to representations as a whole is Theorem 3.4.4, which states that a matrix element involving a product of basis functions for three irreducible representations will vanish if the space whose character is described by the product of the three characters does not contain the identity representation.

More specific information about individual matrix elements and individual partner functions of a space is given by the base vector orthogonality theorem. This theorem states that an operator possessing the full symmetry of a group cannot connect different irreducible representations of the group. Moreover, it can only connect corresponding partners of the same irreducible representation. When a given representation is repeated in the physical space of a problem, this theorem tells us to arrange the basis vectors in each of the repeated spaces so that they transform in an identical manner. The advantage of this procedure is that a large matrix of dimension $n_i l_i$ by $n_i l_i$ is decomposed into a direct sum of l_i identical $n_i \times n_i$ matrices, thus greatly reducing the dynamical problem to the solution of one $n_i \times n_i$ problem.

Methods of projecting into a subspace belonging to a given representation and even into a subspace belonging to a given partner of a given representation are discussed in Section 3.7. These projection techniques are simplified in Section 3.8 to provide an easy way to obtain sets of polynomial basis functions for the irreducible representations of the point groups.

problems

3.1.1. Given a group G of g elements $A_1, A_2, \ldots, A_g$ and a sufficiently *arbitrary* basis function Ψ such that the set of g basis functions $\Psi_i = A_i \Psi$ is an independent set, show that the Ψ_i form a basis for a representation

$$R\Psi_j = \sum \Psi_i D_{ij}^{\text{reg}}(R)$$

[11]This is Noether's theorem. See Weyl (1946), p. 275.

known as the *regular representation*, whose elements are 1 or 0 with

$$D_{ij}^{\text{reg}}(R)=\delta(R,A_jA_i^{-1})$$

3.1.2. (*a*) Show that for the regular representation

$$\chi^{\text{reg}}(E)=g, \qquad \chi^{\text{reg}}(R)=0 \quad \text{for} \quad R\neq E.$$

(*b*) Show that representation Γ_i of dimension l_i is contained l_i times in the regular representation.

(*c*) Prove Burnside's theorem, $\Sigma(l_i)^2=g$ (Eq. 1.5.9).

3.1.3. (*a*) Show, without using the character table for T, that the "polar vector" representation generated by the basis functions x,y,z is irreducible in the tetrahedral point group $T(23)$.

(*b*) Show that there are four classes in T.

(*c*) Use Burnside's theorem to show that the group T must have three one-dimensional representations in addition to the vector representation with basis x,y,z.

3.1.4. Verify that Eq. 2.4.6,

$$\rho(\alpha_1\alpha_2\alpha_3)\,d\alpha_1\,d\alpha_2\,d\alpha_3=\left[\frac{2\rho(0)(1-\cos\alpha)}{\alpha^2}\right]\alpha^2 d\alpha\,d\Omega$$

gives the Hurwitz invariant volume element for the rotation group when $\alpha\mathbf{n}=(\alpha_1,\alpha_2,\alpha_3)$ are the parameters, $d\Omega$ is an element of solid angle in $\alpha_1\alpha_2\alpha_3$ space, and $\alpha=(\alpha_1^2+\alpha_2^2+\alpha_3^2)^{1/2}$ is the rotation angle.

3.1.5. Verify the irreducibility of the representation Γ_j of the full rotation group (see Section 2.4) by showing that

$$\frac{\int_0^{\pi}\left[\sin\alpha(j+\frac{1}{2})/\sin\frac{1}{2}\alpha\right]^2(1-\cos\alpha)\,d\alpha}{\int_0^{\pi}(1-\cos\alpha)\,d\alpha}=1$$

3.2.1. An F center (Mott and Gurney, **1948**) in NaCl is a localized electron which fills a chlorine vacancy and produces color via its absorption lines. The chlorine site has cubic symmetry. How are the *s*, *p*, and *d* levels of the full rotational group separated by cubic symmetry? If the crystal is compressed along the 111 axis, what site symmetry remains? What further reduction of degeneracy takes place in the distorted crystal?

3.3.1. Prove that $\Gamma_j \times \Gamma_k$ is irreducible if Γ_j is irreducible and Γ_k is one dimensional.

3.3.2. Show that, if $\Gamma = \Sigma_j \Gamma_j$,

$$(\Gamma \times \Gamma)_{\mathrm{sym}} = \sum_j (\Gamma_j \times \Gamma_j)_{\mathrm{sym}} + \sum_{i>j} (\Gamma_i \times \Gamma_j)$$

$$(\Gamma \times \Gamma)_{\mathrm{asym}} = \sum_j (\Gamma_j \times \Gamma_j)_{\mathrm{asym}} + \sum_{i>j} (\Gamma_i \times \Gamma_j)$$

whereas

$$\Gamma \times \Gamma = \sum_j (\Gamma_j \times \Gamma_j) + 2 \sum_{i>j} (\Gamma_i \times \Gamma_j)$$

3.3.3. (*a*) Show that, if representation Γ decomposes into

$$\Gamma = \sum a_j \Gamma_j$$

(where the a_j are necessarily integers), then

$$(\Gamma \times \Gamma)_{\mathrm{sym}} = \sum_j \left[a_j (\Gamma_j \times \Gamma_j)_{\mathrm{sym}} + \tfrac{1}{2} a_j (a_j - 1)(\Gamma_j \times \Gamma_j) \right]$$

$$+ \sum_{i>j} a_i a_j (\Gamma_i \times \Gamma_j)$$

(*b*) Show that the corresponding antisymmetric product is simply obtained by replacing "sym" with "asym" in the above equation.

3.3.4. Show that the representations Γ_j and Γ_k of the proper rotation group $R^+(3)$ have the product

$$\Gamma_j \times \Gamma_k = \sum_{l=|j-k|}^{j+k} \Gamma_l$$

where j, k, and l are integers or half-integers. *Hint*: Show that

$$c_{jkl} = \frac{\displaystyle\int_0^{\pi} \frac{\sin \alpha (j+\frac{1}{2})}{\sin \frac{1}{2}\alpha} \frac{\sin \alpha (k+\frac{1}{2})}{\sin \frac{1}{2}\alpha} \frac{\sin \alpha (l+\frac{1}{2})}{\sin \frac{1}{2}\alpha} \sin^2 \tfrac{1}{2}\alpha \, d\alpha}{\displaystyle\int_0^{\pi} \sin^2 \tfrac{1}{2}\alpha \, d\alpha}$$

$$= 1 \quad \text{if} \, |j-k| \leqslant l \leqslant j+k$$

$$= 0 \quad \text{otherwise}$$

3.3.5. Show with the help of Eq. 3.3.30 that, if Γ_j is a half-integral representation of the rotation group the symmetric product is given by

$$\Gamma_{(j\times j)}=\Gamma_1+\Gamma_3+\cdots+\Gamma_{2j}$$

which contains all "odd" representations, whereas the antisymmetric product contains the "even" representations:

$$\Gamma_{[j\times j]}=\Gamma_0+\Gamma_2+\cdots+\Gamma_{2j-1}$$

3.3.6. Under what circumstances does

$$\sum_{\mu\nu}\int[\phi_\mu^i(\mathbf{r})\Psi_\nu^j(\mathbf{r}')]^* R\phi_\mu^i(\mathbf{r})\,\Psi_\nu^j(\mathbf{r}')\,d\mathbf{r}$$

$$=\chi^{i\times j}(R)=\chi^i(R)\chi^j(R)$$

3.3.7. If $F(x)$ is a polynomial with integral coefficients, show that

$$\frac{1}{g}\sum_R F[\chi(R)]=\text{an integer}$$

Hint: $[\chi(R)]^n$ is the character of $D(R)\times D(R)\times\cdot\times D(R)$ (n factors).

3.4.1. Determine the selection rules for magnetic dipole radiation for the group C_{3v}.

3.4.2. For which point groups are the selection rules for magnetic dipole radiation distinct from those for electric dipole radiation?

3.4.3. For which groups are the selection rules for radiation independent of the polarization? For which groups are there two different sets of selection rules, depending on the polarization? Can there be three different sets of selection rules?

3.4.4. Determine the selection rules for electric quadrupole radiation for the group C_{3v}.

3.4.5. Show, using the composition of angular momentum, that a nucleus with total angular momentum j has no multipole moments of order greater than $2j$.

3.4.6. Prove Eq. 3.4.12.

3.4.7. Show, using the composition of angular momentum, that a nucleus can have no magnetic dipole moment unless its spin obeys $j\geq\frac{1}{2}$, and no electric quadrupole moment unless $j\geq 1$.

3.4.8. (*a*) Show that, if a system has a Hamiltonian even under inversion, the eigenfunctions of the Hamiltonian are even or odd and have no electric moment.

(*b*) Polar molecules, however, have a susceptibility appropriate to the presence of a permanent electric moment. Explain.

3.4.9. Show that the form of $(\Phi^i_\mu, V^m_\lambda \Psi^j_\nu)$, that is, the dependence on μ, λ, and ν, is unique if

$$c^{imj} = \frac{1}{g}\sum_R \chi^i(R)^* \chi^m(R) \chi^j(R) = 1$$

Hint: See Section 4.2.

3.4.10. Use the addition theorem of angular momentum derived in Problem 3.3.4 and the uniqueness implied by Problem 3.4.9 to show that, if Ψ^j_m is a state of total angular momemtum j and magnetic quantum number m, then $(\Psi^j_m, \mathbf{A}\Psi^j_m)$ has a *form* independent of the choice of vector **A**. Show, therefore, that the resulting matrix elements can be evaluated by making the replacement $\mathbf{A} \to \mathbf{J}(\mathbf{A}\cdot\mathbf{J})/J^2$. This is the content of the Wigner-Eckart theorem (Condon and Shortley, **1967**).

3.4.11. If V_s is any set of quantities obeying

$$V_s = \sum V_{s'} D_{s's}(R) \qquad (1)$$

for all R in some group, and $D_{s's}(R)$ constitutes a unitary representation of the group, show that a particular element V_s must vanish if

$$N_s \equiv \frac{1}{g}\sum_R D_{ss}(R) = 0 \qquad (2)$$

and all elements $V_s = 0$ if $\sum_R \chi(R) = 0$, where $\chi(R)$ is the trace of the possibly reducible representation $\mathbf{D}(R)$. *Hint*: Average Eq. 1 over R to obtain

$$V_s = \sum V_{s'} M_{s's}, \qquad M_{s's} = \frac{1}{g}\sum_R D_{s's}(R)$$

Show that $\sum_{s'} |M_{s's}|^2 = N_s$. If $N_s = 0$, all $M_{s's}$ must vanish!

3.8.1. Obtain a basis function for the determinantal representation A_2 of the group C_{3v}.

3.8.2. Develop three sets of crystal harmonics for the representation Γ_{15} of the group O.

3.8.3. (*a*) Show that the independent set of basis functions $\psi^{ja}_\mu(\mathbf{r})$, $a = 1, \ldots, l_j$, of Theorem 3.8.1 yields a determinant, $\det \psi_{a\mu}(\mathbf{r})$ (with j suppressed), that forms a basis for the representation $\det D^j_{\mu\nu}(R)$.

(*b*) Apply this, using the two sets of basis functions of E for C_{3v} shown in Eq. 3.8.8 to construct the representation A_2 of C_{3v}.

3.8.4. Consider an atom at a point of $O(432)$ symmetry in the crystal. By projection procedure or otherwise, show that the simplest invariants of group O (other than functions of r) are of fourth degree and take the form $x^4+y^4+z^4$ and $x^2y^2+y^2z^2+z^2x^2$. By making these orthogonal to r^4 and suitably normalizing them, show that the potential must take the form (Von der Lage and Bethe, **1947**).

$$V=\frac{5(21)^{1/2}}{4}\left(x^4+y^4+z^4-\tfrac{3}{5}r^4\right)$$

chapter four

MACROSCOPIC CRYSTAL TENSORS

4.1. MACROSCOPIC POINT GROUP SYMMETRY

Infinite crystals are invariant under symmetry operations that involve pure translation (the lattice translations discussed in Chapter 6), pure rotation (in some cases), and, in general, combinations of rotations (or reflections) with translation:

$$\mathbf{r}' \equiv (\boldsymbol{\alpha}|\mathbf{a})\mathbf{r} = \boldsymbol{\alpha}\mathbf{r} + \mathbf{a} \tag{4.1.1}$$

where $\boldsymbol{\alpha}$ is a 3×3 rotation matrix, and $\mathbf{a}$ is a translation—not necessarily a lattice translation. (See Chapter 8.) The complete set of operations (Eq. 4.1.1) that restore a crystal to itself is referred to as a space group.

Theorem 4.1.1

The macroscopic physical properties of a crystal (conductivity tensor, piezoelectric tensor, etc.) are invariant under the operations $(\boldsymbol{\alpha}|0)$ of the point group P associated with the space group G whose elements are $(\boldsymbol{\alpha}|\mathbf{a})$. (The term macroscopic means these properties are invariant to lattice translations and are sensitive to bulk rather than surface effects.)

Proof: For an *infinite* crystal, *all* operations $(\boldsymbol{\alpha}|\mathbf{a})$ that restore the crystal to itself leave invariant the macroscopic physical properties. If the space group G is *symmorphic*, that is, contains the point group $P=\{(\boldsymbol{\alpha}|0)\}$ as a subgroup, then clearly these operations leave invariant all macroscopic properties.

For nonsymmorphic groups, one can by judicious removal of lattice translations produce a set of operations $(\boldsymbol{\alpha}|\mathbf{v}(\boldsymbol{\alpha}))$, where the $\mathbf{v}(\boldsymbol{\alpha})$ are some proper fractions of a lattice translation and do not all vanish simultaneously. (See Theorem 8.2.2.) It can be argued that on a macroscopic scale such small translations can be ignored. If, say, the electrical conductivity is invariant with respect to a fourfold rotation C_4 combined with a translation of one-fourth a lattice spacing, no macroscopic experiment will detect whether or not such a small translation has in fact been made. A more adequate proof will be supplied in Section 8.9.

4.2. TENSORS OF THE FIRST RANK: FERROELECTRICS AND FERROMAGNETICS

Let us consider a crystal of symmetry $\mathcal{C}_3$, which can possess a permanent total moment $\mathbf{M}$ ($=\mathbf{p}$ or $\mathbf{m}$ for the electric or magnetic case, respectively). How many constants among M_1, M_2, and M_3 are independent?

If R represents an arbitrary proper rotation in three-dimensional space, and $\mathbf{D}(R)$ the corresponding three-dimensional matrix, then, after such a rotation, the vector $\mathbf{M}$, by its definition as a vector, transforms into

$$\mathbf{M}'=\mathbf{D}(R)\mathbf{M} \qquad \text{or} \qquad M'_\mu=\sum_{\nu=1}^{3} D_{\mu\nu}(R)M_\nu \tag{4.2.1}$$

If R is not an arbitrary rotation, but a member of a group G that restores the crystal to itself, then, in fact, the moment must not change:

$$\mathbf{M}'=\mathbf{M} \qquad \text{or} \qquad \mathbf{D}(R)\mathbf{M}=\mathbf{M} \tag{4.2.2}$$

In this case $\mathbf{M}$ is an *invariant vector* of the group of matrices $\mathbf{D}(R)$. In different terms, the vector $\mathbf{M}$ constitutes a basis, or symmetry coordinate, for the identity representation of the group. *The number of independent constants in* $\mathbf{M}$ *is then the number of independent base vectors for the identity representation*; that is, *it is equal to the number of times that the identity representation is contained in the three-dimensional space spanned by the components* M_1, M_2, *and* M_3.

Theorem 4.2.1

If $\chi(R)$ is the character, for a proper or improper rotation R, of the space spanned by the components of a tensor $\mathbf{M}$, the number of independent constants in the tensor is given by

$$N=\frac{1}{g}\sum_R \chi(R) \tag{4.2.3}$$

since by Eq. 3.4.5 this is the number of times that the identity representation is contained in the space spanned by the components of the tensor **M**.

In accord with Eq. 3.1.7 we use

$$\chi(R)=\chi^p(R)=\chi^{1-}(R)$$

$$\chi(R)=\chi^m(R)=\chi^{1+}(R)$$

for polar vectors **p** (= electric polarization) and for magnetic vectors **m** (= pseudovectors or axial vectors), respectively.

For our group $\mathcal{C}_3$, which contains no improper elements, there is no distinction between polar and axial vectors. The three elements E, C_3, and C_3^{-1} have characters (3,0,0), respectively (see Table 3.1.1). Thus $N=\frac{1}{3}(3+0+0)=1$, that is, only one constant is available. Indeed, the relation $\mathbf{D}(R)\mathbf{M}=\mathbf{M}$ for $R=C_3$ implies that the vector $\mathbf{M}=\mathbf{p}$ or $\mathbf{m}$ must be along the threefold axis.

For the group C_{3v}, the m and p representations are given in Table 3.1.3, and the numbers of independent constants are

$$N_m=\tfrac{1}{6}[3+0+3(-1)]=0$$

$$N_p=\tfrac{1}{6}[3+0+3(1)]=1$$

that is, a crystal of symmetry C_{3v} can be ferroelectric but not ferromagnetic. These results may be extended in the following theorems.

Theorem 4.2.2

A crystal can be ferroelectric only if its point group is $\mathcal{C}_1$, $\mathcal{C}_2$, $\mathcal{C}_3$, $\mathcal{C}_4$, or $\mathcal{C}_6$ or $\mathcal{C}_{1v}$, $\mathcal{C}_{2v}$, $\mathcal{C}_{3v}$, $\mathcal{C}_{4v}$, or $\mathcal{C}_{6v}$.

Proof: The ferroelectric direction must be an invariant axis for each element of the group. Thus the inversion must be absent. Also, **p** cannot be simultaneously directed along two distinct axes C_n and C_m. Thus the group can contain only one such axis, which leaves only the groups $\mathcal{C}_n$, $\mathcal{C}_{nh}$, $\mathcal{C}_{nv}$, and $\mathcal{S}_n$. But $\mathcal{C}_{nh}$ can be ruled out because the $\boldsymbol{\sigma}_h$ plane reverses the direction of the axis. Planes $\boldsymbol{\sigma}_v$ containing the axis leave it invariant and are allowed. Similarly, $\mathcal{S}_n$ contains the reversal element S_n and must be ruled out.

Theorem 4.2.3

A ferromagnetic crystal must have one of the point groups $\mathcal{C}_1$, $\mathcal{C}_2$, $\mathcal{C}_3$, $\mathcal{C}_4$, $\mathcal{C}_6$, or, one of the groups formed from these by adding inversion $\mathcal{C}_i$, $\mathcal{C}_{2h}$, $\mathcal{S}_6$,

$\mathcal{C}_{4h}$, and $\mathcal{C}_{6h}$, respectively, or one of the groups isomorphic to $\mathcal{C}_n$: $\mathcal{C}_s \sim \mathcal{C}_2$, $\mathcal{S}_4 \sim \mathcal{C}_4$, and $\mathcal{C}_{3h} \sim \mathcal{C}_6$. (See Table 2.15.1.)

Proof: The group must have only one axis as before, a requirement which means that $\mathcal{C}_1, \mathcal{C}_2, \ldots, \mathcal{C}_6$ are the only allowed proper groups. Inversion by definition has no effect on a pseudovector **m** and is thus allowed, giving $\mathcal{C}_n \times \mathcal{C}_i$. Among the groups not containing the inversion, the *corresponding* isomorphic proper group, Eq. 2.10.8, must also leave **m** invariant. Hence this isomorphic proper group must be among $\mathcal{C}_1, \mathcal{C}_2, \ldots, \mathcal{C}_6$. The corresponding improper groups are listed in the second column of Table 2.15.1. (Alternatively, the group can contain a mirror plane $\boldsymbol{\sigma}_h$ or a reversal axis $\mathcal{S}_n$, since these do not reverse a magnetic vector whereas $\boldsymbol{\sigma}_v$ does and is ruled out.)

These theorems are somewhat puzzling in view of the fact that iron is cubic and becomes ferromagnetic. We conclude that at the ferromagnetic transition the Fe atoms must displace in such a way as to destroy the O symmetry. If the magnetic forces are weak, these displacements will be small. They are in fact unobservable with the X-ray precision presently available. Similar remarks apply to the ferroelectric case, but there the forces are stronger, and the symmetry change on a phase transition can usually be observed.

4.3. SECOND-RANK TENSORS: CONDUCTIVITY, SUSCEPTIBILITY

Current **J** and electric field **E** are polar vectors transforming under a rotation R by the polar vector representation $\mathbf{D}^p(R)$, written briefly as **R**:

$$\mathbf{J}' = \mathbf{R}\mathbf{J}, \qquad \mathbf{E}' = \mathbf{R}\mathbf{E} \tag{4.3.1}$$

The conductivity tensor $\sigma_{\mu\nu}$ which relates them:

$$\mathbf{J} = \boldsymbol{\sigma} \cdot \mathbf{E}, \qquad \mathbf{J}' = \boldsymbol{\sigma}' \cdot \mathbf{E}' \tag{4.3.2}$$

then transforms according to

$$\boldsymbol{\sigma}' = \mathbf{R} \cdot \boldsymbol{\sigma} \cdot \mathbf{R}^{-1} \tag{4.3.3}$$

Using the orthogonality of **R**: $\tilde{\mathbf{R}}^{-1} = \mathbf{R}$, we find

$$\sigma'_{\mu\nu} = D^p_{\mu\mu'}(R) D^p_{\nu\nu'}(R) \sigma_{\mu'\nu'} \tag{4.3.4}$$

that is, the elements of $\boldsymbol{\sigma}$ form a basis for the *product* representation $\Gamma_p \times \Gamma_p = \Gamma_{\mathrm{I}}^- \times \Gamma_{\mathrm{I}}^-$ of $R(3)$ with character

$$\chi(R) = [\chi^p(R)]^2 \tag{4.3.5}$$

However $\boldsymbol{\sigma}$ (in contrast to **J** or **E**) is a matter tensor, that is, if R is a rotation belonging to the point group that restores the crystal to itself, $\boldsymbol{\sigma}$ (but not **J** or **E**) is invariant with respect to the rotation:

$$\boldsymbol{\sigma} = \mathbf{D}^{p\times p}(R)\boldsymbol{\sigma} \tag{4.3.6}$$

and must belong to the identity representation. The number of independent constants in $\boldsymbol{\sigma}$ is still given by Eq. 4.2.3:

$$N = \frac{1}{g}\sum_R \chi(R) \tag{4.3.7}$$

the number of times that the identity representation is present in the space $\Gamma_p \times \Gamma_p$. An identical conclusion applies to the dielectric constant tensor, the thermal conductivity tensor, and the magnetic susceptibility tensor. (Strictly speaking, one uses $\Gamma_m \times \Gamma_m$ for the last-named tensor, but the product is identical to $\Gamma_p \times \Gamma_p$.)

If the tensor is symmetric, Eq. 4.3.7 remains valid but $\chi(R)$ must be replaced by the character of the symmetric product representation $(\Gamma_p \times \Gamma_p)_{\text{sym}}$:

$$\chi(R) = \tfrac{1}{2}[\chi^p(R)]^2 + \tfrac{1}{2}\chi^p(R^2) \tag{4.3.8}$$

This expression is valid even in the magnetic case, since R^2 is always a *proper* rotation. *Thus all second-rank tensors (all second-rank symmetric tensors) relating two polar or two axial vectors have a form and a number of independent constants that are functions of the point group and not of the physical nature of the tensor.*

The static dielectric constant tensor $\epsilon_{\mu\nu}$ must be symmetric because the polarization $P_\mu = \partial W/\partial E_\mu$ is the derivative of some potential energy W. The symmetry of $\epsilon_{\mu\nu}$ follows from

$$\frac{\partial P_\mu}{\partial E_\nu} = \frac{\partial P_\nu}{\partial E_\mu} \tag{4.3.9}$$

that is, the fact that the second derivatives of W may be taken in either order.

Let us illustrate two methods of finding the form of an arbitrary (not necessarily symmetric) second-rank tensor $T_{\mu\nu}$ under the group C_{3v}.

1. *Characters Plus Guesswork.* The classes of C_{3v} are E, $2C_3$, $3\sigma_v$. The corresponding characters of Γ_p are (3,0,1). (See Table 3.1.3.) Thus $\Gamma_p \times \Gamma_p$ has characters (9,0,1), and the number of independent parameters is, by Eq. 4.3.7, $\frac{1}{6}[9+0+3(1)] = 2$. We therefore guess that the matrix has elements $T_{11} = T_{22}$ and T_{33}, where 3 is the direction of the C_3 axis.

(For any of the cubic groups, T, O, T_d, T_h, O_h, only one constant results and we know that $T_{11} = T_{22} = T_{33}$ and all other elements vanish.)

2. *Eigenvalue Arguments.* The basis for the preceding guess is our intuition about quadratic forms. The symmetric part of **T** can be thought of as a quadratic form, that is, an ellipsoid. The invariance of such an ellipsoid under an axis C_n for $n>2$ requires the ellipsoid to be an ellipsoid of revolution about the axis. The two minor axes (i.e., transverse eigenvalues) must be equal because they are mixed by C_n. Thus $(T)_{\text{sym}}$ has the form $T_{11}=T_{22}\neq T_{33}$ with other elements vanishing.

The antisymmetric part of **T** can be thought of as a vector, $V_\mu=\frac{1}{2}e_{\mu\nu\lambda}T_{\nu\lambda}$, using the tensor density $e_{\mu\nu\lambda}$ defined in Eq. 2.2.3. To be invariant under C_n (for $n\geqslant 2$), **V** must be directed along C_n, that is, only $T_{12}=-T_{21}\neq 0$. Thus our tensor must have the form

$$\begin{bmatrix} a & c & 0 \\ -c & a & 0 \\ 0 & 0 & b \end{bmatrix} \tag{4.3.10}$$

Invariant Functions

Let $\mathbf{x}=(x_1,x_2,\ldots,x_n)$ be a vector in an n-dimensional space, and $\mathbf{D}(R)$ be an $n\times n$ matrix representation in that space of a set of operators R. Then $F(\mathbf{x})$ is said to be an invariant function if

$$F[\mathbf{D}(R)\mathbf{x}]=F(\mathbf{x})$$

Theorem 4.3.1

If $F(\mathbf{x})=\mathbf{P}^*\cdot\mathbf{x}$ is an invariant linear form with respect to a set of *unitary* matrices $\mathbf{D}(R)$, then $\mathbf{P}$ is an invariant vector of the set of matrices: $\mathbf{D}(R)\cdot\mathbf{P}=\mathbf{P}$. If the representation $\mathbf{D}(R)$ is real, $\mathbf{P}$ can be taken as real.

The relation that we have intuitively assumed above between tensors and quadratic forms can be summarized in the following theorem.

Theorem 4.3.2

If **T** is invariant under the real orthogonal transformation **R**, that is, $\mathbf{RTR}^{-1}=\mathbf{T}$ as in Eq. 4.3.3, then $F(\mathbf{r},\mathbf{r}')=\mathbf{r}\cdot\mathbf{T}\cdot\mathbf{r}'$ is an invariant bilinear form.

The proofs of these theorems are left to the reader.

We note that inversion yields no information concerning second-rank tensors because $\mathbf{T}'\sim(-\mathbf{r})(-\mathbf{r}')=\mathbf{rr}'=\mathbf{T}$. Conversely, *measurements on tensors of even rank cannot be used to direct whether a crystal has a center of symmetry*. Note, however,

Table 4.3.2. Essential Symmetry Elements in Each Crystal System

Crystal System	Point Groups	Essential Symmetry Elements
Triclinic	C_i, C_1	None or $\bar{1}$ = inversion
Monoclinic	C_{2h}, C_s, C_2	A single 2 or $\bar{2} = m$
Orthorhombic	D_{2h}, C_{2v}, D_2	Three perpendicular 2 or $\bar{2} = m$
Tetragonal	D_{4h}, C_{4h}, D_{2d} C_{4v}, D_4, S_4, C_4	A single 4 or $\bar{4}$
Cubic	O_h, T_h, T_d, O, T	Four threefold axes along body diagonals of a cube
Trigonal	$D_{3d}, D_3, C_{3v}, S_6, C_3$	A single 3 or $\bar{3}$ axis
Hexagonal	$D_{6h}, C_{6h}, D_{3h}, C_{6v}$ D_6, C_{3h}, C_6	A single 6 or $\bar{6}$ axis

that a linear relation between an axial and a polar vector would have the properties of a *pseudotensor* $\mathbf{T} \sim \mathbf{mr}$ or tensor density, that is, it would change sign under inversion: $\mathbf{T}' \sim \mathbf{m}(-\mathbf{r}) = -\mathbf{T}$.

With respect to second-rank tensors (not pseudotensors) the group $H \times \mathcal{C}_i$ contains no more information than the group H. Also the group $G = H + MH$, when H is proper and M is improper, yields the same result as the corresponding proper group $G = H + IMH$. With this information and the methods above, the form of a symmetric second-rank tensor can be constructed for each point group. For purposes of summarizing the results in Table 4.3.1, the 32 point groups are arranged into seven crystal systems such that all point groups within

Table 4.3.1. Nonvanishing Components of a Symmetrical Second-Rank Tensor in Each of the Crystal Systems

Crystal System	Components					
Triclinic	11	22	33	23	31	12
Monoclinic[a]	11	22	33	0	0	12
Orthorhombic	11	22	33	0	0	0
Trigonal	11	11	33	0	0	0
Tetragonal	11	11	33	0	0	0
Hexagonal	11	11	33	0	0	0
Cubic	11	11	11	0	0	0

[a]With twofold axis along the z or 3 direction.

each system have the same form of second-rank symmetric tensor. The point groups within each system and the essential symmetry elements appropriate to the system are listed in Table 4.3.2. In Chapter 6 we shall see that groups (e.g., O_h, T_h, T_d, O, T) are assigned to the same crystal system, since they *force* a lattice to have the same structure (e.g., cubic).

4.4. DIRECT INSPECTION METHODS FOR TENSORS OF HIGHER RANK: THE HALL EFFECT[1]

In this section we discuss the method of determining the form of a tensor by "direct inspection." As an example, let us consider the linear Hall effect:

$$J_\mu = \sigma_{\mu\nu\lambda} E_\nu H_\lambda \tag{4.4.1}$$

The symmetry of $\sigma_{\mu\nu\lambda}$ is restricted by the Onsager relations,

$$\sigma_{\mu\nu}(\mathbf{H}) = \sigma_{\nu\mu}(-\mathbf{H}) \tag{4.4.2}$$

for the conductivity tensor $\boldsymbol{\sigma}(\mathbf{H})$ in the presence of a magnetic field $\mathbf{H}$. (These relations are a consequence of time reversal and are established in Section 10.5.) In view of Eq. 4.4.2, the part of $\sigma_{\mu\nu}(\mathbf{H})$ even in $\mathbf{H}$ is symmetric in $(\mu\nu)$, as indicated by parentheses and the part odd in $\mathbf{H}$, including the Hall effect, is antisymmetric in $[\mu\nu]$, as indicated by brackets. A discussion similar to that in Section 4.3 shows that $\sigma_{\mu\nu\lambda}$ transforms according to

$$\sigma_{\mu\nu\lambda} = \sigma'_{\mu\nu\lambda} = D^p_{\mu\mu'}(R) D^p_{\nu\nu'}(R) D^m_{\lambda\lambda'}(R) \sigma_{\mu'\nu'\lambda'} \tag{4.4.3}$$

that is, according to the representation $(\Gamma_p \times \Gamma_p)_{\text{asym}} \times \Gamma_m$ in view of the antisymmetry (asym) in $[\mu\nu]$. In a vacuum or an isotropic material, we have the full rotation group:

$$(\Gamma_p \times \Gamma_p)_{\text{asym}} \times \Gamma_m = \Gamma_1^+ \times \Gamma_1^+ = \Gamma_2^+ + \Gamma_1^+ + \Gamma_0^+$$

which contains the identity representation once, so that only one arbitrary constant occurs. Indeed, in this case we are familiar with the answer:

$$\mathbf{J} = \sigma_0(\mathbf{E} \times \mathbf{H}) \qquad \text{or} \qquad \sigma_{\mu\nu\lambda} = \sigma_0 e_{\mu\nu\lambda} \tag{4.4.4}$$

where $e_{\mu\nu\lambda}$ is the Levi-Civita tensor density defined by Eq. 2.2.3.

For the group O we have the decomposition

$$\Gamma_2 \to \Delta_{12} + \Delta_{25}, \qquad \Gamma_1 \to \Delta_{15}, \qquad \Gamma_0 \to \Delta_1$$

[1]For the physical meaning of "Hall effect," "conductivity," and other terms see any elementary textbook on solid-state physics, for example, Kittel (**1971**), Harrison (**1970**), or Nye (**1957**).

so that $\Gamma_2+\Gamma_1+\Gamma_0$ contains Δ_1 once. With only one available constant, the form must be the same as in a vacuum. A similar conclusion applies to group T. See, however, Problems 4.4.1–4.4.4.

Direct Inspection

Let us now consider $\sigma_{\mu\nu\lambda}$ for the group D_2. A twofold rotation about the z axis has the matrix

$$\boldsymbol{\delta}_{2z}=\begin{bmatrix} -1 & 0 & 0 \\ 0 & -1 & 0 \\ 0 & 0 & 1 \end{bmatrix} \tag{4.4.5}$$

The use of Eq. 4.4.3, together with the invariance requirement $\boldsymbol{\sigma}'=\boldsymbol{\sigma}$, yields

$$\sigma_{\mu\nu\lambda}=(-1)^{n_1+n_2}\sigma_{\mu\nu\lambda} \tag{4.4.6}$$

where n_1 is the number of 1's and n_2 the number of 2's appearing in μ, ν, and λ. Since

$$n_1+n_2+n_3=3$$

we can conclude that

$$\sigma_{\mu\nu\lambda}=(-1)(-1)^{n_3}\sigma_{\mu\nu\lambda} \tag{4.4.7}$$

Thus $\sigma_{\mu\nu\lambda}$ vanishes unless n_3 is odd. A similar use of $\boldsymbol{\delta}_{2x}$ and $\boldsymbol{\delta}_{2y}$ shows n_1 and n_2 odd for nonvanishing elements. Therefore each possible index, 1, 2, and 3, must appear once, and there are six independent third-rank tensor components, $(\sigma_{123},\sigma_{231},\sigma_{312})$, $(\sigma_{213},\sigma_{321},\sigma_{132})$, in D_2. If $\sigma_{\mu\nu\lambda}$ is symmetric (for piezoelectricity)[2] or antisymmetric (for the Hall effect) in $[\mu\nu]$, the second set of three constants is determined by the first.

The group T can be obtained from D_2 by adding the generator $\boldsymbol{\delta}_{3xyz}$, which takes xyz into yzx. Equation 4.4.3 now leads to

$$\sigma_{123}=\sigma_{231}=\sigma_{312} \quad \text{and} \quad \sigma_{213}=\sigma_{321}=\sigma_{132} \tag{4.4.8}$$

The group O can be obtained from T by adding the twofold generator $\boldsymbol{\delta}_{2xy}$,

[2]Piezoelectricity describes the electric field E induced by a strain $e_{\mu\nu}$, $E_\lambda=S_{\mu\nu\lambda}e_{\mu\nu}$. Since $e_{\mu\nu}$ is symmetric in $(\mu\nu)$ (see Eq. 4.6.2), $S_{\mu\nu\lambda}$ is symmetric in $(\mu\nu)$.

which takes xyz into $yx\bar{z}$ so that Eq. 4.4.3 yields

$$\sigma_{123} = -\sigma_{213} \tag{4.4.9}$$

and $\sigma_{\mu\nu\lambda}$ is *necessarily* antisymmetric in $[\mu\nu]$. Thus piezoelectricity is ruled out by the group O even in the absence of inversion symmetry.

The procedure just used is known as the method of direct inspection (Fumi, **1952**). It is most convenient for group operations (two- and fourfold elements) that simply permute the indices, except for signs.

Theorem 4.4.1

Let a tensor T_p, $p=\mu\nu\lambda\ldots$, be invariant with respect to a group G, containing a subgroup H of all elements that simply permute the indices $\mu\nu\lambda$ (aside from sign). Then all information in H concerning the *vanishing* of a particular element T_p is contained in a subgroup[3] of H that we shall call the *group of the indices*, namely, the elements that leave p invariant.

Proof: If $\mathbf{R}T_p = rT_{p'}$, $\mathbf{S}T_p = sT_{p'}$ where $r=\pm 1$, $s=\pm 1$, then $\mathbf{R}^{-1}\mathbf{S}T_p = r^{-1}sT_p$.

Thus a vanishing that can be produced by combining two elements in H that lead from p to the same end point p can equally well be obtained by using a member of the group of the indices. (Of course, this procedure is not applicable to three- and sixfold elements that produce a linear combination of tensor components. The effect of these elements must be determined separately, using Eq. 4.4.3.)

Theorem 4.4.2

If in Theorem 4.4.1 the matrix $T_{\mu\nu\lambda\ldots}$ is known to have a definite symmetry with respect to the operation $F=$ "flip" of interchanging the indices $\mu\nu$, the group of the indices must be enlarged to include the elements Q that reverse the indices $\mu\nu$, since FQ then leaves the indices invariant. The enlarged group can be referred to as the *reversal group of the indices*.

4.5. METHOD OF INVARIANTS

When three- and sixfold elements are present, the direct use of a set of equations such as 4.4.3 can be annoying. We therefore propose a scheme[4] that makes use of the basis functions that accompany most character tables (see Appendix A) and of the following theorem.

[3]The subgroup techniques based on Theorems 4.4.1, 4.4.2, and related theorems are applied as needed in many places in this book. A different proof of the equivalence of full group and subgroup techniques, with examples, is given in Lax (**1965c**). See also Lax and Birman (**1972**).

[4]A related procedure is used by Erdos (**1964**).

Theorem 4.5.1

If $T_{\mu\nu\lambda\ldots}$ transforms as a vector with respect to indices $\mu\nu\ldots$ and as a pseudovector with respect to indices $\lambda\ldots$, then $F(\mathbf{r},\mathbf{r}',\mathbf{m}\ldots) = \sum T_{\mu\nu\lambda\ldots} r_\mu r'_\nu m_{\lambda\ldots}$ is an invariant under the group G if $\mathbf{T}$ is an invariant tensor under G. Here $\mathbf{r},\mathbf{r}',\ldots$ are vectors and $\mathbf{m},\ldots$ are pseudovectors.

This theorem generalizes Theorems 4.3.1 and 4.3.2.

Our procedure is therefore to write down an invariant polynomial whose degree is equal to the rank of the tensor. The tensor can then be read off as the coefficients in the polynomials.

For example, under the full rotation group $R(3)$ the only invariant that can be formed out of $\mathbf{r},\mathbf{r}'$, and $\mathbf{m}$ is

$$\mathbf{r}\times\mathbf{r}'\cdot\mathbf{m} = \sigma_{\mu\nu\lambda} r_\mu r'_\nu m_\lambda \tag{4.5.1}$$

Thus the Hall effect tensor has the form

$$\sigma_{\mu\nu\lambda} = e_{\mu\nu\lambda}$$

of the Levi-Civita tensor density. In writing Eq. 4.5.1 we used the following theorem.

Theorem 4.5.2

For the *proper* orthogonal group in n dimensions $R^+(n)$, an arbitrary invariant *polynomial* Σ in n vectors $\mathbf{r}_1,\ldots,\mathbf{r}_n$ can be written in the form $\Sigma_1 + J\Sigma_2$, where Σ_1 and Σ_2 are *polynomials* in the scalar products $\mathbf{r}_i\cdot\mathbf{r}_j$ and $J = \det(\mathbf{r}_1,\mathbf{r}_2,\ldots,\mathbf{r}_n)$ is the determinant formed by the vectors.

This is a deep theorem because of its algebraic nature, and we refer to Weyl (**1946**, pp. 31, 53) for proof. It is related, however, to a more elementary theorem.

Theorem 4.5.3

Any arbitrary function Σ of $\mathbf{r}_1,\mathbf{r}_2,\ldots,\mathbf{r}_n$ invariant under the proper orthogonal group $R^+(n)$ can be expressed as some arbitrary function $\Sigma(\mathbf{r}_1\cdot\mathbf{r}_j)$ of the basic invariants $\mathbf{r}_i\cdot\mathbf{r}_j$.

Comment: The n vectors $\mathbf{r}_i$ taken from the origin define an n-dimensional "tetrahedron." An invariant function of $\mathbf{r}_1,\ldots,\mathbf{r}_n$ is a function of this tetrahedron but not of its orientation. The tetrahedron is uniquely defined by the lengths of its edges and the angles between them, that is, by the set of numbers $(\mathbf{r}_i\cdot\mathbf{r}_j)$. Note that

$$\det(\mathbf{r}_1,\mathbf{r}_2,\ldots,\mathbf{r}_n)\det(\mathbf{s}_1,\mathbf{s}_2,\ldots,\mathbf{s}_n) = \det(\mathbf{r}_i\cdot\mathbf{s}_j) \tag{4.5.2}$$

$$\therefore \quad J=[\det(\mathbf{r}_i\cdot\mathbf{r}_j)]^{1/2} \tag{4.5.3}$$

is expressible in terms of $\mathbf{r}_i\cdot\mathbf{r}_j$ but not as a polynomial.

A natural generalization of the idea of forming invariants by taking scalar products is given by the following theorem.

First Shell Theorem 4.5.4

Consider two sets of orthonormal functions $\varphi_\mu(\mathbf{r})$ and $\psi_\mu(\mathbf{r}')$. Each set is assumed to span a space invariant under any real unitary operation S, and the representations $D_{\nu\mu}(S)$ generated by these two sets are assumed to be identical. Then the *shell sum*

$$F(\mathbf{r},\mathbf{r}')\equiv\sum_\mu \varphi_\mu^*(\mathbf{r})\psi_\mu(\mathbf{r}')$$

is invariant under S.

Proof:

$$S[\varphi_\mu^*(\mathbf{r})\psi_\mu(\mathbf{r}')]=\sum_\nu \varphi_\nu^*(\mathbf{r})D_{\nu\mu}^*(S)\sum_{\nu'}\psi_{\nu'}(\mathbf{r}')D_{\nu'\mu}(S) \tag{4.5.4}$$

since φ_μ^* generates the complex conjugate representation for real operators (see Eqs. 3.4.8 and 10.6.6). A sum on μ using the unitarity of $\mathbf{D}(S)$ yields the desired result:

$$SF(\mathbf{r},\mathbf{r}')=F(\mathbf{r},\mathbf{r}') \tag{4.5.5}$$

If the conditions of Theorem 4.5.4 are obeyed for all S in a group G, and if in addition $\varphi_\mu(\mathbf{r})$ and $\psi_\mu(\mathbf{r}')$ constitute bases for identical *irreducible* representations of G, we can sum Eq. 4.5.4 on S, divide by g, and use the orthogonality theorem appropriate to *irreducible* representations to obtain

$$\frac{1}{g}\sum_S S[\varphi_\mu^j(\mathbf{r})^*\psi_\mu^j(\mathbf{r}')]=\frac{1}{l_j}\sum_{\nu=1}^{l_j}\varphi_\nu^j(\mathbf{r})^*\psi_\nu^j(\mathbf{r}') \tag{4.5.6}$$

a result independent of μ. The superscript j reminds us of the need for irreducibility in this second shell theorem, which we have established.

Second Shell Theorem 4.5.5

The "orientation average" of $\varphi_\mu^j(\mathbf{r})^*\psi_\mu^j(\mathbf{r}')$ (left-hand side of Eq. 4.5.6) is equal to the *shell average* (right-hand side of Eq. 4.5.6).

Example 4.5.1: Hall Effect in C_{3v}

The Hall effect tensor belongs to $\Gamma_{[p\times p]}\times\Gamma_m$, where $\Gamma_{[p\times p]}=(\Gamma_1^-\times\Gamma_1^-)_{\text{asym}}=\Gamma_m$. But in the group C_{3v} the pseudovector representation as shown in Eq. 3.1.9 decomposes into A_2+E, that is, into m_z and (m_x,m_y). Similarly, we can write $\Gamma_{[p\times p]} \sim (\mathbf{r}\times\mathbf{r}') = (\mathbf{r}\times\mathbf{r}')_z + [(\mathbf{r}\times\mathbf{r}')_x, (\mathbf{r}\times\mathbf{r}')_y] \sim A_2 + E$. Our invariants can now be constructed using the shell sums $E\cdot E$ and $A_2\cdot A_2$. Using the two sets of basis vectors for E,

$$\varphi_1^E = (\mathbf{r}\times\mathbf{r}')_x, \qquad \varphi_2^E = (\mathbf{r}\times\mathbf{r}')_y$$

$$\psi_1^E = m_x, \qquad \psi_2^E = m_y$$

we obtain for the invariant shell sum $E\cdot E$ the form

$$(yz' - zy')m_x + (zx' - xz')m_y$$

The corresponding invariant $A_2\cdot A_2$ is $(xy' - yx')m_z$. Adding the two invariants with arbitrary coefficients, we arrive at the most general invariant:

$$a(xy' - yx')m_z + b[(yz' - zy')m_x + (zx' - xz')m_y]$$

By Theorem 4.5.1 we can now read off the tensor as the coefficients in this cubic polynomial:

$$\sigma_{123} = -\sigma_{213} = a, \qquad \sigma_{231} = \sigma_{312} = -\sigma_{321} = -\sigma_{132} = b \tag{4.5.7}$$

with all other coefficients vanishing.

Shell Theorem with Spin 4.5.6

With $s = \pm 1$ to indicate the two spin components, the shell sum

$$F(\mathbf{r},\mathbf{r}';s,s') = \sum_{\mu} \varphi_\mu^\dagger(\mathbf{r},s)\psi_\mu(\mathbf{r}',s') \tag{4.5.8}$$

over a space invariant under a set of operations R is invariant under these operations:

$$RF(\mathbf{r},\mathbf{r}';s,s') = F(\mathbf{r},\mathbf{r}';s,s') \tag{4.5.9}$$

Moreover, if R is a rotation, the transition $(\mathbf{r}' \neq \mathbf{r})$ charge density

$$\rho(\mathbf{r},\mathbf{r}') = \sum_{s=1,2} F(\mathbf{r},\mathbf{r}';s,s) \tag{4.5.10}$$

is invariant under rotations:

$$\rho(\mathbf{rR},\mathbf{r}'\mathbf{R}) = \rho(\mathbf{r},\mathbf{r}') \tag{4.5.11}$$

Proof: The dagger on φ serves to remind us that $\varphi^\dagger$ comes from the left-hand side of a scalar product and by Eq. 3.4.8 generates the conjugate representation $\mathbf{D}(R)^*$. The proof is then identical to that of the first shell theorem 4.5.4 and leads to Eq. 4.5.9.

We should not conclude from Eq. 4.5.9, however, that different components of the charge density, for example,

$$\rho_{\text{up}}(\mathbf{r}) = F(\mathbf{r},\mathbf{r};11) \tag{4.5.12}$$

are invariant under rotations, since $\rho_{\text{up}}(\mathbf{rR}) \neq \rho_{\text{up}}(\mathbf{r})$. Equation 4.5.9 *in extenso* is the statement

$$\sum_{s_1 s_1'} \left[\exp\left(\frac{-i\alpha\boldsymbol{\sigma}\cdot\mathbf{n}}{2}\right)\right]^*_{s s_1} \left[\exp\left(-\frac{i\alpha\boldsymbol{\sigma}\cdot\mathbf{n}}{2}\right)\right]_{s' s_1'} F(\mathbf{rR},\mathbf{r}'\mathbf{R};s_1 s_1') = F(\mathbf{r},\mathbf{r}',s,s') \tag{4.5.13}$$

for a rotation of angle α about axis $\mathbf{n}$. If we set $s'=s$, sum over s, and use the unitarity of the 2×2 spin matrices, Eq. 4.5.13 leads directly to the rotational invariance, Eq. 4.5.11, of the complete charge density.

4.6. MEASURES OF INFINITESIMAL AND FINITE STRAIN[5]

A crystal is distorted by carrying points initially at $\mathbf{X}=(X^1,X^2,X^3)$ into $\mathbf{x} = (x^1,x^2,x^3)$, where $x^i = x^i(X^L)$. The displacement field $\mathbf{u}$ is defined by[6]

$$\mathbf{u}=\mathbf{x}-\mathbf{X} \qquad \text{or} \qquad u^\mu = x^\mu - X^\mu \tag{4.6.1}$$

Since a uniform displacement of the crystal as a whole produces no forces ("stress"), we are concerned with nonuniform displacements that are adequately described to lowest order by the displacement gradient $\partial u^\mu/\partial X^\nu \equiv u^\mu{}_{,\nu}$. As in Section 3.3, we can decompose this second-rank tensor into parts that transform as S, P, and D under the full rotation group. The P-wave part is the antisymmetric part of the tensor, $\frac{1}{2}(u^\mu{}_{,\nu} - u^\nu{}_{,\mu})$, that is, the vector curl $\mathbf{u}$. This vector provides a measure of local rotation. No stress or increase in energy is induced by a pure rotation.[7] It is therefore conventional to adopt

$$e_{\mu\nu} = \tfrac{1}{2}(u^\mu{}_{,\nu} + u^\nu{}_{,\mu}) \tag{4.6.2}$$

[5]Murnaghan (1951), Pearson (1959), Truesdell (1952), Truesdell and Toupin (1960), Eringen (1962, 1967), Leigh (1968), Thurston (1974).

[6]The subtraction in Eq. 4.6.1 is meaningful in the second form shown only if $\mathbf{x}$ and $\mathbf{X}$ are vectors in a common Cartesian coordinate system.

[7]An inhomogeneous rotation would require energy, but we shall discard terms such as grad curl $\mathbf{u}$ as producing small effects. See Eq. 4.6.10 and the following discussion.

as a local measure of ("infinitesimal") strain. The S-wave part of $u^\mu{}_{,\nu}$ is $\frac{1}{3}(\mathrm{div}\,\mathbf{u})\delta^\mu{}_\nu$, which is simply the *dilatation* since the Jacobian of the transformation (i.e., the volume change)

$$J=\det\left(\frac{\partial x^\mu}{\partial X^\nu}\right)=\det[\delta^\mu{}_\nu+u^\mu{}_{,\nu}] \tag{4.6.3}$$

expanded to terms linear in the u's is simply[8]

$$J\approx 1+u^\mu{}_{,\mu}=1+\mathrm{div}\,\mathbf{u} \tag{4.6.4}$$

The five components belonging to D, that is, the traceless part $e_{\mu\nu}-\frac{1}{3}(\mathrm{trace}\;\mathbf{e})\delta_{\mu\nu}$, are referred to as components of *shear* strain, since they involve no volume change or rotation.

The above definitions of strain and rotation are in fact valid only if all $u^\mu{}_{,\nu}$ are small. A *finite* rotation followed by an infinitesimal strain leads to different $e_{\mu\nu}$ according to whether Eq. 4.6.2 is used directly or in the rotated coordinate system. To obtain a rotationally invariant definition of the strain, let us consider the *metric* C_{LM} of the *deformed* system in the undeformed reference frame:[9]

$$ds^2=dx^i\,dx_i=C_{LM}\,dX^L\,dX^M \tag{4.6.5}$$

where

$$C_{LM}=x^i{}_{,L}x_{i,M}=\sum_i\left(\frac{\partial x^i}{\partial X^L}\right)\left(\frac{\partial x_i}{\partial X^M}\right) \tag{4.6.6}$$

We can regard the particles as *named* by $\mathbf{X}$, their position in a reference frame that we shall call the *material* reference frame. Then the $\mathbf{x}$ are the *positions* of the particles in a *spatial* reference frame. But ds^2 by construction is a $(\text{length})^2$ that is invariant against spatial rotations, that is, rotations of the spatial reference frame keeping the material reference frame fixed. Similarly C_{12} is a scalar product between two quantities, $\mathbf{x}_{,1}$ and $\mathbf{x}_{,2}$, that are vectors under spatial rotations. Thus C_{LM} and

$$E_{LM}\equiv\tfrac{1}{2}(C_{LM}-\delta_{LM}) \tag{4.6.7}$$

are invariant under spatial rotations. We define E_{LM} as the appropriate (sym-

[8]Repeated indices are understood to be summed over.

[9]For the Cartesian coordinate systems to which we shall confine our remarks there is no distinction between superscripts and subscripts. We follow, however, a notation of Truesdell (1952) and Toupin (1956) that is valid in more general (e.g., spherical-coordinate) systems.

metric in L and M) measure of *finite strain*. If we choose the spatial and material frames to be identical Cartesian frames, we can use Eq. 4.6.1 to write

$$E_{LM} = \tfrac{1}{2}\left(u^M{}_{,L} + u^L{}_{,M}\right) + u^i{}_{,L} u^i{}_{,M} \tag{4.6.8}$$

so that in the limit of "infinitesimal" strain E_{LM} reduces to the conventional definition e_{LM}. Note that infinitesimal $u^M{}_{,L}$ implies infinitesimal E_{LM}, but the converse is not true. A rigid rotation through a large angle yields zero finite strain: $E_{LM}=0$, but $u^M{}_{,L}$ and even $\tfrac{1}{2}(u^L{}_{,M}+u^M{}_{,L})$ need not be small. When all $u^M{}_{,L}$ are small, however, $e_{LM}=\tfrac{1}{2}(u^L{}_{,M}+u^M{}_{,L})$ is a suitable measure of the strain, and $\tfrac{1}{2}(u^L{}_{,M}-u^M{}_{,L})$ of the rotation.

The Potential Energy

Since the energy of a solid is invariant against arbitrary rigid displacements $\mathbf{x}\to\mathbf{x}+\mathbf{d}$ of the solid as a whole, the potential energy must be expressible in terms of differences $\mathbf{x}(\mathbf{X})-\mathbf{x}(\mathbf{Y})$. If we restrict ourselves to materials in which the range $|\mathbf{X}-\mathbf{Y}|$ over which forces are exerted is small compared to the distances important in macroscopic measurements, we can expand in $\mathbf{X}-\mathbf{Y}$ and express the potential energy

$$V = \int \Sigma \, d\mathbf{X} \tag{4.6.9}$$

in terms of a local function of first and higher derivatives:

$$\Sigma \equiv \Sigma\left(\frac{\partial x^i}{\partial X^L}, \frac{\partial^2 x^i}{\partial X^L \partial X^M}, \ldots\right) \tag{4.6.10}$$

where Σ is the potential energy per unit *undeformed* volume. The term in x^i is missing because of displacement invariance. An expansion in terms of increasingly higher derivatives is essentially an expansion in $a\,\partial/\partial\mathbf{X}$, where a is a length that measures the range of the forces, that is, some multiple of the lattice constant. Thus higher derivatives can be important only if the strain is inhomogeneous over distances of the order of a. It is therefore quite a good approximation to keep only the lowest-order contributions and to assume that[10]

[10]The introduction of second derivatives $x^i_{,LM}$ permits the introduction of curvature "twists" grad curl $\mathbf{u}$ and results in *couple stresses*. The latter concept dates back to Cosserat and Cosserat (**1909**) but has received renewed attention recently from Truesdell and Toupin (**1960**), Aero and Kuvshinskii (**1960**), Toupin (**1962**, **1963**, **1964**), and Tiersten and Mindlin (**1962**). Part of this attention is due to a controversy created by Laval (**1951**, **1954**), who claims that the stress tensor need not be symmetric, so that in the absence of all crystalline symmetry the elastic constant matrix has 45 rather than 21 components. The resolution of this rather subtle controversy in favor of the usual theory with 21 constants is discussed by Lax (**1965a**). Rotational effects have been found, however, in photoelasticity (Lax and Nelson, **1971**; Nelson, Lazay, and Lax, **1972**; Lax and Nelson, **1973**).

$$\Sigma \equiv \Sigma(x^i{}_{,L}) \tag{4.6.11}$$

Provided that the crystal is not subject to external fields, its potential energy must be invariant against arbitrary proper rotations of the crystal as a whole, that is, Σ must be invariant under rotations of the spatial frame $\mathbf{x}$ relative to a fixed material frame $\mathbf{X}$. But $\mathbf{x}_{,L}$ for $L=1,2,3$ are three *vectors* under such transformations. By Theorem 4.5.3, then, Σ must be a function of the (six) scalar products:

$$\Sigma \equiv \Sigma(C_{LM}) \tag{4.6.12}$$

$$C_{LM} \equiv \mathbf{x}_{,L} \cdot \mathbf{x}_{,M} \equiv x^i{}_{,L} x_{i,M} = C_{ML} \tag{4.6.13}$$

Indeed, Theorem 4.5.2 asserts more: if $\Sigma(x^i{}_{,L})$ is any (invariant) *polynomial* in the $x^i{}_{,L}$, then

$$\Sigma(x^i{}_{,L}) = \Sigma_1(C_{LM}) + J\Sigma_2(C_{LM}) \tag{4.6.14}$$

where

$$J \equiv \det x^i{}_{,L} = [\det C_{LM}]^{1/2} = \frac{\rho_0}{\rho} \tag{4.6.15}$$

is the change in volume: hence the inverse change in density ρ, during the deformation, and Σ_1 and Σ_2 are polynomials. If the initial configuration $\mathbf{X}$ is an equilibrium configuration (i.e., $\partial\Sigma/\partial E_{LM}=0$ at $E_{LM}=0$), we can write to lowest order (Toupin and Gazis, **1965**)

$$\Sigma = \tfrac{1}{2} C^{MANB} E_{MA} E_{NB} \tag{4.6.16}$$

in the region of infinitesimal strain. The coefficients C^{MANB} represent the *elastic stiffness tensor* or, briefly, in this book the elasticity tensor. Clearly C^{MANB} is symmetric on the pair interchange $(MA) \rightleftarrows (NB)$:

$$C^{MANB} = C^{NBMA} \tag{4.6.17}$$

since only this portion of C contributes to Σ. Similarly, the symmetry of E_{MA} guarantees that C^{MANB} is symmetric in MA (and also in NB):

$$C^{MANB} = C^{AMNB} \tag{4.6.18}$$

By construction C_{LM} and also E_{LM} are scalars under rotations of the spatial frame $\mathbf{x}$ and second-rank tensors under rotations of the material frame $\mathbf{X}$. In order for Σ to be invariant under the name-changing transformations of $\mathbf{X}$, C^{MANB} must transform as a fourth-rank tensor in the material framework. Transformations that restore the crystal to itself must, however, leave C^{MANB} invariant. This provides us with a set of conditions that can be used by direct inspection or by the method of invariants to determine the form of C^{MANB}. The latter procedure is illustrated in Section 4.7.

In the absence of any crystal symmetry requirements the tensor C^{MANB} has $\frac{1}{2}\cdot 6\cdot 7=21$ independent components, since there are six E_{MA}. If, however, E_{MA} had nine independent components, there would be the $\frac{1}{2}\cdot 9\cdot 10=45$ components claimed by Laval (**1951**, **1954**).

In referring to C^{MANB} as the elasticity tensor we have tacitly assumed that the *stress tensor* can be defined by

$$T^{LM} \equiv \frac{\partial \Sigma}{\partial E_{LM}} = T^{ML} \tag{4.6.19}$$

so that

$$T^{MA} \approx C^{MANB} E_{NB} \tag{4.6.20}$$

for small E_{NB}. To show that C^{MANB} does indeed provide a relation between stress and strain, however, it is necessary to give an independent definition of the stress tensor in terms of a force per unit area. Since the components of the force can be measured in the spatial or material reference frame and taken per unit deformed or per unit undeformed area, three related tensors arise: t^{ij} (all spatial—the *Cauchy* stress tensor), T^{LM} (all material—the *Piola* stress tensor), and T^{iM} (mixed—the *Kirchhoff* stress tensor). These tensors can be computed in terms of the potential energy Σ, since the surface forces produce an increase in potential energy during a displacement. Careful analysis (see, e.g., Truesdell, **1952**, Eq. 39.4) shows that T^{LM} is indeed $\partial\Sigma/\partial E_{LM}$ and hence must be symmetric. The Cauchy stress tensor is also found to be

$$t^{ij} = \left(\frac{1}{J}\right) x^i_{,L} T^{LM} x^j_{,M} \tag{4.6.21}$$

which is then also symmetric. However,

$$T^{iM} = x^i_{,L} T^{LM} = \frac{\partial \Sigma}{\partial x_{i,M}} \tag{4.6.22}$$

is not symmetric [except in the limit of (*a*) infinitesimal strain, (*b*) a spatial reference system identical to the material reference system, and (*c*) no initial stress, i.e., stress vanishes when $\mathbf{x}=\mathbf{X}$]. Thus the usual computation of stress as a derivative of the potential energy with respect to a displacement gradient yields the possibly nonsymmetric Kirchhoff tensor.

4.7. THE ELASTICITY TENSOR FOR GROUP C_{3v}

The symmetry of the elasticity tensor on its indices implies that it transforms according to $(L\times L)_{\text{sym}}$, where

$$L=(\Gamma_p\times\Gamma_p)_{\text{sym}}\equiv\Gamma_{(p\times p)}=\Gamma_2+\Gamma_0$$

can be evaluated using the symmetrized product character formula 3.3.30. For classes E, $2C_3$, and 3σ of C_{3v}, we find characters $L=(6,0,2)$ and the decomposition $L=2(A_1+E)$. Furthermore, $(L\times L)_{\text{sym}}$ has characters $(21,0,5)$, leading to $\frac{1}{6}[21+0+5(3)]=6$ arbitrary constants.

Because of the threefold axis, the method of direct inspection is inconvenient. However, this problem, as difficult as any tensor example is likely to be, can be handled readily by the method of invariants. $L=\Gamma_2+\Gamma_0$ contains all six products, x^2,y^2,z^2,xy,yz,zx. Of these

$$A=z^2,\qquad A'=x^2+y^2 \tag{4.7.1}$$

belong to the identity representation, and

$$E=(xz,yz),\qquad E'=(2xy,x^2-y^2) \tag{4.7.2}$$

belong to the two-dimensional representation; both transform precisely as (x,y) under C_{3v}. See Section 3.7.

Our six invariants are now $A\cdot A$, $A'\cdot A'$, $A\cdot A'+A'\cdot A$, $E\cdot E$, $E'\cdot E'$, and $E\cdot E'+E'\cdot E$. We do not regard $A\cdot A'$ and $A'\cdot A$ as separate invariants because it is the symmetrized product of L with itself that is desired. Using lower-case letters for left-hand factors and upper-case for right-hand factors, we have for our complete invariant

$$\begin{aligned}F(x,y,z,X,Y,Z)&=az^2Z^2+b(x^2+y^2)(X^2+Y^2)\\&\quad+c[(x^2+y^2)Z^2+z^2(X^2+Y^2)]\\&\quad+d[(2xz)(2XZ)+(2yz)(2YZ)]\\&\quad+e[(x^2-y^2)(X^2-Y^2)+2xy(2XY)]\\&\quad+f[(2xy)(2XZ)+(x^2-y^2)(2YZ)+(2xz)(2XY)+(2yz)(X^2-Y^2)]\end{aligned} \tag{4.7.3}$$

This expression is already the invariant strain energy, provided that x^2 is interpreted as the strain E_{11}, $2xy$ as $E_{12}+E_{21}$, and so on. The coefficient of x^2X^2 is thus

$$C_{11,11}=b+e$$

whereas that of x^2Y^2 is

$$C_{11,22} = b - e$$

and that of $(2xy)(2XY)$ is

$$C_{12,12} = e = \tfrac{1}{2}(C_{11,11} - C_{11,22})$$

If we make the customary correspondence between double and single indices,

$$\begin{array}{ccccccc} ij & 11 & 22 & 33 & 23 & 31 & 12 \\ m & 1 & 2 & 3 & 4 & 5 & 6 \end{array} \tag{4.7.4}$$

$$C_{ijkl} = C_{mn} \tag{4.7.5}$$

our results can be summarized in the form of a symmetric matrix with the components

$$\begin{array}{cccccc} 11 & 12 & 13 & 14 & 0 & 0 \\ & 11 & 13 & -14 & 0 & 0 \\ & & 33 & 0 & 0 & 0 \\ & & & 44 & 0 & 0 \\ & & & & 44 & 14 \\ & & & & & X \end{array} \tag{4.7.6}$$

where $66 = X = \frac{1}{2}(11 - 12)$ is a relation of the type produced by three- and sixfold axes, but not by the operations that simply permute the indices. The simplicity of the above procedure is much greater than that of the most efficient direct inspection technique used by Hearmon (**1961**).

4.8. SUMMARY

The matter tensors of a crystal are invariant under the operations that restore a crystal to itself. Thus they are bases for identity representations of the group. The number of independent constants in any such invariant tensor is given by the number of times that the identity representation is contained in the space spanned by the independent components of the tensor. Tensors of the same type (i.e., same transformation properties, such as the dielectric tensor and the thermal conductivity tensor) have a form determined only by the group and not by the *physical* nature of the individual tensor.

The form of a tensor can be determined by direct inspection procedures for operations, such as fourfold rotations, that permute the indices (with $\pm$ signs).

When three- or sixfold elements are present, a fast, efficient procedure is to use Theorem 4.5.1: Write down an arbitrary invariant polynomial whose degree is equal to the rank of the tensor, and read off the tensor as the coefficients in this polynomial.

Invariant polynomials are formed under the orthogonal group by using scalar products of vectors and determinants of vectors. Under more general groups, the shell sums of Theorems 4.5.4 and 4.5.5 form invariants that are the analogs of the scalar products.

Invariance against arbitrary displacements and rotations is used to express the potential energy of an elastic continuum in terms of the measure $E_{LM}=E_{ML}$ of finite strain. The elasticity tensor of an arbitrary crystal is then shown to have 21 constants. These constants obey additional restrictions for crystals with nontrivial point group symmetries.

problems

4.3.1. Show that the only tensor of second rank of octahedral $O(432)$ symmetry is an isotropic tensor, that is, a tensor invariant under all rotations.

4.3.2. Show that a measurement of electrical conductivity in a crystal of cubic symmetry is independent of the orientation of the crystal. Is this also true of optical absorption?

4.3.3. Show that the presence of a reflection plane containing the principal axis forces c to equal 0 in Eq. 4.3.10.

4.3.4. Calculate the character of the rotations of the group C_{3v} in the six-dimensional space spanned by the functions $x^2, y^2, z^2, xy, yz, zx$. Show that the representation generated contains $2A_1+2E$ and hence that a second-rank symmetric tensor has two independent components. Compare this procedure with a character computation based on $\Gamma_p \times \Gamma_p$ or on Eq. 4.3.8.

4.3.5. The displacement vector **D** and electric field **E**, when both are plane waves $\exp(i\mathbf{k}\cdot\mathbf{r})$, are related by $D_\mu=\epsilon_{\mu\nu}(\mathbf{k})E_\nu$, where time reversal requires that $\epsilon_{\mu\nu}(\mathbf{k})=\epsilon_{\nu\mu}(-\mathbf{k})$. Show that in the long-wave limit this gives rise to a relationship of the form

$$D_\mu=\epsilon_{\mu\nu}(0)E_\nu - i\epsilon_{\mu\nu\lambda}\left(\frac{\partial E_\nu}{\partial x_\lambda}\right)+\cdots$$

where $\epsilon_{\mu\nu\lambda}=\partial\epsilon_{\mu\nu}(\mathbf{k})/\partial k_\lambda$ is antisymmetric in μ and ν and gives rise to optical rotation. How does $\epsilon_{\mu\nu\lambda}$ differ from the Hall tensor $\sigma_{\mu\nu\lambda}$, which is also antisymmetric in μ and ν? For which point groups does $\epsilon_{\mu\nu\lambda}$ exist?

4.4.1. Show that for second-rank tensors there is no distinction between crystals of tetrahedral (T) symmetry and isotropic materials. Is there a distinction for third-rank tensors? For fourth-rank tensors?

4.4.2. For which tensors are polycrystalline materials equivalent to isotropic materials? Note the Orson Anderson (1963) procedure of replacing a crystal by an "equivalent" isotropic material before computing the Debye constant.

4.4.3. Use a character formula to determine the number of independent parameters in the Hall effect tensor for a material with point group C_{3v}. Compare with Example 4.5.1.

4.4.4. Use the group of the indices (Theorem 4.4.1) to determine the form of a fourth-rank tensor p_{ijkl} in a crystal with tetrahedral symmetry T and octahedral symmetry O.

4.4.5. Extend the results of Problem 4.4.4 to a tensor of the form (1) $p_{ij(kl)}$ [where ( ) denotes symmetry on the enclosed indices]; (2) $p_{(ij)(kl)}$; (3) $p_{(ij)(kl)}$ with symmetry on interchange of the pair $(ij)\rightleftarrows(kl)$. Use the reversal group of the indices.

4.4.6. For what point groups is a longitudinal Hall effect possible?

4.5.1. Show that, if there are $m>n$ vectors $\mathbf{r}_1\ldots\mathbf{r}_m$ in an n-dimensional space, a complete set of algabraic invariants is given by $\mathbf{r}_i\cdot\mathbf{r}_j$ and the set of determinants $\det(\mathbf{r}_a,\mathbf{r}_b,\ldots,\mathbf{r}_s)$ formed from all combinations of n vectors. Is this set redundant?

4.5.2. Show that the first shell theorem 4.5.4 implies that

$$I(S)\equiv\sum_{\mu}\left|\int\psi_\mu(\mathbf{r})f(\mathbf{Sr})\,d\mathbf{r}\right|^2$$

is independent of S.

4.5.3. Show using Problem 4.5.2 that

$$I_\nu^m(S)=\sum_{\lambda\mu}\left|\int\varphi_\lambda(\mathbf{r})^*\chi_\mu(\mathbf{r})\psi_\nu^m(\mathbf{Sr})\,d\mathbf{r}\right|^2$$

is independent of S and thus equals $I_\nu^m(E)$. Average this result over S and use the second shell theorem 4.5.5 to establish that

$$I_\nu^m(S)=\frac{1}{l_m}\sum_{\nu=1}^{l_m}I_\nu^m(E)=I^m(E)$$

that is, $I_\nu^m(S)$ is independent of ν as well as S. This may be referred to as the *polarization theorem*. Attempts to observe a polarization (ν) dependence of, say, a scattered beam will fail unless either the initial state λ or the interaction χ_μ is also polarized.

4.5.4. Prove the *principle of spectroscopic stability*: The total transition probability from a set of degenerate initial states to a set of degenerate final states is unaffected by a weak perturbation that produces negligible mixing between nondegenerate states.

4.6.1. Calculate the number of independent constants in the dielectric tensor, the Hall tensor, the piezoelectric tensor, and the elasticity tensor for crystals such as CdS and ZnO of the wurtzite structure. (See Table 6.1.4 for the nature of this structure and the associated point group C_{6v}.)

4.7.1. Show that bismuth with point group D_{3d} has the elasticity tensor shown in Eq. 4.7.6 for C_{3v}.

4.7.2. Show that for GaAs in the zincblende structure of Fig. 6.1.11, the point group is T_{2d} and the piezolectric tensor has a form described by a single constant $e_{ijk} = e_{123}$ when all indices ijk are different and $e_{ijk} = 0$ otherwise.

chapter five

MOLECULAR VIBRATIONS

In Chapters 3 and 4 we illustrated the power of group theory in dealing with compatibility relations, selection rules, and the form and number of independent constants in tensors of various ranks. The central problem, however, is to exhaust in a systematic manner all symmetry information before attempting the solution of a dynamical problem. The chief point of Chapter 1 is that this is accomplished by diagonalizing the Dirac characters. By this we do not mean simply the construction and use of a character table—steps that can be carried out in an abstract way without knowledge of the physical space of the problem. Rather, we mean that the *diagonalization* of the Dirac characters *in the physical space of the problem will automatically select for us spaces belonging to a given irreducible representation*.

To illustrate this technique we consider a nontrivial problem: the vibrations of the ammonia molecule. The steps involved in the treatment of this problem are numbered 1–7 (they extend through Sections 5.1 and 5.2) to emphasize that they appear in the solution of almost any problem. A discussion of the *form* of the Hamiltonian is left for one of the last steps (although it could equally well be done first) because we wish to show how far it is possible to go using the properties of the space of the problem, with no information as to the dynamics.

5.1. REPRESENTATIONS CONTAINED IN NH_3 VIBRATIONS

1. *Select the Group of Operations Appropriate to the Physical Problem.* With the H_3 forming an equilateral triangle and N directly above the center (see Fig. 5.1.1), the molecular group that restores the equilibrium molecule to itself is clearly C_{3v}. The invariance of the potential energy of the *distorted* molecule is established in Section 5.4. (Can any information be obtained by adding time reversal to the groups? What about infinitesimal translations and rotations?)

2. *Determine the Physical Space of the Problem.* In this case, it is a twelve-dimensional space,

$$\psi=(x^1,y^1,z^1;x^2,y^2,z^2;x^3,y^3,z^3;x^4,y^4,z^4) \tag{5.1.1}$$

consisting of the displacements $\mathbf{u}^m$, $m=1,2,3,4$, of the three H and N atoms from their equilibrium positions $\mathbf{X}^m$, respectively. (In a quantum treatment, however, the wave functions span an infinite Hilbert space.)

3. *Find Out How the Group Operators Act in This Space.* The action of $A=\boldsymbol{\sigma}_x=$a reflection that reverses x was shown in Example 1.2.1 in the two-dimensional space of one particle. However, as shown in Fig. 5.1.2, $\boldsymbol{\sigma}_x$ interchanges particles 1 and 2. Thus we can write

$$A\begin{bmatrix}\mathbf{u}^1\\ \mathbf{u}^2\\ \mathbf{u}^3\\ \mathbf{u}^4\end{bmatrix}=\begin{bmatrix}a\mathbf{u}^2\\ a\mathbf{u}^1\\ a\mathbf{u}^3\\ a\mathbf{u}^4\end{bmatrix}=\begin{bmatrix}0 & a & 0 & 0\\ a & 0 & 0 & 0\\ 0 & 0 & a & 0\\ 0 & 0 & 0 & a\end{bmatrix}\begin{bmatrix}\mathbf{u}^1\\ \mathbf{u}^2\\ \mathbf{u}^3\\ \mathbf{u}^4\end{bmatrix} \tag{5.1.2}$$

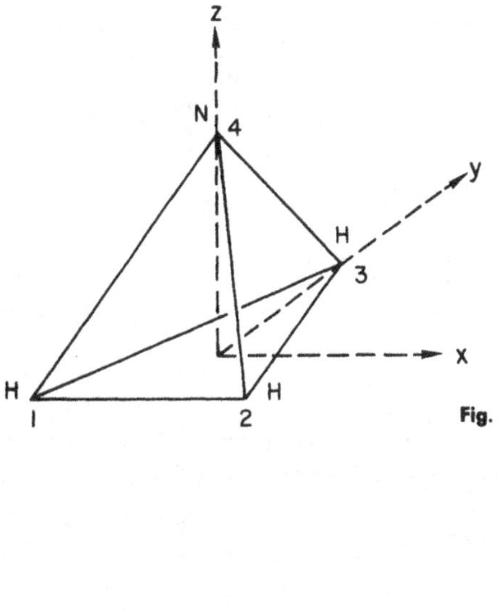

Fig. 5.1.1. Symmetry of the ammonia molecule.

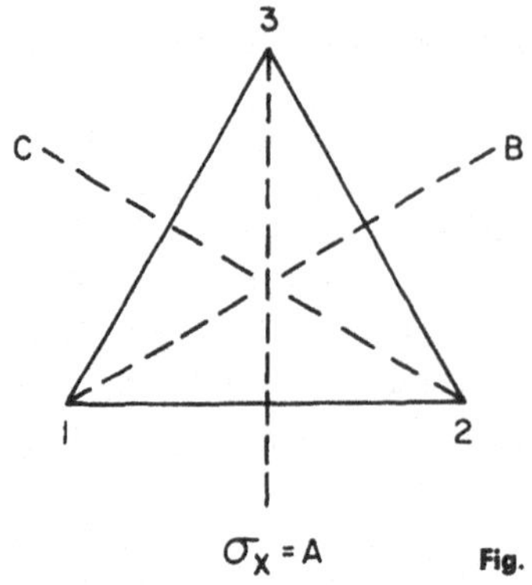

Fig. 5.1.2. Reflection plane perpendicular to the plane of H_3 in NH_3.

where

$$a = \begin{bmatrix} -1 & 0 & 0 \\ 0 & 1 & 0 \\ 0 & 0 & 1 \end{bmatrix}$$

is the usual matrix of A in the space Γ_p of one (polar) vector.

Since D is a 120° counterclockwise rotation, C_3, the new displacement at position 1, is the displacement that was previously at position 3 (see Fig. 5.1.3), rotated counterclockwise by 120°. Thus

$$D\begin{bmatrix} \mathbf{u}^1 \\ \mathbf{u}^2 \\ \mathbf{u}^3 \\ \mathbf{u}^4 \end{bmatrix} = \begin{bmatrix} d\mathbf{u}^3 \\ d\mathbf{u}^1 \\ d\mathbf{u}^2 \\ d\mathbf{u}^4 \end{bmatrix} = \begin{bmatrix} 0 & 0 & d & 0 \\ d & 0 & 0 & 0 \\ 0 & d & 0 & 0 \\ 0 & 0 & 0 & d \end{bmatrix} \begin{bmatrix} \mathbf{u}^1 \\ \mathbf{u}^2 \\ \mathbf{u}^3 \\ \mathbf{u}^4 \end{bmatrix} \tag{5.1.3}$$

where

$$d = \begin{bmatrix} -\frac{1}{2} & -\frac{1}{2}\sqrt{3} & 0 \\ +\frac{1}{2}\sqrt{3} & -\frac{1}{2} & 0 \\ 0 & 0 & 1 \end{bmatrix} \tag{5.1.4}$$

The remaining elements of the groups can be obtained from A and D by applying the multiplication table 1.2.4, by using A and D as generators:

$$A^2 = D^3 = E; \qquad F = D^2, \quad B = AD, \quad C = DA = AD^2 \tag{5.1.5}$$

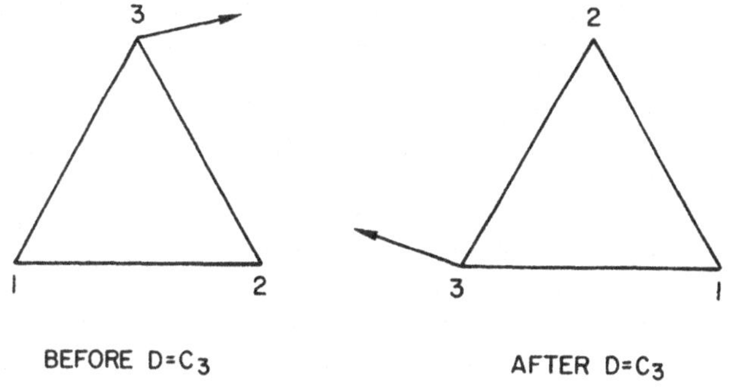

Fig. 5.1.3. Threefold rotation of the H_3 plane in NH_3.

or by inspection of Figs. 5.1.1 and 5.1.2 with the result:

$$B\begin{bmatrix}\mathbf{u}^1\\ \mathbf{u}^2\\ \mathbf{u}^3\\ \mathbf{u}^4\end{bmatrix}=\begin{bmatrix}b\mathbf{u}^1\\ b\mathbf{u}^3\\ b\mathbf{u}^2\\ b\mathbf{u}^4\end{bmatrix};\quad C\begin{bmatrix}\mathbf{u}^1\\ \mathbf{u}^2\\ \mathbf{u}^3\\ \mathbf{u}^4\end{bmatrix}=\begin{bmatrix}c\mathbf{u}^3\\ c\mathbf{u}^2\\ c\mathbf{u}^1\\ c\mathbf{u}^4\end{bmatrix};\quad F\begin{bmatrix}\mathbf{u}^1\\ \mathbf{u}^2\\ \mathbf{u}^3\\ \mathbf{u}^4\end{bmatrix}=\begin{bmatrix}f\mathbf{u}^2\\ f\mathbf{u}^3\\ f\mathbf{u}^1\\ f\mathbf{u}^4\end{bmatrix} \tag{5.1.6}$$

with the elements b,c,f readily deducible from Example 1.2.1 or Eq. 5.1.5

4. *Determine the Character of a Symmetry Operation in the Space.* The character of an operation such as A in Eq. 5.1.2 is given by

$$\chi(A)=N(A)\chi^{P}(A) \tag{5.1.7}$$

where $N(A)$ is the number of times that a appears on the main diagonal, that is, *the number of atoms whose position is unchanged by* A, and

$$\chi^{P}(A)=\text{trace } a \tag{5.1.8}$$

is the trace of the rotation a in the three-dimensional polar vector representation Γ_p. Indeed, the supermatrix notation of Eq. 5.1.2 properly describes the fact that the complete operation A involves the *direct product* of two operations: a permutation P_{12} of atoms 1 and 2, and the mirror operation $a=\sigma_x$. The character of the product is then the product of $N(A)$, the character of A as a permutation operation, and $\chi^P(A)$, its character as a reflection.

For the classes E, C_3, σ_v the characters of $\chi^P(R)$ are $(3,0,1)$, as shown in Table 3.1.3, those of $N(R)$ are $(4,1,2)$, and those of the product $N(R)\chi^P(R)$ are $(12,0,2)$.

5. *Find Out What Irreducible Representations Are Contained in the Space.* Using the characters (12,0,2) and the character table 3.1.3, we can decompose the complete space into

$$\Gamma_{tot} = 3A_1 + A_2 + 4E \tag{5.1.9}$$

However, we are concerned only with the *internal* vibrations of NH_3. Therefore we must eliminate the 3 degrees of freedom associated with translation of the molecule as a whole, which transform as (x,y,z) (i.e., as Γ_p), and the 3 rotational degrees of freedom of the molecule as a whole, which transform as $\mathbf{r} \times \mathbf{p}$ (i.e., as the pseudovector representation Γ_m). The characters of Γ_m can be written as

$$\chi^m(R) = \det R \chi^p(R) \tag{5.1.10}$$

since det R introduces a factor of $+1$ for the proper rotations and -1 for the improper rotations in agreement with Eq. 3.1.8. Alternatively, the characters for Γ_m in C_{3v} are shown in Table 3.1.3.

Since the total space of molecular motion is the space of internal motions plus the space of motions of the molecule as a whole, and characters are additive, the character for the internal motions is

$$\chi^{int}(R) = \chi^{tot}(R) - \chi^p(R) - \det R \chi^p(R) = [N(R) - 1 - \det R]\chi^p(R) \tag{5.1.11}$$

Thus

$$\chi^{int}(E) = (4-2)\cdot 3 = 6, \qquad \chi^{int}(D) = (1-2)\cdot 0 = 0$$

$$\chi^{int}(A) = (2-1+1)\cdot 1 = 2$$

and

$$\Gamma_{int} = 2A_1 + 2E \tag{5.1.12}$$

The result $\chi^{int}(E) = 6$ verifies that there are 6 degrees of freedom remaining in the internal modes $2A_1 + 2E$ (12 total degrees of freedom minus 6 degrees of freedom for the external motion).

6. *Determine the Size of Equations and the Number of Parameters.* The original NH_3 problem requires the diagonalization of a 12×12 matrix. The 6 external degrees of freedom are uninteresting. (They describe free motion of the molecule as a whole, i.e., vibrations with zero restoring force or zero eigenfrequencies.) By imposing zero external motion on the symmetry coordinates that we seek to find, we shall automatically project ourselves into the space of internal motions Γ_{int}. In this space no representation appears more than twice. By using identically transforming symmetry vectors, we can then ensure via Eq. 3.4.15 that the dynamical problem can be solved by diagonalizing nothing worse than a 2×2 matrix! (Indeed, $\Gamma_{int} = 2A_1 + 2E$ informs us that the 6×6 matrix that describes

the internal motion problem *can* be decomposed into a 2×2 matrix for the A_1 modes and a pair of *identical* 2×2 matrices for the E modes.)

The potential energy is an invariant linear combination of displacement products $\mathbf{u}^m\mathbf{u}^n$. The number of independent invariant quadratic forms is given by the number of times that the identity representation is contained in the space spanned by $\mathbf{u}^m\mathbf{u}^n$. The character of this space is that of the direct product $\Gamma_{tot}\times\Gamma_{tot}$, or, more properly, since $\mathbf{u}^m\mathbf{u}^n$ cannot be distinguished from $\mathbf{u}^n\mathbf{u}^m$, it is the symmetrized direct product $(\Gamma_{tot}\times\Gamma_{tot})_{sym}$. If we omit the (vanishing) constants associated with external motion, the number of independent parameters is given by Eq. 4.2.3 as

$$N=\frac{1}{g}\sum_R \tfrac{1}{2}\left\{\left[\chi_{int}(R)\right]^2+\chi_{int}(R^2)\right\} \tag{5.1.13}$$

The external constants vanish, since there are no restoring forces against uniform displacements or rotations of the molecule as a whole.

For $R=E,D,A$ the characters from Eq. 5.1.12 are $[\chi_{int}(R)]^2=6^2,0^2,2^2=(36,0,4)$, whereas those of $\chi_{int}(R^2)$ are $(6,0,6)$, with the averages $(21,0,5)$, so that

$$N=\tfrac{1}{6}[21+0+3(5)]=6 \tag{5.1.14}$$

(The factor 3 enters since the class of A contains three mirror planes.)

5.2. DETERMINATION OF THE SYMMETRY VECTORS FOR NH_3

7a. *Project into the Space of a Given Symmetry Vector by Guesswork. Symmetry vectors* are vectors (distortions of NH_3) that *transform as* φ_μ^i, the μth partner of the ith irreducible representation. By starting with an arbitrary vector ψ in the twelve-dimensional space Eq. 5.1.1 of the problem, one can apply Eq. 3.7.13 or Eq. 3.7.14 to project into a subspace that belongs to a given symmetry vector, that is, a given φ_μ^i. Formal projection procedures are cumbersome, however, and should be used as a last resort. For one-dimensional representations it is usually possible to guess the answer. For example, the identity representation must have the full symmetry of the figure. For the motion of the H_3 triangle alone, two possibilities come to mind—the radial breathing mode, and the vertical motion of the triangle as a whole. (See Fig. 5.2.1).The N atom can also be given an arbitrary motion in the z direction, holding the H_3 triangle fixed. These last two vertical motions can be combined to give an internal vibration of N against the H_3 plane, in which the center of mass of the molecule as a whole remains fixed, and a uniform motion of the molecule as a whole.

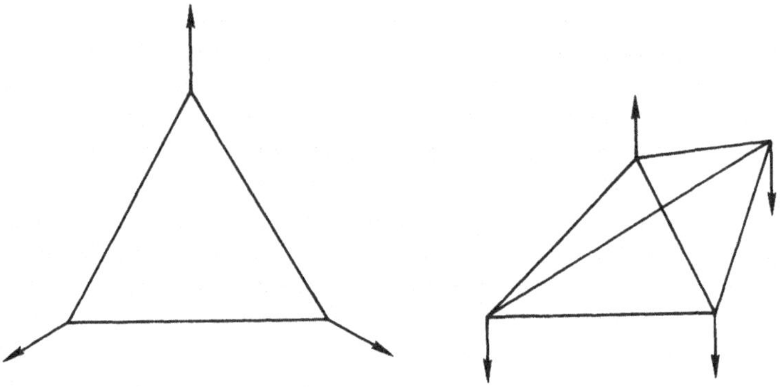

Fig. 5.2.1. Breathing mode of H_3 triangle and vertical motion of H_3 triangle are modes belonging to the identity representation.

Our symmetry coordinates in the 12 component notation

$$(x^1,y^1,z^1;x^2,y^2,z^2;x^3,y^3,z^3;x^4,y^4,z^4)$$

are given respectively by the unnormalized vectors

$$\begin{aligned}
\mathbf{b}^1 &= (-\sqrt{3},\,-1,0; \quad \sqrt{3},\,-1,0; \quad 0,2,0; \quad 0,0,\ \ 0)\\
\mathbf{b}^{1'} &= (\quad 0,\quad 0,1; \quad\ 0,\quad 0,1; \quad 0,0,1; \quad 0,0,-\mu)\\
\mathbf{b}^{1''} &= (\quad 0,\quad 0,1; \quad\ 0,\quad 0,1; \quad 0,0,1; \quad 0,0,\ \ 1)
\end{aligned} \qquad (5.2.1)$$

where $\mu = 3M_1/M_4 = 3$(mass of H)/(mass of N) is chosen so that the constraint (of no translational motion of the center of mass)

$$\sum M_n \mathbf{u}^n = 0 \qquad (5.2.2)$$

is obeyed. That there are two internal and one external A_1 modes agrees with Eqs. 5.1.9 and 5.1.12.

7b. *Project by the Use of Group Generators.* Usually one-dimensional representations can be guessed. If not, we can use the fact that the basis vectors for such representations are eigenvectors of all elements of the group, with each eigenvalue equal to the character of the element. It is sufficient to impose this requirement on the group generators, in this case $A = \sigma_x$ and $D = C_3$, since it will then hold automatically for all other elements of the group. Let us illustrate the procedure for the external mode A_2, which transforms as R_z, a rotation around the z axis. Although the form of the answer shown in Fig. 5.2.2 can almost be guessed, let us perform the projection by using the group generators.

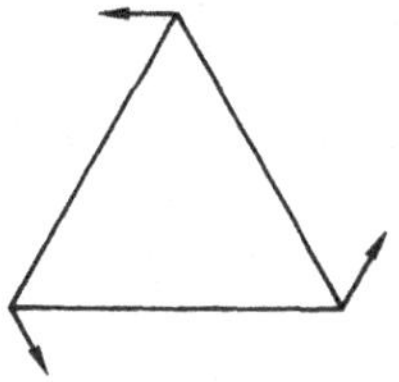

Fig. 5.2.2. Mode $\mathbf{b}^2$ belonging to irreducible representation A_2 is a simple rotation of the triangle.

In the A_2 representation, $A=-1$ and $D=+1$. Since $A=-1$, we find using Eq. 5.1.2 that

$$(-x^2,y^2,z^2;-x^1,y^1,z^1;-x^3,y^3,z^3;-x^4,y^4,z^4)$$
$$=-(x^1,y^1,z^1;x^2,y^2,z^2;x^3,y^3,z^3;x^4,y^4,z^4) \quad (5.2.3)$$

Therefore

$$x^2=x^1\equiv x;\qquad -y^2=y^1\equiv y;\qquad -z^2=z^1\equiv z;\qquad y^4=z^4=0 \quad (5.2.4)$$

which implies that the general form of the symmetry vector is

$$(x,y,z;x,-y,-z;x^3,0,0;x^4,0,0) \quad (5.2.5)$$

Since $D=+1$, Eq. 5.1.3 implies that we must have

$$\left(-\tfrac{1}{2}x^3,\tfrac{1}{2}\sqrt{3}\,x^3,0;-\tfrac{1}{2}x-\tfrac{1}{2}\sqrt{3}\,y,\tfrac{1}{2}\sqrt{3}\,x-\tfrac{1}{2}y,z;\right.$$
$$\left.-\tfrac{1}{2}x+\tfrac{1}{2}\sqrt{3}\,y,\tfrac{1}{2}\sqrt{3}\,x+\tfrac{1}{2}y,-z;-\tfrac{1}{2}x^4,\tfrac{1}{2}\sqrt{3}\,x^4,0\right)$$
$$=(x,y,z;x,-y,-z;x^3,0,0;x^4,0,0) \quad (5.2.6)$$

which in turn implies that

$$x^3=-2x;\qquad y=-\sqrt{3}\,x;\qquad z=x^4=0 \quad (5.2.7)$$

This allows us to remove scale factor x and obtain another symmetry vector,

$$\mathbf{b}^2=(1,-\sqrt{3},0;1,\sqrt{3},0;-2,0,0;0,0,0) \quad (5.2.8)$$

This symmetry coordinate corresponds to a displacement of the type shown in Fig. 5.2.2. As we had anticipated, $\mathbf{b}^2$ describes pure rotation about the z axis.

7c. *Project by the Use of Dirac Characters*. For the two-dimensional representation E, the preceding methods are less effective. We can project into the representation E by requiring that the Dirac characters obey

$$\Omega_2=D+F=-1,\qquad \Omega_3=A+B+C=0 \quad (5.2.9)$$

(See representation $E=\Gamma_3$ in Eq. 1.3.7.) However, in accord with our discussion in Sections 3.2 and 3.4, complete reduction of the matrix representation is obtained only if repetitions of a given symmetry coordinate are chosen so that they transform in precisely the same way. Since E has basis functions (x,y), let us require all symmetry coordinates to transform as these do.

Since x is odd under the reflection of σ_x and y is even, we can project into the subspace of odd partners of E by requiring that

$$A=-1 \tag{5.2.10}$$

The diagonalization of an *individual group operator* such as A is usually *more restrictive* than the diagonalization of the Ω's and *should be done first*. The operator to be diagonalized should be chosen (if possible) so that it has a *different* eigenvalue for each basis vector of the representation, as J_z does for the full rotation group. Otherwise two or more operators are needed to complete a set of commuting observables. As in Eqs. 5.2.3–5.2.5, the condition $A=-1$ restricts us to motions of the form

$$(x,y,z;x,-y,-z;x^3,0,0;x^4,0,0) \tag{5.2.11}$$

The condition $D+F=-1$ using Eqs. 5.1.3 and 5.1.6 leads to twelve equations that contain only one new requirement:

$$x^3=x-\sqrt{3}\,y \tag{5.2.12}$$

The condition $\Omega_3=0$ is automatically satisfied in view of the group multiplication table 1.2.4 for

$$\Omega_3=A+B+C=A+FA+DA=(1+D+F)A=(1+\Omega_2)A=0 \tag{5.2.13}$$

since $\Omega_2=-1$. When Eqs. 5.2.11 and 5.2.12 are combined, the result of projecting into the odd partner of E is

$$\left(x,y,z;x,-y,-z;x-\sqrt{3}\,y,0,0;x^4,0,0\right) \tag{5.2.14}$$

The coefficients of y, z, x, and x^4 constitute four[1] (unnormalized) symmetry coordinates:

$$\begin{matrix}(0,1,0; & 0,-1,\ \ 0; & -\sqrt{3},0,0; & 0,0,0)\\ (0,0,1; & 0,\ \ 0,-1; & 0,0,0; & 0,0,0)\\ (1,0,0; & 1,\ \ 0,\ \ 0; & 1,0,0; & 0,0,0)\\ (0,0,0; & 0,\ \ 0,\ \ 0; & 0,0,0; & 1,0,0)\end{matrix} \tag{5.2.15}$$

[1]That there are four coordinates which transform as x (the odd component of E) is evident, since the representation E appears four times in the complete space. See Eq. 5.1.9.

The last two symmetry vectors can be combined as before to give translation of the molecule as a whole and a motion $(1,1,1,-\mu)$ of the x components. Such a motion, however, contains a rotation about the y axis. But so do the first two symmetry coordinates above. By combining these motions, it is possible to construct two symmetry coordinates that do not contain any rotations. It is simpler, however, to return to form 5.2.14 and impose the constraint

$$\sum_n M_n(\mathbf{X}^n \times \mathbf{u}^n) = 0 \tag{5.2.16}$$

where $\mathbf{X}^n$ is the equilibrium position, $\mathbf{u}^n$ the displacement, and M_n the mass of the particle n. (The time derivative of this constraint represents the vanishing of the total angular momentum.) The $\mathbf{X}^n$ can be measured from any origin as long as Eq. 5.2.2 is imposed. Using the center of the triangle as origin, we obtain the following conditions for the vanishing of rotations about x and y, respectively:

$$R_x: \quad \frac{M_1 L}{\sqrt{3}}\left[z^3 - \tfrac{1}{2}(z^1 + z^2)\right] - M_4 H y^4 = 0 \tag{5.2.17}$$

$$R_y: \quad \frac{M_1 L}{2}(z^1 - z^2) + M_4 H x^4 = 0 \tag{5.2.18}$$

where L is the side of the triangle, and H is the height of the N atom above the plane. Equation 5.2.17 is satisfied automatically by Eq. 5.2.14 because we have projected into the subspace odd under σ_x, whereas R_x is even under the σ_x reflection. But Eq. 5.2.18 leads to

$$z \equiv (z^1 - z^2) = \frac{3H}{L}\left(\frac{-x^4}{\mu}\right) \equiv \beta \cdot X \tag{5.2.19}$$

where $\beta = 3H/L$, and

$$X \equiv -\frac{x^4}{\mu} = -\frac{M_4 x^4}{3M_1} \tag{5.2.20}$$

Similarly, displacement invariance (Eq. 5.2.2) leads to

$$x = \frac{y}{\sqrt{3}} + X \equiv Y + X \tag{5.2.21}$$

Expressing Eq. 5.2.14 in terms of X and Y leads to

$$\left(Y+X, \sqrt{3}\,Y, \beta X; Y+X, -\sqrt{3}\,Y, -\beta X; X-2Y, 0, 0; -\mu X, 0, 0\right) \tag{5.2.22}$$

and to the two internal symmetry coordinates, the coefficients of Y and X in Eq. 5.2.22:

$$\begin{aligned}\mathbf{b}_1^3 &= (1, \sqrt{3}, 0; \quad 1, -\sqrt{3}, \; 0; \quad -2, 0, 0; \quad 0, 0, 0)\\ \mathbf{b}_1^{3\prime} &= (1, \; 0, \beta; \quad 1, \; 0, -\beta; \quad 1, 0, 0; \quad -\mu, 0, 0)\end{aligned} \tag{5.2.23}$$

The even partners can be obtained in the same manner as the odd partners, by imposing $A=1$ (instead of $A=-1$). One must also impose $\Omega_2=-1$, $\Omega_3=0$ to project into the representation Γ_3 and displacement and rotational invariance as before. This procedure will determine $\mathbf{b}_2^3$ and $\mathbf{b}_2^{3\prime}$ except for an unknown sign (in other problems, an unknown phase factor).

Alternatively, one could apply the transfer operator 3.7.2:

$$\mathbf{b}_2^3 = P_{21}^3 \mathbf{b}_1^3 \tag{5.2.24}$$

but this is tedious, since P_{21}^3 involves a sum over six group operations. The simplest procedure is to make use of the transformation properties 5.3.1 of the $\mathbf{b}$'s in the form

$$\begin{aligned} D\mathbf{b}_1^3 &= \mathbf{b}_1^3 D_{11}^{(3)}(D) + \mathbf{b}_2^3 D_{21}^{(3)}(D)\\ &= -\tfrac{1}{2}\mathbf{b}_1^3 + \tfrac{1}{2}\sqrt{3}\,\mathbf{b}_2^3 \end{aligned} \tag{5.2.25}$$

or

$$\mathbf{b}_2^3 = \frac{2}{\sqrt{3}}\left(\tfrac{1}{2} + D\right)\mathbf{b}_1^3 \tag{5.2.26}$$

with D given by Eq. 5.1.3. This procedure has the advantages that (1) it automatically obeys displacement and rotational invariance, and (2) it determines the even partners uniquely and ensures that they transform alike. The corresponding even partners are found to be

$$\begin{aligned}\mathbf{b}_2^3 &= (-\sqrt{3}, 1, 0; \quad \sqrt{3}, 1, 0; \quad 0, -2, \; 0; \quad 0, \; 0, 0)\\ \mathbf{b}_2^{3\prime} &= (\quad 0, 1, \alpha; \quad 0, 1, \alpha; \quad 0, \; 1, -2\alpha; \quad 0, -\mu, 0)\end{aligned} \tag{5.2.27}$$

where $\alpha = \beta/\sqrt{3} = \sqrt{3}\, H/L$. These even modes, by the base vector orthogonality theorem 3.4.3, are orthogonal to odd modes. They happen also to be orthogonal to each other, or we would orthogonalize them. The partners $\mathbf{b}_1^3$ and $\mathbf{b}_2^3$ are shown in Fig. 5.2.3. The mode $\mathbf{b}_1^{3\prime}$ is a shearing motion of the H_3 triangle in the x direction relative to the N, combined with enough rotation about the y axis (drawn through the center of the triangle) to cancel the rotation contained in the shearing motion. The mode $\mathbf{b}_2^{3\prime}$ is a similar combination of y shear and x rotation.

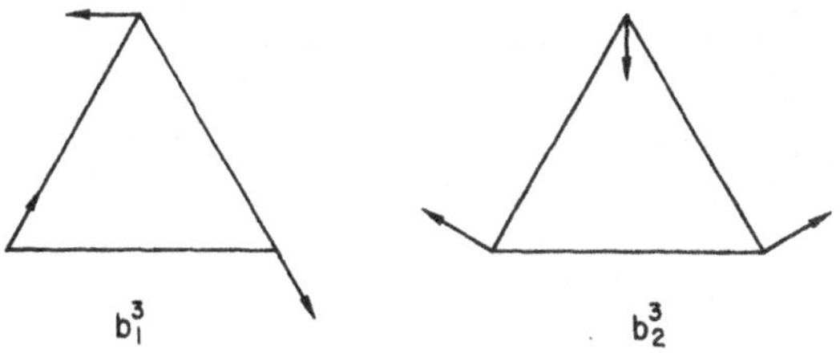

b_1^3 b_2^3

Fig. 5.2.3. Symmetry vectors for representation E of C_{3v}.

To determine the vibration frequencies of symmetry type E, one need only insert form 5.2.22 into the set of twelve similar equations to be solved. Two of these equations will determine X and Y, and the remaining ten equations will then be satisfied automatically. The particular admixture $Y\mathbf{b}_1^3 + X\mathbf{b}_1^{3'}$ which constitutes the actual normal mode of vibration will depend on details of the force constants not determined by symmetry.

No new information is gained by using the other partners of the representation; the assumed form $Y\mathbf{b}_2^3 + X\mathbf{b}_2^{3'}$ leads to the same frequencies (because E is degenerate) and to the same ratio Y/X.

In this section, we used the operator $A = \boldsymbol{\sigma}_x$ to distinguish between the even and odd partners of the representation Γ_3. In the process, we found it more efficient to diagonalize A first, and the Dirac characters later, because the diagonalization of an individual operator is more restrictive than that of a Dirac character.

How can we choose the best operator or operators to diagonalize? The following is a useful rule: *Find the largest Abelian subgroup and diagonalize the generators of this subgroup*. For the ammonia molecule, this subgroup is the group C_3. If the representations of C_3 that transform as z(or R_z), $x+iy$, and $x-iy$ are denoted as Δ_1, Δ_ϵ, and Δ_{ϵ^2} respectively, the compatability relations are

$$A_1 = \Delta_1, \qquad A_2 = \Delta_1, \qquad E = \Delta_\epsilon + \Delta_{\epsilon^2} \tag{5.2.28}$$

Thus diagonalization of the element C_3 with eigenvalue $\epsilon = \exp(2\pi i/3)$ or $\epsilon^2 = \exp(4\pi i/3)$ immediately projects us into the modes that transform as $x+iy$ and $x-iy$. The real part of these symmetry coordinates must yield the modes that transform as x, that is, the four modes shown in Eq. 5.2.15 or linear combinations thereof. Similarly, the imaginary part will yield four modes that transform as y.

More generally, since C_{3v} is constructed from C_3 by adding $\boldsymbol{\sigma}_x$, the space $\Delta_\epsilon \oplus \boldsymbol{\sigma}_x \Delta_\epsilon$ is the space invariant under C_{3v}; indeed, $\frac{1}{2}(1-\boldsymbol{\sigma}_x)\Delta_\epsilon$ will yield the odd (x) partner of E, and $-(i/2)(1+\boldsymbol{\sigma}_x)\Delta_\epsilon$ the even (y) partner of E. This procedure of generating a representation of a group C_{3v} from that of a subgroup C_3 is known as the method of *induced representations*. It was used in Section 3.8 to construct the polynomial basis functions for the point group character tables. In dealing with space groups in Chapter 8, the translation operators will constitute the Abelian subgroup that is diagonalized as the first step.

The representations A_1 and A_2 cannot be distinguished by C_3 (see Eq. 5.2.28). Hence we must provide additional information by diagonalizing Dirac characters. For these one-dimensional representations, it is sufficient to diagonalize $\boldsymbol{\sigma}_x$, which is what we did in the first place.

5.3. SYMMETRY COORDINATES, NORMAL COORDINATES, INTERNAL COORDINATES, AND INVARIANTS

Symmetry Coordinates

In Fig. 5.2.3 we showed the distortions of the NH_3 molecule asssociated with $\mathbf{b}_1^3$ (of Eq. 5.2.23) and $\mathbf{b}_2^3$ (of Eq. 5.2.27). These distortions are referred to as *symmetry vectors* because they have been constructed so as to transform into one another as (odd and even) partners of irreducible representation Γ_3 or, more generally,

$$R\mathbf{b}_\nu^{ia} = \sum_\mu \mathbf{b}_\mu^{ia} D_{\mu\nu}^i(R) \tag{5.3.1}$$

where i labels the representations, μ or ν denotes the partner function, and $a=1,2,\ldots,n_i$ is a label that we have introduced to differentiate between the n_i different repetitions of a given representation in the space. For example, we can write $\mathbf{b}_1^3$ as $\mathbf{b}_1^{31}$, and $\mathbf{b}_1^{3\prime}$ as $\mathbf{b}_1^{32}$. Each $\mathbf{b}_\mu^{ia}$ is, in this case, a vector with twelve components that can be denoted by an index $r=1,2,\ldots,12$. An *arbitrary* (possibly time-dependent) distortion $\{u^r\}=(u^1,u^2,\ldots,u^{12})=(x^1,y^1,z^1;x^2\ldots z^4)$ can be expressed as a linear combination of *all* the symmetry vectors:

$$u^r(t) = \sum_{ia\mu} b_{\mu r}^{ia} q_\mu^{ia}(t) \tag{5.3.2}$$

where the q_μ^{ia} are (the possibly time-dependent) amplitudes describing the amount of distortion of type $\mathbf{b}_\mu^{ia}$ present at time t.

Under a symmetry operation R of the group we can write

$$Ru^r = \sum_{ia\nu} (Rb_{\nu r}^{ia}) q_\nu^{ia}(t) = \sum b_{\mu r}^{ia} D_{\mu\nu}^i(R) q_\nu^{ia}(t) \tag{5.3.3a}$$

that is, a rotated distortion has the same amplitudes when taken with respect to the rotated symmetry vectors, but the modified amplitudes

$$(q_\mu^{ia})' = \sum_\nu D_{\mu\nu}^i(R) q_\nu^{ia} \tag{5.3.3b}$$

when taken with respect to the unrotated symmetry vectors.

The initial symmetry vectors such as those of Eq. 5.2.15 involve only motion of a set of identical particles, since only these are mixed by group operators. And these vectors have been chosen orthogonal to one another. Symmetry vectors of the same type are then mixed in such a way as to eliminate external translation or rotation. Such a procedure can be arranged to leave the kinetic energy as a diagonal form and therefore constitutes a rotation that preserves $\Sigma M_r(u^r)^2$, so that the final *symmetry vectors* obey the orthogonality requirement

$$\sum_r M_r(b_{\mu r}^{ia})^* b_{\nu r}^{jb} = M\,\delta(i,j)\,\delta(a,b)\,\delta(\mu,\nu) \tag{5.3.4}$$

where M is arbitrary and can be set equal to unity or to some mass in the problem. A formal proof is given in Appendix D.

For example, if we set $M = M_1$, the normalized eigenvectors of Eq. 5.2.23 become

$$\begin{aligned}
\mathbf{b}_1^{31} &= (1, \sqrt{3}, 0; \quad 1, -\sqrt{3}, 0; \quad -2, 0, 0; \quad 0,0,0)/(12)^{1/2} \\
\mathbf{b}_1^{32} &= (1, 0, \beta; \quad 1, 0, -\beta; \quad 1, 0, 0; \quad -\mu, 0, 0)/(3+2\beta^2+3\mu)^{1/2}
\end{aligned} \tag{5.3.5}$$

whereas the even partners become

$$\begin{aligned}
\mathbf{b}_2^{31} &= (-\sqrt{3}, 1, 0; \quad \sqrt{3}, 1, 0; \quad 0, -2, 0; \quad 0, 0, 0)/(12)^{1/2} \\
\mathbf{b}_2^{32} &= (0, 1, \alpha; \quad 0, 1, \alpha; \quad 0, 1, -2\alpha; \quad 0, -\mu, 0)/(3+6\alpha^2+3\mu)^{1/2}
\end{aligned} \tag{5.3.6}$$

and the identity representation vectors of Eq. 5.2.1 are

$$\begin{aligned}
\mathbf{b}^{11} &= (-\sqrt{3}, -1, 0; \quad \sqrt{3}, -1, 0; \quad 0, 2, 0; \quad 0, 0, 0)/(12)^{1/2} \\
\mathbf{b}^{12} &= (0, 0, 1; \quad 0, 0, 1; \quad 0, 0, 1; \quad 0, 0, -\mu)/(3+3\mu)^{1/2}
\end{aligned} \tag{5.3.7}$$

When $ia\mu$ for a moment is replaced by the single index s, Eq. 5.3.4 is equivalent to requiring that

$$\left(\frac{M_r}{M}\right)^{1/2} \cdot b_r^s \equiv U_{rs} \tag{5.3.8}$$

be a unitary matrix. Thus Eq. 5.3.2 can be inverted easily to obtain the *symmetry coordinates*:

$$q^s = \sum_r \left(\frac{M_r}{M}\right)(b_r^s)^* u^r$$

or

$$q_\mu^{ia} = \sum_r \left(\frac{M_r}{M}\right)(b_{\mu r}^{ia})^* u^r \tag{5.39}$$

Thus Eqs. 5.3.5–5.3.7 result in

$$\begin{aligned}
q_1^{31} &= \frac{x^1 + x^2 - 2x^3 + \sqrt{3}\,(y^1 - y^2)}{(12)^{1/2}} \\
q_1^{32} &= \frac{x^1 + x^2 + x^3 - 3x^4 + \beta(z^1 - z^2)}{(3 + 2\beta^2 + 3\mu)^{1/2}} \\
q_2^{31} &= \frac{y^1 + y^2 - 2y^3 + \sqrt{3}\,(x^2 - x^1)}{(12)^{1/2}} \\
q_2^{32} &= \frac{y^1 + y^2 + y^3 - 3y^4 + \alpha(z^1 + z^2 - 2z^3)}{(3 + 6\alpha^2 + 3\mu)^{1/2}} \\
q^{11} &= \frac{\sqrt{3}\,(x^2 - x^1) + 2y^3 - y^1 - y^2}{(12)^{1/2}} \\
q^{12} &= \frac{z^1 + z^2 + z^3 - 3z^4}{(3 + 3\mu)^{1/2}}
\end{aligned} \tag{5.3.10}$$

There are, of course, six external symmetry coordinates. Aside from normalization, the three translational coordinates are

$$\mathbf{q}_{\text{transl}} = \sum M_n \mathbf{u}^n \tag{5.3.11}$$

and the three rotational coordinates are

$$\mathbf{q}_{\text{rot}} = \sum M_n (\mathbf{X}^n \times \mathbf{u}^n) \tag{5.3.12}$$

The transformation to symmetry coordinates Eq. 5.3.2 reduces the potential energy to a form

$$V = \frac{1}{2} \sum A_{ab}^i (q_\mu^{ia})^* q_\mu^{ib} \tag{5.3.13}$$

coupling only *symmetry coordinates* belonging to the same partner of the same irreducible representation (see Eq. 3.4.15). In Eq. 5.3.13 only the internal symmetry coordinates appear, as V must be invariant against translations and rotations of the molecule as a whole.

Our orthonormality requirement 5.3.4 was chosen so that the kinetic energy would retain the diagonal form,

$$T=\tfrac{1}{2}M\sum|\dot{q}_{\mu}^{ia}|^2+T_{\text{ext}} \tag{5.3.14}$$

where T_{ext} is the kinetic energy associated with the external degrees of freedom and can be omitted from our discussions of the vibration problem.

Normal Coordinates

The normal coordinates $q_{ti\mu}$ are obtained by diagonalizing the Hermitian matrix A_{ab}^{i}:

$$\sum_b A_{ab}^{i}C_t^{ib}=M\omega_{ti}^2C_t^{ia} \tag{5.3.15}$$

The normalized eigenvectors C_t^{ia} then produce a unitary transformation:

$$q_{\mu}^{ia}=\sum_t C_t^{ia}q_{ti\mu} \tag{5.3.16}$$

The solution of Eq. 5.3.15 and the resulting transformation are simpler than the profusion of indices indicates because all of the work is done within a given partner μ of a given representation i. If we suppress indices $i\mu$, needed only to remind us of this fact, Eqs. 5.3.15 and 5.3.16 take the simpler forms

$$\sum_b A_{ab}C_t^b=M\omega_t^2C_t^a$$

and $q^a=\sum C_t^a q_t$, with A_{ab} an $n_i\times n_i$ matrix and C_t^a an n_i component eigenvector.

Transformation 5.3.16 diagonalizes V,

$$V=\tfrac{1}{2}M\sum\omega_{ti}^2|q_{ti\mu}|^2 \tag{5.3.17}$$

while leaving T diagonal:

$$T_{\text{int}}=\tfrac{1}{2}M\sum|\dot{q}_{ti\mu}|^2 \tag{5.3.18}$$

The complete transformation from the original displacements u^r to the normal coordinates can be written as

$$u^r=\sum b_{\mu r}^{ti}q_{ti\mu} \tag{5.3.19}$$

with

$$b_{\mu r}^{ti} = \sum_{a=1}^{n_i} b_{\mu r}^{ia} C_t^{ia} \qquad (5.3.20)$$

that is, the *correct eigenvectors are linear combinations of symmetry vectors of a given representation i and a given partner* μ.

Number of Independent Parameters in the Potential Energy Matrix

In terms of the symmetry coordinates, the potential energy matrix of Eq. 5.3.13 has the reduced form

$$\left[\begin{array}{c|c} A_{ab}^{1} & 0 \\ \hline 0 & A_{ab}^{3} \end{array}\right]$$

where the ith block on the diagonal is the $n_i \times n_i$ Hermitian matrix A_{ab}^{i}. Group theory places no restriction on the elements of the matrix A_{ab}^{i}, so that the ith block contributes, in general, $(n_i)^2$ arbitrary parameters to the potential energy matrix. If the symmetry coordinates q_{μ}^{ia} are real, the potential energy, Eq. 5.3.13, is a quadratic form and A_{ab}^{i} must then be symmetric. In this case A_{ab}^{i} contributes $\frac{1}{2}n_i(n_i+1)$ arbitrary parameters to the potential energy matrix. If A_{ab}^{i} were antisymmetric, it would contribute $\frac{1}{2}n_i(n_i-1)$ constants.

We can always insert the real character $\chi_{\text{int}}(R)=\Sigma_i n_i \chi^i(R)$ into Eq. 5.1.13 to decompose the total number of parameters into parts appropriate to each i:

$$N = \frac{1}{2}\sum_i n_i(n_i + b_i) \qquad (5.3.21)$$

where

$$b_i = \frac{1}{g}\sum_R \chi^i(R^2) \qquad (5.3.22)$$

is a parameter that can take only the values $+1, -1, 0$ (see Theorem 10.6.9). Thus there is some group-theoretical property b_i of representation i that determines whether the matrix A_{ab}^{i} is to be symmetric, antisymmetric, or Hermitian. We shall see in Chapter 10 that the three cases $b_i = +1, -1, 0$ will correspond, respectively, to cases in which (1) $D^i(R)^*$ is equivalent to $D^i(R)$ and can be made real, (2) $D^i(R)^*$ is equivalent to $D^i(R)$ but cannot be made real, (3) $D^i(R)^*$ is not equivalent to $D^i(R)$ and is necessarily complex. The representations are referred to as real, pseudoreal, and complex, respectively. (Only the last has complex *characters*.) Equation 5.3.22 is known as the Frobenius-Schur (**1906**) criterion.

Table 5.3.1. Character Table for Group C_3 Showing the Polar Vector Representation

		E	C_3	C_3^2	Basis Vectors
Δ_1	Γ_1	1	1	1	z
Δ_ϵ	Γ_2	1	ϵ	ϵ^2	$x+iy$
Δ_{ϵ^2}	Γ_3	1	ϵ^2	ϵ	$x-iy$
	Γ_p	3	0	0	(x,y,z)

Note. $\epsilon=\exp(-2\pi i/3)$.

The pseudoreal case does not occur for any (single-valued) representation of any point group, although it does occur in space groups. The real case is the most common. All three representations of C_{3v} are real.

The meaning of complex representations can be illustrated simply by reference to an anisotropic oscillator in three dimensions bound to the origin with symmetry C_3. The threefold axis causes the potential energy

$$V=Az^2+Bx^2+Cy^2+Dxy+Exz+Fyz \tag{5.3.23}$$

to reduce to the form

$$V=Az^2+B(x^2+y^2) \tag{5.3.24}$$

leaving only two independent constants, even though the space (x,y,z) contains all three representations,

$$\Gamma_p=\Gamma_1+\Gamma_2+\Gamma_3 \tag{5.3.25}$$

(as is clear from the character table, Table 5.3.1), and we would expect three parameters for the three 1×1 matrices A^i_{ab}, with $i=1,2,3$.

What has happened? The condition that V be real and hence symmetric (which we shall show in Chapter 11 corresponds to time reversibility) has forced the two complex conjugate representations $\Gamma_2, \Gamma_3=\Gamma_2^*$ described by symmetry coordinates $x+iy$ and $x-iy$ to be degenerate. In physical terms, the right and left circularly polarized normal motions are time inverses of one another and must have the same frequency. This time-reversal degeneracy guarantees that *real* normal coordinates (in this case, x and y) can always be constructed. For representations such as Γ_2 and Γ_3, for which $b=0$, this is done at the expense of mixing symmetry coordinates from nonequivalent representations.

Indeed, we show in Section 10.6 that when $b_i=-1$ or 0 time reversal causes extra degeneracy, and the representation can be made real only at the expense of being made reducible.

Internal Coordinates

Molecular physics (e.g., Wilson, Decius, and Cross, **1955**) often introduces internal coordinates: bond stretchings, angle bendings, bending of a bond relative to a plane, and so on. These internal coordinates have the merit of automatically eliminating motions of the molecule as a whole. But they have the disadvantage that one must re-express them in terms of Cartesian displacements to compute the kinetic energy. What is worse, the mass-weighted orthonormality condition is usually violated and the kinetic energy matrix acquires off-diagonal terms.[2] Conversely, however, a particular set of internal coordinates can often be chosen so that most of the potential energy is stored in one or two of these coordinates. This is an aid in visualization as well as in the final numerical solutions.

As an example, the $3n-6=6$ internal coordinates of the NH_3 molecule can be described completely by the six bond stretches

$$\Delta r^{12}, \Delta r^{23}, \Delta r^{31}; \quad \Delta r^{41}, \Delta r^{42}, \Delta r^{43} \tag{5.3.26}$$

A potential energy can be written down immediately with the desired symmetry:

$$\begin{aligned} V = a\left[(\Delta r^{12})^2 + (\Delta r^{23})^2 + (\Delta r^{31})^2\right] + b(\Delta r^{12}\Delta r^{23} + \Delta r^{23}\Delta r^{31} + \Delta r^{31}\Delta r^{12}) \\ + c\left[(\Delta r^{41})^2 + (\Delta r^{42})^2 + (\Delta r^{43})^2\right] + d(\Delta r^{41}\Delta r^{12} + \Delta r^{42}\Delta r^{23} + \Delta r^{43}\Delta r^{31}) \\ + e(\Delta r^{41}\Delta r^{23} + \Delta r^{42}\Delta r^{31} + \Delta r^{43}\Delta r^{12}) + f(\Delta r^{41}\Delta r^{42} + \Delta r^{42}\Delta r^{43} + \Delta r^{43}\Delta r^{41}) \end{aligned} \tag{5.3.27}$$

where Δr^{ij} is the change in the "bond length" connecting particles i and j, and for NH_3 the index 4 represents the N atom. This procedure is undoubtedly the simplest, but it has two pitfalls:

1. One may forget to write down all the invariants.
2. For more than four particles in three dimensions, the number of bond lengths will be an overcomplete set, that is to say,

$$\tfrac{1}{2}(n)(n-1) > 3n-6 \quad \text{for} \quad n>4 \tag{5.3.28}$$

Therefore it will, in general, be possible to write down invariants that are not independent of other invariants.

Methods of eliminating redundant variables and of recomputing the kinetic

[2]This difficulty is clearly avoidable by using appropriate combinations of internal coordinates.

energy in the internal coordinates are given by Wilson, Decius, and Cross **(1955)**.

We next indicate a method of writing down invariant potential energies directly in Cartesian coordinates.

Invariants

Since the potential energy is invariant under the group operations and has form 5.3.13 with constants A^i_{ab} that are arbitrary, in the sense that they can be altered without affecting the symmetry of the molecule, each of the coefficients of A^i_{ab}, namely,

$$\sum_{\mu} (q_{\mu}^{ia})^* q_{\mu}^{ib} \tag{5.3.29}$$

must be an invariant. This is simply the shell theorem 4.5.4. [Since the q_{μ}^{ia} of Eq. 5.3.9 are expressed via Eq. 5.3.10 in terms of the displacements u^r with coefficients $(b_{\mu r}^{ia})^*$ that are now regarded as numbers, the q_{μ}^{ia} transform under rotations in accord with Eq. 5.3.3b.] Thus the NH_3 molecule problem (whose symmetry coordinates are real) has the *six* invariants

$$(q^{11})^2, (q^{12})^2, q^{11}q^{12}, (q_1^{31})^2 + (q_2^{31})^2, (q_1^{32})^2 + (q_2^{32})^2, q_1^{31}q_1^{32} + q_2^{31}q_2^{32}$$

where the q_{μ}^{ia} are given in Eq. 5.3.10. The total potential is a linear combination of these invariants with six arbitrary coefficients as predicted in Eq. 5.1.14.

The procedure just described is the same as the method of invariants used in Section 4.5 to construct various tensors. It is particularly useful when three-and sixfold elements are present. Also, by suitable choice of symmetry coordinates, the external motions are eliminated. When no three-or sixfold elements are present, it is usually simpler to use the direct inspection method discussed in the next section.

5.4. POTENTIAL ENERGY AND FORCE CONSTANTS

If $\mathbf{u}^m$ denotes the vector displacement of atom m from its equilibrium position $\mathbf{X}^m$, the potential energy of a *molecule or solid* takes, in the harmonic approximation, the form

$$V(\mathbf{u}) = \frac{1}{2}\sum_{m,n} \mathbf{u}^m \cdot \mathbf{K}^{mn} \cdot \mathbf{u}^n = \frac{1}{2}\sum_{m,n,\mu\nu} u_{\mu}^m K_{\mu\nu}^{mn} u_{\nu}^n \tag{5.4.1}$$

where the second form, which explicitly displays Cartesian components $\mu\nu = 1,2,3$, will be avoided whenever possible.

We may as well assume **K** to be symmetric, since its antisymmetric part does not contribute to the potential energy. Thus **K** is invariant under the operation F ("flip") that interchanges (flips) all indices:

$$K^{mn}_{\mu\nu} = K^{nm}_{\nu\mu} \equiv FK^{mn}_{\mu\nu}$$

or

$$F\mathbf{K}^{mn} \equiv \tilde{\mathbf{K}}^{nm} = \mathbf{K}^{mn} \tag{5.4.2}$$

where $\tilde{\mathbf{K}}$ indicates the transpose of **K** with respect to its (suppressed) Cartesian indices. The relation of F to time reversal is discussed in Section 10.9.

If the displacements $\mathbf{u}^m$ are altered by the amount $\Delta\mathbf{u}^m$, the change in potential energy is

$$\Delta V = \frac{1}{2} \sum_{m,n} (\mathbf{u}^m \cdot \mathbf{K}^{mn} \cdot \Delta\mathbf{u}^n + \Delta\mathbf{u}^m \cdot \mathbf{K}^{mn} \cdot \mathbf{u}^n + \Delta\mathbf{u}^m \cdot \mathbf{K}^{mn} \cdot \Delta\mathbf{u}^n)$$

By reversing the order of the factors and then interchanging the names of the dummy indices, the middle term can be made to take two new forms:

$$\sum \mathbf{u}^n \cdot \tilde{\mathbf{K}}^{mn} \cdot \Delta\mathbf{u}^m = \sum \mathbf{u}^m \cdot \tilde{\mathbf{K}}^{nm} \cdot \Delta\mathbf{u}^n$$

The symmetry of **K**, Eq. 5.4.2, then shows that the second term is identical to the first; hence

$$\Delta V = \sum_{m,n} (\mathbf{u}^m \cdot \mathbf{K}^{mn} \cdot \Delta\mathbf{u}^n + \tfrac{1}{2} \Delta\mathbf{u}^m \cdot \mathbf{K}^{mn} \cdot \Delta\mathbf{u}^n) \tag{5.4.3}$$

If the changes $\Delta\mathbf{u}^m$ are such as to leave the potential energy invariant for an arbitrary initial configuration $\{\mathbf{u}^m\}$, the coefficient of each $\mathbf{u}^m$ and the term independent of all $\mathbf{u}$ in ΔV must vanish:

$$\sum_n \mathbf{K}^{mn} \cdot \Delta\mathbf{u}^n = 0 \tag{5.4.4}$$

$$\sum_{m,n} \Delta\mathbf{u}^m \cdot \mathbf{K}^{mn} \cdot \Delta\mathbf{u}^n = 0 \tag{5.4.5}$$

If the first condition is obeyed for *all m*, the second is automatically satisfied. The symmetry of **K** permits the first condition to be written equally well in the form

$$\sum_m \Delta\mathbf{u}^m \cdot \mathbf{K}^{mn} = 0 \tag{5.4.6}$$

A molecule or solid with only internal forces is invariant against arbitrary infinitesimal (or finite) displacements or rotations. The former with $\Delta\mathbf{u}^n = \mathbf{d}$ yields

$$\sum_n \mathbf{K}^{mn} = 0 \qquad \text{or} \qquad \sum_m \mathbf{K}^{mn} = 0 \tag{5.4.7}$$

The latter with the infinitesimal rotation angle ω,

$$\Delta\mathbf{u}^n = \omega \times (\mathbf{X}^n - \mathbf{X}_0) \tag{5.4.8}$$

[where $\mathbf{X}^n$ is the equilibrium position of the atom n, and $\mathbf{X}_0$ is the (arbitrary) origin of the rotation] yields

$$\sum_n \mathbf{K}^{mn} \times (\mathbf{X}^n - \mathbf{X}_0) = 0 \tag{5.4.9}$$

where the cross product is taken using the second (spatial) index of $\mathbf{K}$ with the first treated as a parameter. In view of the displacement invariance, Eq. 5.4.7, $\mathbf{X}_0$ can be chosen arbitrarily and is taken to be 0 or, more frequently, to be equal to $\mathbf{X}^m$:

$$\sum_n \mathbf{K}^{mn} \times (\mathbf{X}^n - \mathbf{X}^m) = 0 \tag{5.4.10}$$

Let us now consider a member $(\mathbf{S}|\mathbf{v})$ of the molecular or crystal group that restores the crystal to itself:

$$(\mathbf{X}^m)' = \mathbf{X}^{m'} = \mathbf{S}\cdot\mathbf{X}^m + \mathbf{v} \tag{5.4.11}$$

The total vector $\mathbf{X}^m + \mathbf{u}^m$ is transformed into

$$(\mathbf{X}^m + \mathbf{u}^m)' = \mathbf{S}\cdot(\mathbf{X}^m + \mathbf{u}^m) + \mathbf{v} \tag{5.4.12}$$

Since $\mathbf{S}$ is a finite rotation, we can think of $(\mathbf{X}^m + \mathbf{u}^m)' - \mathbf{X}^m$ as a new *large* displacement from atom m, that is, from position $\mathbf{X}^m$. This would simply generalize the infinitesimal rotational invariance of Eq. 5.4.9 to invariance against an arbitrary finite rotation: $V' = V$ as before. But we can equally well consider, after subtracting Eq. 5.4.11 from 5.4.12, that we now have a small displacement,

$$(\mathbf{X}^m + \mathbf{u}^m)' - \mathbf{X}^{m'} \equiv (\mathbf{u}^m)' = \mathbf{S}\cdot\mathbf{u}^m \tag{5.4.13}$$

from the position $\mathbf{X}^{m'}$ of the renamed atom,

$$m' = S(m) \tag{5.4.14}$$

defined by Eq. 5.4.11. In this case, the potential energy can be computed by giving each atom $m' = S(m)$ the displacement $\mathbf{S}\cdot\mathbf{u}^m$:

$$V' = \frac{1}{2}\sum (\mathbf{S}\cdot\mathbf{u}^m)\cdot\mathbf{K}^{m'n'}\cdot(\mathbf{S}\cdot\mathbf{u}^n) \tag{5.4.15}$$

Since, aside from the renaming of the atoms, we have merely rotated the distorted molecule, we can again argue that $V' = V$ if the crystal or molecule is subject only to internal forces.

Equation 5.4.15 is equivalent to introducing a new force constant matrix,

$$V' = \frac{1}{2}\sum \mathbf{u}^m\cdot\mathbf{K}'^{mn}\cdot\mathbf{u}^n \tag{5.4.16}$$

where

$$\mathbf{K}'^{mn} = \tilde{\mathbf{S}}\cdot\mathbf{K}^{m'n'}\cdot\mathbf{S} \tag{5.4.17}$$

or

$$(K')^{mn}_{\mu\nu} = \sum \tilde{S}_{\mu\mu'}K^{m'n'}_{\mu'\nu'}S_{\nu'\nu} = \sum K^{m'n'}_{\mu'\nu'}S_{\mu'\mu}S_{\nu'\nu} \tag{5.4.18}$$

The invariance of the potential energy under the group operations, $V' = V$ or $\mathbf{K}' = \mathbf{K}$, leads to a series of conditions

$$K^{mn}_{\mu\nu} = \sum_{\mu'\nu'} K^{m'n'}_{\mu'\nu'}S_{\mu'\mu}S_{\nu'\nu} \equiv \sum_{\mu'\nu'} K^{S(m)S(n)}_{\mu'\nu'}S_{\mu'\mu}S_{\nu'\nu} \tag{5.4.19}$$

that limit the number of independent constants among the $\mathbf{K}$'s. If we denote a given pair (m,n) by p, most of the above equations yield relationships between $\mathbf{K}^p$ and $\mathbf{K}^{p'}$, that is, *all the force constants between the set of equivalent bonds can be expressed in terms of those of a prototype bond.*

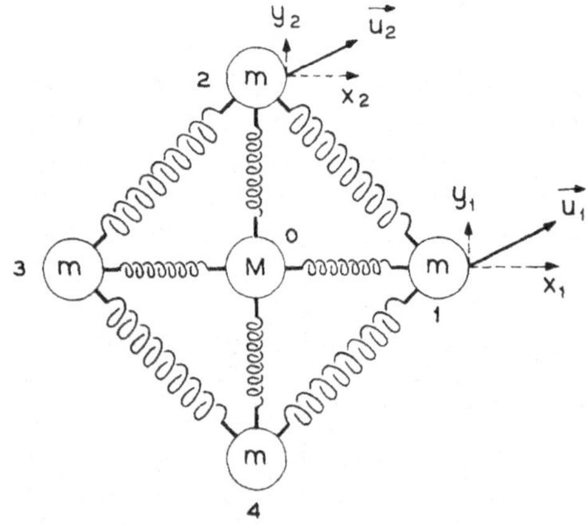

Fig. 5.4.1. Planar vibrations of a molecular with symmetry C_{4v}.

Theorem 5.4.1

The conditions obeyed by the force constants of a given (prototype) bond are determined by the elements of the group that leave the bond invariant (the "group of the bond"). When the symmetry of **K** is included, one must include the elements that reverse the bond (interchange m and n). Together these form the "reversal group of the bond."

The proof of Theorem 5.4.1 is similar to the proofs of the analogous theorems 4.4.1-4.4.2 about tensors. A detailed discussion is given by Lax (**1965c**).

Example 5.4.1

Find the force constants appropriate to the planar vibrations of planar molecules AB_4 with the two-dimensional symmetry C_{4v}. See Fig. 5.4.1.

Bond (01): The only element of C_{4v} that leaves this bond invariant is the $\boldsymbol{\sigma}_y$ reflection plane, which contains the bond. Equation 5.4.19 can be written as

$$\mathbf{K}^{01} = \tilde{\boldsymbol{\sigma}}_y \cdot \mathbf{K}^{01} \cdot \boldsymbol{\sigma}_y$$

$$\begin{bmatrix} \alpha & \gamma \\ \delta & \beta \end{bmatrix} = \begin{bmatrix} 1 & 0 \\ 0 & -1 \end{bmatrix} \begin{bmatrix} \alpha & \gamma \\ \delta & \beta \end{bmatrix} \begin{bmatrix} 1 & 0 \\ 0 & -1 \end{bmatrix} = \begin{bmatrix} \alpha & -\gamma \\ -\delta & \beta \end{bmatrix}$$

$$\therefore \gamma = \delta = 0 \qquad \text{or} \qquad \mathbf{K}^{01} = \begin{bmatrix} \alpha & 0 \\ 0 & \beta \end{bmatrix} \tag{5.4.20}$$

More briefly, $\boldsymbol{\sigma}_y$ leaves invariant all the 11 and 22 components of the quadratic form but reverses the sign of the 12 or 21 components (i.e., x^0y^1 or y^0x^1 components), causing the terms linear in the y's to vanish.

Bond (12): There is no element of C_{4v} that leaves bond (12) invariant. Thus, if we ignore flip, Eq. 5.4.2, $\mathbf{K}^{12}$ has four arbitrary constants. The diagonal reflection $\boldsymbol{\sigma}_d \equiv \boldsymbol{\sigma}_{xy}$ that interchanges x and y changes $\mathbf{K}^{12}$ into $\mathbf{K}^{21}$. This reversal element can be combined with the symmetry requirement 5.4.2 to yield

$$K_{11}^{12} = K_{22}^{21} = K_{22}^{12}, \qquad K_{12}^{12} = K_{21}^{21} \tag{5.4.21}$$

Thus the combined operation $\boldsymbol{\sigma}_d F$ has reduced $\mathbf{K}^{12}$ to the form

$$\mathbf{K}^{12} = \begin{bmatrix} a & b \\ c & a \end{bmatrix} \tag{5.4.22}$$

with three independent constants. *Note that* $\mathbf{K}^{12}$ *is not symmetric* (in the suppressed spatial indices).

Bond (13): The $\boldsymbol{\sigma}_y$ symmetry element guarantees that

$$\mathbf{K}^{13}=\begin{bmatrix}\alpha' & 0\\ 0 & \beta'\end{bmatrix} \tag{5.4.23}$$

The operation $\boldsymbol{\sigma}_x F$ restores $\mathbf{K}^{13}$ to itself but adds no further constraint.

"*Bond*" *(00):* This is not really a bond, but $\mathbf{x}^0\cdot\mathbf{K}^{00}\cdot\mathbf{x}^0$ does contribute to the potential energy and must be evaluated. All elements C_{4v} leave site 0 invariant. Element $\boldsymbol{\sigma}_y$ kills the off-diagonal terms as usual. And a fourfold rotation C_4 guarantees that the x^2 and y^2 terms have equal coefficients; thus

$$\mathbf{K}^{00}=\begin{bmatrix}f & 0\\ 0 & f\end{bmatrix} \tag{5.4.24}$$

"*Bond*" *(11):* Element $\boldsymbol{\sigma}_y$ guarantees that

$$\mathbf{K}^{11}=\begin{bmatrix}g & 0\\ 0 & h\end{bmatrix} \tag{5.4.25}$$

We have evaluated all the "prototype" matrices and have arrived at a total of ten constants. Have we made use of all the possible symmetries of the point group? We can check our results by a direct calculation of the total number of independent constants by the method discussed in Section 5.1. The character of the ten-dimensional space of the molecule is given by

$$\chi_{\text{tot}}(R)=N(R)\chi(R) \tag{5.4.26}$$

where $\chi(R)$ is the character of the two-dimensional representation E generated by (x,y), and $N(R)$ is the number of points of the molecule left fixed by R. The calculation is summarized in Table 5.4.1. Thus

$$\frac{1}{g}\sum_R[\chi_{\text{tot}}(R)]^2=\tfrac{1}{8}(100+4)=13$$

$$\frac{1}{g}\sum_R\chi_{\text{tot}}(R^2)=\tfrac{1}{8}[6\chi_{\text{tot}}(E)+2\chi_{\text{tot}}(C_2)]=7$$

Table 5.4.1. Calculation of Number of Force Constants for Molecules with Symmetry C_{4v}

Representation		E	C_2	$2C_4$	σ_x, σ_y	$2\sigma_d$
		Character Table for C_{4v}				
z	A_1 Δ_1	1	1	1	1	1
R_z	A_2 Δ_1'	1	1	1	−1	−1
x^2-y^2	B_1 Δ_2	1	1	−1	1	−1
xy	B_2 Δ_2'	1	1	−1	−1	1
(x,y)	E Δ_5	2	−2	0	0	0
	$N(R)$	5	1	1	3	1
	$N(R)\chi^E(R)$	10	−2	0	0	0
	$[\chi_{tot}(R)]^2$	100	4	0	0	0
	$\chi_{tot}(R^2)$	10	10	−2	10	10

and the number of independent constants in the quadratic form is obtained from the symmetrized product, Eq. 3.3.30, that is, $\frac{1}{2}(13+7)=10$, which agrees with our direct calculation. (See, however, Problem 5.4.1.)

The three constants f, g, and h in the diagonal terms $\mathbf{K}^{00}$ and $\mathbf{K}^{11}$ can be eliminated by making use of displacement invariance (Eq. 5.4.7):

$$\mathbf{K}^{00} = -(\mathbf{K}^{01}+\mathbf{K}^{02}+\mathbf{K}^{03}+\mathbf{K}^{04}) \tag{5.4.27}$$

$$\mathbf{K}^{11} = -(\mathbf{K}^{10}+\mathbf{K}^{12}+\mathbf{K}^{13}+\mathbf{K}^{14}) \tag{5.4.28}$$

Equation 5.4.17 can be written as

$$\mathbf{K}^{m'n'} = \mathbf{S}\mathbf{K}^{mn}\tilde{\mathbf{S}} \tag{5.4.29}$$

Using $S = C_4 =$ a 90° counterclockwise rotation, we obtain

$$\mathbf{K}^{02} = \begin{bmatrix} 0 & -1 \\ 1 & 0 \end{bmatrix}\begin{bmatrix} \alpha & 0 \\ 0 & \beta \end{bmatrix}\begin{bmatrix} 0 & 1 \\ -1 & 0 \end{bmatrix} = \begin{bmatrix} \beta & 0 \\ 0 & \alpha \end{bmatrix}$$

$$\mathbf{K}^{03} = \begin{bmatrix} \alpha & 0 \\ 0 & \beta \end{bmatrix}, \quad \mathbf{K}^{04} = \begin{bmatrix} \beta & 0 \\ 0 & \alpha \end{bmatrix} \tag{5.4.30}$$

$$\mathbf{K}^{00} = -\begin{bmatrix} 2\alpha+2\beta & 0 \\ 0 & 2\alpha+2\beta \end{bmatrix}$$

Similarly,

$$\mathbf{K}^{14}=\sigma_y\cdot\mathbf{K}^{12}\cdot\tilde{\sigma}_y=\begin{bmatrix} a & -b \\ -c & a \end{bmatrix}$$

$$\mathbf{K}^{11}=-\begin{bmatrix} \alpha+\alpha'+2a & 0 \\ 0 & \beta+\beta'+2a \end{bmatrix} \tag{5.4.31}$$

Our number of independent constants has now been reduced from ten to seven. A further reduction can be accomplished with the help of rotational invariance. When atom 0 is used as center, Eq. 5.4.9 yields

$$\mathbf{K}^{00}\times 0+\mathbf{K}^{01}\times\mathbf{X}^1+\mathbf{K}^{02}\times\mathbf{X}^2+\mathbf{K}^{03}\times\mathbf{X}^3+\mathbf{K}^{04}\times\mathbf{X}^4=0$$

or

$$(\mathbf{K}^{01}-\mathbf{K}^{03})\times\mathbf{X}^1+(\mathbf{K}^{02}-\mathbf{K}^{04})\times\mathbf{X}^2=0 \tag{5.4.32}$$

which is satisfied automatically. Let us conjecture that this automatic cancellation is not an accident of this particular example but will often occur in a crystal, shell by shell because the atoms in a shell and the force constants are related by appropriate group elements.

The conditions near the boundary of a finite crystal are different, however, and are typified by the results obtained using atom 1 as origin (but for convenience $\mathbf{X}_0=0$). Rotational invariance then yields

$$\sum\mathbf{K}^{1j}\times\mathbf{X}^j=0 \tag{5.4.33}$$

or

$$(\mathbf{K}^{11}-\mathbf{K}^{13})\times\mathbf{X}^1+(\mathbf{K}^{12}-\mathbf{K}^{14})\times\mathbf{X}^2=0$$

$$-\begin{bmatrix} \alpha+2\alpha'+2a & 0 \\ 0 & \beta+2\beta'+2a \end{bmatrix}\times\begin{bmatrix} 1 \\ 0 \end{bmatrix}+\begin{bmatrix} 0 & 2b \\ 2c & 0 \end{bmatrix}\times\begin{bmatrix} 0 \\ 1 \end{bmatrix}=0$$

The cross product is to be taken as if each row is a vector:

$$\begin{bmatrix} 0 \\ \beta+2\beta'+2a \end{bmatrix}+\begin{bmatrix} 0 \\ 2c \end{bmatrix}=0$$

or

$$\beta+2(\beta'+a+c)=0 \tag{5.4.34}$$

which provides one constraint among our seven constants $\alpha,\beta,\alpha',\beta',a,b,c$, leaving six independent parameters. As a check we note that

$$\begin{aligned}\Gamma^{\text{tot}} &= A_1+A_2+B_1+B_2+3E\\ \Gamma^{\text{transl}} &= E\,[\text{transforms as } (x,y)]\\ \Gamma^{\text{rot}} &= A_2(\text{transforms as } R_z)\end{aligned} \tag{5.4.35}$$

Since the representations are all real, the number of constants can be checked using Eq. 5.3.21, with all $b_i=1$; the results are 10 for Γ^{tot}, 7 for $\Gamma^{\text{tot}}-\Gamma^{\text{transl}}$, and 6 for $\Gamma^{\text{tot}}-\Gamma^{\text{transl}}-\Gamma^{\text{rot}}$. On the other hand, the use of Eqs. 5.1.11 and 5.1.13 leads to only five constants. What is wrong with the use of these equations here?

5.5. THE NUMBER OF FORCE CONSTANTS

In this section we provide an alternative derivation for the number of force constants in a given system that is more readily generalized to crystals and to finding the number of force constants in a particular bond.

We rewrite Eq. 5.4.17 in the form

$$SK^{mn}_{\mu\nu}=\sum K^{S(m),S(n)}_{\mu'\nu'}S_{\mu'\mu}S_{\nu'\nu} \tag{5.5.1}$$

Interchanging m and n and also μ and ν leads to

$$SFK^{mn}_{\mu\nu}=SK^{nm}_{\nu\mu}=\sum K^{S(n),S(m)}_{\mu'\nu'}S_{\mu'\nu}S_{\nu'\mu} \tag{5.5.2}$$

but in view of the symmetry of **K**, Eq. 5.4.2, the left-hand sides are identical. As in Eqs. 3.3.26–3.3.29, the symmetric and antisymmetric parts of **K** generate separate representations:

$$S(K^{mn}_{\mu\nu})^{\pm}=\sum (K^{m'n'}_{\mu'\nu'})^{\pm}\Delta^{\pm}_{m'n'\mu'\nu';mn\mu\nu}(S) \tag{5.5.3}$$

$$\begin{aligned}\Delta^{\pm}_{m'n'\mu'\nu';mn\mu\nu}(S) &= \tfrac{1}{2}\delta[m',S(m)]\,\delta[n',S(n)]S_{\mu'\mu}S_{\nu'\nu}\\ &\quad\pm\tfrac{1}{2}\delta[m',S(n)]\,\delta[n',S(m)]S_{\mu'\nu}S_{\nu'\mu}\end{aligned} \tag{5.5.4}$$

and we shall henceforth be concerned only with the symmetric representation (+ sign), since $K^{mn}_{\mu\nu}$ is symmetric.

The trace of this representation provides us with a character for the system, namely,

$$\chi_{\text{sys}}(S)=\frac{1}{2}\sum_{m,n}\delta[m,S(m)]\delta[n,S(n)][\chi(S)]^2$$
$$+\frac{1}{2}\sum_{m,n}\delta[m,S(n)]\delta[n,S(m)]\chi(S^2) \tag{5.5.5}$$

Let us first verify that this expression is equivalent to our previous procedure by noting that

$$N(S)=\sum_n \delta[n,S(n)] \tag{5.5.6}$$

is the number of atoms in the system left invariant by S, so that

$$\sum_{m,n}\delta[m,S(m)]\delta[n,S(n)]=[N(S)]^2$$
$$\sum_{m,n}\delta[m,S(n)]\delta[n,S(m)]=\sum_n \delta\{n,S[S(n)]\}=N(S^2) \tag{5.5.7}$$

or

$$\chi_{\text{sys}}(S)=\tfrac{1}{2}[N(S)\chi(S)]^2+\tfrac{1}{2}N(S^2)\chi(S^2) \tag{5.5.8}$$

is the symmetric product of $N(S)\chi(S)$ with itself, as expected.

We can now split Eq. 5.5.5 into terms $m<n$ that correspond to bonds, and terms $m=n$ that correspond to atoms:

$$\chi_{\text{sys}}(S)=\sum_{m<n}\chi_{mn}(S)+\sum_n \chi_n(S) \tag{5.5.9}$$

where

$$\chi_{mn}(S)=[\chi(S)]^2 J_{mn}(S)+\chi(S^2)J'_{mn}(S)$$

$$\chi_n(S)=\tfrac{1}{2}\{[\chi(S)]^2+\chi(S^2)\}J_{nn}(S)$$

$$J_{mn}(S)=\delta[m,S(m)]\delta[n,S(n)]$$

$$J'_{mn}(S)=\delta[m,S(n)]\delta[n,S(m)]$$

that is, $J_{mn}(S)=1$ if S belongs to the group of the bond (mn), and $J'_{mn}(S)=1$ if S is a reversal element for bond (mn).

Group of the Atom

Equation 5.5.9 suggests that the total number of force constants in a molecule is readily decomposable into the number of constants in $\mathbf{K}^{mn}$ associated with each bond (mn) and the number of constants in $\mathbf{K}^{nn}$ associated with each "atom" n. We can concentrate our attention on a single atom p by using only the elements of the group G_p of the atom, that is, the elements in G that restore p to itself. Since $\mathbf{K}^{pp}$ is a symmetric tensor in its suppressed indices, the number of independent parameters in $\mathbf{K}^{pp}$ is, according to Eq. 3.3.30 or 4.3.8,

$$N_p = \frac{1}{g_p} \sum_{S \in G_p} \tfrac{1}{2}\{[\chi(S)]^2 + \chi(S^2)\} \tag{5.5.10}$$

$$= \frac{1}{g_p} \sum_{S \in G} \tfrac{1}{2}\{[\chi(S)]^2 + \chi(S^2)\} J_{pp}(S) \tag{5.5.11}$$

since $J_{pp}(S) = 1$ for each of the g_p elements of G_p and $= 0$ otherwise. To establish the consistency of Eq. 5.5.11 with 5.5.9, we note that p can be regarded as a prototype atom for a set of atoms $[p]$ that are all equivalent to p (i.e., obtainable from p by an operation in G). If X_j is a group element that carries p into atom j, then so does each element in the set of elements $X_j G_p$. Thus the group G is decomposed into

$$G = \sum_j X_j G_p \tag{5.5.12}$$

that is, into g/g_p sets of elements (called *cosets*; see Section 8.5), one for each of the g/g_p atoms in $[p]$. Since N_p takes the same value for any p in $[p]$, we can rewrite Eq. 5.5.11 as

$$N_p = \frac{1}{g} \frac{g}{g_p} \sum_{S \in G} \tfrac{1}{2}\{[\chi(S)]^2 + \chi(S^2)\} J_{pp}(S)$$

$$= \frac{1}{g} \sum_{p \in [p]} \sum_{S \in G} \tfrac{1}{2}\{[\chi(S)]^2 + \chi(S^2)\} J_{pp}(S) \tag{5.5.13}$$

If we sum this last result over all prototypes p and compare with Eq. 5.5.9 we get

$$N_{\text{atoms}} = \frac{1}{g} \sum_n \sum_S \chi_n(S) \tag{5.5.14}$$

This is the same as the contribution of the second term of 5.5.9 to the total number of force constants.

Group of the Bond

Similarly, let G_{pq} be the group of all elements that leave the bond (pq) invariant or reverse it. Then the number of force constants associated with this bond is

$$N_{pq}=\frac{1}{g_{pq}}\sum_{S\in G_{pq}}\left\{[\chi(S)]^2 J_{pq}(S)+\chi(S^2)J'_{pq}(S)\right\} \tag{5.5.15}$$

The restriction to $S\in G_{pq}$ is unnecessary, since $J_{pq}(S)$ selects the ordinary elements and $J'_{pq}(S)$ selects the reversal elements (if any) of G_{pq}.

If $[pq]$ denotes the set of bonds equivalent to (pq), then g/g_{pq} must be the number of bonds in this set. Writing

$$\frac{1}{g_{pq}}\sum_{S}=\frac{1}{g}\frac{g}{g_{pq}}\sum_{S}=\frac{1}{g}\sum_{(pq)\in[pq]}\sum_{S}$$

and summing over all prototype bonds (pq), we find that the total number of force constants associated with all bonds is

$$N_{\text{bonds}}=\frac{1}{g}\sum_{m<n}\sum_{S}\chi_{mn}(S) \tag{5.5.16}$$

that is, the contribution of the first term of Eq. 5.5.9.

Example 5.5.1

Calculate the number of force constants associated with "atom 1," that is, bond (11), of Example 5.4.1.

Solution: Apply Eq. 5.5.10, using the group of atom 1 consisting of the identity ϵ and σ_y. Use $\chi(\epsilon)=2, \chi(\sigma_y)=0$ to obtain

$$N_1=\tfrac{1}{2}\cdot\tfrac{1}{2}\left\{[\chi(\epsilon)]^2+[\chi(\sigma_y)]^2+\chi(\epsilon^2)+\chi(\sigma_y^2)\right\}$$

$$=\tfrac{1}{4}(4+0+2+2)=2$$

in agreement with Eq. 5.4.25.

If there are no reversal elements and we use G_b to denote the group of the bond, Eq. 5.5.15 reduces to the usual formula,

$$N_{pq}=\frac{1}{g_b}\sum_{S\in G_b}[\chi(S)]^2\equiv N_b \tag{5.5.17}$$

since $g_b=g_{pq}$.

If there are reversal elements and Q is one of them, $G=G_b+QG_b, g_{pq}=2g_b$, and Eq. 5.5.15 can be rewritten as

$$N_{pq}=\tfrac{1}{2}(N_b+N_b') \tag{5.5.18}$$

$$N_b \equiv \frac{1}{g_b}\sum[\chi(S)]^2 J(S)=\frac{1}{g_b}\sum_{S\in G_b}[\chi(S)]^2$$

$$N_b' \equiv \frac{1}{g_b}\sum \chi(S^2)J(QS)=\frac{1}{g_b}\sum_{S\in QG_b}\chi(S^2) \tag{5.5.19}$$

when we write $J(S)$ briefly for $J_{pq}(S)$.

Example 5.5.2

Find the number of force constants in bond (01) of Example 5.4.1.

Solution: There are no reversal elements. The only nontrivial element that leaves (01) alone is $\boldsymbol{\sigma}_y$. We use Eq. 5.5.17 to obtain

$$N_{01}=\tfrac{1}{2}\{[\chi(\epsilon)]^2+[\chi(\sigma_y)]^2\}=\tfrac{1}{2}\{2^2+0\}=2$$

which agrees with Eq. 5.4.20.

Example 5.5.3

Find the number of force constants in bond (12) of Example 5.4.1.

Solution: G_b contains only the identity element, and $\boldsymbol{\sigma}_d$ is the only reversal element. Thus

$$N_b=[\chi(\epsilon)]^2=4, \qquad N_b'=\chi(\sigma_d^2)=2$$

and

$$N_{12}=\tfrac{1}{2}(4+2)=3$$

in agreement with Eq. 5.4.22.

5.6. SUMMARY

A dynamical problem is reduced to its simplest form by introducing the symmetry vectors as a basis. These are constructed by projecting into the subspace appropriate to a given ϕ_μ^i, that is, a given partner μ of a given irreducible representation i. This projection may be accomplished by using the transfer operator $P_{\mu\nu}^i$ of Eq. 3.7.4, but it is simpler to proceed by first diagonalizing the elements of an Abelian subgroup (preferably the largest), and then diagonalizing the Dirac characters.

We show how to separate the external motion of a molecule from its internal motion. The total number of independent parameters required to describe this internal motion, as well as the number of such parameters in each bond, is obtained using character formulas. The form and the number of constants for a given bond are shown to be determined by the *reversal group* of the bond, namely, the elements that either leave the bond invariant or reverse it.

problems

5.3.1. Show that the orthogonality requirement, Eq. 5.3.4, is automatically obeyed as far as (ij) or $(\mu\nu)$ is concerned because of the base vector orthogonality theorem 3.4.3 with H as the mass operator M_r.

5.3.2. Show that, if the symmetry vectors of Eq. 5.3.4 are normalized in accord with

$$\sum M_r (b_r^{ia})^* b_r^{ia} = M^{ia}$$

the kinetic energy takes the form

$$T = \frac{1}{2} \sum M^{ia} |\dot{q}^{ia}|^2$$

5.3.3. Determine the *symmetry* coordinates for the planar vibrations of the planar molecule AB_4 of Example 5.4.1 and Fig. 5.4.1.

5.3.4. Determine the symmetry coordinates for the purely transverse vibrations of the planar molecule AB_4 of Fig. 5.4.1. These are motions perpendicular to the equilibrium plane of the molecule. (First ignore translational and rotational invariance.)

5.3.5. Use *internal* coordinates directly in describing the NH_3 molecule. Construct the character of the space spanned by these internal coordinates, and find what symmetry coordinates are contained. Construct the symmetry coordinates themselves.

5.3.6. Show that the transformation (Eq. 5.3.16) from symmetry coordinates to normal coordinates can be chosen to be a unitary transformation. Under what circumstances would a nonunitary transformation arise?

5.4.1. Show that the use of Eq. 5.1.13 with 5.1.11 yields five independent internal constants for the AB_4 molecule of Example 5.4.1, whereas six constants are found in that example. What is wrong?

5.4.2. Find the *normal* coordinates for the AB_4 molecule of Example 5.4.1 and Fig. 5.4.1.

5.4.3. Determine the form of the potential energy matrix $\mathbf{K}^{mn}$ for the NH_3 problem.

5.4.4. Find the reciprocal of the matrix

$$\mathbf{M}=\begin{bmatrix} A & C & B & C & D \\ C & A & B & D & C \\ B & B & A & B & B \\ C & D & B & A & C \\ D & C & B & C & A \end{bmatrix}$$

(Levitas and Lax, **1958**). *Hint*: This matrix has the symmetry of Fig. 5.4.1 and could describe vibrations of such a molecule transverse to the plane (see Problem 5.3.4). Show that, if S_{jp} and m_p, $p=1,2,\ldots,5$, are the eigenvectors and eigenvalues of $\mathbf{M}$ in the sense of Problem 1.4.1, then

$$(\mathbf{M}^{-1})_{ij}=\sum_p S_{ip}(m_p)^{-1}S_{jp}^*$$

5.4.5. Discuss the vibrations of methane, CH_4 (or CCl_4), which has the complete symmetry T_d of the tetrahedron.

5.4.6. Discuss the vibrations of uranium hexafluoride, UF_6, which possesses a structure of cubic (O_h) symmetry.

5.4.7. Consider the set of spin-$\frac{1}{2}$ particles situated at the points 1,2,3,4 of a square with spin Hamiltonian

$$H=J(\boldsymbol{\sigma}_1\cdot\boldsymbol{\sigma}_2+\boldsymbol{\sigma}_2\cdot\boldsymbol{\sigma}_3+\boldsymbol{\sigma}_3\cdot\boldsymbol{\sigma}_4+\boldsymbol{\sigma}_4\cdot\boldsymbol{\sigma}_1)+J'(\boldsymbol{\sigma}_1\cdot\boldsymbol{\sigma}_3+\boldsymbol{\sigma}_2\cdot\boldsymbol{\sigma}_4)$$

(*a*) What is the space of this problem? What is its dimension?
(*b*) What are the possible symmetry operations that commute with H?
(*c*) What is the group of these symmetry operations? Is it a single or double group?
(*d*) How is the problem modified if there is a central spin $\boldsymbol{\sigma}_0$ and an additional term $J''\boldsymbol{\sigma}_0\cdot(\boldsymbol{\sigma}_1+\boldsymbol{\sigma}_2+\boldsymbol{\sigma}_3+\boldsymbol{\sigma}_4)$ in the Hamiltonian?
(*e*) What representations are contained in the problem space?
(*f*) Find the symmetry eigenvectors, the energies, and the eigenstates.

5.5.1. Sometimes a particular representation does not appear among the internal modes of vibration of a molecule or solid, for example, Γ_2 in the ammonia molecule or Γ_{25}^- in the diamond crystal. Show that this is a property of the molecule or crystal and not of the group. *Hint*: Construct a molecule (or crystal) sufficiently complicated that its vibrations contain all representations.

5.5.2. Let $\mathbf{M} = \Sigma \mathbf{B}^n \cdot \mathbf{u}^n$ be the electric moment induced in a molecule by a displacement $\mathbf{u}^n$ of its atoms.

(*a*) Show that

$$\mathbf{B}^{S(n)} = \mathbf{S} \cdot \mathbf{B}^n \cdot \mathbf{S}^{-1}$$

(cf. Eq. 9.1.7).

(*b*) Show how to determine the number of independent parameters in $\mathbf{B}^n$ by using the "group of the atom n." Find the number of parameters and the form of the effective charge matrices $\mathbf{B}^n$ for the ammonia molecule.

5.5.3. Transform the electric moment expression of Problem 5.5.2 into a linear form in the normal coordinates. Examine the resulting effective charge coefficients and determine which symmetry coordinates are "infrared active," that is, have nonvanishing charges.

5.5.4. Consider the nonlinear interaction energy

$$V = \sum_{\substack{mnp \\ \mu\nu\lambda}} K_{\mu\nu\lambda}^{mnp} u_\mu^m u_\nu^n u_\lambda^p$$

Show that

$$RK_{\mu\nu\lambda}^{mnp} = \sum K_{\mu'\nu'\lambda'}^{R(m)R(n)R(p)} R_{\mu'\mu} R_{\nu'\nu} R_{\lambda'\lambda}$$

If R belongs to the group of the mnp triangle (i.e., it restores the triangle to itself modulo some permutation of the indices), show (Lax, **1965c**) that the character of the representation generated by K^{mnp} is

$$\Phi^{mnp}(R) = J_1(R)[\Phi(R)]^3 + J_2(R)\Phi(R^2)\Phi(R) + J_3(R)\Phi(R^3)$$

where $\Phi(R) = \Sigma_\mu R_{\mu\mu}$ is the character of the 3×3 (ordinary) vector representation,

$$J_1(R) = \delta[m, R(m)]\delta[n, R(n)]\delta[p, R(p)]$$

selects the ordinary elements that leave the triangle invariant,

$$\begin{aligned} J_2(R) = {} & \delta[m, R(n)]\delta[n, R(m)]\delta[p, R(p)] \\ & + \delta[m, R(p)]\delta[n, R(n)]\delta[p, R(m)] \\ & + \delta[m, R(m)]\delta[n, R(p)]\delta[p, R(n)] \end{aligned}$$

selects the elements that interchange any pair of atoms, and

$$\begin{aligned} J_3(R) = {} & \delta[m, R(n)]\delta[n, R(p)]\delta[p, R(m)] \\ & + \delta[m, R(p)]\delta[n, R(m)]\delta[p, R(n)] \end{aligned}$$

selects the elements that produce a cyclic interchange.

chapter six

TRANSLATIONAL PROPERTIES OF CRYSTALS

6.1. CRYSTAL SYSTEMS, BRAVAIS LATTICES, AND CRYSTAL CLASSES

A crystal can have an infinite number of possible structures. It may belong to any of 230 space groups, 32 crystal classes, 14 Bravais lattices, and 7 crystal systems. Each classification in this hierarchy is more specific than all those that follow it and includes them. We now define these and some related terms.

CRYSTAL: A periodic array of physical objects—atoms or molecules or charge and current distributions.

CRYSTAL STRUCTURE: A statement of the repetition pattern (Bravais lattice) plus a complete description of the contents of the unit cell (repetition unit).

SPACE GROUP: The set of all operations $(\alpha|\mathbf{a})$ defined in Eq. 4.1.1 (where α is a proper or improper rotation, and $\mathbf{a}$ is a translation) that restore the crystal to itself.

CRYSTAL CLASS: The point group of the crystal, that is, the operations $(\alpha|0)$, even if these point operations do not restore the crystal to itself. [The crystal class can be determined from the shape (morphology) of the crystal

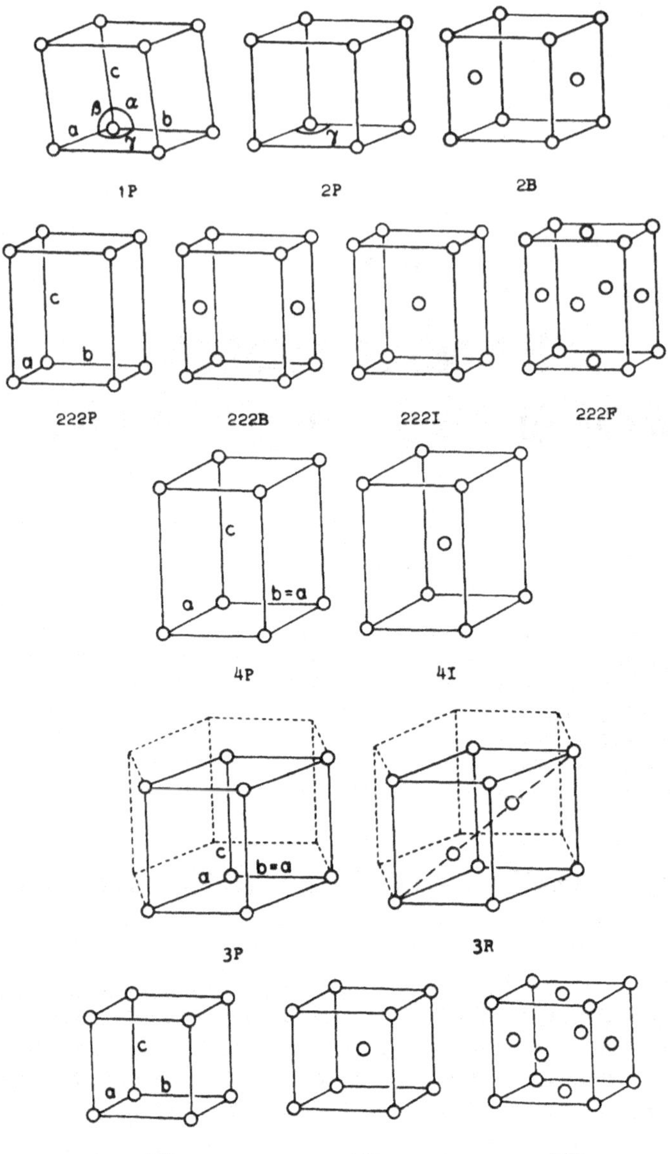

Fig. 6.1.1. The Bravais lattices: $1P$ = triclinic, $2P$ = monoclinic primitive, $2B$ = B-centered monoclinic, $222P$ = primitive orthorhombic, $222B$ = B-centered orthorhombic, $222I$ = body-centered orthorhombic, $222F$ = face-centered orthorhombic, $4P$ = primitive tetragonal, $4I$ = body-centered tetragonal, $3P$ = primitive hexagonal, $3R$ = rhombohedral, $23P$ = simple cubic, $23I$ = body-centered cubic, $23F$ = face-centered cubic. The numbers are International notation for the smallest proper group that enforces the given Bravais lattice. In all cases, the nonprimitive cell is shown. Thus $3R$ = rhombohedral is displayed as a centered hexagonal lattice. See Figs. 6.1.2–6.1.5.

and the symmetry of its matter tensors (conductivity, piezoelectricity, etc.).]

BRAVAIS LATTICE: An array of points that (in three-dimensional space) are expressable as an integral linear combination of three ("primitive") basis vectors:

$$\mathbf{t} = t_1\mathbf{a}_1 + t_2\mathbf{a}_2 + t_3\mathbf{a}_3, \qquad t_i \text{ integers} \tag{6.1.1}$$

See Fig. 6.1.1 and Problem 6.1.1. A crystal is said to belong to a given Bravais lattice if the crystal is restored to itself by the set (6.1.1) of pure translations $(\epsilon|\mathbf{t})$, where ϵ = the identity. The set $\{\epsilon|\mathbf{t}\}$ constitutes the

TRANSLATION GROUP of the crystal.

HOLOHEDRY: The point group that describes the rotational symmetry of the Bravais lattice. Bravais lattices automatically have inversion symmetry as an element of the holohedry. Moreover, if a lattice possesses an n-fold

Table 6.1.1. Crystal Systems and Bravais Lattices

Crystal System	Description of Nonprimitive Cell[a]	Crystal Classes	Bravais Lattice[b]
Triclinic	$a \neq b \neq c$	C_i, C_1	$1P$
	$\alpha \neq \beta \neq \gamma$		
Monoclinic	$a \neq b \neq c$	C_{2h}, C_s, C_2	$2P$
	$\alpha = \beta = 90°$		$2B$
Orthorhombic	$a \neq b \neq c$	D_{2h}, C_{2v}, D_2	$222P$
	$\alpha = \beta = \gamma = 90°$		$222C$
			$222F$
			$222I$
Tetragonal	$a = b \neq c$	$[D_{4h}, C_{4h}, D_{2d},$	$4P$
	$\alpha = \beta = \gamma = 90°$	$C_{4v}, D_4, S_4, C_4]$	$4I$
Trigonal	Same as hexagonal	$[D_{3d}, D_3, C_{3v},$	$3R$
(rhombohedral)		$S_6, C_3]$	$3P$
Hexagonal	$a = b \neq c$	$[D_{6h}, C_{6h}, D_{3h},$	$3P$
	$\alpha = \beta = 90°, \gamma = 120°$	$C_{6v}, D_6, C_{3h}, C_6]$	
Cubic	$a = b = c$	$[O_h, T_h, T_d,$	$23P$
	$\alpha = \beta = \gamma = 90°$	$O, T]$	$23I$
			$23F$

[a]The not equal symbol ($\neq$) implies that equality is not imposed by symmetry, although it could occur accidentally.

[b]Since the point group $D_2 = 222$ (is the smallest proper point group that) imposes an orthorhombic lattice, read $222C$ as an orthorhombic C-centered lattice. (See Table 6.1.2 for the nature of the centerings.) Lattices $222A$ and $222B$ are not listed separately since they are reorientations of $222C$.

Table 6.1.2. Space Lattice Designations

Symbol	Name	Total Number of Lattice Points per Cell	Location of Extra Points[a]
P	Primitive	1	—
I	Body-centered	2	$\frac{1}{2}(\mathbf{a}+\mathbf{b}+\mathbf{c})$
A	Center of A-face	2	$\frac{1}{2}(\mathbf{b}+\mathbf{c})$
B	B-centered	2	$\frac{1}{2}(\mathbf{c}+\mathbf{a})$
C	C-centered	2	$\frac{1}{2}(\mathbf{a}+\mathbf{b})$
F	Face-centered	4	$\frac{1}{2}(\mathbf{a}+\mathbf{b})$
			$\frac{1}{2}(\mathbf{b}+\mathbf{c})$
			$\frac{1}{2}(\mathbf{c}+\mathbf{a})$
R	Rhombohedral centered (obverse)	3	$\frac{2}{3}\mathbf{a}+\frac{1}{3}\mathbf{b}+\frac{1}{3}\mathbf{c}$
			$\frac{1}{3}\mathbf{a}+\frac{2}{3}\mathbf{b}+\frac{2}{3}\mathbf{c}$

[a] Relative to any lattice point.

rotation C_n, for $n>2$, it also possesses a mirror plane containing the axis (or equivalently by inversion a twofold rotation perpendicular to the axis). (For proof see Lyubarskii, **1960**.) Thus there are only seven holohedral groups: $C_i, C_{2h}, D_{2h}, D_{4h}, D_{3d}, D_{6h}, O_h$. The classification of lattices and crystal classes into seven systems in Table 6.1.1 is based on these holohedral groups.

CRYSTAL SYSTEMS: Bravais lattices that possess the same holohedry are said to belong to the same crystal system. Thus primitive (P), body-centered (I), and face-centered (F) cubic lattices all have O_h holohedry and hence belong to the cubic system. (For the lattice notation P,A,B,C,I,F,R see Table 6.1.2 and Fig. 6.1.1).

CENTERED LATTICES: The face-centered and body-centered cubic lattices can be described (see Figs. 6.1.2 and 6.1.3) in terms of primitive rhombohedral axes. (A primitive rhombohedral lattice has three basis vectors of equal length, making equal angles with one another.) The rhombohedral cell does not display the full O_h symmetry. We can, however, construct a larger nonprimitive cell with basis vectors **a**, **b**, **c** and more than one lattice point per cell that does display the full symmetry of the lattice holohedry. The nonprimitive cells are described in Table 6.1.1, the locations of the extra

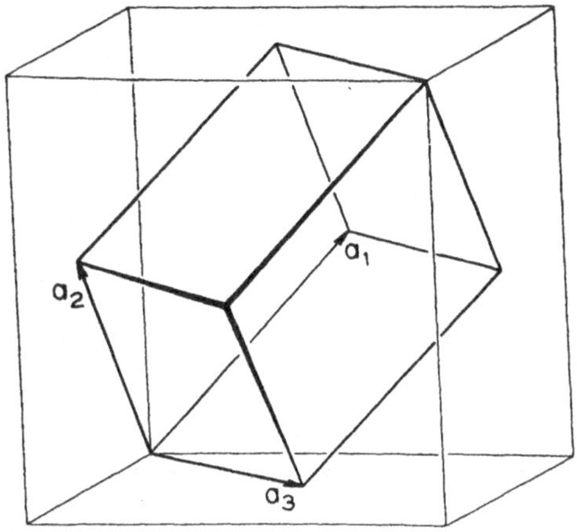

Fig. 6.1.2. The face-centered cubic lattice, $23F$, in relation to its primitive cell spanned by $\mathbf{a}_1, \mathbf{a}_2, \mathbf{a}_3$.

points (the "centering positions") are given in Table 6.1.2, and the relation between primitive and nonprimitive basis vectors is shown in Table 6.1.3. The primitive rhombohedral lattice R with trigonal symmetry is unusual in that the corresponding nonprimitive cell has sixfold hexagonal symmetry. See Figs. 6.1.4 and 6.1.5.

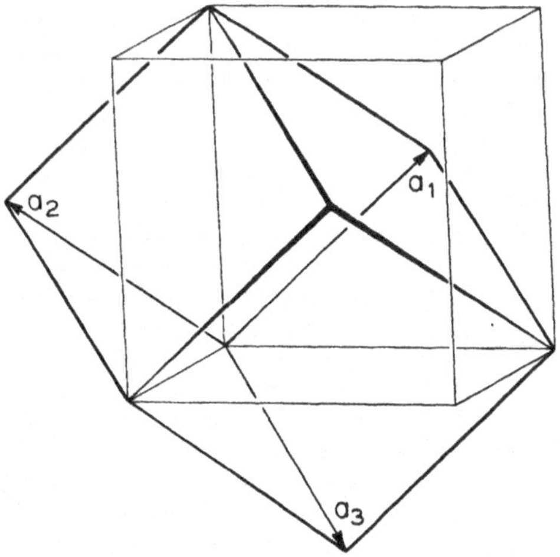

Fig. 6.1.3. The body-centered cubic lattice, $23I$, in relation to its primitive cell spanned by $\mathbf{a}_1$, $\mathbf{a}_2$, $\mathbf{a}_3$.

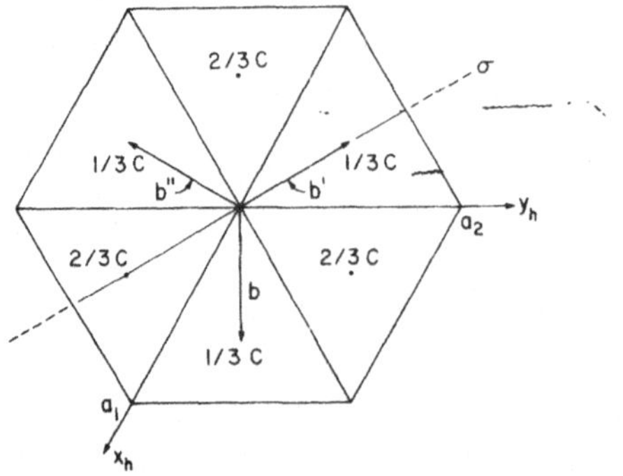

Fig. 6.1.4. Top view of a rhombohedral lattice. The primitive rhombohedral vectors, $\mathbf{b}+\frac{1}{3}\mathbf{c}$, $\mathbf{b}'+\frac{1}{3}\mathbf{c}$, $\mathbf{b}''+\frac{1}{3}\mathbf{c}$, are of equal length and make equal angles with each other. Vector **c** is out of the paper.

In Fig. 6.1.6 we see that every point group is a subgroup of some holohedral group. Each point group is associated with the holohedral group of lowest order of which it is a subgroup and is assigned to the corresponding crystal system in Table 6.1.1. The groups within one crystal system have the same *essential* symmetry elements (see Table 4.3.2). Thus they enforce Bravais lattices within the same crystal system; for example, O_h, T_h, T_d, O, and T are all "cubic" groups. The *first* crystal class listed in Table 6.1.1 is the holohedry; the *last* is the

Table 6.1.3. Relation Between Primitive And Nonprimitive Cells

The standard choice of the International Tables for X-ray Crystallography (**1952**):

I:	$\mathbf{a}_1=-\frac{1}{2}\mathbf{a}+\frac{1}{2}\mathbf{b}+\frac{1}{2}\mathbf{c}$	F:	$\mathbf{a}_1=\frac{1}{2}(\mathbf{b}+\mathbf{c})$	$R\equiv R_{\text{obverse}}$:	$\mathbf{a}_1=\frac{2}{3}\mathbf{a}+\frac{1}{3}\mathbf{b}+\frac{1}{3}\mathbf{c}$
	$\mathbf{a}_2=\frac{1}{2}\mathbf{a}-\frac{1}{2}\mathbf{b}+\frac{1}{2}\mathbf{c}$		$\mathbf{a}_2=\frac{1}{2}(\mathbf{a}+\mathbf{c})$		$\mathbf{a}_2=-\frac{1}{3}\mathbf{a}+\frac{1}{3}\mathbf{b}+\frac{1}{3}\mathbf{c}$
	$\mathbf{a}_3=\frac{1}{2}\mathbf{a}+\frac{1}{2}\mathbf{b}-\frac{1}{2}\mathbf{c}$		$\mathbf{a}_3=\frac{1}{2}(\mathbf{a}+\mathbf{b})$		$\mathbf{a}_3=-\frac{1}{3}\mathbf{a}-\frac{2}{3}\mathbf{b}+\frac{1}{3}\mathbf{c}$

A: $\mathbf{a}_1=\mathbf{a}$,	$\mathbf{a}_2=\mathbf{b}$,	$\mathbf{a}_3=\frac{1}{2}(\mathbf{b}+\mathbf{c})$
B: $\mathbf{a}_1=\mathbf{a}$,	$\mathbf{a}_2=\mathbf{b}$,	$\mathbf{a}_3=\frac{1}{2}(\mathbf{c}+\mathbf{a})$
C: $\mathbf{a}_1=\mathbf{a}$,	$\mathbf{a}_2=\frac{1}{2}(\mathbf{a}+\mathbf{b})$,	$\mathbf{a}_3=\mathbf{c}$

A choice with $\mathbf{a}_1,\mathbf{a}_2$ in a plane perpendicular to the principal axis (useful in constructing three-dimensional lattices from plane lattices):

I:	$\mathbf{a}_1=\mathbf{a}$,	$\mathbf{a}_2=\mathbf{b}$,		$\mathbf{a}_3=\frac{1}{2}(\mathbf{a}+\mathbf{b}+\mathbf{c})$
F:	$\mathbf{a}_1=\frac{1}{2}(\mathbf{a}+\mathbf{b})$,	$\mathbf{a}_2=\frac{1}{2}(\mathbf{b}-\mathbf{a})$,		$\mathbf{a}_3=\frac{1}{2}(\mathbf{b}+\mathbf{c})$
R:	$\mathbf{a}_1=\mathbf{a}$,	$\mathbf{a}_2=\mathbf{b}$	R_{obverse}:	$\mathbf{a}_3=\frac{2}{3}\mathbf{a}+\frac{1}{3}\mathbf{b}+\frac{1}{3}\mathbf{c}$
			R_{reverse}:	$\mathbf{a}_3=\frac{1}{3}\mathbf{a}+\frac{2}{3}\mathbf{b}+\frac{1}{3}\mathbf{c}$

A,B,C same as above

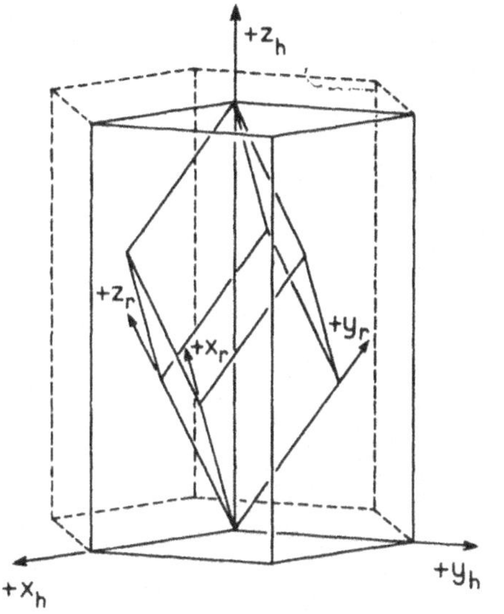

Fig. 6.1.5. The primitive rhombohedral cell is shown embedded in its (nonprimitive) hexagonal cell. See Tables 6.1.2 and 6.1.3.

smallest group (of proper elements) within the system and is used (in International notation) to help *label* the Bravais lattices.

Crystals with a given Bravais lattice occur in nature with *less than holohedral symmetry* because the atom or molecule that is repeated has a structure that is not *holosymmetric* (Fig. 6.1.7) or because there are several atoms in the cell whose distribution is not holosymmetric (Fig. 6.1.8).

The holohedral groups themselves form a hierarchy. In following the arrows in Fig. 6.1.9, one passes from a group to a subgroup by an *infinitesimal* distortion of the lattice. If a crystal in class D_2 (orthorhombic) appeared to have a hexagonal lattice, we would regard this as *accidental* since the hexagonal form would not be imposed; the crystal would tend to distort (e.g., under temperature changes) to orthorhombic form to minimize its cohesive energy. However, the arrow from D_{6h} to the subgroup D_{3d} is missing in Fig. 6.1.9, since a hexagonal lattice cannot distort into a rhombohedral lattice (see Figs. 6.1.4 and 6.1.5) by an infinitesimal distortion. Hence in Table 6.1.1 and also in nature we find that the five trigonal classes $D_{3d}, D_3, C_{3v}, \mathcal{S}_6, \mathcal{C}_3$ are compatible with the primitive hexagonal lattice (named $3P$ for this reason), as well as with the primitive rhombohedral lattice $3R$. The fifth crystal system in Table 6.1.1 should then be called trigonal or rhombohedral according to whether we are specifying the crystal class or the Bravais lattice.

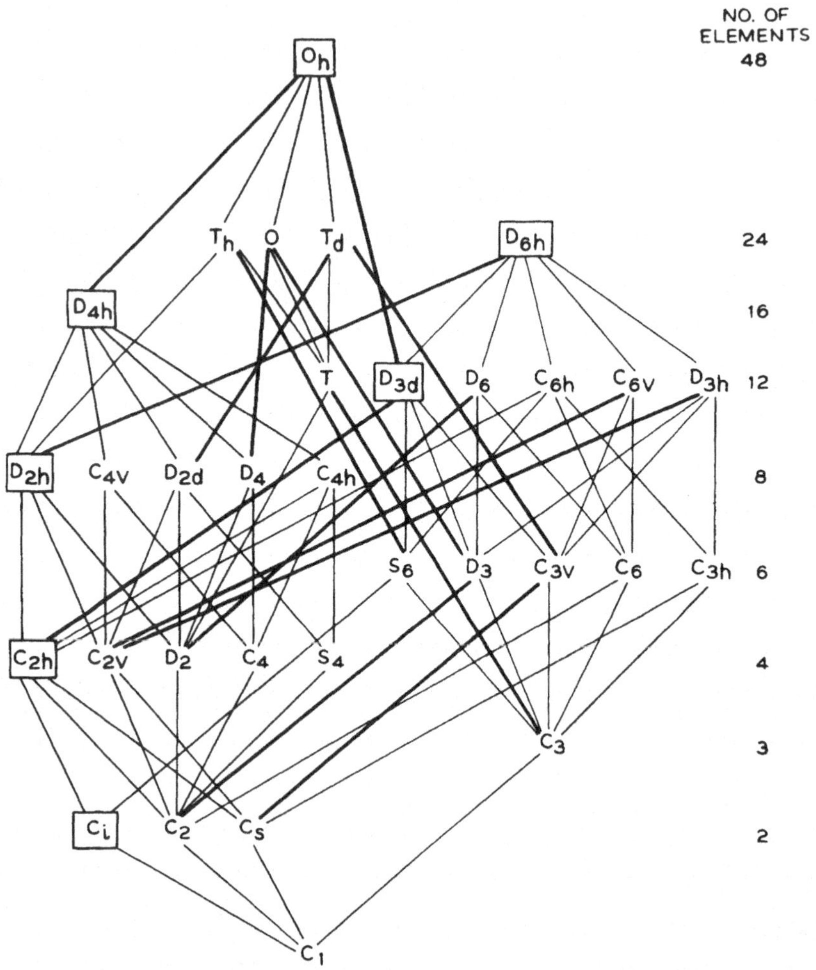

Fig. 6.1.6. Holohedral groups and their subgroups. The seven holohedral groups are enclosed in boxes. Every group is connected by a line to each of its subgroups. If the subgroup is not invariant, the line is heavy.

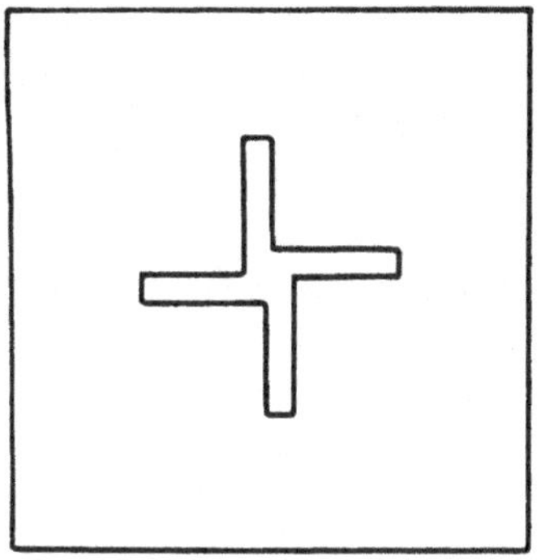

Fig. 6.1.7. A "molecule" whose symmetry, C_4, is less than the holohedral symmetry C_{4v} of the square Bravais lattice.

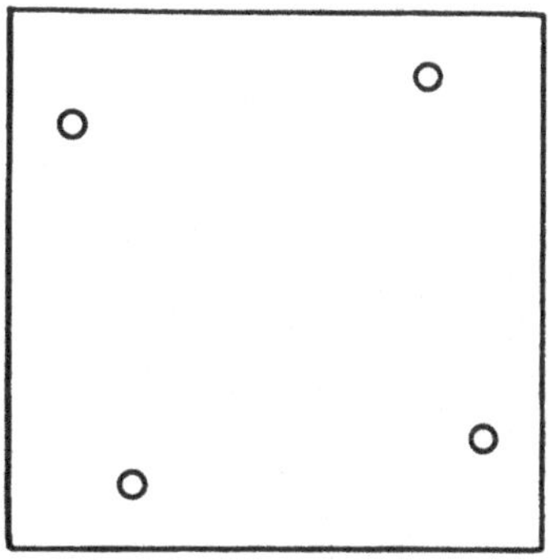

Fig. 6.1.8. A set of points whose collective symmetry, C_4, is less than the holohedral symmetry C_{4v} of the square Bravais lattice.

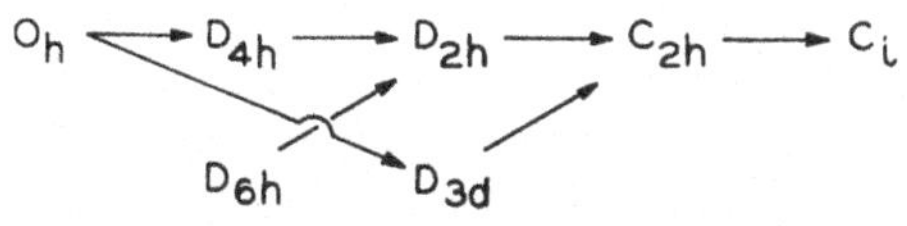

Fig. 6.1.9. The hierarchy of holohedral groups.

Wigner-Seitz or Proximity Cell

It is also possible to construct a new type of primitive cell, the Wigner-Seitz or proximity cell, with the following properties:

1. It has a lattice point at its center.

2. It contains all space points closer to the center than to any other lattice point.
3. The cells associated with different lattice points can be translated into one another and hence are identical.
4. Every point in space is nearest some lattice point; hence the cells fill space.
5. The cell has the full symmetry of the "site" operations that leave a lattice point invariant. (The crystal is restored to itself by such operations and also by property 2 above; so is the cell.)
6. The cell is primitive in that its volume is the volume per point of the lattice.
7. The cell can be constructed by drawing lines from a given lattice point to all of the nearby lattice points and forming the perpendicular bisector planes of each line. The cell is the polyhedron (not necessarily regular) consisting of all points on the near side of each plane. [In Section 6.3 we define the (first) Brillouin zone as the proximity cell of the reciprocal lattice.]

Crystal Structures

An X-ray crystallographer specifies a *crystal structure* by stating the Bravais lattice (i.e., the repetition pattern) *and* the location of the atoms within the cell (what is repeated). Thus the rock salt (NaCl) structure has a face-centered cubic Bravais lattice with one Na and one Cl in each primitive cell. (See Fig. 6.1.10 and Table 6.1.4.) If we place the Na at the cell corner (0,0,0), the Cl can be

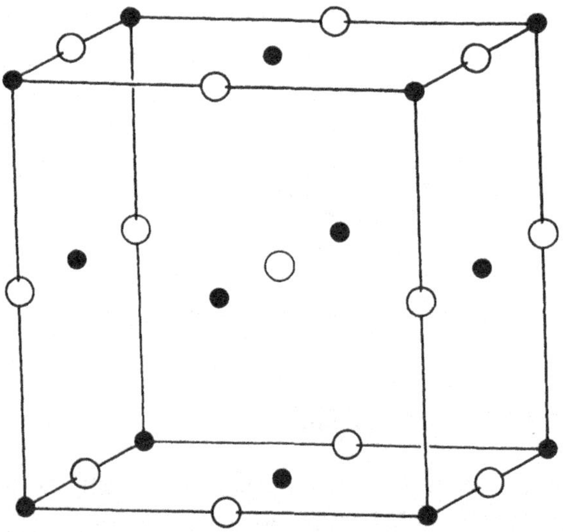

Fig. 6.1.10. The face-centered cubic rock salt (NaCl) structure.

Table 6.1.4. Common Crystal Structures R_n and RX

Structure	Bravais Lattice	Extra **R** Positions[a]	**X** Positions	Space Group	Comments
Close-packed cubic	$23F$	—	—	$O_h^5 = F\frac{4}{m}\bar{3}\frac{2}{m}$	
Body-centered cubic	$23I$	—	—	$O_h^9 = I\frac{4}{m}\bar{3}\frac{2}{m}$	
Hexagonal close-packed (h.c.p.)	$3P$	$\frac{1}{3}, \frac{2}{3}, \frac{1}{2}$	—	$D_{6h}^4 = P\frac{6_3}{m}\frac{2}{m}\frac{2}{c}$	$\frac{c}{a} \sim \left[\frac{8}{3}\right]^{1/2}$
Diamond	$23F$	$\frac{1}{4}, \frac{1}{4}, \frac{1}{4}$	—	$O_h^7 = F\frac{4_1}{d}\bar{3}\frac{2}{m}$	
Graphite	$3P$	$0, 0, \frac{1}{2}$ $\frac{1}{3}, \frac{2}{3}, v$ $\frac{2}{3}, \frac{1}{3}, v + \frac{1}{2}$		$C_{6v}^4 = P6_3mc$	$v < 0.05$
Rock salt (NaCl)	$23F$		$\frac{1}{2}, \frac{1}{2}, \frac{1}{2}$	$O_h^5 = F\frac{4}{m}\bar{3}\frac{2}{m}$	
Zinc blende (ZnS, InSb)	$23F$		$\frac{1}{4}, \frac{1}{4}, \frac{1}{4}$	$T_d^2 = F\bar{4}3m$	
Cesium chloride	$23I$		$\frac{1}{2}, \frac{1}{2}, \frac{1}{2}$	$O_h^1 = P\frac{4}{m}\bar{3}\frac{2}{m}$	
Wurtzite (ZnS, ZnO)	3P	$\frac{1}{3}, \frac{2}{3}, \frac{1}{2}$	$0, 0, u$ $\frac{1}{3}, \frac{2}{3}, \frac{1}{2} + u$	$C_{6v}^4 = P6_3mc$	$\frac{c}{a} \sim \left[\frac{8}{3}\right]^{1/2}$. This is two h.c.p. lattices from (000) and $(0,0,u)$ $u \approx 3/8$.

[a] One R atom is assumed at (0,0,0), and x, y, z means $x\mathbf{a} + y\mathbf{b} + z\mathbf{c}$.

placed at the body center of either the primitive or the cubic cells, $\frac{1}{2}(\mathbf{a}_1+\mathbf{a}_2+\mathbf{a}_3) \equiv \frac{1}{2}(\mathbf{a}+\mathbf{b}+\mathbf{c})$. Of course, there are three additional Cl positions, $(\frac{1}{2},0,0),(0,\frac{1}{2},0),(0,0,\frac{1}{2})$, in the cubic cell, but these are obtainable from $(\frac{1}{2},\frac{1}{2},\frac{1}{2})$ by the usual face-centering translations and therefore we need not mention them.

The zinc blende structure (ZnS) is also face-centered cubic with two atoms per primitive cell. The second atom is located at $\boldsymbol{\tau}=\frac{1}{4}(\mathbf{a}+\mathbf{b}+\mathbf{c})$ relative to the first, that is, one-quarter of the way up the body diagonal. (See Fig. 6.1.11 and Table 6.1.4.)

If the two types of atoms in the zinc blende structure are made identical, the structure reduces to the diamond type. The two atoms per cell in the diamond structure are *equivalent*, since a center of inversion placed halfway between them interchanges not only these two atoms but also the entire face-centered cubic sublattice generated from each atom. However, these two atoms are not translationally equivalent: there is no way of choosing a new set of primitive base vectors that can reduce the diamond structure to a primitive Bravais lattice with one atom per cell.

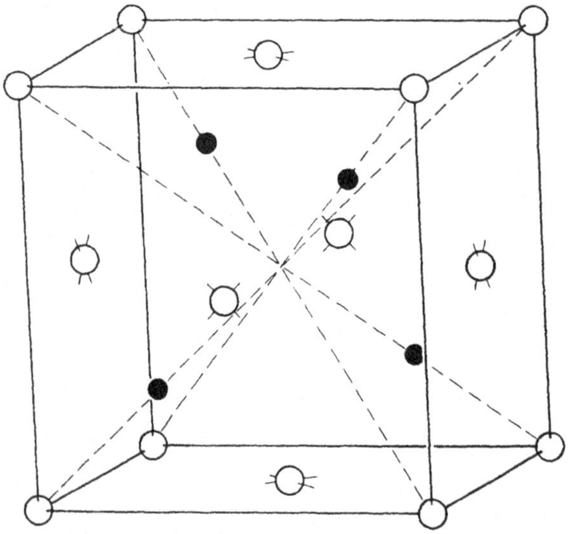

Fig. 6.1.11. The face-centered cubic zinc blende (ZnS) structure. If Zn is at $(0,0,0)$, S is at $(\frac{1}{4},\frac{1}{4},\frac{1}{4})$. If the Zn and S atoms are made identical, the diamond structure is obtained.

6.2. REPRESENTATIONS OF THE TRANSLATION GROUP

Let $(\epsilon|\mathbf{t})$ denote a pure translation $\mathbf{t}$ of a body. Then

$$(\epsilon|\mathbf{t})\psi(\mathbf{r})=\psi(\mathbf{r}-\mathbf{t})=\psi\left[(\epsilon|\mathbf{t})^{-1}\mathbf{r}\right] \tag{6.2.1}$$

describes the effect on the wave function (cf. Eq. 2.5.17). If

$$\mathbf{t}=t_1\mathbf{a}_1+t_2\mathbf{a}_2+t_3\mathbf{a}_3 \tag{6.2.2}$$

where t_1, t_2, t_3 are integers, the $(\epsilon|\mathbf{t})$ constitute the (discrete) translation group of a space lattice. This group is Abelian. Indeed, it is the direct product of three cyclic groups with elements $(\epsilon|t_1\mathbf{a}_1)$, $(\epsilon|t_2\mathbf{a}_2)$, and $(\epsilon|t_3\mathbf{a}_3)$, respectively. In this sense all Bravais lattices are isomorphic, and representations of three-dimensional lattices can be compounded out of representations of one-dimensional lattices.

Theorem 6.2.1

The irreducible representations of finite Abelian groups are one dimensional.

Proof: Each element of an Abelian group is in a class by itself. By Theorem 1.5.6 the number of representations is then g (the number of elements), and this is consistent with $\Sigma l_i^2=g$ (Theorem 1.5.5) only if each $l_i=1$.

Alternatively, we could simply use the quantum-mechanical theorem that a set of diagonalizable operators can be simultaneously diagonalized if (and only if) the operators commute. And if all group elements are represented by diagonal matrices, the representations are clearly one-dimensional. *Note*: The eigenvalues of all the operators in an Abelian group are determined by those of the group generators.

One-Dimensional Case

For the translation group, these generators are $(\epsilon|\mathbf{a}_1)$, $(\epsilon|\mathbf{a}_2)$, and $(\epsilon|\mathbf{a}_3)$. The diagonalization of one of these is essentially a one-dimensional problem. Let us therefore require

$$(\epsilon|a)\psi(x)\equiv\psi(x-a)=\lambda\psi(x) \tag{6.2.3}$$

We set

$$\lambda=\exp(-ika) \tag{6.2.4}$$

where k can as yet be complex, and

$$\psi(x)=\exp(ikx)u(x) \tag{6.2.5}$$

Then Eq. 6.2.3 implies that

$$u(x-a)=u(x) \tag{6.2.6}$$

that is, $u(x)$ is periodic. It is customary to restrict eigenfunctions on an *infinite* lattice to be bounded, which requires that k be real.

Bloch's Theorem 6.2.2

The eigenfunctions of a one-dimensional periodic Hamiltonian can be required to take the form $\psi(x)=\exp(ikx)u(x)$, a plane wave times a periodic function.

Proof: $(\epsilon|a)$ commutes with the Hamiltonian. Diagonalize it first.

Note: Complex values of k can arise in considering a semi-infinite lattice ("surface states" that decay as one leaves the crystal surface). They arise in treatments of tunneling from one band to another (Kane, **1961**) and in dispersion-theoretic arguments that necessarily enter the complex k plane (Kaus and Watson, **1960**; Blount, **1962a**).

Corollary 6.2.3

The values of k in Bloch's theorem can be restricted to the domain

$$-\frac{\pi}{a}<k\leqslant\frac{\pi}{a} \tag{6.2.7}$$

referred to as the "*first Brillouin zone.*"

Proof: A factor $\exp(i2\pi x/a)$ can always be absorbed into the periodic factor $u(x)$.

Alternative Proof: Since $(\epsilon|t_1a)\exp(ikx)u(x)=\exp[ik(x-t_1a)]u(x-t_1a)$ and $u(x-t_1a)=u(x)$ for t_1 an integer, the value and character of $(\epsilon|t)$ for lattice displacement $t=t_1a$ in representation k are

$$\chi^k(\epsilon|t)=\exp(-ikt) \tag{6.2.8}$$

Thus the irreducible representations of our *infinite discrete* group are labeled by a continuous parameter k. If we set

$$K=2\pi\left(\frac{h}{a}\right), \qquad h=\text{integer} \tag{6.2.9}$$

then $\exp(-iKt)=1$ and

$$\chi^{k+K}(\epsilon|t)=\chi^{k}(\epsilon|t)=\exp(-ikt) \tag{6.2.10}$$

Thus representations outside the first Brillouin zone are identical to those inside.

If we wish to limit ourselves to the domain of finite groups, we can add the Born-von Karman "cyclic" or periodic boundary condition; we regard the Nth lattice point and the zeroth lattice point as identical. The ends of the one-dimensional chain can be thought of as being joined. Thus a translation through N lattice points returns us to our starting point:

$$(\epsilon|Na)=(\epsilon|0) \tag{6.2.11}$$

This restricts k to the discrete values

$$\exp(ikNa)=1 \qquad \text{or} \qquad k=\left(\frac{2\pi}{Na}\right)m$$

where $m=$ an integer. These values are uniformly spaced in k. Furthermore, the group is now isomorphic to the cyclic group $A, A^2, \ldots, A^{N-1}, A^N=E$ of order N. Thus there are precisely N representations, with $A=\omega^{-m}=\exp(-2\pi im/N)$ in the mth representation,

$$\chi^{m}(\epsilon|t_1a)=\chi^{m}(A^{t_1})=\omega^{-mt_1}=\exp(-ikt) \tag{6.2.12}$$

The representation $m+N$ is identical to m since $\omega^N=1$. Therefore we can restrict m to $0\leqslant m\leqslant N-1$ or, more symmetrically (if N is even), to

$$-\frac{N}{2}<m\leqslant\frac{N}{2} \tag{6.2.13}$$

The "zone boundary" representations at $N/2$ and $-N/2$ are equivalent. One or the other, but not both, may be used. The same remarks then apply to $k=\pi/a$ and $-\pi/a$.

Three-Dimensional Case

The generalization to the three-dimensional case is immediate. It is necessary to use a three-dimensional Bloch function $\psi(\mathbf{k},\mathbf{r})=\exp(i\mathbf{k}\cdot\mathbf{r})u(\mathbf{r})$ obeying

$$(\epsilon|\mathbf{t})\exp(i\mathbf{k}\cdot\mathbf{r})\,u(\mathbf{r})=\exp(-i\mathbf{k}\cdot\mathbf{t})\exp(i\mathbf{k}\cdot\mathbf{r})\,u(\mathbf{r}) \tag{6.2.14}$$

where

$$u(\mathbf{r}-\mathbf{t})=u(\mathbf{r}) \tag{6.2.15}$$

so that the character of (one-dimensional) representation **k** is

$$\chi^{\mathbf{k}}[(\epsilon|\mathbf{t})] = \exp(-i\mathbf{k}\cdot\mathbf{t}) \tag{6.2.16}$$

The periodicity conditions now take the form

$$(\epsilon|N_1\mathbf{a}_1) = (\epsilon|N_2\mathbf{a}_2) = (\epsilon|N_3\mathbf{a}_3) = (\epsilon|0) \tag{6.2.17}$$

that is, the **k**'s are restricted by the requirements

$$\mathbf{k}\cdot\mathbf{a}_i = \frac{2\pi m_i}{N_i}, \qquad i = 1,2,3 \tag{6.2.18}$$

For an orthorhombic lattice ($\mathbf{a}_1 \perp \mathbf{a}_2 \perp \mathbf{a}_3$) we can immediately write

$$\mathbf{k} = 2\pi\left[\left(\frac{m_1}{N_1}\right)\mathbf{b}_1 + \left(\frac{m_2}{N_2}\right)\mathbf{b}_2 + \left(\frac{m_3}{N_3}\right)\mathbf{b}_3\right] \tag{6.2.19}$$

where $\mathbf{b}_i$ is parallel to $\mathbf{a}_i$ and has magnitude $1/a_i$. For lattices with nonorthogonal basis vectors, we merely introduce the reciprocal lattice vectors $\mathbf{b}_1, \mathbf{b}_2, \mathbf{b}_3$ such that

$$\mathbf{b}_1 = \frac{\mathbf{a}_2 \times \mathbf{a}_3}{\mathbf{a}_1\cdot\mathbf{a}_2 \times \mathbf{a}_3}, \qquad \mathbf{a}_i\cdot\mathbf{b}_j = \delta_{ij} \tag{6.2.20}$$

If an assumed solution $\mathbf{k} = \Sigma\lambda_i\mathbf{b}_i$ is inserted into Eq. 6.2.18, the biorthogonality conditions 6.2.20 show that our previously quoted result (Eq. 6.2.19) *remains correct with this more general definition of the* $\mathbf{b}_i$. Thus we find a total of $N = N_1N_2N_3$ irreducible representations whose labels **k** are distributed uniformly over a zone (cell) in **k** space of volume

$$(2\pi)^3(\mathbf{b}_1\cdot\mathbf{b}_2 \times \mathbf{b}_3) = \frac{(2\pi)^3}{\Omega_0} \tag{6.2.21}$$

where

$$\Omega_0 = \mathbf{a}_1\cdot\mathbf{a}_2 \times \mathbf{a}_3 \tag{6.2.22}$$

is the volume of a primitive cell in the original lattice. As in the one-dimensional case, the zone can be bounded by $0 \leqslant m_i \leqslant N_i - 1$, that is, the zone is the interior of the parallelepiped generated by the vectors $2\pi\mathbf{b}_1, 2\pi\mathbf{b}_2, 2\pi\mathbf{b}_3$.

6.3. RECIPROCAL LATTICES AND BRILLOUIN ZONES

It is convenient to introduce a reciprocal lattice, the points in **k** space defined by

$$\mathbf{K} = 2\pi(h_1\mathbf{b}_1 + h_2\mathbf{b}_2 + h_3\mathbf{b}_3) \tag{6.3.1}$$

where the h_i are integers. These vectors obey

$$\mathbf{K}\cdot\mathbf{a}_i = 2\pi h_i = 2\pi(\text{integer}) \tag{6.3.2}$$

so that

$$u(\mathbf{r}) = \sum_{\mathbf{K}} C(\mathbf{K}) \exp(i\mathbf{K}\cdot\mathbf{r}) \tag{6.3.3}$$

is an arbitrary function with the periodicity of the lattice, and we write **K** briefly for $K_{h_1h_2h_3}$.

The zone in **k** space corresponding to our irreducible representations then consists of any primitive cell of this reciprocal lattice. It is clear, however, that any zone of the same volume, $(2\pi)^3/\Omega_0$, chosen so that by repetition it can fill reciprocal space, will contain the identical representations. The most common choice, the *Brillouin zone*, is the *proximity cell* in *reciprocal space*: it is the set of points closer to the lattice point $\mathbf{K}=0$ than to any other reciprocal lattice point **K**. The Brillouin zone is constructed by drawing perpendicular bisector planes to each of the nearby reciprocal lattice vectors **K** and is the interior enclosed by these planes. This zone is sometimes referred to as the *first* Brillouin zone. It may be translated to any other equivalent zone by adding a reciprocal lattice vector **K**. Since

$$\exp(-i\mathbf{k}\cdot\mathbf{t}) = \exp[-i(\mathbf{k}+\mathbf{K})\cdot\mathbf{t}] \tag{6.3.4}$$

it is clear that vectors in **k** space that differ by a reciprocal lattice vector label the same representation and are referred to as *equivalent*. Since the holohedral group of a space lattice restores the lattice (and its Wigner-Seitz cell) to itself, it must also restore the reciprocal lattice (and its Brillouin zone) to itself. Thus the *reciprocal lattice* of a given space lattice must belong to the *same crystal system* as the original, but it need not have the identical repetition pattern. Hence a face-centered cubic lattice has a body-centered reciprocal lattice, and vice versa. The three and sixfold axes for rhombohedral and hexagonal lattices ensure that the reciprocal lattices have Bravais characters identical to those of the originals. Appendix E shows the Brillouin zones for all Bravais lattices with points of special symmetry labeled.

6.4. CHARACTER ORTHONORMALITY THEOREMS

The first and second character orthonormality theorems (Eqs. 1.5.6 and 1.5.10) for the translation group in three dimensions can be written as

$$\Delta(\mathbf{k}-\mathbf{k}') \equiv \frac{1}{N}\sum_{\mathbf{t}} \exp\left[i(\mathbf{k}-\mathbf{k}')\cdot\mathbf{t}\right] = \delta(\mathbf{k},\mathbf{k}') \tag{6.4.1}$$

$$\sum_{\mathbf{k}\subset \mathrm{BZ}} \exp\left[i\mathbf{k}\cdot(\mathbf{t}-\mathbf{t}')\right] = N\delta(\mathbf{t},\mathbf{t}') \tag{6.4.2}$$

where $N = N_1 N_2 N_3$ is the number of cells, and $\delta(\mathbf{a},\mathbf{b})$ is a Kronecker delta function. These theorems follow from general group-theoretical principles or directly from the Nth roots of unity theorem.

Nth Roots of Unity Theorem 6.4.1

$$\sum_{s=0}^{N-1} \omega^s = N\delta(\omega,1) = N\delta(m,0)$$

where $\omega^N = 1$ or $\omega = \exp(2\pi i m/N)$.

Proof: It is easy to sum the geometric progression, but it is simpler to visualize the ω^s as vectors in the complex plane (see Fig. 6.4.1) distributed uniformly over angles, and adding up to zero, except in the special case when $m=0$ or $\omega=1$.

It is often convenient to have such theorems transcribed to the form appropriate when $\mathbf{k}$ is continuous. If $f(\mathbf{k})$ is an arbitrary smooth function, the transcription must be such that

$$\sum f(\mathbf{k})|\Delta\mathbf{k}| = \int f(\mathbf{k})\, d\mathbf{k} \tag{6.4.3}$$

Fig. 6.4.1. The Nth roots of unity theorem 6.4.1 is illustrated for $N=6$, $\omega = \exp(2\pi i/6)$. The arrows are ω^s for $s = 0,1,2,3,4,5$, and the sum is zero.

where $|\Delta\mathbf{k}|$ is the volume per point of reciprocal space:

$$|\Delta\mathbf{k}| = \frac{(2\pi)^3/\Omega_0}{N} \equiv \frac{2\pi}{N_1}\mathbf{b}_1 \cdot \frac{2\pi}{N_2}\mathbf{b}_2 \times \frac{2\pi}{N_3}\mathbf{b}_3 \tag{6.4.4}$$

Thus we arrive at the rule

$$\sum f(\mathbf{k}) = \frac{\Omega}{(2\pi)^3} \int f(\mathbf{k})\, d\mathbf{k} \tag{6.4.5}$$

where $\Omega = N\Omega_0$. This rule is equivalent to the usual statement that there is one state for each volume h^3 in phase space:

$$\frac{\Omega\, d\mathbf{p}}{h^3} = \frac{\Omega\, d\mathbf{k}}{(2\pi)^3} \tag{6.4.6}$$

The second character orthonormality theorem can now be given the form

$$\int_{\mathrm{BZ}} \exp[i\mathbf{k}\cdot(\mathbf{t}-\mathbf{t}')]\, d\mathbf{k} = (2\pi)^3 n\, \delta(\mathbf{t},\mathbf{t}') \tag{6.4.7}$$

where

$$n = \frac{N}{\Omega} = \frac{1}{\Omega_0} \tag{6.4.8}$$

is the number of cells per unit volume in the space lattice. (For an orthorhombic lattice, Eq. 6.4.7 is a trivial statement of the orthonormality of Fourier series.)

To rewrite Eq. 6.4.1 in continuous form, we note that the Fourier expansion of a function $F(\mathbf{k})$ periodic in reciprocal space, $F(\mathbf{k}+\mathbf{K}) = F(\mathbf{k})$, is given by

$$F(\mathbf{k}) = \sum_{\mathbf{t}} C_{\mathbf{t}} \exp(i\mathbf{k}\cdot\mathbf{t}) \tag{6.4.9}$$

where the Fourier coefficients are given by

$$C_{\mathbf{t}} = \frac{\int_{\mathrm{BZ}} F(\mathbf{k}') \exp(-i\mathbf{k}'\cdot\mathbf{t})\, d\mathbf{k}'}{\int_{\mathrm{BZ}} d\mathbf{k}'} \tag{6.4.10}$$

The notation indicates that the integral is to be carried over a single Brillouin zone. The denominator is the volume $(2\pi)^3/\Omega_0$ of this zone. These results can be combined to yield

$$F(\mathbf{k}) = \int_{\mathrm{BZ}} F(\mathbf{k}')\, d\mathbf{k}' \left\{ \frac{\Omega_0}{(2\pi)^3} \sum_{\mathbf{t}} \exp[i(\mathbf{k}-\mathbf{k}')\cdot\mathbf{t}] \right\}$$

For **k** within the first Brillouin zone, the quantity in braces must be the Dirac delta function $\delta(\mathbf{k}-\mathbf{k}')$. Since $F(\mathbf{k})$ is periodic, this integral yields the correct result for $F(\mathbf{k})$ everywhere only if

$$\Delta(\mathbf{k}-\mathbf{k}') = \frac{1}{N}\sum_{\mathbf{t}} \exp[i(\mathbf{k}-\mathbf{k}')\cdot\mathbf{t}] = \frac{(2\pi)^3}{\Omega}\sum_{\mathbf{K}} \delta(\mathbf{k}-\mathbf{k}'-\mathbf{K}) \quad (6.4.11)$$

This is to be compared with the discrete case (Eq. 6.4.1) rewritten in a form appropriate to the full reciprocal lattice:

$$\Delta(\mathbf{k}-\mathbf{k}') = \frac{1}{N}\sum_{\mathbf{t}} \exp[(\mathbf{k}-\mathbf{k}')\cdot\mathbf{t}] = \sum_{\mathbf{K}} \delta(\mathbf{k},\mathbf{k}'+\mathbf{K}) \quad (6.4.12)$$

If we now set $\mathbf{k}'=0$ in Eq. 6.4.11, multiply by an arbitrary nonperiodic $F(\mathbf{k})$, and integrate over **k**, we obtain

$$\sum_{\mathbf{t}} f(\mathbf{t}) = (2\pi)^3 n \sum_{\mathbf{K}} F(\mathbf{K}) \quad (6.4.13)$$

where

$$f(\mathbf{r}) = \int F(\mathbf{k}) \exp(i\mathbf{k}\cdot\mathbf{r})\, d\mathbf{k} \quad (6.4.14)$$

Equation 6.4.13 is simply a generalization of the Poisson sum formula (Titchmarsh, **1937**) from one dimension to a three-dimensional nonorthogonal lattice.

6.5. CONSERVATION OF CRYSTAL MOMENTUM

Let us consider a matrix element of the form

$$\begin{aligned} M &= \int \varphi(\mathbf{k}',\mathbf{r})^* V(\mathbf{r}) \psi(\mathbf{k},\mathbf{r})\, d\mathbf{r} \\ &= \int_{\text{crystal}} \exp[i(\mathbf{k}-\mathbf{k}')\cdot\mathbf{r}]\, U(\mathbf{r})\, d\mathbf{r} \end{aligned} \quad (6.5.1)$$

where $\psi(\mathbf{k},\mathbf{r})$, $\varphi(\mathbf{k}',\mathbf{r})$ are Bloch functions, and $V(\mathbf{r})$ has the full periodicity of the lattice. Then $U(\mathbf{r})$ is the product of all the periodic factors. By translating each cell to the origin, we can convert an integral over the entire crystal into an integral over one primitive cell Ω_0:

$$M = \sum_{\mathbf{t}} \exp[i(\mathbf{k}-\mathbf{k}')\cdot\mathbf{t}] J \quad (6.5.2)$$

where

$$J \equiv \int_{\Omega_0} \exp[i(\mathbf{k}-\mathbf{k}')\cdot(\mathbf{r})]\, U(\mathbf{r})\, d\mathbf{r} \quad (6.5.3)$$

and $\mathbf{r}-\mathbf{t}$ is used as a new variable of integration. Thus M contains a factor $\Delta(\mathbf{k}-\mathbf{k}')$, and the matrix element exists only if

$$\mathbf{k}-\mathbf{k}'=\mathbf{K} \tag{6.5.4}$$

This conservation of "crystal momentum" is a consequence of homogeneity, that is, translational invariance under (usually discrete) lattice *translations*, whereas conservation of ordinary momentum is a consequence of invariance of energy under arbitrary infinitesimal *displacements*, which requires freedom from external forces but not homogeneity.

If $V(\mathbf{r})$ itself behaves as $\exp(-i\mathbf{k}''\cdot\mathbf{r})$ times a periodic function, the selection rule becomes

$$\mathbf{k}=\mathbf{k}'+\mathbf{k}''+\mathbf{K} \tag{6.5.5}$$

A transition with $\mathbf{K}\neq 0$ is referred to as an Umklapp-process. Peierls (**1955**, p. 128) was the first to emphasize that the attainment of thermal equilibrium in heat conduction *requires* the use of Umklapp-processes.

6.6. LAUE-BRAGG X-RAY DIFFRACTION

There is a close relation between the reciprocal lattice and X-ray diffraction. Consider a plane wave of propagation vector $\mathbf{k}$ impinging on two atoms at 0 and $\mathbf{t}$, respectively. Examine the outgoing wave in direction $\mathbf{k}'$.

Figure 6.6.1 shows that the atom at $\mathbf{t}$ "sees" the incoming wave with a phase lag $\mathbf{k}\cdot\mathbf{t}$ (compared to $\mathbf{t}=0$) and emits in direction $\mathbf{k}'$ with a phase lead $\mathbf{k}'\cdot\mathbf{t}$. Thus the net phase change is $\exp[i(\mathbf{k}-\mathbf{k}')\cdot\mathbf{t}]$. With $\Delta\mathbf{k}=\mathbf{k}-\mathbf{k}'$, the diffracted X-ray intensity is proportional to

$$\left|\sum_{\mathbf{t}}\exp(i\Delta\mathbf{k}\cdot\mathbf{t})\right|^2=\sum_{\mathbf{t},\mathbf{t}'}\exp[i\Delta\mathbf{k}\cdot(\mathbf{t}-\mathbf{t}')]=N\sum_{\mathbf{t}}\exp(i\Delta\mathbf{k}\cdot\mathbf{t}) \tag{6.6.1}$$

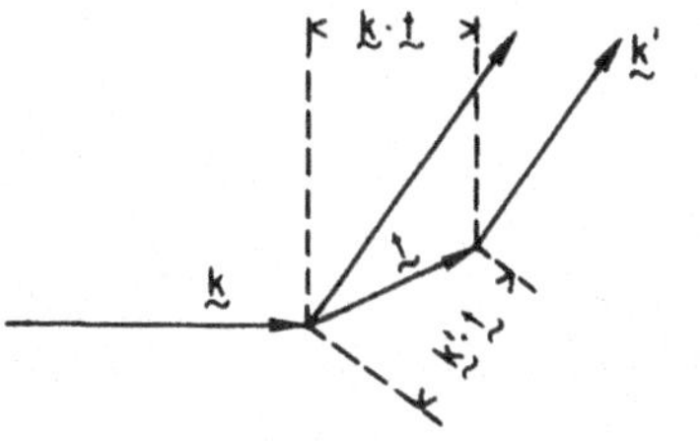

Fig. 6.6.1. The incident wave with wave vector $\mathbf{k}$ hits the scatterer at $\mathbf{t}$ with a phase lag $\mathbf{k}\cdot\mathbf{t}$ compared to the scatterer at the origin. The emitted wave from $\mathbf{t}$ leads that at 0 with the phase lead $\mathbf{k}'\cdot\mathbf{t}$. Thus the net phase change is $\exp[i(\mathbf{k}-\mathbf{k}')\cdot\mathbf{t}]$.

where the last result is obtained by summing over $\mathbf{t}$, using $\mathbf{t}'$ as origin. The final sum over $\mathbf{t}'$ then yields a factor N. In view of Eq. 6.4.11 the diffracted X-ray intensity is proportional to

$$\sum_{\mathbf{K}} \delta(\mathbf{k}-\mathbf{k}'-\mathbf{K}) \tag{6.6.2}$$

Thus there is a Laue diffraction spot associated with each reciprocal lattice vector $\mathbf{K}$. The conditions

$$(\mathbf{k}-\mathbf{k}')\cdot\mathbf{a}_i=\mathbf{K}\cdot\mathbf{a}_i=2\pi h_i, \qquad i=1,2,3 \tag{6.6.3}$$

are known as the *Laue-Bragg diffraction conditions*, and h_1, h_2, h_3 are referred to as the *Miller indices* of the reflecting crystal planes. These are a set of planes normal to $\mathbf{K}$ and defined by

$$\mathbf{K}\cdot\mathbf{t}=2\pi n \tag{6.6.4a}$$

or

$$h_1t_1+h_2t_2+h_3t_3=n \tag{6.6.4b}$$

Every lattice point is in one of these planes, since Σh_it_i is *some* integer. The closest distance to the origin of one of these planes is the component parallel to $\mathbf{K}$ of any $\mathbf{t}$ in the plane:

$$\frac{\mathbf{t}\cdot\mathbf{K}}{|\mathbf{K}|}=\frac{2\pi n}{|\mathbf{K}|} \tag{6.6.5}$$

Hence the separation between planes n and $n+1$ is

$$d_{\mathbf{K}}=\frac{2\pi}{|\mathbf{K}|} \tag{6.6.6}$$

The density of points in each plane must be proportional to $d_{\mathbf{K}}$, since (density of points in a plane)$\times$(density of planes)=(number of lattice points per unit volume) is fixed.

The intercept of such a plane with the $\mathbf{a}_1$ axis, obtained by setting $t_2=t_3=0$ in Eq. 6.6.4, is given by

$$t_1=\frac{n}{h_1} \tag{6.6.7}$$

The three intercepts are then n/h_1, n/h_2, and n/h_3, and n must be large enough to make these ratios all integers. [These intercepts are often used to define the set of Miller indices (h_1,h_2,h_3).]

We must add to the Laue-Bragg condition

$$\mathbf{k}-\mathbf{k}'=\mathbf{K} \tag{6.6.8}$$

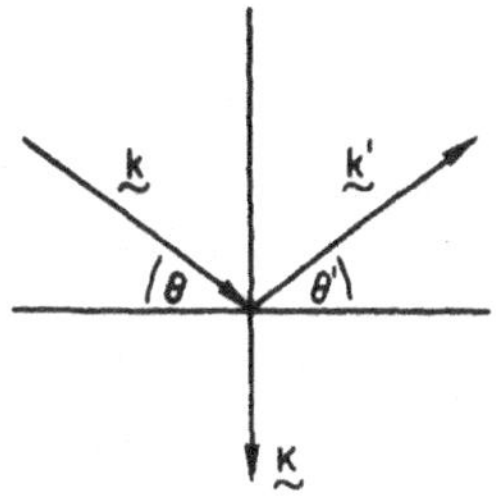

Fig. 6.6.2. The Laue-Bragg condition $\mathbf{k}-\mathbf{k}'=\mathbf{K}$, where $\mathbf{K}$ is a reciprocal lattice vector normal to a scattering plane, guarantees (when $k=k'$) that the angles of incidence and reflection are equal.

the condition that the scattering is elastic:

$$|\mathbf{k}'|=|\mathbf{k}| \tag{6.6.9}$$

For $\mathbf{k}-\mathbf{k}'$ to be equal to $\mathbf{K}$, that is, normal to the Bragg plane (with $k'=k$), the angle of reflection θ' must be equal to the angle of incidence θ. (See Fig. 6.6.2.) Also,

$$(\mathbf{k}-\mathbf{k}')^2=k^2+k^2-2k^2\cos 2\theta=|\mathbf{K}|^2 \tag{6.6.10}$$

which can be simplified to

$$2k\sin\theta=|\mathbf{K}| \tag{6.6.11}$$

or

$$\lambda=2d\sin\theta \tag{6.6.12}$$

which is the well-known Bragg condition for planes whose spacing is $d\equiv d_{\mathbf{K}}=2\pi/|\mathbf{K}|$ by Eq. 6.6.6. These conditions are used in Fig. 6.6.3 to construct the Laue diffraction spots.

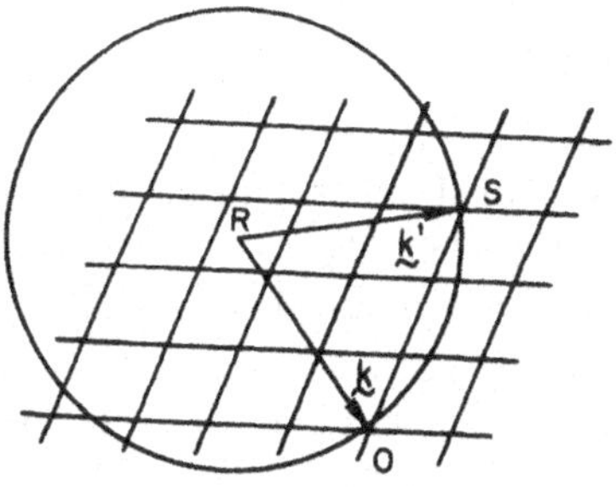

Fig. 6.6.3. The Ewald sphere method is a graphical procedure for determining the Laue diffraction spots. From the origin O of the reciprocal lattice draw the vector $-\mathbf{k}$. From the end point R of $-\mathbf{k}$, which is not in general a point of the lattice, draw a sphere of radius $|\mathbf{k}|$. If the sphere intersects a point S of the reciprocal lattice, OS is some $\mathbf{K}$ and RS is an allowed $\mathbf{k}'$ that obeys the conditions $|\mathbf{k}'|=RS=|\mathbf{k}|$ and $\mathbf{k}'-\mathbf{k}=OS=$ some $\mathbf{K}$.

6.7. SUMMARY

The point group symmetry of a crystal controls its Bravais lattice. Only seven point groups (the holohedral groups) can be symmetry groups of a lattice of points. The fourteen Bravais lattices are thus divided into seven systems. Several lattices can be associated with one holohedry (e.g., simple cubic, face-centered cubic, and body-centered cubic with the group O_h) because of the possibility of centered lattices.

The symmetry properties of the translation group are completely exhausted by diagonalizing the three generators $(\epsilon|\mathbf{a}_1)$, $(\epsilon|\mathbf{a}_2)$, and $(\epsilon|\mathbf{a}_3)$ or the general element $(\epsilon|\mathbf{t})$. This leads immediately to Bloch's theorem that an electronic energy band $\psi(\mathbf{k},\mathbf{r})$ must be a product of a plane wave $\exp(i\mathbf{k}\cdot\mathbf{r})$ times a function $u(\mathbf{r})$ with the periodicity of the lattice. It is shown that representations belonging to $\mathbf{k}$ and to $\mathbf{k}+\mathbf{K}$, where $\mathbf{K}$ is an arbitrary vector of the reciprocal lattice, are equivalent. The allowed values of $\mathbf{k}$ are shown to fill (uniformly) any one cell in reciprocal space. The Brillouin zone or proximity cell in reciprocal space is the most useful choice, since it has the full point group symmetry of the crystal.

The character orthonormality theorems are derived and are presented in continuous as well as discrete forms. They are shown to lead immediately to conservation of crystal momentum and to the Laue-Bragg conditions for X-ray diffraction.

problems

6.1.1. Show that any array of points equivalent under translations can be expressed in the form

$$\mathbf{t}=t_1\mathbf{a}_1+t_2\mathbf{a}_2+t_3\mathbf{a}_3$$

where t_1,t_2,t_3 are integers (Wannier, **1959**).

6.1.2. Why is T_h not a holohedral group?

6.1.3. Why is the group C_{3h} assigned to the hexagonal rather than the trigonal system? The converse question can be asked about S_6.

6.1.4. Why are there no face-centered lattices in the tetragonal system?

6.1.5. Show that monoclinic I (body-centered) lattices are possible but not new; that is, show that $2C$ is not a distinct lattice but that $2A$ or $2B$ are.

6.1.6. Construct the five plane Bravais lattices.

6.1.7. Show that for proper rotations the only possible angles of rotation in a crystallographic point group are $\varphi=2\pi/n$ with $n=1,2,3,4,6$.

6.3.1. Prove that the reciprocal lattice of a face-centered cubic lattice is body centered, and vice versa.

6.3.2. If a nonprimitive basic cell is used in ordinary space instead of a primitive basic cell, show (1) that the cell in reciprocal space is reduced in size but (2) that the extra points in the reciprocal lattice do not contribute to X-ray diffraction because the cell structure factor $\Sigma \exp(i\mathbf{K}\cdot\mathbf{X}^{\alpha})$ (where $\mathbf{X}^{\alpha}$ are the locations of the sites within the nonprimitive cell) vanishes.

6.4.1. Evaluate $\Sigma_{\mathbf{t}} \exp(i\mathbf{k}\cdot\mathbf{t})$ by summation over a finite lattice. Take the limit as $N\rightarrow\infty$ and verify Eq. 6.4.11.

chapter seven

ELECTRONIC ENERGY BANDS

In this chapter we discuss the concept of an electronic energy band, using only the most elementary of symmetry considerations (i.e., individual operations but not groups). The set of Bloch waves $\psi_n\,(\mathbf{k},\mathbf{r})$ that belong to a given band index n is decided (1) by naming the bands in order of energy, $E_1\,(\mathbf{k}) \leqslant E_2\,(\mathbf{k}) \leqslant E_3\,(\mathbf{k})$ and so on, (2) by analytical continuation from $\mathbf{k}$ to $\mathbf{k}+\Delta\mathbf{k}$ by "$\Delta\mathbf{k}\cdot\mathbf{v}$ perturbation theory," and (3) by imposing periodicity $\psi_n\,(\mathbf{k}+\mathbf{K},\mathbf{r}) = \psi_n\,(\mathbf{k},\mathbf{r})$. In Section 7.8 we prove that these requirements can in fact be simultaneously satisfied for a nondegenerate band. Also, we discuss the ambiguities introduced by degeneracy.

In Section 7.1 we show that many-electron wave functions have the same symmetry properties as one-electron wave functions. Thereafter, our discussion is confined to the latter. In Sections 7.3 and 7.4 we start with an empty lattice, add a weak periodic perturbation (to describe "almost-free electrons"), and show how this perturbation introduces gaps between the energy bands. Symmetry considerations are invoked in Sections 7.5 and 7.6 to show that (contrary to the almost-free-electron argument) such energy gaps do not always appear, and that energy bands need not always approach the zone boundary with zero slope. Section 7.9 discusses the velocity and acceleration of wave packets of electrons and holes in solids, and Section 7.10 applies these considerations to an effective Hamiltonian treatment of donor states.

7.1. RELATION BETWEEN THE MANY-ELECTRON AND ONE-ELECTRON VIEWPOINTS

For the purposes of this chapter we regard the *nuclei* as *fixed in a periodic array*. The electrons interact with the nuclei and with each other. It is customary to make a Hartree-Fock approximation, that is, to take as the many-electron wave function a determinant of one-electron wave functions each of which obeys a one-electron equation

$$\left(-\frac{\hbar^2}{2m}\nabla^2+V\right)\psi(\mathbf{r})=E\psi(\mathbf{r}) \tag{7.1.1}$$

The potential V consists of a Coulomb part that can be expressed as a function of $\mathbf{r}$, $V_c(\mathbf{r})$, and an exchange part that is nonlocal:

$$V_{\mathrm{ex}}\psi(\mathbf{r})=\int A(\mathbf{r},\mathbf{r}')\,d\mathbf{r}'\psi(\mathbf{r}') \tag{7.1.2}$$

We shall, however, think of $V=V(\mathbf{r})$ simply as a periodic function of $\mathbf{r}$ because at the moment we are interested not in calculations, but in symmetry properties.

The qualitative nature of a one-electron energy band in a solid and its symmetry properties accurately reflect corresponding properties of the correct many-electron wave function. For example, the translation operator $(\epsilon|\mathbf{t})$ in a many-electron problem can be taken to mean

$$(\epsilon|\mathbf{t})\psi(\mathbf{r}_1,\ldots,\mathbf{r}_N)=\psi(\mathbf{r}_1-\mathbf{t},\mathbf{r}_2-\mathbf{t},\ldots,\mathbf{r}_N-\mathbf{t}) \tag{7.1.3}$$

Since $(\epsilon|\mathbf{t})$ restores the nuclear array to itself, it commutes with the exact Hamiltonian. We can now choose to diagonalize this operator and obtain

$$\psi(\mathbf{r}_1+\mathbf{t},\ldots,\mathbf{r}_N+\mathbf{t})=\exp\,(i\mathbf{k}\cdot\mathbf{t})\psi(\mathbf{r}_1,\ldots,\mathbf{r}_N) \tag{7.1.4}$$

so that $\mathbf{k}$ now plays the role of the crystal momentum in the many-electron case. If we define

$$\psi(\mathbf{r}_1,\ldots,\mathbf{r}_N)=\exp(i\mathbf{k}\cdot\mathbf{r}_1)\,u(\mathbf{r}_1,\ldots,\mathbf{r}_N) \tag{7.1.5}$$

then u will be periodic in the sense that

$$u(\mathbf{r}_1+\mathbf{t},\ldots,\mathbf{r}_N+\mathbf{t})=u(\mathbf{r}_1,\ldots,\mathbf{r}_N) \tag{7.1.6}$$

Thus Bloch's theorem is trivially extended to the many-electron case.

The space group element $(\alpha|\mathbf{a})$ converts a plane wave with vector $\mathbf{k}$ into

$$(\alpha|\mathbf{a})\exp(i\mathbf{k}\cdot\mathbf{r})=\exp\left[i\mathbf{k}\cdot(\alpha|\mathbf{a})^{-1}\mathbf{r}\right]$$

But $(\alpha|\mathbf{a})^{-1}=(\alpha^{-1}|-\alpha^{-1}\mathbf{a})$ by Eq. 8.1.9, so that

$$\begin{aligned}(\alpha|\mathbf{a})\exp(i\mathbf{k}\cdot\mathbf{r})&=\exp\left[i\mathbf{k}\cdot(\alpha^{-1}\mathbf{r}-\alpha^{-1}\mathbf{a})\right]\\&=\exp\left[i\alpha\mathbf{k}\cdot(\mathbf{r}-\mathbf{a})\right]\end{aligned}$$

a wave with vector $\alpha\mathbf{k}$; then $(\alpha|\mathbf{a})$ does the same to the many-electron wave function:

$$(\alpha|\mathbf{a})\psi_n(\mathbf{k},\mathbf{r}_1,\ldots,\mathbf{r}_N)=\psi_{n'}(\alpha\mathbf{k},\mathbf{r}_1,\ldots,\mathbf{r}_N) \tag{7.1.7}$$

(See Problem 7.1.1.). *Note that* $(\alpha|\mathbf{a})$ *takes* $\mathbf{k}$ *into* $\alpha\mathbf{k}$ *regardless of the value of* $\mathbf{a}$. Since $(\alpha|\mathbf{a})$ commutes with the Hamiltonian, the energy is unchanged:

$$E_n(\mathbf{k})=E_{n'}(\alpha\mathbf{k}) \tag{7.1.8}$$

If the energy E_n ($\mathbf{k}$) is not degenerate at $\mathbf{k}$, we can identify n' and n as referring to the same band and write

$$E_n(\mathbf{k})=E_n(\alpha\mathbf{k}) \tag{7.1.9}$$

Thus, even in the many-electron case, the energy of a particular band is the same at all points $\alpha\mathbf{k}$. These points are called the "star of $\mathbf{k}$" (see Fig. 8.6.1).

Example 7.1.1

Show that the *set* of energy functions $\{E_n(\mathbf{k})\}$ is even in $\mathbf{k}$:

$$\{E_n(\mathbf{k})\}=\{E_n(-\mathbf{k})\} \tag{7.1.10}$$

Solution: Let us restrict ourselves to the case in which the Hamiltonian can be written in the form

$$H=-\tfrac{1}{2}\hbar^2\sum\left(\frac{1}{m_i}\right)\nabla_i^2+V(\mathbf{r}_1,\mathbf{r}_2,\ldots,\mathbf{r}_N) \tag{7.1.11}$$

Then H cannot be Hermitian unless it is also real: $H^*=H$. Let us now take the complex conjugate of

$$H\psi_n(\mathbf{k},\mathbf{r}_1,\ldots,\mathbf{r}_N)=E_n(\mathbf{k})\psi_n(\mathbf{k},\mathbf{r}_1,\mathbf{r}_2,\ldots,\mathbf{r}_N)$$

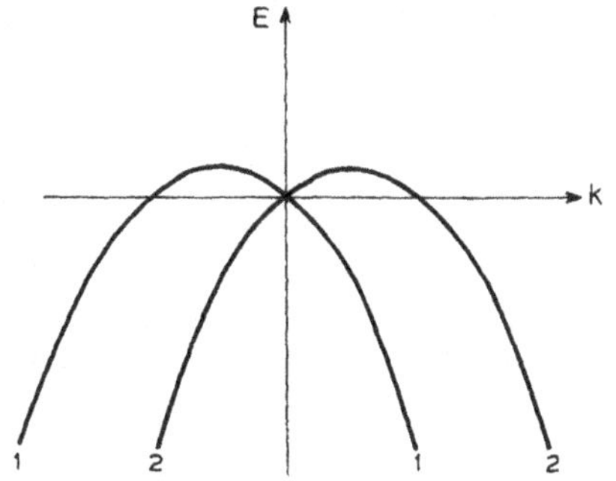

Fig. 7.1.1. The set of $E(\mathbf{k})$ functions is even in **k**. If we choose 1-1 (or 2-2) to be the bands, the individual bands are smooth, but not even in **k**. If we label the bands in order of energy, each one is even in **k** but possesses a cusp at $\mathbf{k}=0$.

to obtain

$$H\psi_n(\mathbf{k},\mathbf{r}_1,\ldots,\mathbf{r}_N)^* = E_n(\mathbf{k})\psi_n(\mathbf{k},\mathbf{r}_1,\ldots,\mathbf{r}_N)^* \tag{7.1.12}$$

so that $\psi_n(\mathbf{k})^*$ is an eigenfunction of H with the same eigenvalue. But the Bloch form (7.1.5) guarantees that $\psi_n(\mathbf{k},\cdots)^*$ is some Bloch wave n' with crystal momentum $-\mathbf{k}$. Thus

$$\psi_n(\mathbf{k},\ldots)^* = \psi_{n'}(-\mathbf{k},\ldots) \tag{7.1.13}$$

$$E_n(\mathbf{k}) = E_{n'}(-\mathbf{k}) \tag{7.1.14}$$

which establishes Eq. 7.1.10. If the bands are named in order of energy, we will always have

$$E_n(\mathbf{k}) = E_n(-\mathbf{k}) \tag{7.1.15}$$

but the bands may then have a cusp at $\mathbf{k}=0$, as in Fig. 7.1.1.

We shall see in Chapter 10 that it is appropriate to regard $\psi_n^* \equiv K_0\psi_n$ as the time-reversed wave function, where K_0, the symbol for time reversal in the absence of spin, is simply complex conjugation. The proof can then be immediately generalized to the case of spin including spin-orbit forces by introducing the general time-reversal operator K of Chapter 10 and noting that $K\psi_n$ is a solution with the same energy and reversed **k** vector because K also involves complex conjugation.

The reader should keep in mind that all of the symmetry arguments and most of the others can be generalized to the many-electron case. We now return to the one-electron viewpoint so as to obtain the results of this chapter in the most elementary way possible.

7.2. CONCEPT OF AN ENERGY BAND

Bloch's theorem tells us that the energies and wave functions in a periodic potential can be written in the forms $E_n(\mathbf{k})$ and

$$\psi_n(\mathbf{k},\mathbf{r}) = \exp(i\mathbf{k}\cdot\mathbf{r})\, u_n(\mathbf{k},\mathbf{r}) \tag{7.2.1}$$

where $u_n(\mathbf{k},\mathbf{r})$ is periodic in $\mathbf{r}$. For the purpose of classifying our states, we can restrict ourselves to the first Brillouin zone, since any factor of the form $\exp[i(\mathbf{k}+\mathbf{K})\cdot\mathbf{r}]$, where $\mathbf{K}$ is a reciprocal lattice vector, can be reduced to the form $\exp(i\mathbf{k}\cdot\mathbf{r})$ by absorbing $\exp(i\mathbf{K}\cdot\mathbf{r})$ into the periodic part of the wave function. (This is referred to as the "reduced zone scheme.")

One possible way to label the states is simply to arrange them in order of energy, so that for each $\mathbf{k}$

$$E_1(\mathbf{k}) \leqslant E_2(\mathbf{k}) \leqslant E_3(\mathbf{k}) < \cdots \tag{7.2.2}$$

Another possible scheme is to try to choose the band indices n so that $\psi_n(\mathbf{k},\mathbf{r})$ and $E_n(\mathbf{k})$ are analytic, or at least continuous functions of $\mathbf{k}$. Outside the first Brillouin zone, since no new functions can arise, one would also like to impose the condition of periodicity in reciprocal space,

$$\psi_n(\mathbf{k}+\mathbf{K},\mathbf{r}) = \psi_n(\mathbf{k},\mathbf{r}); \qquad E_n(\mathbf{k}+\mathbf{K}) = E_n(\mathbf{k}) \tag{7.2.3}$$

Are these two classification schemes possible? And are they consistent with one another? These questions will be answered as this chapter unfolds.

We may remark immediately that in one dimension the two schemes necessarily lead to the same classification: Kramers (**1935**) has established[1] that a periodic potential in one dimension permits the eigenvalue $\lambda = \exp(-ika)$ of the translation operator $(\epsilon|\mathbf{a})$ to be in the range $|\lambda| \leqslant 1$ only if the energy is in a series of regions separated by energy gaps. Thus the spectrum looks qualitatively as shown in Fig. 7.2.1. The presence of the energy gaps ensures that the highest point of one band is separated by a finite energy gap from the lowest point of the next band. Thus degeneracy, or a crossing of $E(\mathbf{k})$ curves, is impossible, and classification by order of increasing energy and classification by analytic continuation must coincide. Periodicity of $\psi(\mathbf{k},\mathbf{r})$ is not obvious, however. A proof for the one-dimensional case when inversion symmetry is present is given by Kohn (**1959a**). A proof for the three-dimensional nondegenerate case is presented in Section 7.8. Since the one-dimensional case has been shown by Kramers to be nondegenerate, our proof automatically covers the one-dimensional case, whether or not inversion is present.

We shall not repeat Kramer's proof because it is not applicable in three

[1] See also Wilson (**1953**), p. 23.

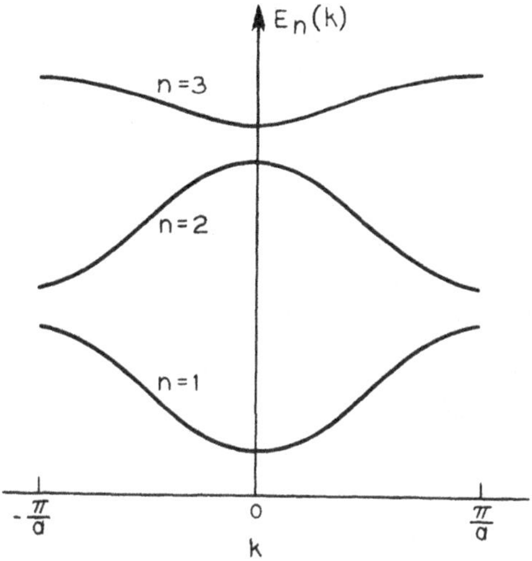

Fig. 7.2.1. The extremal energies in band n occur at $k=0$ and $k=\pi/a$.

dimensions. Instead we shall review the theory of almost-free electrons to obtain a qualitative understanding of why energy gaps originate and how they differ in one and in three dimensions.

7.3. THE EMPTY LATTICE

Free electrons are described by plane waves $\exp(i\mathbf{k}\cdot\mathbf{r})$ extending over all of $\mathbf{k}$ space. Electrons in a periodic potential, even a very weak one, are described by Bloch waves $\psi(\mathbf{k},\mathbf{r})$ in one zone, and as periodic functions of $\mathbf{k}$ outside that zone. This radical change in description is due partly to the perturbation and partly to the use of a reduced zone description. We can clarify this point by describing free electrons ("the empty lattice") in the reduced zone scheme.

In Fig. 7.3.1 the free-electron relation $E=\hbar^2k^2/2m$ is shown as $ABCD$. The piece BC is shifted by a reciprocal lattice vector $2\pi/a$ to bring it to $B'C'$ inside the first zone. Similarly CD is shifted to $C''D'$. The dashed lines show other pieces shifted in from the negative k region.

Crossings or "degeneracies" occur at $k=0$ and $k=\pm\pi/a$. We shall show that the effect of a weak periodic potential is to lift these degeneracies, giving rise to *energy gaps*, and to cause the $E_n(k)$ curves to have zero slope at these points. Figure 7.3.2 shows the resulting energy bands plotted in a conventional way as periodic functions of k, whereas in Fig. 7.3.3 the resulting E versus k relation is plotted against the original k vector, that is, in the *extended zone scheme*.

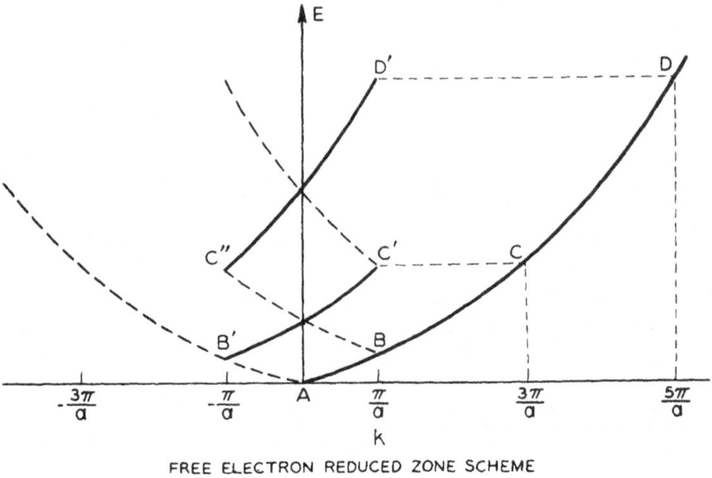

Fig. 7.3.1.

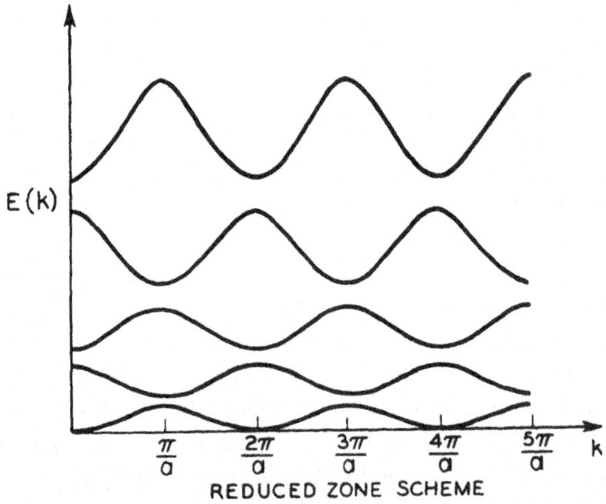

Fig. 7.3.2

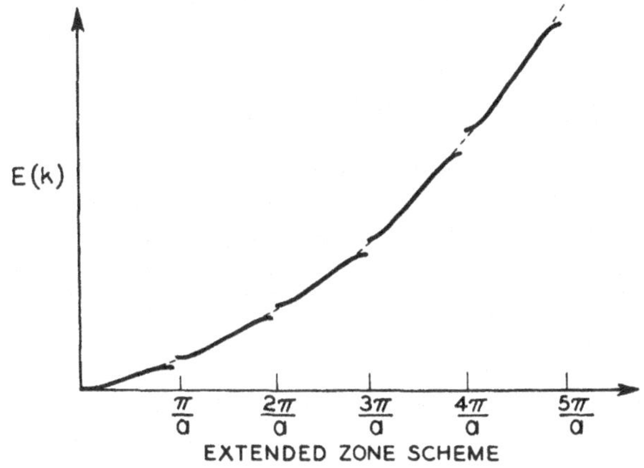

Fig. 7.3.3

The transition from the extended to the reduced zone scheme is a trivial matter in one dimension, and the basic idea is the same in three dimensions. Nevertheless, the geometry of reduction to the first Brillouin zone is considerably more complicated in three dimensions. Considerable insight has been obtained into the properties of some metals by regarding the electrons as free and occupying a Fermi surface that extends in some directions beyond the first zone. Mapping these outside portions into the first zone, one obtains "pockets" of electrons with properties different from those of the majority of electrons (Harrison, **1960**, **1966**).

7.4. ALMOST-FREE ELECTRONS

The periodic potential can be expressed as a Fourier series

$$V(\mathbf{r}) = \sum_{\mathbf{K}} V_{\mathbf{K}} \exp(i\mathbf{K}\cdot\mathbf{r}) \tag{7.4.1}$$

where the **K** are reciprocal lattice vectors. By Bloch's theorem, the wave function can be written in the form

$$\psi(\mathbf{k},\mathbf{r}) = \sum_{\mathbf{K}} c(\mathbf{K}) \exp[i(\mathbf{k}+\mathbf{K})\cdot\mathbf{r}] \tag{7.4.2}$$

where $c(\mathbf{K}) = c(\mathbf{K},\mathbf{k})$ depends on $\mathbf{k}$, but this dependence is not explicitly shown.

Schrödinger's equation for the coefficients takes the form

$$\left[E-\left(\frac{\hbar^2}{2m}\right)(\mathbf{k}+\mathbf{K})^2\right]c(\mathbf{K})=\sum_{\mathbf{K}'}c(\mathbf{K}')V_{\mathbf{K}-\mathbf{K}'} \tag{7.4.3}$$

For $\mathbf{K}\neq 0$ we can write

$$c(\mathbf{K})=\frac{\sum_{\mathbf{K}'}c(\mathbf{K}')V_{\mathbf{K}-\mathbf{K}'}}{E-(\hbar^2/2m)(\mathbf{k}+\mathbf{K})^2} \tag{7.4.4}$$

whereas the $\mathbf{K}=0$ equation can be written as an equation for the energy

$$E=\left(\frac{\hbar^2}{2m}\right)k^2+V_0+\sum_{\mathbf{K}\neq 0}V_{-\mathbf{K}}\left[\frac{c(\mathbf{K})}{c(0)}\right] \tag{7.4.5}$$

In the almost-free-electron approximation, $E\gg V$ and $\psi\approx\exp(i\mathbf{k}\cdot\mathbf{r})$, that is, $c(\mathbf{K})/c(0)\ll 1$. Retaining only the dominant term in Eq. 7.4.4, we find

$$\frac{c(\mathbf{K})}{c(0)}\approx\frac{V_{\mathbf{K}}}{E-(\hbar^2/2m)(\mathbf{k}+\mathbf{K})^2} \tag{7.4.6}$$

and

$$\begin{aligned} E&\approx\left(\frac{\hbar^2}{2m}\right)k^2+V_0+\sum_{\mathbf{K}\neq 0}\frac{|V_{\mathbf{K}}|^2}{E-(\hbar^2/2m)(\mathbf{k}+\mathbf{K})^2}\\ &\approx\left(\frac{\hbar^2}{2m}\right)k^2+V_0+\sum_{\mathbf{K}\neq 0}\frac{|V_{\mathbf{K}}|^2}{(\hbar^2/2m)\left[\mathbf{k}^2-(\mathbf{k}+\mathbf{K})^2\right]} \end{aligned} \tag{7.4.7}$$

This weak perturbation approximation breaks down for $\mathbf{k}$, obeying or nearly obeying

$$\mathbf{k}^2=(\mathbf{k}+\mathbf{K})^2 \tag{7.4.8}$$

for some reciprocal lattice vector $\mathbf{K}$. In the rest of this section $\mathbf{K}$ is not a summation index, but the particular value of $\mathbf{K}$ for which Eq. 7.4.8 is nearly obeyed. This condition is (1) the Laue-Bragg condition for X-ray scattering (cf. Eqs. 6.6.8–6.6.12) and (2) the equation for a set of planes in reciprocal space. These planes are the boundary planes of first and higher Brillouin zones.

We shall now show that in the almost-free approximation a jump occurs in crossing the zone boundary plane defined by Eq. 7.4.8 if $V_{\mathbf{K}}$ does not vanish. At a general point $\mathbf{k}_0$ of such a plane, the denominator of Eq. 7.4.4 will be small. Thus the almost-free-electron approximation must now consist in regarding both $c(\mathbf{K})$ and $c(0)$ as large, and all other coefficients as small. The pair of equations for these two amplitudes leads to a secular determinant of the form

$$\begin{vmatrix} E_1 - E & V_{-\mathbf{K}} \\ V_{\mathbf{K}} & E_2 - E \end{vmatrix} = 0 \tag{7.4.9}$$

where

$$\begin{aligned} E_1 &= \left(\frac{\hbar^2}{2m}\right)k^2 + V_0 = E_b + \left(\frac{\hbar^2}{2m}\right)\left[(\Delta\mathbf{k})^2 + 2\mathbf{k}_0\cdot\Delta\mathbf{k}\right] \\ E_2 &= \left(\frac{\hbar^2}{2m}\right)(\mathbf{k}+\mathbf{K})^2 + V_0 = E_b + \left(\frac{\hbar^2}{2m}\right)\left[(\Delta\mathbf{k})^2 + 2(\mathbf{k}_0+\mathbf{K})\cdot\Delta\mathbf{k}\right] \end{aligned} \tag{7.4.10}$$

and

$$\begin{gathered} \mathbf{k} = \mathbf{k}_0 + \Delta\mathbf{k}, \qquad k_0^2 = (\mathbf{k}_0+\mathbf{K})^2 \\ E_b = \left(\frac{\hbar^2}{2m}\right)k_0^2 + V_0 \end{gathered} \tag{7.4.11}$$

so that E_b is the (first-order) energy at the boundary point, and $\Delta\mathbf{k}$ is the deviation from the boundary. The new eigenvalues are given to this order by

$$E = E_b + \left(\frac{\hbar^2}{2m}\right)\left[(\Delta\mathbf{k})^2 + (2\mathbf{k}_0+\mathbf{K})\cdot\Delta\mathbf{k}\right] \pm \left\{\left[\left(\frac{\hbar^2}{2m}\right)\mathbf{K}\cdot\Delta\mathbf{k}\right]^2 + |V_{\mathbf{K}}|^2\right\}^{1/2} \tag{7.4.12}$$

At the boundary, $\Delta\mathbf{k} = 0$ and the energy displays a *gap* $2|V_{\mathbf{K}}|$:

$$E = E_b \pm |V_{\mathbf{K}}| \tag{7.4.13}$$

If we plot the energy versus $\Delta\mathbf{k}$ normal to the boundary, we find the result shown in Fig. 7.4.1.

For $V_{\mathbf{K}} \neq 0$, the radical can be expanded and yields $\pm|V_{\mathbf{K}}|$ plus a correction term of order $(\Delta k)^2$: since $\mathbf{k}_0$ is a zone boundary point obeying Eq. 7.4.8,

$$(2\mathbf{k}_0+\mathbf{K})\cdot\mathbf{K} = 0 \tag{7.4.14}$$

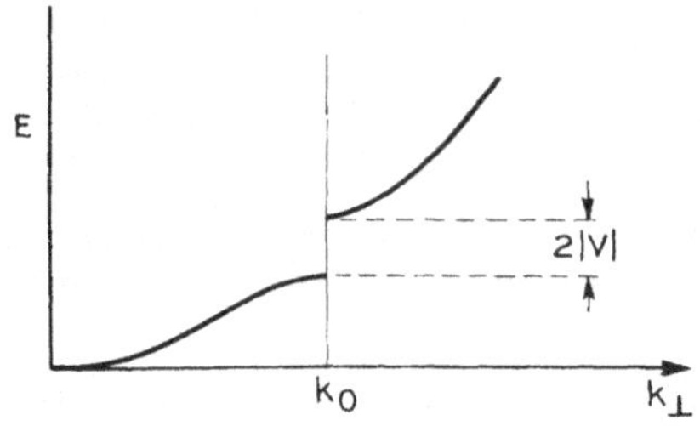

Fig. 7.4.1. Jump in energy and vanishing slope as one crosses a zone boundary in a perpendicular direction.

Thus, if we choose $\Delta\mathbf{k}\|\mathbf{K}$, that is, perpendicular to the plane, the only linear term in $\Delta\mathbf{k}$ in Eq. 7.4.12 vanishes, so that

$$\nabla_{\mathbf{k}} E(\mathbf{k})\cdot\mathbf{n}=0 \tag{7.4.15}$$

where $\mathbf{n}=\mathbf{K}/|\mathbf{K}|$ is normal to the boundary.

7.5. ENERGY GAPS AND SYMMETRY CONSIDERATIONS

How generally valid are the conclusions derived above concerning an energy gap at, and zero slope normal to, a zone boundary? Can $V_{\mathbf{K}}$ vanish other than accidentally? Let us consider the symmetry of the $V_{\mathbf{K}}$'s. The space group operation $(\boldsymbol{\alpha}|\mathbf{a})^{-1}$ requires that

$$V(\mathbf{r})=V((\boldsymbol{\alpha}|\mathbf{a})\mathbf{r})\equiv V(\boldsymbol{\alpha}\cdot\mathbf{r}+\mathbf{a})$$

or

$$V_{\mathbf{K}}=\exp[i(\boldsymbol{\alpha}\cdot\mathbf{K})\cdot\mathbf{a}]V_{\boldsymbol{\alpha}\cdot\mathbf{K}} \tag{7.5.1}$$

Symmorphic space groups contain the elements $(\boldsymbol{\alpha}|0)$ as a subgroup. Hence, for these, we can set $\mathbf{a}=0$ to obtain a set of relations

$$V_{\mathbf{K}}=V_{\boldsymbol{\alpha}\cdot\mathbf{K}} \tag{7.5.2}$$

between different Fourier coefficients that does not cause any one of them to vanish. For a nonsymmorphic space group we can get a vanishing of $V_{\mathbf{K}}$ provided that an $(\boldsymbol{\alpha}|\mathbf{a})$ exists such that

$$\boldsymbol{\alpha}\cdot\mathbf{K}=\mathbf{K}, \qquad \exp(i\mathbf{K}\cdot\mathbf{a})=-1 \tag{7.5.3}$$

For example, in the diamond structure we can select from Table 9.1.1 the reflection plane $(\boldsymbol{\alpha}|\mathbf{a})=(\boldsymbol{\rho}_z|\boldsymbol{\tau})$, where $\boldsymbol{\rho}_z$ takes z into $-z$ and $\boldsymbol{\tau}=(\mathbf{a}/4)(1,1,1)$.

Using $\mathbf{K}=(2\pi/a)(P_1,P_2,P_3)$, we obtain

$$V_{P_1P_2P_3}=\exp\left[i\left(\frac{\pi}{2}\right)(P_1+P_2-P_3)\right]V_{P_1P_2\bar{P}_3} \tag{7.5.4}$$

Thus, for $P_3=0$, $P_1+P_2=$twice an odd integer, $V_{P_1P_2P_3}$ vanishes. In particular, V_{200} and V_{420} vanish.[2] [Of course, V_{110} also vanishes, but this adds no new information, since diamond has a face-centered cubic Bravais lattice, and its body-centered cubic reciprocal lattice is restricted to triplets $(P_1P_2P_3)$ that are *all even* or *all odd*. Such "trivial" vanishings do not occur in the reciprocal of the original *primitive* lattice.]

An approximate vanishing can also occur for structure factor reasons if several identical atoms appear at particular positions in the cell. If the potential can be represented approximately as a sum of identical but displaced terms,

$$V(\mathbf{r})=\sum_{\alpha i} v(\mathbf{r}-\mathbf{R}^{\alpha i}) \tag{7.5.5}$$

Then the Fourier coefficient will contain a structure factor

$$V_{\mathbf{K}}\propto\sum_{\alpha}\exp(i\mathbf{K}\cdot\mathbf{R}^{\alpha}) \tag{7.5.6}$$

In the diamond structure again, with $\mathbf{R}^1=0$, $\mathbf{R}^2=(a/4)(1,1,1)$, we find that $V_{222}=0$. This (222) line is seen in X-ray diffraction [Göttlicher and Wölfel (**1959**)] as a weak line (1/100 of the intensity of other lines), indicating that the decomposition of Eq. 7.5.6 is approximate but not exact.

Although the vanishing of a $V_{\mathbf{K}}$ *may* mean the absence of an energy gap across the corresponding plane, it does *not necessarily* have this significance. The degeneracy between $\exp(i\mathbf{k}\cdot\mathbf{r})$ and $\exp[i(\mathbf{k}+\mathbf{K})\cdot\mathbf{r}]$ will *usually* be lifted in second order. To second order, we have the same 2×2 degeneracy problem, with $V_{\mathbf{K}}$ replaced by the equivalent Hamiltonian of Problem 11.10.7:

$$V_{\mathbf{K}}^{\text{eff}}=V_{\mathbf{K}}+\frac{1}{2}\sum_{\mathbf{K}'}V_{\mathbf{K}-\mathbf{K}'}V_{\mathbf{K}'}\left[\frac{1}{\mathbf{k}^2-(\mathbf{k}+\mathbf{K}')^2}+\frac{1}{(\mathbf{k}+\mathbf{K})^2-(\mathbf{k}+\mathbf{K}')^2}\right] \tag{7.5.7}$$

[2] Jones (**1960**) has chosen to divide reciprocal space into cells bounded by planes of energy discontinuity, that is, zone boundaries for which $V_{\mathbf{K}}$ is not zero or is not anomalously small. These zones usually possess a nonintegral number of states per atom and are useful in discussing electronic motion and changes in occupancy of different parts of **k** space with changes in number of free carriers (e.g., because of alloying). We shall refer to them as Jones zones to distinguish them from Brillouin zones, which are more directly related to the translation group.

where $(\mathbf{k}+\mathbf{K})^2=\mathbf{k}^2$ on the zone boundary plane. When $V_{\mathbf{K}}=0$, then, we expect a *small* gap to occur (of second order in the strength of the potential). In diamond, the combined use of V_{111} and V_{1-1-1}, both allowed, produces the same effect as V_{200} would have produced if Eq. 7.5.4 had not caused V_{200} to vanish.

The case of a twofold screw axis normal to a zone boundary plane (such as occurs in hexagonal close-packed lattices) provides us with one example, however, in which degeneracy is not lifted to any order over an entire plane. See Example 10.8.2.

7.6. POINTS OF ZERO SLOPE

The almost-free-electron theory predicts that the energy gradient normal to a zone boundary plane vanishes for all points in the plane. To what extent is this statement generally true? If we calculate the energy using Eq. 7.4.12 with $V_{\mathbf{K}}$ replaced by $V_{\mathbf{K}}^{\text{eff}}$ of Eq. 7.5.7, the energy gradients will not vanish, in general, since $V_{\mathbf{K}}^{\text{eff}}$ depends on $\Delta\mathbf{k}$, including its component normal to the surface. Our conclusion, then, is that the normal energy gradient will not vanish unless forced to do so by some symmetry requirement. A complete discussion is postponed until Section 10.4, but we can give here an elementary treatment, omitting time reversal, of the case in which no degeneracy occurs at a general point in the plane.

The only symmetry element $\boldsymbol{\alpha}$ that can leave a general vector $\mathbf{k}$ of some plane invariant, $\boldsymbol{\alpha}\mathbf{k}=\mathbf{k}$ (or equivalent, $\boldsymbol{\alpha}\mathbf{k}=\mathbf{k}+\mathbf{K}$), is a reflection plane parallel to the plane in question. Suppose for simplicity of notation that these planes are perpendicular to the x axis. The presence of a reflection plane $\boldsymbol{\rho}_x$ then guarantees that

$$E(-k_x,k_y,k_z)=E(k_x,k_y,k_z) \tag{7.6.1}$$

Hence E is an even function of k_x and

$$\frac{\partial E}{\partial k_x}=0 \quad \text{at} \quad k_x=0 \tag{7.6.2}$$

Thus the surface $E(\mathbf{k})=$ constant cuts the symmetry plane $k_x=0$ at right angles. [If there is more than one reflection plane, the surface $E(\mathbf{k})=$ constant will cut each symmetry plane at right angles.]

Now let us consider the $E(\mathbf{k})$ surfaces near the zone boundary. Consider a reciprocal lattice vector $\mathbf{K}$ oriented in the x direction. Then $k_x=+\frac{1}{2}|\mathbf{K}|$ and $k_x=-\frac{1}{2}|\mathbf{K}|$ are the zone boundary planes. It follows from Eq. 7.1.15 that

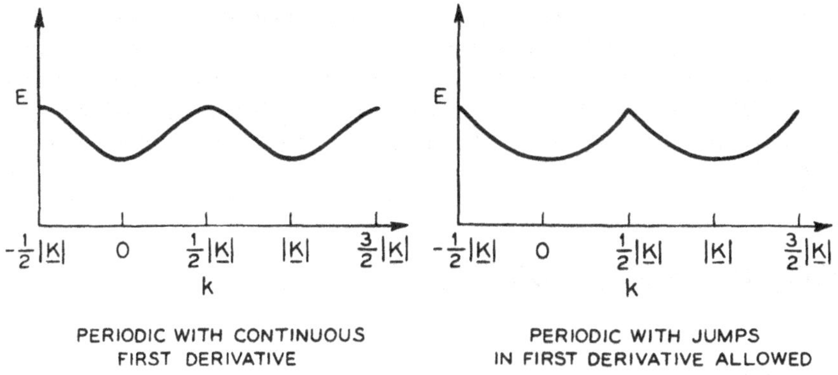

Fig. 7.6.1

$\partial E/\partial k_x$ is an odd function of k_x:

$$\left.\frac{\partial E}{\partial k_x}\right|_{k_x=-\frac{1}{2}|\mathbf{K}|} = -\left.\frac{\partial E}{\partial k_x}\right|_{k_x=+\frac{1}{2}|\mathbf{K}|} \tag{7.6.3}$$

However, these points are equivalent in the sense that they differ by a reciprocal lattice vector $\mathbf{K}$. We shall show in the next section that this implies

$$E_n(\mathbf{k}+\mathbf{K}) = E_{n'}(\mathbf{k}) \tag{7.6.4}$$

In the absence of degeneracy, we can set $n'=n$ and require periodicity. If the gradient of E is a continuous function. $\partial E/\partial k_x$ must take the same value at $k_x = \pm\frac{1}{2}|\mathbf{K}|$ and hence vanishes. (See Fig. 7.6.1, which illustrates the discontinuous as well as the continuous case.)

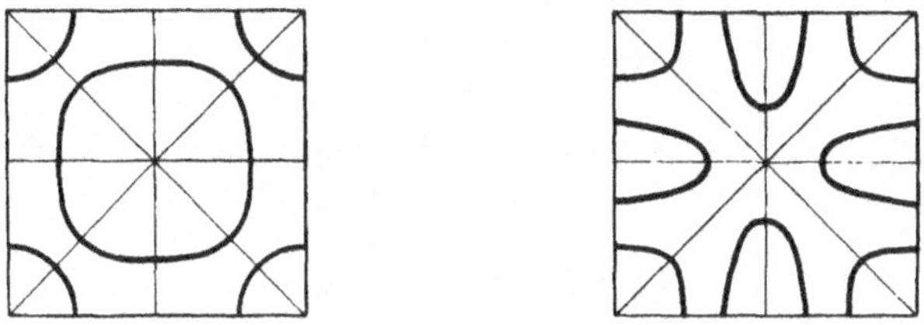

THE FIGURES SHOW ALL THE SYMMETRY OF A SQUARE AND CROSS ALL LINES (SYMMETRY PLANES) AT RIGHT ANGLES

Fig. 7.6.2

When all the conditions mentioned above are satisfied, the constant energy surfaces $E(\mathbf{k})=\text{constant}$ must cut all symmetry planes and all zone boundary planes at right angles. This places an appreciable limitation on the energy contours, and knowledge of a few key values permits them to be sketched out roughly. Two possibilities in the $k_z=0$ plane of a simple cubic lattice are shown in Fig. 7.6.2.

7.7. PERIODICITY IN RECIPROCAL SPACE

Tight Binding Approximation

Bloch, in his original treatment of energy bands, sought a solution in the form of a linear combination of atomic orbitals:

$$\psi(\mathbf{k},\mathbf{r})=\sum_{\mathbf{R}}F(\mathbf{R})\varphi(\mathbf{r}-\mathbf{R}) \tag{7.7.1}$$

centered at the atomic positions $\mathbf{R}$ located on a Bravais lattice. We can minimize the expectation value of the energy by requiring the coefficients $F(\mathbf{R})$ to obey

$$\sum_{\mathbf{R}'}H(\mathbf{R},\mathbf{R}')F(\mathbf{R}')=E\sum_{\mathbf{R}'}S(\mathbf{R},\mathbf{R}')F(\mathbf{R}') \tag{7.7.2}$$

where

$$\begin{aligned} S(\mathbf{R},\mathbf{R}')&=S(\mathbf{R}'-\mathbf{R})=\int\varphi(\mathbf{r}-\mathbf{R})^*\varphi(\mathbf{r}-\mathbf{R}')\,d\mathbf{r}\\ H(\mathbf{R},\mathbf{R}')&=H(\mathbf{R}'-\mathbf{R})=\int\varphi(\mathbf{r}-\mathbf{R})^*H\varphi(\mathbf{r}-\mathbf{R}')\,d\mathbf{r} \end{aligned} \tag{7.7.3}$$

and H is the complete periodic Hamiltonian. The (lattice) translational symmetry that makes $H(\mathbf{R},\mathbf{R}')$ and $S(\mathbf{R},\mathbf{R}')$ functions only of $\mathbf{R}'-\mathbf{R}$ guarantees that

$$F(\mathbf{R})=\exp(i\mathbf{k}\cdot\mathbf{R}) \tag{7.7.4}$$

is a solution, appropriate to wave vector $\mathbf{k}$ with eigenvalue $E=E(\mathbf{k})$ determined from

$$H(\mathbf{k})=E(\mathbf{k})S(\mathbf{k}) \tag{7.7.5}$$

where

$$\begin{aligned} H(\mathbf{k})&=\sum_{\mathbf{R}}H(\mathbf{R})\ \exp(i\mathbf{k}\cdot\mathbf{R})\\ S(\mathbf{k})&=\sum_{\mathbf{R}}S(\mathbf{R})\ \exp(i\mathbf{k}\cdot\mathbf{R}) \end{aligned} \tag{7.7.6}$$

For any reciprocal lattice vector $\mathbf{K}$,

$$\exp[i(\mathbf{k}+\mathbf{K})\cdot\mathbf{R}]=\exp(i\mathbf{k}\cdot\mathbf{R}) \tag{7.7.7}$$

Hence it follows automatically that we have

$$\begin{aligned}\psi(\mathbf{k}+\mathbf{K},\mathbf{r})&=\psi(\mathbf{k},\mathbf{r})\\ E(\mathbf{k}+\mathbf{K})&=E(\mathbf{k})\end{aligned} \tag{7.7.8}$$

The tight binding approximation, which makes use of only one atomic orbital, is valid only for well separated atoms, so that the energy bands are narrow compared to the spacing between such bands. Otherwise interactions between the orbitals that form these bands become important. The *generalized tight binding* approach uses a set $\varphi_m(\mathbf{r}-\mathbf{R})$ of atomic orbitals:

$$\psi(\mathbf{k},\mathbf{r})=\sum_{\mathbf{R},n}F_n(\mathbf{R})\varphi_n(\mathbf{r}-\mathbf{R}) \tag{7.7.9}$$

This approach becomes rigorous as the set of atomic orbitals is enlarged to a complete set. Our equations retain the same structure, but H and S become matrices in the band indices, namely,

$$\begin{aligned}H_{mn}(\mathbf{R})&=\int\varphi_m(\mathbf{r})^*H\varphi_n(\mathbf{r}-\mathbf{R})\,d\mathbf{r},\quad H_{mn}(\mathbf{k})=\sum_{\mathbf{R}}H_{mn}(\mathbf{R})\exp(i\mathbf{k}\cdot\mathbf{R})\\ S_{mn}(\mathbf{R})&=\int\varphi_m(\mathbf{r})^*\varphi_n(\mathbf{r}-\mathbf{R})\,d\mathbf{r},\quad S_{mn}(\mathbf{k})=\sum_{\mathbf{R}}S_{mn}(\mathbf{R})\exp(i\mathbf{k}\cdot\mathbf{R})\end{aligned} \tag{7.7.10}$$

and the solution takes the form

$$F_n(\mathbf{R})=f_n(\mathbf{k})\exp(i\mathbf{k}\cdot\mathbf{R}) \tag{7.7.11}$$

where $f_n(\mathbf{k})$ obeys

$$\sum_n H_{mn}(\mathbf{k})f_n(\mathbf{k})=E(\mathbf{k})\sum_n S_{mn}(\mathbf{k})f_n(\mathbf{k}) \tag{7.7.12}$$

Since $H_{mn}(\mathbf{k})$ and $S_{mn}(\mathbf{k})$ are periodic functions of $\mathbf{k}$, we can *choose* our solutions $f_n(\mathbf{k})$ to be periodic in $\mathbf{k}$:

$$f_n(\mathbf{k}+\mathbf{K})=f_n(\mathbf{k}) \tag{7.7.13}$$

so that our Bloch functions $\psi(\mathbf{k},\mathbf{r})$ and their energies $E(\mathbf{k})$ will likewise obey the periodicity conditions 7.7.8.

Plane Wave Treatment

The problem of periodicity can be discussed rigorously in the plane wave representation of Eq. 7.4.2. If in Eq. 7.4.3 for the coefficients $c(\mathbf{k},\mathbf{K})$ of the Fourier representation of $\psi(\mathbf{k},\mathbf{r})$:

$$\left[E-\left(\frac{\hbar^2}{2m}\right)(\mathbf{k}+\mathbf{K})^2\right]c(\mathbf{k},\mathbf{K})=\sum_{\mathbf{K}'}V_{\mathbf{K}-\mathbf{K}'}c(\mathbf{k},\mathbf{K}') \qquad (7.7.14)$$

we make the replacement $\mathbf{k}\to\mathbf{k}+\mathbf{K}_0$, we see that the resulting equation is the same as if we had made the replacements $\mathbf{K}\to\mathbf{K}+\mathbf{K}_0$, $\mathbf{K}'\to\mathbf{K}'+\mathbf{K}_0$, keeping $\mathbf{k}$ fixed. Comparing these two sets of equations, we learn that

$$c(\mathbf{k}+\mathbf{K}_0,\mathbf{K})=c(\mathbf{k},\mathbf{K}+\mathbf{K}_0) \qquad (7.7.15)$$

with the same eigenvalue, $E(\mathbf{k}+\mathbf{K}_0)=E(\mathbf{k})$.

To be precise, all that we really know is that a solution $c_n(\mathbf{k}+\mathbf{K}_0,\mathbf{K})$ for $c(\mathbf{k}+\mathbf{K}_0,\mathbf{K})$ obeys the same equations as *some* solution for the quantities $c(\mathbf{k},\mathbf{K}+\mathbf{K}_0)$:

$$c_n(\mathbf{k}+\mathbf{K}_0,\mathbf{K})=c_{n'}(\mathbf{k},\mathbf{K}+\mathbf{K}_0) \qquad (7.7.16)$$

If this is substituted into

$$\psi_n(\mathbf{k},\mathbf{r})=\sum c_n(\mathbf{k},\mathbf{K})\exp[i(\mathbf{k}+\mathbf{K})\cdot\mathbf{r}] \qquad (7.7.17)$$

we obtain

$$\psi_n(\mathbf{k}+\mathbf{K}_0,\mathbf{r})=\psi_{n'}(\mathbf{k},\mathbf{r}) \qquad (7.7.18)$$

and hence

$$E_n(\mathbf{k}+\mathbf{K}_0)=E_{n'}(\mathbf{k})$$

In other words, the change $\mathbf{k}\to\mathbf{k}+\mathbf{K}_0$ produces a shift in the rows of the secular determinant:

$$\det\left\{\left[E-\left(\frac{\hbar^2}{2m}\right)(\mathbf{k}+\mathbf{K})^2\right]\delta_{\mathbf{K}\mathbf{K}'}-V_{\mathbf{K}-\mathbf{K}'}\right\}=0 \qquad (7.7.19)$$

which does not change the *set* of solutions $E_n(\mathbf{k})$ but *may* rearrange them. Thus as a *set* we have periodicity

$$\{E_n(\mathbf{k}+\mathbf{K}_0)\}=\{E_n(\mathbf{k})\} \qquad (7.7.20)$$

The individual band functions $E_n(\mathbf{k})$ must then be periodic *if* we *name* the bands in order of energy. [See, however, Section 7.8 for a proof that $\psi_n(\mathbf{k},\mathbf{r})$ can be made periodic in the nondegenerate case.]

7.8. THE k·p METHOD OF ANALYTICAL CONTINUATION

We have already examined the properties of electronic energy bands in the large: periodicity under $\mathbf{k}\to\mathbf{k}+\mathbf{K}$ or symmetry under $\mathbf{k}\to\alpha\mathbf{k}$. But we wish our energy bands to be continuous, and preferably also differentiable in $\mathbf{k}$. Therefore we shall investigate the properties of electronic energy bands in moving from $\mathbf{k}$ to $\mathbf{k}+\Delta\mathbf{k}$. [See, for example, Kane (1966).] Let us set

$$\psi(\mathbf{k}+\Delta\mathbf{k},\mathbf{r}) = \exp(i\Delta\mathbf{k}\cdot\mathbf{r})\,\chi(\mathbf{k},\mathbf{r}) \tag{7.8.1}$$

Then $\chi(\mathbf{k},\mathbf{r})$ has the translational symmetry of a Bloch wave of crystal momentum $\mathbf{k}$ and is therefore expandable in terms of the complete set of Bloch waves at $\mathbf{k}$:

$$\chi(\mathbf{k},\mathbf{r}) = \sum a_n\psi_n(\mathbf{k},\mathbf{r}) \tag{7.8.2}$$

The equation for $\psi(\mathbf{k}+\Delta\mathbf{k},\mathbf{r})$ can be rewritten as an equation for χ:

$$H(\Delta\mathbf{k})\chi(\mathbf{k},\mathbf{r}) = E(\mathbf{k}+\Delta\mathbf{k})\chi(\mathbf{k},\mathbf{r}) \tag{7.8.3}$$

where

$$H(\Delta\mathbf{k}) = \exp(-i\Delta\mathbf{k}\cdot\mathbf{r})H\exp(i\Delta\mathbf{k}\cdot\mathbf{r}) \tag{7.8.4}$$

This unitary transformation commutes with $\mathbf{r}$, but it takes $\mathbf{p}=(\hbar/i)\nabla$ into

$$\exp(-i\Delta\mathbf{k}\cdot\mathbf{r})\mathbf{p}\exp(i\Delta\mathbf{k}\cdot\mathbf{r}) = (\mathbf{p}+\hbar\Delta\mathbf{k}) \tag{7.8.5}$$

[To see this, let Eq. 7.8.5 act on an arbitrary $f(\mathbf{r})$.] The Hamiltonian $H(\mathbf{p},\mathbf{r})$ becomes

$$H(\Delta\mathbf{k}) = H(\mathbf{p}+\hbar\Delta\mathbf{k},\mathbf{r}) \tag{7.8.6}$$

For the three most frequently used Hamiltonians:

$$H=\frac{p^2}{2m}+V \qquad \text{(Schrödinger)}$$

$$H=\frac{p^2}{2m}+V+\left(\frac{\hbar}{4m^2c^2}\right)[\nabla V(\mathbf{r})\times\mathbf{p}\cdot\boldsymbol{\sigma}] \qquad \text{(Pauli)} \tag{7.8.7}$$

$$H=c\boldsymbol{\alpha}\cdot\mathbf{p}+V \qquad \text{(Dirac)}$$

we can write

$$H(\Delta\mathbf{k}) = H + \hbar\Delta\mathbf{k}\cdot\mathbf{v} + \frac{\hbar^2(\Delta\mathbf{k})^2}{2m} \tag{7.8.8}$$

[In the Dirac case, the term in $(\mathbf{\Delta k})^2$ does not occur.] Here, $\mathbf{v}$ is the velocity operator:

$$\mathbf{v} = \dot{\mathbf{r}} = \frac{[\mathbf{r}, H]}{i\hbar} \tag{7.8.9}$$

which takes the respective forms

$$\mathbf{v} = \frac{\mathbf{p}}{m}$$

$$\mathbf{v} = \frac{\mathbf{p}}{m} + \left(\frac{\hbar}{4m^2c^2}\right)\boldsymbol{\sigma}\times\nabla V \tag{7.8.10}$$

$$\mathbf{v} = c\boldsymbol{\alpha}$$

When Eq. 7.8.2 is inserted into 7.8.3, we obtain the set of simultaneous equations:

$$\sum H(\Delta\mathbf{k})_{mn}a_n = E(\mathbf{k}+\Delta\mathbf{k})a_m \tag{7.8.11}$$

where

$$H(\Delta\mathbf{k})_{mn} = \left[E_m(\mathbf{k}) + \frac{\hbar^2(\Delta\mathbf{k})^2}{2m}\right]\delta_{mn} + \hbar\Delta\mathbf{k}\cdot\mathbf{v}^{mn}$$

and

$$\mathbf{v}^{mn} = \int \psi_m(\mathbf{k},\mathbf{r})^*\mathbf{v}\psi_n(\mathbf{k},\mathbf{r})\,d\mathbf{r} \tag{7.8.12}$$

where the integral is over the volume of normalization of the ψ_n. See Eq. 7.8.24 and the following discussion.

If at the point $\mathbf{k}$ there is a finite energy separation between the energy $E_m(\mathbf{k})$ and all other energies $E_n(\mathbf{k})$, Schrödinger perturbation theory yields the new eigenvalue correct to terms of second order in $\Delta\mathbf{k}$:

$$E_m(\mathbf{k}+\Delta\mathbf{k}) = E_m(\mathbf{k}) + \frac{\hbar^2(\Delta\mathbf{k})^2}{2m} + \hbar\Delta\mathbf{k}\cdot\mathbf{v}^{mm} + \hbar^2\sum_{n\neq m}\frac{(\Delta\mathbf{k}\cdot\mathbf{v}^{mn})(\Delta\mathbf{k}\cdot\mathbf{v}^{nm})}{E_m(\mathbf{k}) - E_n(\mathbf{k})} \tag{7.8.13}$$

In this nondegenerate case the energy is unique, and by Eq. 7.8.14 it is continuous and differentiable in **k** space. Indeed, the first two derivatives are given by

$$\nabla E_m(\mathbf{k}) \equiv \frac{\partial E_m(\mathbf{k})}{\partial \mathbf{k}} = \hbar \mathbf{v}^{mm}$$
$$\frac{\partial^2 E_m(\mathbf{k})}{\partial \mathbf{k}\,\partial \mathbf{k}} = 2\hbar^2 \sum_{n\neq m} \frac{\mathbf{v}^{mn}\mathbf{v}^{nm}}{E_m(\mathbf{k}) - E_n(\mathbf{k})} + \frac{\hbar^2}{m}\mathbf{1} \qquad (7.8.14)$$

An interesting by-product of Eq. 7.8.14 is that $\nabla \mathbf{E}_m(\mathbf{k})$ vanishes at any point **k** [at which $E_m(\mathbf{k})$ is nondegenerate] for which $\psi_m(\mathbf{k},\mathbf{r})$ is real, for (with $\mathbf{v} \propto i\nabla$) the scalar product

$$(\psi_m(\mathbf{k},\mathbf{r}), i\nabla\psi_m(\mathbf{k},\mathbf{r})) = \text{purely imaginary}$$

whereas the diagonal elements $\mathbf{v}^{mm}$ of a Hermitian operator must be real. This proof can be extended immediately to the Pauli and Dirac cases by using the oddness of **v** under time reversal and the techniques of Chapter 10.

But $\psi(\mathbf{k},\mathbf{r})$ can be made real only if (1) $-\mathbf{k}$ is equivalent to **k** (i.e., at $\mathbf{k}=0$ and at the center of a zone boundary plane, at the center of an edge, or at a corner point of the Brillouin zone), and if (2) time reversal produces no added degeneracy. (See Section 10.7.)

We now turn to the perturbation expression for $\psi_m(\mathbf{k},\mathbf{r})$. Since the eigenstate is always ambiguous to the extent of an arbitrary **k**-dependent phase factor, we can write the perturbation expression for it, correct to first order in $\Delta\mathbf{k}$, as:

$$\psi_m(\mathbf{k}+\Delta\mathbf{k},\mathbf{r}) = \exp[i\Delta\mathbf{k}\cdot(\boldsymbol{\alpha}+\mathbf{r})]\left[\psi_m(\mathbf{k},\mathbf{r}) - i\sum_{n\neq m}\psi_n(\mathbf{k},\mathbf{r})\xi_{nm}(\mathbf{k})\cdot\Delta\mathbf{k}\right] \qquad (7.8.15)$$

where $\xi_{nm}(\mathbf{k})$ is an abbreviation for

$$\xi_{nm}(\mathbf{k}) = \frac{\hbar}{i}\frac{\mathbf{v}^{nm}}{E_n(\mathbf{k}) - E_m(\mathbf{k})} \qquad (7.8.16)$$

and $\exp(i\boldsymbol{\alpha}\cdot\Delta\mathbf{k})$ is an arbitrary phase factor at our disposal. In principle, $\boldsymbol{\alpha}$ can depend on the *direction* of $\Delta\mathbf{k}$. In this case, however, $\psi_m(\mathbf{k},\mathbf{r})$ would not possess a unique derivative in **k** space. Let us therefore set $\boldsymbol{\alpha} \equiv -\xi_{mm}(\mathbf{k})$, independent of $\Delta\mathbf{k}$, but an as yet arbitrary function of **k**. In terms of the periodic parts of the Bloch functions, Eq. 7.8.15 can be rewritten as

$$\frac{i\,\partial u_m(\mathbf{k},\mathbf{r})}{\partial \mathbf{k}} = \sum_n u_n(\mathbf{k},\mathbf{r})\xi_{nm}(\mathbf{k}) \qquad (7.8.17)$$

where the sum over n includes m. Using the orthonormality

$$\int u_n(\mathbf{k},\mathbf{r})^* u_m(\mathbf{k},\mathbf{r})\,d\mathbf{r} = \delta_{nm} \tag{7.8.18}$$

(which follows from the orthonormality of the corresponding Bloch waves), we obtain

$$\xi_{nm}(\mathbf{k}) = \int u_n(\mathbf{k},\mathbf{r})^* \left(i\frac{\partial}{\partial \mathbf{k}} \right) u_m(\mathbf{k},\mathbf{r})\,d\mathbf{r} \tag{7.8.19}$$

The volume of integration in Eq. 7.8.19 must be the same as in 7.8.18, and both *can* be taken equal to the volume of the primitive cell. The ξ_{nm} were first introduced by Adams (**1952**, **1953**) via Eq. 7.8.17 in discussing the crystal momentum representation. See also Blount (**1962a**).

Thus, given a set of solutions $u_m(\mathbf{k},\mathbf{r})$ differentiable in $\mathbf{k}$, we can construct the $\xi_{nm}(\mathbf{k})$. Conversely, given a set of $\xi_{nm}(\mathbf{k})$, we can use Eq. 7.8.17 to integrate step by step from, say, $\mathbf{k}=0$ to $\mathbf{k}$ to produce a set of functions $u_m(\mathbf{k},\mathbf{r})$.

Will these functions be independent of the path of integration from $\mathbf{k}=0$ to $\mathbf{k}$? The path integral of a gradient

$$\int_0^{\mathbf{k}} \nabla_{\mathbf{k}} u_m(\mathbf{k},\mathbf{r})\cdot d\mathbf{k} \tag{7.8.20}$$

will depend on the end points, and not on the path. However, the right-hand side of Eq. 7.8.17 can represent a gradient only if its curl vanishes. Since

$$\operatorname{curl}(u\xi) = u \operatorname{curl}\xi + \nabla_{\mathbf{k}} u \times \xi$$

we obtain the condition

$$\sum_n \left(u_n \operatorname{curl}\xi_{nm} + \frac{\partial u_n}{\partial \mathbf{k}} \times \xi_{nm} \right) = 0$$

Let us now eliminate $\partial u_n/\partial \mathbf{k}$ using Eq. 7.8.17. Multiply the preceding equation by $u_s(\mathbf{k},\mathbf{r})$, and integrate over $\mathbf{r}$. After changing the name s to n, we obtain the set of conditions

$$\operatorname{curl}\xi_{nm} = i\sum_p \xi_{np} \times \xi_{pm} \tag{7.8.21}$$

for each n. For $n=m$ we obtain a restriction on our previously arbitrary ξ_{mm}:

$$\begin{aligned} \operatorname{curl}\xi_{mm}(\mathbf{k}) &= i\sum_p \xi_{mp} \times \xi_{pm} \\ &= i\sum_{p\neq m} \xi^*_{pm} \times \xi_{pm} \end{aligned} \tag{7.8.22}$$

In obtaining the last form, we omit the term $\xi_{mm} \times \xi_{mm}$, which vanishes since ξ_{mm} is real. To establish this reality, we use the Hermiticity condition

$$\xi_{mn}(\mathbf{k}) = \xi^*_{nm}(\mathbf{k}) \tag{7.8.23}$$

which is an immediate consequence of Eqs. 7.8.17 and 7.8.18 and

$$\frac{\partial}{\partial \mathbf{k}} \int u^*_m(\mathbf{k},\mathbf{r})\, u_n(\mathbf{k},\mathbf{r})\, d\mathbf{r} = 0 \tag{7.8.24}$$

Phase Factor Transformations

Solutions $\psi_n(\mathbf{k},\mathbf{r})$ of the Schrödinger equation are unique (in the nondegenerate case), except for an arbitrary phase factor $\exp[i\varphi_n(\mathbf{k})]$. Under the transformation

$$\psi'_n(\mathbf{k},\mathbf{r}) = \exp[\,i\varphi_n(\mathbf{k})\,]\psi_n(\mathbf{k},\mathbf{r})$$

we have

$$\begin{aligned} \xi'_{nm}(\mathbf{k}) &= \xi_{nm}(\mathbf{k})\exp[\,i\varphi_m(\mathbf{k}) - i\varphi_n(\mathbf{k})\,], \qquad n \neq m \\ \xi'_{mm}(k) &= \xi_{mm}(\mathbf{k}) - \frac{\partial \varphi_m(\mathbf{k})}{\partial \mathbf{k}} \end{aligned} \tag{7.8.25}$$

Thus the off-diagonal elements of ξ transform in exactly the same way as those of any operator A. However, for an operator, A_{mm} is unique, whereas ξ_{mm} is not. A specification of $\xi_{mm}(\mathbf{k})$ is thus equivalent to a specification of the phase of the Bloch waves.

Since by Eq. 7.8.22 curl$\,\xi_{mm}$ is independent of phase factor transformations, it is in general impossible to make a transformation that reduces $\xi_{mm}(\mathbf{k})$ to zero at all $\mathbf{k}$. (This is possible, however, when inversion is an element of the space group. See Problems 7.8.5–7.8.8.)

The standard application of $\mathbf{k}\cdot\mathbf{p}$ perturbation theory seems to use $\xi_{mm} = 0$, that is, $\alpha = 0$ in Eq. 7.8.15. This is deceptive. All that is really used is $\xi_{mm}\cdot \Delta\mathbf{k} = 0$ along the path of integration, and the path (in, e.g., Eq. 7.8.20) is usually understood to be a straight line, say from $\mathbf{k} = 0$ to $\mathbf{k}$. With the path specified, a unique function $u_m(\mathbf{k},\mathbf{r})$, differentiable in $\mathbf{k}$, is obtained, with $\xi_{mm}(\mathbf{k})\cdot\mathbf{k} = 0$. The components of $\xi_{mm}(\mathbf{k})$ transverse to $\mathbf{k}$ were not used in the construction of $u_m(\mathbf{k},\mathbf{r})$ and cannot be read from Eq. 7.8.15 or 7.8.17. When curl $\xi_{mm} \neq 0$, these transverse components will automatically assume nonvanishing values, since $\int \xi_{mm}\cdot d\mathbf{k}$ around a closed path including two radial rays plus a small $\Delta\mathbf{k}$ cannot vanish by Stokes's theorem, even though the radial contributions vanish by construction.

Construction of a Periodic $\xi_{mm}(\mathbf{k})$

If band m is not degenerate with other bands at $\mathbf{k}$, then $\psi_m(\mathbf{k}+\mathbf{K},\mathbf{r})$ can differ from $\psi_m(\mathbf{k},\mathbf{r})$ only by a phase factor λ:

$$\psi_m(\mathbf{k}+\mathbf{K},\mathbf{r}) = \lambda\psi_m(\mathbf{k},\mathbf{r}) \tag{7.8.26}$$

Quantities such as the energy and curl ξ_{mm}, which by Eq.7.8.25 is also independent of phase factor transformations, are then automatically periodic:

$$E_m(\mathbf{k}+\mathbf{K}) = E_m(\mathbf{k}) \tag{7.8.27}$$

$$\operatorname{curl}\xi_{mm}(\mathbf{k}+\mathbf{K}) = \operatorname{curl}\xi_{mm}(\mathbf{k}) \tag{7.8.28}$$

We shall now display an explicit "special solution" $\xi'_{mm}(\mathbf{k})$ that is (*a*) periodic, and (*b*) consistent with given $\operatorname{curl}\xi_{mm}$. Let us express $\Omega \equiv \operatorname{curl}\xi$ as a Fourier series (see Eq. 6.4.9).

$$\Omega(\mathbf{k}) = \operatorname{curl}\xi_{mm}(\mathbf{k}) = \sum \Omega_{\mathbf{t}} \exp(i\mathbf{k}\cdot\mathbf{t}) \tag{7.8.29}$$

where $\mathbf{t}$ are the vectors of the Bravais lattice, and

$$\Omega_{\mathbf{t}}\cdot\mathbf{t} = 0 \tag{7.8.30}$$

since $\operatorname{div}\operatorname{curl}\xi = 0$. It is now easy to verify that

$$\xi'_{mm}(\mathbf{k}) = i\sum_{t\neq 0}\left(\frac{\mathbf{t}\times\Omega_{\mathbf{t}}}{t^2}\right)\exp(i\mathbf{k}\cdot\mathbf{t}) \tag{7.8.31}$$

is *periodic* in $\mathbf{k}$ and has a curl identical to the right-hand side of Eq. 7.8.29, provided we assume that $\Omega_0 = 0$, a result that can be demonstrated using the relation $\Omega(-\mathbf{k}) = -\Omega(\mathbf{k})$, which follows from time reversal considerations. (See Problem 7.8.7.)

Let us suppose now that we have obtained a particular set of (differentiable) functions $u_m(\mathbf{k},\mathbf{r})$ by any method whatever—plane waves, atomic orbitals, or $\mathbf{k}\cdot\mathbf{p}$ along a specified path. If we compute $\xi_{mm}(\mathbf{k})$ and $\operatorname{curl}\xi_{mm}(\mathbf{k})$ for these functions and determine ξ' for this $\operatorname{curl}\xi$, we have

$$\operatorname{curl}(\xi_{mm} - \xi'_{mm}) = 0 \tag{7.8.32}$$

But any vector whose curl vanishes can be expressed as a gradient:

$$\xi'_{mm}(\mathbf{k}) = \xi_{mm}(\mathbf{k}) - \frac{\partial\varphi_m(\mathbf{k})}{\partial\mathbf{k}} \tag{7.8.33}$$

Thus there exists a phase factor transformation $\exp(i\varphi_m)$ that converts $\xi_{mm}(\mathbf{k})$ to $\xi'_{mm}(\mathbf{k})$. In other words, in addition to the mandatory requirement 7.8.22 on curlξ, it is possible to require that the phases of our Bloch waves be so chosen that

$$\xi_{mm}(\mathbf{k}+\mathbf{K})=\xi_{mm}(\mathbf{k}) \tag{7.8.34}$$

the periodicity property obeyed by $\xi'_{mm}(\mathbf{k})$.

Moreover, our choice of special solution (Eq. 7.8.31) obeys the condition

$$\operatorname{div}\xi'_{mn}=0$$

which is useful in minimizing the range of the Wannier functions constructed from $\psi_m(\mathbf{k},\mathbf{r})$. [See Wannier (**1937**) or Blount (**1962a**)].

The ξ_{mm} that we have obtained is not unique. It is still subject to phase transformations where $\nabla\varphi_m(\mathbf{k})$ is periodic. If we retain $\operatorname{div}\xi=0$, however, $\nabla^2\varphi_m=0$ and only φ_m linear in $\mathbf{k}$ are permitted.

Construction of Bloch Functions Periodic in k

We can rewrite Eq.7.8.26 in the form

$$u_m(\mathbf{k}+\mathbf{K},\mathbf{r})=\lambda\exp(-i\mathbf{K}\cdot\mathbf{r})\,u_m(\mathbf{k},\mathbf{r}) \tag{7.8.35}$$

where $\lambda=\exp[i\theta_m(\mathbf{K},\mathbf{k})]$ is a phase factor that can depend on $\mathbf{K}$ and the band index m, as well as on $\mathbf{k}$. Using Eqs. 7.8.35 and 7.8.19, we find that

$$\begin{aligned}\xi_{mm}(\mathbf{k}+\mathbf{K})&=\xi_{mm}(\mathbf{k})+\lambda(\mathbf{k})^*\frac{i\,\partial\lambda(\mathbf{k})}{\partial\mathbf{k}}\\&=\xi_{mm}(\mathbf{k})-\nabla_{\mathbf{k}}\theta_m(\mathbf{K},\mathbf{k})\end{aligned} \tag{7.8.36}$$

The periodicity of ξ_{mm} by Eq. 7.8.34 then requires that $\theta_m(\mathbf{K},\mathbf{k})$ be independent of $\mathbf{k}$. Moreover, the use of successive displacements in $\mathbf{K}$ space requires θ to be an additive function:

$$\theta(\mathbf{K}_1+\mathbf{K}_2)=\theta(\mathbf{K}_1)+\theta(\mathbf{K}_2) \tag{7.8.37}$$

The only solution of this functional equation is the linear function

$$\theta_m(\mathbf{K})=\mathbf{l}\cdot\mathbf{K} \tag{7.8.38}$$

Thus the transformation

$$\psi'_m(\mathbf{k},\mathbf{r}) = \exp(-i\mathbf{l}\cdot\mathbf{k})\psi_m(\mathbf{k},\mathbf{r}) \tag{7.8.39}$$

makes

$$\psi'_m(\mathbf{k}+\mathbf{K},\mathbf{r}) = \psi'_m(\mathbf{k},\mathbf{r}) \tag{7.8.40}$$

and

$$\xi'_{mm} = \mathbf{l} + \xi_{mm} \tag{7.8.41}$$

If we wish to retain periodicity in the Bloch wave, and div $\xi_{mm} = 0$, *no further phase factor transformations are allowed* except trivial **k**-independent transformations. A phase factor $\exp(-i\mathbf{k}\cdot\mathbf{l})$ takes $\exp(i\mathbf{k}\cdot\mathbf{r})$ into $\exp[i\mathbf{k}\cdot(\mathbf{r}-\mathbf{l})]$ and is related, therefore, to a shift in the origin of the unit cell. If any symmetry is present, the possibilities for such a shift will probably be limited to certain natural choices of origin.

The Naming Procedure in the Presence of Degeneracy

Suppose that we try to name our bands by the requirement that it be possible to go continuously by $\Delta\mathbf{k}\cdot\mathbf{v}$ perturbation theory from any general point of the band to any other such point (avoiding points of degeneracy). If there are points or lines of degeneracy, but *no planes of degeneracy*, we can always find a path from any general point **k** to any general point **k**′ that avoids all points and lines of degeneracy. This procedure results in a *unique assignment* of bands which of necessity leads to $E(\mathbf{k})$ and $\psi(\mathbf{k},\mathbf{r})$ that are continuous, and even differentiable in **k** at all *general* points. Points (or lines) of degeneracy have wave functions that are *not* uniquely assigned to particular bands but belong to several bands simultaneously.

Since the assignment scheme described above avoids all points of degeneracy, if one starts on the lowest energy band, one must surely remain on that band. In other words, *this analytic continuation scheme is entirely equivalent to naming the bands in order of energy!*

Furthermore, since the *complete patterns of energy bands are periodic, the lowest energy band must be periodic, as must the second and the nth*:

$$E_n(\mathbf{k}+\mathbf{K}) = E_n(\mathbf{k}) \tag{7.8.42}$$

That these conclusions are not entirely trivial can be seen by referring to a counterexample, the hexagonal close-packed structure, which contains a twofold screw axis $(\boldsymbol{\delta}_2|\frac{1}{2}\mathbf{c})$ along the hexagonal axis. The Brillouin zone boundary planes normal to this axis are planes containing a time-reversal degeneracy. (See Example 10.8.2.) Thus the energy bands cross over an entire, unavoidable plane,

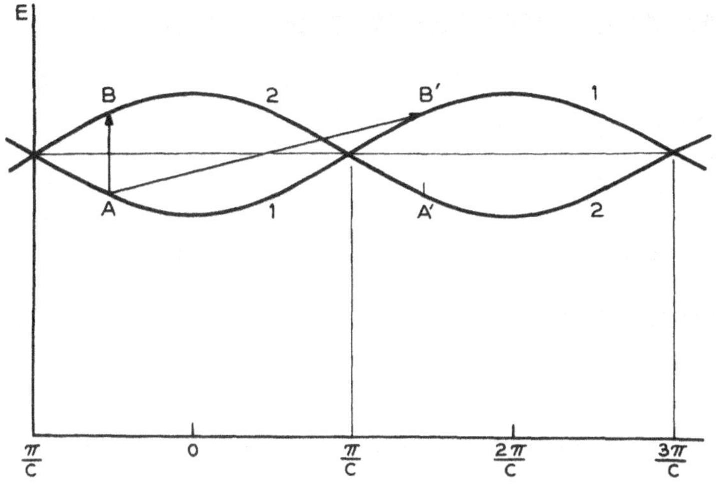

Fig. 7.8.1. Energy is plotted versus k_z, that is, along the c axis of a hexagonal close-packed lattice. The double degeneracy shown at $k_z = \pi/c$ persists for all k_x, k_y, that is, across the entire zone boundary plane.

and the analytical continuation procedure no longer guarantees a unique assignment. Figure 7.8.1 shows energy bands plotted along the hexagonal **c** direction in a hexagonal close-packed structure. If bands 1 and 2 are assigned as shown, this is equivalent to doubling the period in the **c** direction, that is, to using a Brillouin zone doubled in the hexagonal direction. If the bands are assigned in order of energy, the original period is maintained, but each band has a cusp at the π/c zone boundary plane.

Whether to use the single or double Brillouin zone is a question that we should not settle without a specific problem in mind. For example, the use of the double zone implies that direct optical transitions from A to B' do not occur because they do not conserve crystal momentum. However, these transitions are in fact equivalent to the AB transition, which does occur. Indeed, the wave function at B' can differ from the wave function at B only by a phase factor, since there is no degeneracy at a general point **k** of the zone. As another example, in constructing Wannier orbitals [Wannier (**1937**), Slater (**1949**), Koster (**1953**), Kohn (**1959b**)], the use of the single zone with its cusp leads to extended (i.e., poorly localized) Wannier orbitals [Des Cloizeaux (**1963**, **1964a**, **1964b**), Kohn and Onffroy (**1973**)]. The use of the double zone, on the other hand, leads to nicely localized orbitals at half the previous spacing, that is, to two orbitals per primitive cell. This is desirable, since the twofold screw axis that caused the planar degeneracy in fact implies the existence of at least two (or another even number) of atoms per cell.

7.9. DYNAMICS OF ELECTRON MOTION IN CRYSTALS

We are concerned in this section with the motion of an electron whose wave function at some initial time is expressible in terms of the Bloch functions of a single band.

Group Velocity

The group velocity is defined as the spatial velocity of the maximum of a wave packet, which we take in the form

$$\psi(\mathbf{r},t)=\int d\mathbf{k}\,a(\mathbf{k})\,u_m(\mathbf{k},\mathbf{r})\exp\left[i\left(\mathbf{k}\cdot\mathbf{r}-\frac{E_m(\mathbf{k})t}{\hbar}\right)\right] \tag{7.9.1}$$

where $a(\mathbf{k})$ has a maximum near $\mathbf{k}_0$. Thus we can approximate:

$$E_m(\mathbf{k})\approx E_m(\mathbf{k}_0)+\Delta\mathbf{k}\cdot\frac{\partial E_m}{\partial\mathbf{k}_0} \tag{7.9.2}$$

$$|\psi(\mathbf{r},t)|^2\approx|u_m(\mathbf{k}_0,\mathbf{r})|^2\left|\int d\mathbf{k}\,a(\mathbf{k})\exp\left[i\Delta\mathbf{k}\cdot(\mathbf{r}-\mathbf{v}t)\right]\right|^2 \tag{7.9.3}$$

where

$$\mathbf{v}=\frac{1}{\hbar}\frac{\partial E_m}{\partial\mathbf{k}_0} \tag{7.9.4}$$

Thus $\mathbf{v}$ is the group velocity of the packet centered at $\mathbf{k}_0$, that is, the velocity of the moving envelope, disregarding the wiggles in $|u_m(\mathbf{k}_0,\mathbf{r})|^2$. But $\partial E_m/\partial(\hbar\mathbf{k}_0)$, the spatial (group) velocity, is by Eq. 7.8.14 precisely equal to $\mathbf{v}^{mm}$, the mean value in the quantum state $\psi_m(\mathbf{k}_0,\mathbf{r})$ of the operator 7.8.10 representing the velocity.

Acceleration

Suppose that an electron is subject to an external force $\mathbf{F}$ sufficiently small that it cannot cause the electron to jump the gap from one band to another. The only effect of this force will then be to change the $\mathbf{k}$ vector of the electron. The requirement that the rate of increase of energy be equal to the work done per unit time yields the condition

$$\frac{\partial E}{\partial \mathbf{k}} \cdot \frac{d\mathbf{k}}{dt} = \mathbf{v} \cdot \mathbf{F} = \frac{1}{\hbar} \frac{\partial E}{\partial \mathbf{k}} \cdot \mathbf{F} \tag{7.9.5}$$

or

$$\frac{d(\hbar \mathbf{k})}{dt} = \mathbf{F} \tag{7.9.6}$$

Thus in a solid we have an equation analogous to Newton's law, in which the crystal momentum $\hbar\mathbf{k}$ plays the role of the momentum even though the crystal momentum is *not* the same as the actual momentum $(\psi(\mathbf{k},\mathbf{r}), \mathbf{p}\psi(\mathbf{k},\mathbf{r}))$.

If we differentiate the group velocity equation, we obtain

$$\frac{d\mathbf{v}}{dt} = \frac{1}{\hbar} \frac{\partial^2 E}{\partial \mathbf{k} \partial \mathbf{k}} \cdot \frac{d\mathbf{k}}{dt} \tag{7.9.7}$$

or

$$\frac{d\mathbf{v}}{dt} = \mathbf{M}^{-1} \cdot \mathbf{F} \tag{7.9.8}$$

where

$$\mathbf{M}^{-1} = \left(\frac{1}{\hbar}\right)^2 \frac{\partial^2 E}{\partial \mathbf{k} \partial \mathbf{k}} \tag{7.9.9}$$

or

$$M_{\mu\nu}^{-1} = \left(\frac{1}{\hbar}\right)^2 \frac{\partial^2 E}{\partial k_\mu \partial k_\nu} \tag{7.9.10}$$

According to Eq. 7.9.8, the electron accelerates in a solid under an applied force as if it possessed an (anisotropic) mass differing from its vacuum value. We shall refer to $\mathbf{M}$ as the *effective mass tensor* in a solid.

Near an energy minimum, say at $\mathbf{k}_0$, we can write

$$E(\mathbf{k}) \approx E(\mathbf{k}_0) + \left(\frac{\hbar^2}{2}\right)(\mathbf{k}-\mathbf{k}_0) \cdot \mathbf{M}^{-1} \cdot (\mathbf{k}-\mathbf{k}_0) \tag{7.9.11}$$

Masses of Electrons and Holes in a Tight Binding Example

Consider, for example, a simple cubic lattice formed from atomic s states in the tight binding approximation 7.7.5. Neglect overlap integrals in $S(\mathbf{k})$ but retain them out to first neighbors in $H(\mathbf{k})$. Application of the cubic point group operations to $H(\mathbf{R})$ demonstrates that

$$H(\pm 1,0,0) = H(0,\pm 1,0) = H(0,0,\pm 1) = -H_1 \qquad (7.9.12)$$

where the sign appropriate to an attractive potential is taken. The energy can then be written in the form

$$E(\mathbf{k}) = H(0) - 2H_1(\cos k_1 a + \cos k_2 a + \cos k_3 a) \qquad (7.9.13)$$

This band has a width of $12H_1$. Its minimum occurs at $\mathbf{k}=0$, that is, Γ, where

$$E(\mathbf{k}) \approx H(0) - 6H_1 + H_1 a^2 k^2 \qquad (7.9.14)$$

so that the effective mass tensor is isotropic with the value

$$\mathbf{M} = m^* \mathbf{1}, \qquad m^* = \frac{\hbar^2}{2H_1 a^2} \qquad (7.9.15)$$

inversely proportional to the band width.

The maximum energy occurs at $k_1 = k_2 = k_3 = \pi/a$, the point R in the Brillouin zone. (See Appendix E.) In the vicinity of this point

$$E(\mathbf{k}) \approx H(0) + 6H_1 - H_1 a^2 (\Delta\mathbf{k})^2 \qquad (7.9.16)$$

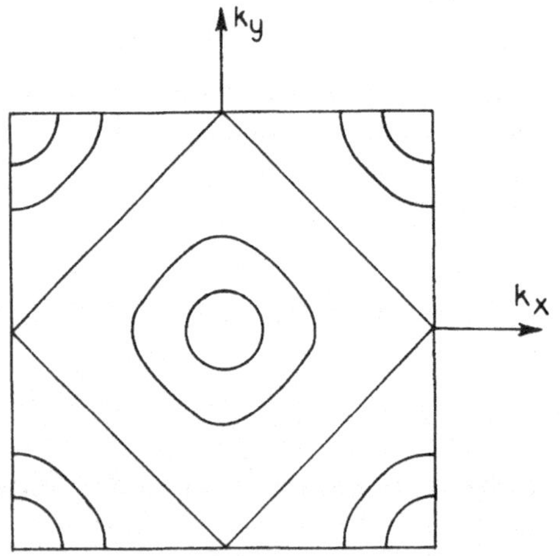

Fig. 7.9.1. Energy bands in k_x, k_y plane of simple cubic lattice in tight-binding approximation.

A sketch of the constant energy contours is given in Fig. 7.9.1.

It is clear from Eqs. 7.9.10 and 7.9.13 that the effective mass will be negative in some portions of an energy band, in particular at the top of the band. Observations of such a negative mass near the top of a band can be carried out only in a band that is full nearly to the top.

A completely full band carries no current, because a filled band that contains the state $\psi(\mathbf{k},\mathbf{r})$ also contains the time-reversed state $\psi(-\mathbf{k},\mathbf{r})$. These two states of necessity carry opposite current, since the evenness of $E(\mathbf{k})$ due to time reversal (see Example 7.1.1) makes the mean velocity $\equiv \nabla E(\mathbf{k})$ odd in $\mathbf{k}$. (This is nothing more than the statement that the velocity operator is odd under time reversal.) With the convention $e = -|e| =$ charge on an electron, the current carried by an empty state $\mathbf{k}$, that is, a *hole* at $\mathbf{k}$, is

$$\mathbf{j}^{\text{hole}}(\mathbf{k}) = -e\mathbf{v}(\mathbf{k}) \tag{7.9.17}$$

But the motion of the state is the same as if it were occupied:

$$\begin{aligned}\frac{d}{dt}\mathbf{j}^{\text{hole}}(\mathbf{k}) &= -e\frac{d\mathbf{v}(\mathbf{k})}{dt}\\ &= -e\left(\frac{1}{\hbar}\right)^2\frac{\partial^2 E}{\partial\mathbf{k}\,\partial\mathbf{k}}\cdot\mathbf{F}\end{aligned} \tag{7.9.18}$$

In an electric and magnetic field, then,

$$\mathbf{F} = e\left(\mathbf{E} + \frac{\mathbf{v}}{c}\times\mathbf{B}\right)$$

$$\frac{d}{dt}\mathbf{j}^{\text{hole}}(\mathbf{k}) = |e|\mathbf{M}_h^{-1}|e|\cdot\left(\mathbf{E} + \frac{\mathbf{v}}{c}\times\mathbf{B}\right)$$

where

$$\mathbf{M}_h^{-1} = -\mathbf{M}^{-1} = -\left(\frac{1}{\hbar}\right)^2\frac{\partial^2 E}{\partial\mathbf{k}\,\partial\mathbf{k}} \tag{7.9.19}$$

is the positive mass tensor of a hole. Thus the hole current can be calculated as if we were dealing with a positive charge and a positive mass.

Of course, there will be regions of $\mathbf{k}$ space that cannot be characterized by a definite sign of the effective mass, since $E(\mathbf{k})$ necessarily has saddle points (Van Hove, **1953**). In our single cubic tight binding example, the mass tensor at $k_1 = k_2 = 0, k_3 = \pi/a$ has two positive components (in the x and y directions) and one negative component (in the z direction). Conversely, at $k_1 = k_2 = \pi/a, k_3 = 0$ the situation is reversed.

Effective Mass Tensor and Oscillator Strength Sum Rule

Since the effective mass is a property of the local environment of a particular **k**, it can be obtained exactly by $\Delta\mathbf{k}\cdot\mathbf{p}$ perturbation theory. Indeed the answer is immediately obtainable from Eq. 7.8.14:

$$(\mathbf{M})^{-1} = \frac{1}{\hbar^2}\frac{\partial^2 E}{\partial\mathbf{k}\,\partial\mathbf{k}} = \frac{1}{m}\mathbf{1} + 2\sum_{n'\neq n}\frac{\mathbf{v}^{nn'}\mathbf{v}^{n'n}}{E_n(\mathbf{k}) - E_{n'}(\mathbf{k})} \tag{7.9.20}$$

This result (Eq. 7.9.20) includes, as a special case, the oscillator strength sum rule appropriate to solids:

$$\sum_{n'\neq n} f_{n'n\mu} = 1 - m(M^{-1})_{\mu\mu} \tag{7.9.21}$$

where

$$f_{n'n\mu} = \frac{2m|v_\mu^{nn'}|^2}{E_{n'}(\mathbf{k}) - E_n(\mathbf{k})} \tag{7.9.22}$$

is the conventional form of oscillator strength. This rule differs from the usual atomic rule by the subtractive term $m(M^{-1})_{\mu\mu}$. This term diminishes the amount of optical absorption to other bands, $n'\neq n$, presumably because of the presence of "free carrier" absorption (Rosenberg and Lax, **1958**) within band n.

7.10. EFFECTIVE HAMILTONIANS AND DONOR STATES

The equations of motion for an electron confined to one (isolated) band subject to an electrostatic potential $\varphi(\mathbf{R})$:

$$\frac{d\mathbf{R}}{dt} = \mathbf{v} = \frac{\partial E(\mathbf{k})}{\partial(\hbar\mathbf{k})} \tag{7.10.1}$$

$$\frac{d(\hbar\mathbf{k})}{dt} = \mathbf{F} = e\mathbf{E} = -e\nabla\varphi(\mathbf{R}) \tag{7.10.2}$$

are simply the Hamiltonian equations of motion for a system with Hamiltonian

$$H = E(\mathbf{k}) + e\varphi(\mathbf{R}) \tag{7.10.3}$$

with **R** and $\hbar\mathbf{k}$ treated as the position and conjugate momentum variables. We now make the extrapolation that it is permissible to regard H as a quantum-mechanical Hamiltonian provided that $\hbar\mathbf{k}$ and **R** are regarded as *op*erators obeying the standard commutation rules

$$[\mathbf{R}_{op}, \hbar\mathbf{k}_{op}] = i\hbar \qquad \text{or} \qquad [\mathbf{R}_{op}, \mathbf{k}_{op}] = i \tag{7.10.4}$$

Indeed, Eqs. 7.10.1 and 7.10.2 remain valid as quantum-mechanical equations and follow from [3]

$$\dot{\mathbf{R}}_{op} = \frac{[\mathbf{R}_{op}, H]}{i\hbar}, \qquad \hbar\dot{\mathbf{k}}_{op} = \frac{[\hbar\mathbf{k}_{op}, H]}{i\hbar} \tag{7.10.5}$$

The action of $\mathbf{R}_{op}$ on a space wave function $F(\mathbf{R})$ is simply

$$\mathbf{R}_{op}F(\mathbf{R}) = \mathbf{R}F(\mathbf{R}) \tag{7.10.6}$$

If we define a momentum space wave function $f(\mathbf{k})$ by

$$F(\mathbf{R}) = \int f(\mathbf{k}) \exp(i\mathbf{k}\cdot\mathbf{R})\, d\mathbf{k} \tag{7.10.7}$$

the action of $\mathbf{k}_{op}$ on $f(\mathbf{k})$ is simply

$$\mathbf{k}_{op}f(\mathbf{k}) = \mathbf{k}f(\mathbf{k}) \tag{7.10.8}$$

and its action on $F(\mathbf{R})$ is given by

$$\begin{aligned} \mathbf{k}_{op}F(\mathbf{R}) &= \int \mathbf{k}f(\mathbf{k}) \exp(i\mathbf{k}\cdot\mathbf{R})\, d\mathbf{k} \\ &= -i\frac{\partial}{\partial \mathbf{R}} \int f(\mathbf{k}) \exp(i\mathbf{k}\cdot\mathbf{R})\, d\mathbf{k} \\ &= \left(-i\frac{\partial}{\partial \mathbf{R}}\right)F(\mathbf{R}) \equiv (-i\nabla)F(\mathbf{R}) \end{aligned} \tag{7.10.9}$$

Equation 7.10.9 is simply a derivation of the standard quantum-mechanical result that a momentum operator is represented by $(\hbar/i)\nabla$ in the space representation. A similar proof shows that $\mathbf{R}$ is represented by $i\partial/\partial\mathbf{k}$ in the momentum representation. Thus the Schrödinger equation associated with our Hamiltonian 7.10.3 takes, in the space and momentum representations, the forms

$$\begin{aligned} [E(-i\nabla) + U(\mathbf{R})]F(\mathbf{R}) &= \frac{i\hbar\,\partial F(\mathbf{R})}{\partial t} \\ \left[E(\mathbf{k}) + U\left(i\frac{\partial}{\partial \mathbf{k}}\right)\right]f(\mathbf{k}) &= \frac{i\hbar\,\partial f(\mathbf{k})}{\partial t} \end{aligned} \tag{7.10.10}$$

where $U(\mathbf{R}) = e\varphi(\mathbf{R})$.

[3]See Landau and Lifschitz (1958), Section 14.

Let us consider the most frequent application of the present "effective" Hamiltonian method: the hydrogenic states of a donor atom in a crystal. If, for example, a phosphorus atom (valence 5) is inserted into silicon (valence 4), an excess charge appears, and the potential can be taken in the form

$$U(\mathbf{R}) = -\frac{e^2}{\epsilon R} \tag{7.10.11}$$

where ϵ is the dielectric constant of the host lattice. At $\mathbf{k}=0$ in a cubic crystal, the effective mass tensor must be isotropic, and the kinetic energy can be written as $-(\hbar^2/2m^*)\nabla^2$. When Eq. 7.9.11 is used with $\mathbf{k}_0=0$, the Schrödinger equation takes the form

$$\left[-\left(\frac{\hbar^2}{2m^*}\right)\nabla^2 - \frac{e^2}{\epsilon R}\right]F(\mathbf{R}) = [E - E(0)]F(\mathbf{R}) \tag{7.10.12}$$

Thus the usual hydrogenic solutions can be used with $m \to m^*$ and $e \to e\epsilon^{-1/2}$. In particular, the Bohr radius is now

$$a_B = \frac{\epsilon m}{m^*}\frac{\hbar^2}{me^2} \tag{7.10.13}$$

larger by a factor $\epsilon m/m^*$ than the usual hydrogenic value. In materials of sufficiently high dielectric constant and small effective mass, then a_B will be significantly larger than the lattice spacing a. Thus $F(\mathbf{R})$ is a slowly varying function, that is, the corresponding momentum wave function $f(\mathbf{k})$ is large only for k less than about $1/a_B$, which is small compared to $1/a$, the size of the Brillouin zone. This is the a postiori justification for having expanded $E(\mathbf{k})$ in powers of $\mathbf{k}$. The binding energy

$$E_B = \left(\frac{m^*}{m}\right)\epsilon^{-2}\left(\frac{me^4}{2\hbar^2}\right) \tag{7.10.14}$$

is significantly less than the free hydrogen atom binding, and measures the extent to which the bound-state energy E is below the energy $E(0)$ corresponding to the minimum energy of the conduction band.

If the minimum occurs at $\mathbf{k}_0$ rather than at the zone center, the combination $\mathbf{k}-\mathbf{k}_0 = -i\nabla - \mathbf{k}_0$ appears in the equation. But the $\mathbf{k}_0$ term can be eliminated by the transformation

$$F(\mathbf{R}) = \exp(i\mathbf{k}_0\cdot\mathbf{R})G(\mathbf{R}) \tag{7.10.15}$$

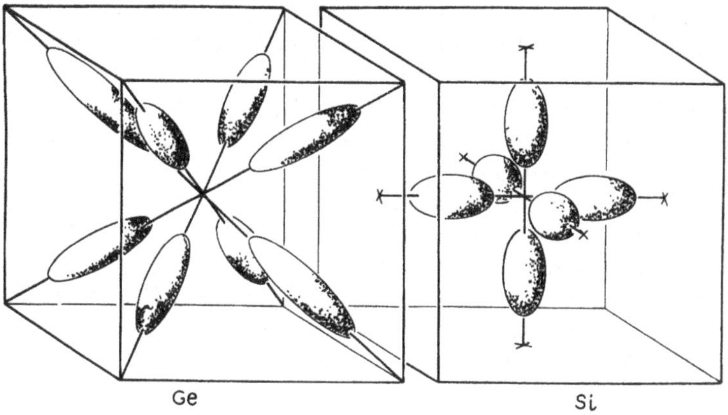

Fig. 7.10.1. Constant energy ellipsoids in **k** space for the conduction bands in germanium and silicon.

where, using Eq. 7.9.11, we find that $G(\mathbf{R})$ obeys

$$\left[-\left(\frac{\hbar^2}{2}\right)\nabla\cdot\mathbf{M}^{-1}\cdot\nabla-\frac{e^2}{\epsilon R}\right]G=[E-E(\mathbf{k}_0)]G \qquad (7.10.16)$$

In silicon, one of the conduction band minima occurs along the (1,0,0) axis, about 85% of the way to the zone boundary (Feher, **1958**). The effective mass tensor **M** is no longer isotropic: the behavior of $E(\mathbf{k})$ in the vicinity of $\mathbf{k}_0$ is restricted only by those group operations $\boldsymbol{\alpha}$ such that $\boldsymbol{\alpha}\mathbf{k}_0 \doteq \mathbf{k}_0$. The fourfold symmetry axis guarantees that the surfaces $E(\mathbf{k})=$constant will be ellipsoids of revolution about the axis (see Section 4.3) shown in Fig. 7.10.1. Thus the mass tensor has two eigenvalues, a longitudinal mass along (1,0,0) and a transverse mass. Moreover, there are altogether six equivalent ellipsoids in the directions $(\pm 1,0,0),(0,\pm 1,0),(0,0,\pm 1)$.

Figure 7.10.1 also shows the eight ellopsoids for germanium at $(\pi/a)(\pm 1, \pm 1, \pm 1)$. These ellipsoids are actually centered at the zone boundary, and the ellipsoid at $(\pi/a)(1,1,1)$ is equivalent to that at $(\pi/a)(-1,-1,-1)$, so that there are indeed only four such ellipsoids (Lax and Hopfield, **1961**).

7.11. SUMMARY

It is commonly assumed (1) that $\nabla E(\mathbf{k})\cdot\mathbf{n}=0$ at the zone boundary, (2) that a periodic potential always introduces gaps, (3) that the bands can be named in order of energy, and (4) that the Bloch waves $\psi_n(\mathbf{k},\mathbf{r})$ possess the periodicity of the reciprocal lattice. We have learned, however, the following:

1. The normal component of the energy gradient is not necessarily zero unless the space group contains a mirror plane parallel to the zone boundary.[4]

2. In nonsymmorphic groups, symmetry elements can force a particular Fourier coefficient $V_{\mathbf{K}}$ of the potential to vanish, thus eliminating the direct source of a particular gap. Even indirect effects of the periodic potential are sometimes prevented by symmetry from introducing a gap.

3. Bands can be named in order of energy but only at the expense of introducing cusps or other nonanalytic behavior.

4. The periodicity of $\mathbf{k}$ is demonstrated for a single nondegenerate band, but this cannot be generalized to the degenerate case. The hexagonal close-packed lattice provides a case (see Fig. 7.8.1) in which the period in $\mathbf{k}$ space along the z axis is double that expected from the usual arguments.

problems

7.1.1. Show that $(\alpha|\mathbf{a})u(\mathbf{k},\mathbf{r})$ is a periodic function of $\mathbf{r}$ if $u(\mathbf{k},\mathbf{r})$ is periodic.

7.4.1. Consider the energy bands of a simple cubic crystal in the plane wave representation. Show that the six plane waves $\exp(i\mathbf{K}\cdot\mathbf{r})$ having $\mathbf{K} = (2\pi/a)(h,k,l)$, with $(hkl) = (100)$, $(\bar{1}00)$, (010), $(0\bar{1}0)$, (001), $(00\bar{1})$, are all degenerate in the absence of a periodic perturbation. Show that the latter splits the six-dimensional space [100] into three irreducible spaces, $[100] \rightarrow \Gamma_1^+ + \Gamma_{12}^+ + \Gamma_{15}^-$, associated with the irreducible representations of the point group O_h. Show that the basis functions ("symmetry orbitals") can be written as

$$\Gamma_1^+ = \cos 2\pi x + \cos 2\pi y + \cos 2\pi z$$

$$\Gamma_{15}^- = [\sin 2\pi x, \sin 2\pi y, \sin 2\pi z]$$

$$\Gamma_{12}^+ = [\cos 2\pi x - \cos 2\pi y, \cos 2\pi z - \tfrac{1}{2}(\cos 2\pi x + \cos 2\pi y)]$$

where x is an abbreviation for x/a, and so forth.

7.4.2. Add the constant term (000) and the set of twelve plane waves [110] to the six plane waves of [100]. Show that the resulting 19×19 determinant can be decomposed by the use of symmetry orbitals so that no matrix larger than 3×3 need be diagonalized.

7.4.3. (*a*) Show that the symmetrized combination Γ_1^+ of the simple cubic crystal

$$\Gamma_1^+ \sim (100) + (\bar{1}00) + (010) + (0\bar{1}0) + (001) + (00\bar{1})$$

[4] Or, if time reversal is used, a twofold axis normal to the plane; see Table 10.8.1.

is replaced by

$$\Gamma_1^+ = (\epsilon + \delta_{2x} + \delta_{2y} + \delta_{2z})(\epsilon + \mathbf{i})(111)$$

$$= (111) + (1\,\overline{1}\,\overline{1}) + (\overline{1}1\overline{1}) + (\overline{1}\,\overline{1}\,1) + (\overline{1}\,\overline{1}\,\overline{1}) + (\overline{1}11) + (1\overline{1}1) + (11\overline{1})$$

for a face-centered cubic crystal.

(*b*) Show for the diamond structure with (hkl) an abbreviation for $\exp(2\pi i/a)(\mathbf{h}\cdot\mathbf{s})$, where $\mathbf{s} = \mathbf{r} - (a/8)(111)$ is a vector measured from the center of inversion, that

$$\Gamma_1^+ = (111) - (1\,\overline{1}\,\overline{1}) - (\overline{1}1\overline{1}) - (\overline{1}\,\overline{1}\,1) + (\overline{1}\,\overline{1}\,\overline{1}) - (\overline{1}11) - (1\overline{1}1) - (11\overline{1})$$

7.4.4. Why is the plane wave expansion method of Section 7.4 almost never used in practical calculations of electronic energy bands? See Williams, Janak, and Moruzzi (**1972**), and Williams and Morgan (**1972**).

7.8.1. Show that for a nondegenerate band

$$u_m(-\mathbf{k}, \mathbf{r}) = \lambda_m u_m(\mathbf{k}, \mathbf{r})^*$$

and that

$$\xi_{mn}(-\mathbf{k}) = \lambda_n \lambda_m^* \xi_{mn}(\mathbf{k})^* = \lambda_n \lambda_m^* \xi_{nm}(\mathbf{k})$$

7.8.2. Show, using Eq. 7.8.29, that the phase invariant $\Omega(\mathbf{k})$ obeys

$$\operatorname{curl} \xi_{mm}(-\mathbf{k}) = -\operatorname{curl} \xi_{mm}(\mathbf{k})$$

7.8.3. With $\theta = \theta_m$ given in Eq. 7.8.36, show that for arbitrary reciprocal vectors $\mathbf{K}_1$ and $\mathbf{K}_2$

$$\theta(\mathbf{K}_1 + \mathbf{K}_2) = \theta(\mathbf{K}_1) + \theta(\mathbf{K}_2)$$

7.8.4. (*a*) Prove Eq. 7.8.38: $\theta(\mathbf{K}) = \mathbf{l} \cdot \mathbf{K}$.

(*b*) Show that the vector $\mathbf{l}$ (with dimensions of length) is given by

$$\mathbf{l} = \frac{1}{2\pi}[\theta(\mathbf{K}_1)\mathbf{a}_1 + \theta(\mathbf{K}_2)\mathbf{a}_2 + \theta(\mathbf{K}_3)\mathbf{a}_3]$$

where $\mathbf{K}_i = 2\pi\mathbf{b}_i =$ a primitive vector of the reciprocal lattice in $\mathbf{K}$ space.

7.8.5. Show that the action of the space group operation $(\alpha|\mathbf{v})$ defined in Section 8.1 on the periodic part of a Bloch function yields

$$(\alpha|\mathbf{v})u_n(\mathbf{k}, \mathbf{r}) = C_n(\alpha, \mathbf{k})u_n(\alpha\mathbf{k}, \mathbf{r})$$

in the absence of degeneracy where C_n is a phase factor (c.f. Eq. 8.6.2).

7.8.6. Using Problem 7.8.5 show that the matrix elements ξ_{mn} obey the symmetry requirement

$$\xi^{\mu}_{mn} = C_m(\alpha,\mathbf{k})^* C_n(\alpha,\mathbf{k}) \xi^{\nu}_{mn}(\alpha,\mathbf{k}) \alpha_{\nu\mu}$$
$$+ \delta_{mn} C_m(\alpha,\mathbf{k})^* i\partial C_n(\alpha,\mathbf{k})/\partial k^{\mu}$$

where μ and ν are Cartesian components. Show that $\Omega(\mathbf{k}) = \operatorname{curl}\xi_{mm}(\mathbf{k})$ obeys the simpler symmetry requirement

$$\Omega(\mathbf{k}) = \det\alpha\ \Omega(\alpha\mathbf{k})\cdot\alpha$$

7.8.7. Under the assumption that time reversal K, (here complex conjugation, see Chapter 10) yields

$$K u_m(\mathbf{k},\mathbf{r}) = D_m(\mathbf{k}) u_m(-\mathbf{k}, r)$$

where $D_m(\mathbf{k})$ is a phase factor, show that

$$\xi_{mm}(\mathbf{k}) = \xi_{mm}(-\mathbf{k})^* = \xi_{mm}(-\mathbf{k})$$

and

$$\Omega(\mathbf{k}) = -\Omega(-\mathbf{k})^* = -\Omega(-\mathbf{k})$$

where reality is established using Eq. 7.8.23.

7.8.8. Combine the result of Problem 7.8.7 with that of Problem 7.8.6 for the inversion operator to show that in the presence of inversion $\Omega(\mathbf{k}) = 0$. If the condition $\operatorname{div}\xi_{mm} = 0$ is imposed ξ_{mm} will vanish. Is this always possible? (Blount, **1962a**).

7.8.9. See additional Problem on page 499.

7.10.1. For silicon, consider the six donor wave functions $\exp(i\mathbf{k}_i\cdot\mathbf{R})G_i(\mathbf{R})$ of Eq. 7.10.16 as basis functions for a representation of the group O_h. What irreducible representations are contained in this space? Construct the basis functions for these representations. When the effective mass approximation is not assumed to be valid in the central cell (where the donor atom is located), what shifts would you expect in the energies of the six previously degenerate states?

***7.10.2.**[5] It has been suggested (Luttinger, **1951**, **1956**; Wannier, **1962**; Wannier and Fredkin **1962**; Roth **1962**) that the addition of a magnetic field via a vector potential $\mathbf{A}$ can be absorbed into the effective mass approximation merely by replacing $\mathbf{k}$ or $-i\nabla$ by $\mathbf{k} - e\mathbf{A}/c$. Discuss the validity of this approximation (Kohn, **1959a**; Blount, **1962b**).

[5]Problems of greater than average difficulty are indicated by an *.

chapter eight

SPACE GROUPS

8.1. SCREW AXES AND GLIDE PLANES

The symmetry transformations that leave a body invariant can be constructed from three basic operations: (1) rotation, (2), reflection, and (3) translation. If we write $\mathbf{r}=(x_1,x_2,x_3)$, then

$$\mathbf{r}'=\boldsymbol{\alpha}\cdot\mathbf{r}+\mathbf{a} \tag{8.1.1}$$

or

$$x'_i=\sum \alpha_{ij}x_j+a_i \tag{8.1.2}$$

describes an arbitrary inhomogeneous transformation, written briefly as $(\boldsymbol{\alpha}|\mathbf{a})$. The matrix α_{ij} will be a real orthogonal matrix of determinant $+1$ or -1 for proper or improper rotations, respectively.

It may appear that whether or not a transformation is inhomogeneous depends on the choice of origin. If we let $\mathbf{s}=\mathbf{r}+\mathbf{b}$, $\mathbf{s}'=\mathbf{r}'+\mathbf{b}$, then

$$\mathbf{s}'=\boldsymbol{\alpha}\cdot\mathbf{s}+\mathbf{a}+\mathbf{b}-\boldsymbol{\alpha}\cdot\mathbf{b} \tag{8.1.3}$$

Thus the transformation can be made homogeneous if we can choose **b** so that

$$(\boldsymbol{\alpha}-\mathbf{1})\cdot\mathbf{b}=\mathbf{a} \tag{8.1.4}$$

If we denote the three eigenvectors of $\boldsymbol{\alpha}$ by $\mathbf{e}_i$ so that

$$\boldsymbol{\alpha}\mathbf{e}_i=\lambda_i\mathbf{e}_i \tag{8.1.5}$$

and write

$$\mathbf{a}=u\mathbf{e}_1+v\mathbf{e}_2+w\mathbf{e}_3 \tag{8.1.6}$$

Eq. 8.1.4 can be solved immediately to yield

$$\mathbf{b}=(\lambda_1-1)^{-1}u\mathbf{e}_1+(\lambda_2-1)^{-1}v\mathbf{e}_2+(\lambda_3-1)^{-1}w\mathbf{e}_3 \tag{8.1.7}$$

But by Corollary 2.9.3 every proper rotation (in an odd number of dimensions) has an axis, that is, an eigenvalue $\lambda_1=1$. Thus, for proper rotations, this solution breaks down unless $u=0$. In other words, translations normal to a rotation axis can be eliminated. Those parallel to the axis cannot, and we refer to these as *screw axes*.

By Corollary 2.9.2 improper rotations automatically have an eigenvalue -1. If they also have an eigenvalue $+1$, the third eigenvalue must also be $+1$ to keep the determinant -1. But $(+1,+1,-1)$ describes a mirror plane, with eigenvalue $+1$ for any direction in the plane. Thus translations perpendicular to a mirror plane can be eliminated (by shifting the origin of the mirror), but translations in the plane cannot. These result in the only other type of element that involves translation in an intrinsic way: the *glide plane*.

The multiplication rule for inhomogeneous transformations is

$$(\boldsymbol{\beta}|\mathbf{b})(\boldsymbol{\alpha}|\mathbf{a})=(\boldsymbol{\beta}\boldsymbol{\alpha}|\mathbf{b}+\boldsymbol{\beta}\mathbf{a}) \tag{8.1.8}$$

so that

$$(\boldsymbol{\alpha}|\mathbf{a})^{-1}=(\boldsymbol{\alpha}^{-1}|-\boldsymbol{\alpha}^{-1}\mathbf{a}) \tag{8.1.9}$$

and

$$(\boldsymbol{\alpha}|\mathbf{a})^{-1}(\boldsymbol{\beta}|\mathbf{b})(\boldsymbol{\alpha}|\mathbf{a})=(\boldsymbol{\alpha}^{-1}\boldsymbol{\beta}\boldsymbol{\alpha}|\boldsymbol{\alpha}^{-1}(\mathbf{b}+\boldsymbol{\beta}\mathbf{a}-\mathbf{a})) \tag{8.1.10}$$

8.2. RESTRICTIONS ON SPACE GROUP ELEMENTS

Space Group Elements

Definition. The *space group* of a crystal has been defined as the set of all inhomogeneous linear transformations $(\boldsymbol{\alpha}|\mathbf{a})$ that restore the crystal to itself.

By using the word crystal, we automatically imply the existence of a translation subgroup T whose elements are $(\boldsymbol{\epsilon}|\mathbf{t})$, with $\boldsymbol{\epsilon}=$identity and

$$\mathbf{t}=t_1\mathbf{a}_1+t_2\mathbf{a}_2+t_3\mathbf{a}_3 \tag{8.2.1}$$

where t_1, t_2, t_3 are integers and $\mathbf{a}_1, \mathbf{a}_2, \mathbf{a}_3$ constitute a set of primitive base vectors; that is, T is the *translation group* of some *Bravais lattice.*

This translation group is, of necessity, an *invariant* subgroup, for Eq. 8.1.10 specializes to

$$(\alpha|\mathbf{a})^{-1}(\epsilon|\mathbf{t})(\alpha|\mathbf{a})=(\epsilon|\alpha^{-1}\mathbf{t}) \tag{8.2.2}$$

Since $(\boldsymbol{\epsilon}|\boldsymbol{\alpha}^{-1}\mathbf{t})$ is a pure translation, it must be a member of the group T of all translations that leave the lattice invariant. Then $\boldsymbol{\alpha}^{-1}\mathbf{t}$ is also expressible as an integral combination of primitive vectors. In short, $\boldsymbol{\alpha}^{-1}$ (and hence also $\boldsymbol{\alpha}$) must be a member of the holohedry, the point group that restores the *lattice* to itself.

The multiplication rule, Eq. 8.1.8, shows that the rotational parts $\boldsymbol{\alpha}$ of $(\boldsymbol{\alpha}|\mathbf{a})$ must separately form a group. This group is referred to as the *point group* of the crystal.

Theorem 8.2.1

The point group of a space group must be a subgroup of the holohedry group appropriate to the crystal system or space lattice.

We have deliberately written $(\boldsymbol{\alpha}|\mathbf{a})$ rather than $(\boldsymbol{\alpha}|\mathbf{t})$ to emphasize that $\mathbf{a}$ need not be one of the translation vectors $\mathbf{t}$ of the space lattice. Since

$$(\epsilon|\mathbf{t})(\alpha|\mathbf{v})=(\alpha|\mathbf{v}+\mathbf{t}) \tag{8.2.3}$$

we see that the translational part of an arbitrary inhomogeneous transformation can be expressed as a lattice translation $\mathbf{t}$ plus a translation $\mathbf{v}$ that is within the Wigner-Seitz (proximity) cell. Indeed, we can show that $\mathbf{v}$ must be unique. If for a given $\boldsymbol{\alpha}$ there were two vectors $\mathbf{v}, \mathbf{v}'$, both in the proximity cell, then

$$(\alpha|\mathbf{v}')^{-1}(\alpha|\mathbf{v})=(\epsilon|\alpha^{-1}(\mathbf{v}-\mathbf{v}')) \tag{8.2.4}$$

implies that $\boldsymbol{\alpha}^{-1}$ $(\mathbf{v}-\mathbf{v}')$ and hence $\mathbf{v}-\mathbf{v}'$ are lattice translations. Therefore $\mathbf{v}$ and $\mathbf{v}'$ must be in different cells. Thus we state another theorem.

Theorem 8.2.2

The elements of a space group can be written in the form $(\boldsymbol{\alpha}|\mathbf{v}(\boldsymbol{\alpha})+\mathbf{t})$, where $\mathbf{v}(\boldsymbol{\alpha})$ is a translation within the proximity cell, and $\mathbf{t}$ is a lattice translation, that is, $(\boldsymbol{\epsilon}|\mathbf{t})$ is a member of the invariant subgroup T, the Bravais translation group of the lattice.

Table 8.2.1. Symbols Used For Glide Planes

Symbol	Description	Translation $\mathbf{v}$
m	Mirror	0
a		$\frac{1}{2}\mathbf{a}$
b	Axial glide	$\frac{1}{2}\mathbf{b}$
c		$\frac{1}{2}\mathbf{c}$
n	Diagonal glide	$\frac{1}{2}(\mathbf{a}+\mathbf{b})$, $\frac{1}{2}(\mathbf{b}+\mathbf{c})$, or $\frac{1}{2}(\mathbf{c}+\mathbf{a})$ or $(\mathbf{a}+\mathbf{b}+\mathbf{c})/2$ (tetragonal and cubic)
d	Diamond glide	$\frac{1}{4}(\mathbf{a}+\mathbf{b})$, $\frac{1}{4}(\mathbf{b}+\mathbf{c})$, or $\frac{1}{4}(\mathbf{c}+\mathbf{a})$ or $(\mathbf{a}+\mathbf{b}+\mathbf{c})/4$ (tetragonal and cubic)

Restrictions on Screw Axes and Glide Planes

If $\boldsymbol{\alpha}^r=\boldsymbol{\epsilon}$, we must have

$$(\boldsymbol{\alpha}|\mathbf{v})^r=(\boldsymbol{\alpha}^r|\mathbf{v}+\boldsymbol{\alpha}\mathbf{v}+\boldsymbol{\alpha}^2\mathbf{v}+\cdots+\boldsymbol{\alpha}^{r-1}\mathbf{v})=(\boldsymbol{\epsilon}|\mathbf{t}) \tag{8.2.5}$$

where $\mathbf{t}$ is a lattice translation. If $\mathbf{v}$ is expanded in terms of the eigenvectors of $\boldsymbol{\alpha}$, the component parallel to the axis (with eigenvalue unity) must obey

$$r\mathbf{v}_{\parallel}=\mathbf{t} \tag{8.2.6}$$

whereas a component associated with any other rth root of unity as eigenvalue automatically yields 0:

$$(1+\boldsymbol{\alpha}+\boldsymbol{\alpha}^2+\cdots+\boldsymbol{\alpha}^{r-1})\mathbf{v}_{\perp}=0 \tag{8.2.7}$$

Thus the $\mathbf{v}_{\perp}$ part of the translation (the nonintrinsic part that could be transformed away for any one element of the group by a new choice of origin) has no restriction placed on it in this way. On the other hand, the $\mathbf{v}_{\parallel}$ part of the translation is restricted, by Eq. 8.2.6, to be a submultiple of a lattice translation. A fourfold screw axis, for example, must have a displacement $\mathbf{v}$ of $1/4$, $2/4$, or $3/4(=-1/4)$ a lattice translation and is designated as 4_1, 4_2, or 4_3 in International notation. Similarly, a glide plane must have $\mathbf{v}=1/2$ a lattice displacement in the plane and is described by the symbols of Table 8.2.1.

Space Group Notation

The notation for a space group starts with the symbol for a point group, and prefixes a letter from Table 6.1.1 for the type of lattice (P, I, A, B, C, F, R for primitive, body-centered, A-centered, B-centered, C-centered, face-centered, or rhombohedral respectively). A knowledge of the point group and the type of centering uniquely determines the Bravais lattice. Thus (see Appendix B for the point group notation)

$$F\frac{4}{m}\bar{3}\frac{2}{m}$$

is a face-centered lattice with the cubic point group O_h; hence the lattice is face-centered cubic. The symbols such as $(4/m)$ describe the generators of the point group with possible redundancy. The International Tables for X-Ray Crystallography [Henry and Lonsdale (**1952**)] list exactly three other space groups with the same point group and Bravais lattice:

$$F\frac{4}{m}\bar{3}\frac{2}{c}, \qquad F\frac{4_1}{d}\bar{3}\frac{2}{m}, \qquad \text{and} \qquad F\frac{4_1}{d}\bar{3}\frac{2}{c}$$

These differ from $F(4/m)\bar{3}(2/m)$ since their generators must contain screw axes (e.g., 4_1) or glide planes (e.g., c). Thus we can see from the notation that the first group, $F(4/m)\bar{3}\,(2/m)$, is symmorphic and the remaining three are nonsymmorphic. Since diamond must be one of these four groups and has a fourfold screw axis (see Table 9.1.1) but simple 110 mirror planes, we can conclude that the diamond structure belongs to the space group

$$O_h^7 = F\frac{4_1}{d}\bar{3}\frac{2}{m}$$

The Schoenflies notation O_h^7 contains no information other than that this is the seventh space group derived by Schoenflies from the point group O_h.

8.3. EQUIVALENCE OF SPACE GROUPS

The inhomogeneous transformation 8.1.1 can be written in the matrix form:

$$\begin{bmatrix} \mathbf{r}' \\ 1 \end{bmatrix} = \begin{bmatrix} \boldsymbol{\alpha} & \mathbf{v} \\ 0 & 1 \end{bmatrix} \cdot \begin{bmatrix} \mathbf{r} \\ 1 \end{bmatrix} \tag{8.3.1}$$

The four-dimensional matrices

$$(\boldsymbol{\alpha}|\mathbf{v}) \rightarrow \begin{bmatrix} \alpha_{11} & \alpha_{12} & \alpha_{13} & v_1 \\ \alpha_{21} & \alpha_{22} & \alpha_{23} & v_2 \\ \alpha_{31} & \alpha_{32} & \alpha_{33} & v_3 \\ 0 & 0 & 0 & 1 \end{bmatrix} \tag{8.3.2}$$

automatically obey the group multiplication rule, Eq. 8.1.8, and constitute the *defining* representation of the group, just as the three-dimensional representations of a set of rotations can be used to define a rotation group. Such a defining representation must be *faithful*, that is, no two operations can be represented by the same matrix if it is to contain all the information[1] in the original group.

Two space groups are equivalent (and in the abstract sense identical) if their defining elements $(\boldsymbol{\alpha}|\mathbf{u})$ or $(\boldsymbol{\alpha}'|\mathbf{u}')$ are related by an equivalence transformation:

$$(\boldsymbol{\beta}|\mathbf{v})^{-1}(\boldsymbol{\alpha}|\mathbf{u})(\boldsymbol{\beta}|\mathbf{v}) = (\boldsymbol{\alpha}'|\mathbf{u}') \tag{8.3.3}$$

where $(\boldsymbol{\beta}|\mathbf{v})$ is not necessarily a member of either group but is an arbitrary proper $(\det\boldsymbol{\beta} > 0)$ inhomogeneous transformation. Equation 8.3.3 states that after a proper rotation, stretching, and shift of origin one defining representation becomes identical to the other.

With the above definition of equivalence there are 230 distinct space groups. Among them, however, are 11 isomorphic pairs (see Table 8.3.1), referred to as enantiomorphic pairs, that is, mirror images of one another. These space groups of necessity contain no mirror planes or inversion and possess screw axes of opposite sense. These pairs would be equivalent if we permitted $\boldsymbol{\beta}$ in Eq. 8.3.3 to include improper rotations.

8.4. CONSTRUCTION OF SPACE GROUPS

Symmorphic Space Groups

If the operations $(\boldsymbol{\alpha}|0)$ constitute a crystallographic point group, and $(\boldsymbol{\epsilon}|\mathbf{t})$ denotes a Bravais lattice consistent with that point group, according to Table

[1]Indeed, a *faithful* representation Γ does contain all the information in the group in the sense (*a*) that it uniquely determines the group multiplication table, and (*b*) that the set of representations Γ, $\Gamma\times\Gamma$, $\Gamma\times\Gamma\times\Gamma$, $\Gamma\times\Gamma\times\Gamma\times\Gamma$, and so forth contains all the irreducible representations of the group (Wigner, **1955**).

Table 8.3.1. Enantiomorphic Pairs

First Member		Second Member	
Schoenflies Symbol	International Symbol	Schoenflies Symbol	International Symbol
C_4^2	$P4_1$	C_4^4	$P4_3$
D_4^3	$P4_122$	D_4^7	$P4_322$
D_4^4	$P4_12_12$	D_4^8	$P4_32_12$
C_3^2	$P3_1$	C_3^3	$P3_2$
D_3^3	$P3_112$	D_3^5	$P3_212$
D_3^4	$P3_121$	D_3^6	$P3_221$
C_6^2	$P6_1$	C_6^3	$P6_5$
C_6^4	$P6_2$	C_6^5	$P6_4$
D_6^2	$P6_122$	D_6^3	$P6_522$
D_6^5	$P6_422$	D_6^4	$P6_222$
O^6	$P4_332$	O^7	$P4_132$

6.1.1, then the set of operations $(\boldsymbol{\alpha}|\mathbf{t})$ constitutes a symmorphic space group. Such a group can be *generated* without using screw axes or glide planes.[2]

This method of combining the translation and rotation groups may be called a *semidirect product* (Lomont, **1959**, p. 29). It is *not* the direct product of the two groups, for the latter would require the multiplication rule

$$(\boldsymbol{\alpha}|\mathbf{a})(\boldsymbol{\beta}|\mathbf{b})=(\boldsymbol{\alpha\beta}|\mathbf{a}+\mathbf{b}) \tag{8.4.1}$$

as opposed to Eq. 8.1.8. In different terms, although a *symmorphic* space group is *characterized* by the fact that it has its *point group* as a *subgroup*, this subgroup is *not invariant* for $(\boldsymbol{\epsilon}|\mathbf{t})^{-1}(\boldsymbol{\alpha}|0)(\boldsymbol{\epsilon}|\mathbf{t})=(\boldsymbol{\alpha}|\boldsymbol{\alpha}\mathbf{t}-\mathbf{t})\neq(\boldsymbol{\alpha}|0)$ for some $\mathbf{t}$.

[2]Symmorphic groups often contain screw axes or glide planes, since (with $\boldsymbol{\delta}_n$= an n-fold rotation) the operator $(\boldsymbol{\epsilon}|\mathbf{t})(\boldsymbol{\delta}_n|0)=(\boldsymbol{\delta}_n|\mathbf{t})$ will be a screw axis if the lattice translation $\mathbf{t}$ has a component along the n-fold axis shorter than the shortest lattice translation in that direction. Consider the symmorphic tetragonal group $P422=D_4^1$: a twofold screw axis is generated by combining the twofold axis $\boldsymbol{\delta}_{2xy}$ that bisects the x and y axes with a displacement in the x direction. [For all centered lattices, the translation $\mathbf{t}$ to the centering point yields a screw axis $(\boldsymbol{\delta}_n|\mathbf{t})$ (or a glide plane).]

There are 73 symmorphic space groups. Of these, 66 can be obtained directly from Table 6.1.1 by multiplying each crystal class by each possible Bravais lattice and adding up the total. The 7 additional symmorphic groups arise because a point group may have two possible nonequivalent orientations with respect to the space lattice that it leaves invariant. For example, the group $D_2=222$ can have an A-centered or a C-centered lattice. This will lead to two space groups $A222$ and $C222$ that are in fact equivalent. On the other hand, for $C_{2v}=mm2$ the space groups $Amm2$ and $Cmm2$ are distinct, since in one case the centering is along a twofold axis and in the other it is not. Similarly, the tetragonal group $D_{2d}=\bar{4}\,2m$ has four instead of the expected two space groups. Finally, the hexagonal group $D_{3h}=\bar{6}m2$ and the trigonal groups $D_3=32$, $C_{3v}=3m$, and $D_{3d}=\bar{3}\ (2/m)$ have two hexagonal space groups instead of one (there are two possible orientations of a mirror plane relative to the hexagonal Bravais lattice).

Nonsymmorphic Space Groups

There are 230-73 = 157 nonsymmorphic space groups, that is, space groups containing operations of the form $(\boldsymbol{\alpha}|\mathbf{v}(\boldsymbol{\alpha}))$, where the $\mathbf{v}(\alpha)$ are not lattice translations, and no shift of origin can reduce all the $\mathbf{v}(\boldsymbol{\alpha})$ simultaneously to zero or a lattice vector. In other words, one starts with a crystal point group $(\boldsymbol{\alpha}|0)$ and an allowed space lattice $(\boldsymbol{\epsilon}|\mathbf{t})$, as for the symmorphic groups, but converts some axes to screw axes, or reflection planes to glide planes. Equation 8.2.5 requires that the nth power of an n-fold screw axis or the square of a glide plane be a lattice translation. Moreover, any product of the form $(\boldsymbol{\alpha}|\mathbf{u})(\boldsymbol{\beta}|\mathbf{v})$ gives rise to new screws or glide planes, and all of these must also be possible elements. The number of possibilities is clearly finite and is determined by enumeration in Buerger (**1956**) and by matrix techniques in Seitz (**1934**, **1935a**, **1935b**, **1936**). A recent translation (Federov, **1971**) compares Federov's classic work with that of Schoenflies on the construction of space groups. See also Problem 8.4.1.

8.5. FACTOR GROUPS OF SPACE GROUPS

The translational properties of a space group G are uniquely specified by its Bravais lattice, that is, the translation group

$$T \equiv \{\,(\boldsymbol{\epsilon}|\mathbf{t})\,\} \tag{8.5.1}$$

where {} means "the set of." The rotational properties of the group are contained in the operators $(\boldsymbol{\alpha}|\mathbf{v}(\boldsymbol{\alpha}))$. Observe that

$$(\boldsymbol{\alpha}|\mathbf{v}(\boldsymbol{\alpha})+\mathbf{t})(\boldsymbol{\beta}|\mathbf{v}(\boldsymbol{\beta})+\mathbf{t}')=(\boldsymbol{\alpha\beta}|\mathbf{v}(\boldsymbol{\alpha\beta})+\mathbf{t}'') \tag{8.5.2}$$

where $\mathbf{t}''$ must be a lattice translation. Even if $\mathbf{t}=\mathbf{t}'=0$, $\mathbf{t}''$ is not necessarily zero. The *representative elements* $(\alpha|\mathbf{v}(\alpha))$ then *fail* to form a *group*. Therefore, to deal with the rotational aspects above, a procedure to divide out or remove the translational effects is needed. The result will be the formation of a *factor group* G/T. We shall discuss this group-theoretical procedure because it is prevalent in the literature, but we shall soon see that it is possible and much simpler to deal directly with the *algebra* of the set $\{(\alpha|\mathbf{v}(\alpha))\}$.

The usual group-theoretical procedure starts by defining multiplication of sets of group elements. If $A\equiv\{A_i\}$, $B\equiv\{B_j\}$, the set product $A\cdot B$ is defined as the collection of elements $\{A_iB_j\}$ with duplicates omitted. With this definition, Eq. 8.5.2 implies the set product relationship

$$(\alpha|\mathbf{v}(\alpha))T\cdot(\beta|\mathbf{v}(\beta))T=(\alpha\beta|\mathbf{v}(\alpha\beta))T \tag{8.5.3}$$

Equation 8.5.3 is the multiplication table of a group whose *elements* are the sets $\{(\alpha|\mathbf{v}(\alpha)+\mathbf{t})\}$. This group is called the *factor group* G/T because it describes multiplication modulo T. The factor group G/T is isomorphic to the point group P. Thus it has g elements, not gN, where $N=N_1N_2N_3\sim10^{23}$ is the order of T.

For proper understanding, we must provide a definition and a discussion of factor groups that do not make use of the special properties of the space groups.

If a group G has an arbitrary subgroup H, it can be decomposed into left cosets with respect to H:

$$G=\sum_j B_jH \tag{8.5.4}$$

where $B_j\in G$. Each collection of elements B_jH is a single left coset. Alternatively, there is a right coset decomposition

$$G=\sum_j HB_j \tag{8.5.5}$$

If H is an *invariant* subgroup, by which is meant $B_jHB_j^{-1}=H$, the left- and right-hand cosets are identical:

$$B_jH=HB_j \tag{8.5.6}$$

In this *invariant* case, the cosets form a *group* under set multiplication:

$$(B_iH)(B_jH)=B_iB_jH^2=(B_iB_j)H \tag{8.5.7}$$

The group formed by the cosets is called the *factor group* or quotient group G/H because, in it, multiplication is performed modulo H, that is, as if the elements of H were set equal to unity.

The only point not fully established in the above discussion is the following theorem.

Coset Decomposition Theorem 8.5.1

A group G has a *unique* left coset decomposition with respect to an arbitrary subgroup H.

This theorem is an immediate consequence of a lemma.

Coset Lemma 8.5.2

If $R \in G$ and $S \in G$, either $S^{-1}R \in H$ and the two cosets RH and SH are *identical*, or $S^{-1}R$ is not in H and the two cosets RH and SH *have no element in common* (i.e., overlapping cosets are impossible).

Proof: If $S^{-1}R \in H$, then $R = SH_1$ (H_1 is some element in H). Thus $RH = SH_1H = SH$, and the cosets are identical. Conversely, if there is an element in common, $RH_1 = SH_2$ and $S^{-1}R$ belongs to H.

The coset decomposition can now be constructed as follows: Choose $B_1 = E$, the identity, B_2 from $G - H$, B_3 from $G - H - B_2H$, B_{j+1} from $G - (B_1H + B_2H + \cdots + B_jH)$. By the coset lemma, $B_{j+1}H$ will be distinct from all previous cosets. This process can be continued until the cosets exhaust the group, leading to the following theorem.

Lagrange's Theorem 8.5.3

If H is a subgroup of G, its order h is a factor of the order g of G.

The generalization of Eq. 8.5.4 is that any representation of a factor group G/H "engenders" a representation of the full group G in which all the elements of H are represented by the identity, and all the elements of any one coset B_jH are represented by one and the same matrix $D(\{B_jH\})$, the matrix that represents the coset B_jH in the factor group.

8.6. GROUP $G_{\mathbf{k}}$ OF THE WAVE VECTOR k

Since the element $(\alpha|\mathbf{a})$ of a space group G converts a plane wave with wave vector $\mathbf{k}$ into

$$(\alpha|\mathbf{a}) \exp(i\mathbf{k}\cdot\mathbf{r}) = \exp\left[i\mathbf{k}\cdot(\alpha|\mathbf{a})^{-1}\mathbf{r}\right]$$

$$= \exp[i\mathbf{k}\cdot(\alpha^{-1}\mathbf{r} - \alpha^{-1}\mathbf{a})] = \exp[i\alpha\mathbf{k}\cdot(\mathbf{r}-\mathbf{a})] \tag{8.6.1}$$

a wave with vector $\alpha\mathbf{k}$, it does the same to a Bloch wave:

$$(\alpha|\mathbf{a})\psi_n(\mathbf{k},\mathbf{r}) = \psi_{n'}(\alpha\mathbf{k},\mathbf{r}) \tag{8.6.2}$$

In the solution of a lattice vibration problem (see Chapter 11) or an electronic energy band problem (see Chapter 7) we start by diagonalizing the $(\epsilon|\mathbf{t})$, that is, picking a **k**. The dynamical problem then separates into uncoupled problems, one for each **k**. To simplify the problem appropriate to a particular **k** we therefore restrict our attention to the subgroup $G_{\mathbf{k}}$ of elements $(\alpha|\mathbf{v}(\alpha)+\mathbf{t})$ whose rotations α leave **k** invariant (or carry it into an equivalent point):

$$\alpha\mathbf{k} \doteq \mathbf{k} \quad \text{that is, } \alpha\mathbf{k} = \mathbf{k} + \mathbf{K} \tag{8.6.3}$$

The addition of a reciprocal lattice vector to the right-hand side of Eq. 8.6.3 is permitted, since $\mathbf{k}+\mathbf{K}$ belongs to the same representation as (is "equivalent to") **k**. The elements $(\epsilon|\mathbf{t})$ of the translation group T obey Eq. 8.6.3 and are included in $G_{\mathbf{k}}$.

If the wave vector **k** is on the zone boundary or at some rational point in the interior, there will be some translations **t** for which $\mathbf{k}\cdot\mathbf{t} = 2\pi$ (integer) or

$$\exp(-i\mathbf{k}\cdot\mathbf{t}) = 1 \tag{8.6.4}$$

These translations constitute an invariant subgroup $T_{\mathbf{k}}$ of T and of $G_{\mathbf{k}}$. Any element of $T_{\mathbf{k}}$ acting on $\psi(\mathbf{k},\mathbf{r})$ yields a factor of unity. Thus, if $R \in G_{\mathbf{k}}$, all elements of the coset $RT_{\mathbf{k}}$ or $T_{\mathbf{k}}R$ yield the same result on $\psi(\mathbf{k},\mathbf{r})$. Hence we are concerned with a representation of $G_{\mathbf{k}}$ in which all $T_{\mathbf{k}}$ are unity, that is, a representation *engendered* by the factor group $G_{\mathbf{k}}/T_{\mathbf{k}}$. The group $G_{\mathbf{k}}/T_{\mathbf{k}}$ is generally one of small finite order whose multiplication table can be evaluated using Eq. 8.5.2. Its representations can then be obtained by means of the class multiplication theorem or the multiplier algebra technique discussed in the remainder of this chapter and in Appendix F.

One word of caution is needed, however: the above group-theoretical procedure *also* produces some *irrelevant* representations of the group $G_{\mathbf{k}}/T_{\mathbf{k}}$ in which one of the added translations $(\epsilon|\mathbf{t})$ is not represented by $\exp(-i\mathbf{k}\cdot\mathbf{t})$. The group $G_{\mathbf{k}}$, just like the translation group T, with which it coincides at a general **k**, possesses representations of arbitrary propagation constant not necessarily equal to **k**. The group $G_{\mathbf{k}}/T_{\mathbf{k}}$ contains all the representations of $G_{\mathbf{k}}$ for which $T_{\mathbf{k}}$ (i.e., each of its elements) is represented by unity. If, however, $(\epsilon|\mathbf{t}_{\mathbf{k}})$ is an arbitrary element of $T_{\mathbf{k}}$, the equation $\exp(i\mathbf{k}\cdot\mathbf{t}_{\mathbf{k}}) = 1$ defines $t_{\mathbf{k}}$. But then

$$\exp(in\mathbf{k}\cdot t_{\mathbf{k}}) = 1 \tag{8.6.5}$$

so that representations belonging to $n\mathbf{k}$, where n is an integer, also map $T_{\mathbf{k}}$ into the identity and are found among the representations of $G_{\mathbf{k}}/T_{\mathbf{k}}$.

Only a small number of possible values of n lead to distinct representations, however. For example, if $(\epsilon|\mathbf{t}) \in T - T_{\mathbf{k}}$ and $\exp(-i\mathbf{k}\cdot\mathbf{t}) = -1$, then $\exp(-in\mathbf{k}\cdot\mathbf{t}) = (-1)^n$ and only two distinct representations occur, those in which the added translation $(\epsilon|\mathbf{t})$ is represented by $+1$ or -1.

When the character table of $Q_{\mathbf{k}} \equiv G_{\mathbf{k}}/T_{\mathbf{k}}$ is constructed by the method of class multiplication using Dirac characters, it is possible to seek only solutions of the algebraic equations that are consistent with the character for $(\epsilon|\mathbf{t})$ taking the value $\exp(-i\mathbf{k}\cdot\mathbf{t})$ and not, for example, $\exp(-in\mathbf{k}\cdot\mathbf{t})$.

The other representations with $n \neq 1$ are needed to provide a "square" character table for $G_{\mathbf{k}}/T_{\mathbf{k}}$, but we shall show that all of the physically relevant information—orthonormality theorems, degeneracies, selection rules—can be obtained from an abbreviated character table containing only the *relevant* representations. Indeed, we shall show that the number of *relevant representations* is equal to the number of *relevant classes* (defined in Section 8.10), so that a smaller "square" character table can be constructed. Furthermore, all symmetry information of other than a purely translational character will be contained in this abbreviated table.

To what extent can we neglect the operations in $G - G_{\mathbf{k}}$? Elements in $G - G_{\mathbf{k}}$ take $\mathbf{k}$ into $\boldsymbol{\beta}\mathbf{k}$, not equivalent to $\mathbf{k}$. The set of such $\boldsymbol{\beta}\mathbf{k}$ is referred to as the *star* of $\mathbf{k}$ (see Fig. 8.6.1). The use of such operators $\boldsymbol{\beta}$ provides relations between wave functions or representations belonging to *different* points of the star. But $G_{\mathbf{k}}$ provides all the necessary information about $\mathbf{k}$ itself. The situation is precisely analogous to our treatment in Theorem 5.4.1 of molecular force constants by restricting attention to the group of the bond and can be justified in a similar way. [Indeed, time reversal is handled in an analogous way by adding the elements that take $\mathbf{k}$ into $-\mathbf{k}$ (or $-\mathbf{k}+\mathbf{K}$) as the *reversal* elements.] This elementary justification is in fact closely related to a deeper justification[‖]: *If basis vectors for a representation of $G_{\mathbf{k}}$ are known, the various $\boldsymbol{\beta}$ acting on these basis vectors provide us with a complete set of basis vectors for a representation of the complete space group G. If the initial representation of $G_{\mathbf{k}}$ is irreducible, the representation of G (using the entire star of* **k**) *is irreducible. Furthermore, all the irreducible representations of G can be found in this way.* [See, for example, Birman (**1974**), Lax (**1965c**), Section 3.]

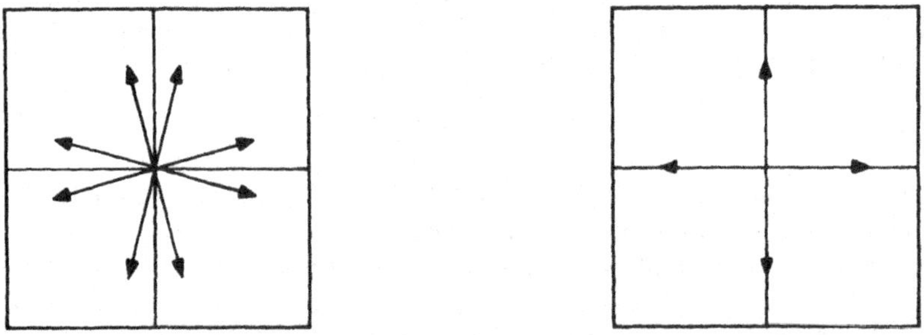

(a) *(b)*

Fig. 8.6.1. (*a*) Square lattice with point group C_{4v} showing the points of a star for a general wave vector. (*b*) Same lattice C_{4v}, special wave vector.

8.7. SPACE GROUP ALGEBRA

The rotational symmetry of the space group is contained in the algebra

$$(\alpha|\mathbf{v}(\alpha))(\beta|\mathbf{v}(\beta)) = (\epsilon|\mathbf{t})(\alpha\beta|\mathbf{v}(\alpha\beta)) \tag{8.7.1}$$

where

$$(\epsilon|\mathbf{t}) = (\epsilon|\mathbf{v}(\alpha) + \alpha\mathbf{v}(\beta) - \mathbf{v}(\alpha\beta))$$

is a lattice translation. We can exhaust the translational information by first diagonalizing $(\epsilon|\mathbf{t})$, that is, by allowing Eq. 8.7.1 to act on a Bloch wave $\psi(\mathbf{k},\mathbf{r})$. It then takes the form

$$(\alpha|\mathbf{v}(\alpha))(\beta|\mathbf{v}(\beta))\psi(\mathbf{k},\mathbf{r}) = \exp(-i\alpha\beta\mathbf{k}\cdot\mathbf{t})(\alpha\beta|\mathbf{v}(\alpha\beta))\psi(\mathbf{k},\mathbf{r}) \tag{8.7.2}$$

since $\boldsymbol{\alpha\beta}$ takes $\psi(\mathbf{k},\mathbf{r})$ into some $\psi(\boldsymbol{\alpha\beta}\mathbf{k},\mathbf{r})$. The elements of $G_{\mathbf{k}}$ will not take us out of a space $\sigma_{\mathbf{k}}$ spanned by Bloch waves of the form $\psi_\mu(\mathbf{k},\mathbf{r})$. Within $\sigma_{\mathbf{k}}$, for elements of $G_{\mathbf{k}}$, Eq. 8.7.2 reduces to

$$(\alpha|\mathbf{v}(\alpha))(\beta|\mathbf{v}(\beta)) = \exp\{-i\mathbf{k}\cdot[\mathbf{v}(\alpha)+\alpha\cdot\mathbf{v}(\beta)-\mathbf{v}(\alpha\beta)]\}(\alpha\beta|\mathbf{v}(\alpha\beta)) \tag{8.7.3}$$

which happens to constitute a multiplier group. The use of multiplier groups or multiplier representations has been emphasized by Lyubarskii (**1960**) and Döring (**1959**). We prefer to follow a choice of Lyubarskii's and to use

$$O(\alpha) \equiv (\alpha|\mathbf{v}(\alpha))\exp[i\mathbf{k}\cdot\mathbf{v}(\alpha)] = (\alpha|\mathbf{v}(\alpha)+\mathbf{t})\exp\{i\mathbf{k}\cdot[\mathbf{v}(\alpha)+\mathbf{t}]\} \tag{8.7.4}$$

as the elements of our algebra, since Eq. 8.7.4 demonstrates that *all members of a given coset* $(\boldsymbol{\alpha}|\mathbf{v}(\boldsymbol{\alpha})+\mathbf{t})$ *are in fact represented by a single element* $O(\boldsymbol{\alpha})$ *in the algebra*:

$$O(\alpha)O(\beta) = \lambda(\alpha,\beta)O(\alpha\beta) \tag{8.7.5}$$

where

$$\lambda(\alpha,\beta) = \exp\{i\mathbf{k}\cdot[\mathbf{v}(\beta)-\alpha\mathbf{v}(\beta)]\} \tag{8.7.6}$$

At one stroke, all the nonsense about cosets and factor groups is avoided!

Since $\mathbf{k}\cdot\alpha\mathbf{v} = (\tilde{\alpha}\mathbf{k})\cdot\mathbf{v} = \alpha^{-1}\mathbf{k}\cdot\mathbf{v}$ and for $\alpha \in G_{\mathbf{k}}$,

$$\alpha^{-1}\mathbf{k} = \mathbf{k} - \mathbf{K}_\alpha \tag{8.7.7}$$

where $\mathbf{K}_\alpha$ is some reciprocal lattice vector (possibly zero), we can write

$$\lambda(\alpha,\beta) = \exp[i\mathbf{K}_\alpha\cdot\mathbf{v}(\beta)] \tag{8.7.8}$$

We now construct the representations of space groups by the "divide and conquer" method: Divide the problem into two parts, and show that the first part is easy. The first division is into symmorphic and nonsymmorphic space groups.

8.8. REPRESENTATIONS OF SYMMORPHIC SPACE GROUPS

When $\lambda(\alpha,\beta)=1$, the $O(\alpha)$ obey the multiplication table of the point group $P(\mathbf{k})$, that is, the rotational parts of the operators in $G_{\mathbf{k}}$. Using Eq. 8.7.4, we can therefore write the representations and characters of the space group elements $(\alpha|\mathbf{v}(\alpha))$ in terms of known point group representations:

$$D^{\mathbf{k},j}(\alpha|\mathbf{v}(\alpha))=\exp[-i\mathbf{k}\cdot\mathbf{v}(\alpha)]D^{P(\mathbf{k}),j}(\alpha)$$
$$\chi^{\mathbf{k},j}(\alpha|\mathbf{v}(\alpha))=\exp[-i\mathbf{k}\cdot\mathbf{v}(\alpha)]\chi^{P(\mathbf{k}),j}(\alpha) \tag{8.8.1}$$

where the index j labels these irreducible representations. In symmorphic space groups it is possible to select an origin of coordinates so that all $\mathbf{v}(\alpha)=0$. Thus $\lambda(\alpha,\beta)=1$, and the representations of all 73 symmorphic space groups, including zone boundary representations, are given by Eq. 8.8.1 with $\mathbf{v}(\alpha)=0$.

8.9. REPRESENTATIONS OF NONSYMMORPHIC SPACE GROUPS

We now divide into interior and zone boundary representations.

Interior Points of the Brillouin Zone

In this case $\mathbf{k}$ and $\alpha^{-1}\mathbf{k}$ must *both* be inside the first Brillouin zone. Hence they cannot differ by a reciprocal lattice vector, that is, $\mathbf{K}_\alpha=0$ and $\lambda(\alpha,\beta)=1$. The representations are again obtainable from point groups via Eq. 8.8.1.

Comment: Since the macroscopic tensors of Chapter 4 are invariant under translations, they belong to $\mathbf{k}=0$ and by Eq. 8.8.1 they *transform* under $(\alpha|\mathbf{v}(\alpha)+\mathbf{t})$ *exactly as they* would under the point group operation $(\alpha|0)$.

We now divide the zone boundary points according to whether a one-dimensional relevant representation exists.

Zone Boundary Points Containing a One-Dimensional Relevant Representation

Call this representation $\nu(\alpha)$. Then

$$\nu(\alpha)\nu(\beta)=\lambda(\alpha,\beta)\nu(\alpha\beta) \tag{8.9.1}$$

Divide Eq. 8.7.5 by 8.9.1 to obtain

$$\frac{O(\alpha)}{\nu(\alpha)}\cdot\frac{O(\beta)}{\nu(\beta)}=\frac{O(\alpha\beta)}{\nu(\alpha\beta)} \tag{8.9.2}$$

that is, the set of operators $O(\alpha)/\nu(\alpha)$ obeys the *point group* multiplication table. Thus

$$D^{\mathbf{k}}[O(\alpha)]=\nu(\alpha)D^{P(\mathbf{k})}(\alpha) \tag{8.9.3}$$

or

$$D^{\mathbf{k},j}(\alpha|\mathbf{v})=\exp[-i\mathbf{k}\cdot\mathbf{v}(\alpha)]\nu(\alpha)D^{P(\mathbf{k}),j}(\alpha) \tag{8.9.4}$$

and the representations and characters are all known. Representations of a multiplier group that differ by a factor transformation $D(\alpha)'=\nu(\alpha)D(\alpha)$ are said to be p-equivalent (projective-equivalent).

The content of the present section can now be summarized in a theorem.

Theorem 8.9.1

If for a given factor system $\lambda(\alpha,\beta)$ a multiplier group has *even one* one-dimensional representation, *all the multiplier representations* are p-equivalent to the corresponding vector (ordinary) representations of the underlying (ordinary) group.

Zone Boundary Points Containing No One-Dimensional Representations

Whether a one-dimensional representation exists can be readily ascertained by seeing whether the algebra is obeyed by a set of numbers. This is not always the case. If $\alpha\beta=\beta\alpha$ but

$$\lambda(\alpha,\beta)\neq\lambda(\beta,\alpha) \tag{8.9.5}$$

then

$$O(\alpha)O(\beta)=\left[\frac{\lambda(\alpha,\beta)}{\lambda(\beta,\alpha)}\right]O(\beta)O(\alpha)\neq O(\beta)O(\alpha) \tag{8.9.6}$$

and no one-dimensional representation is possible. [See Kovalev and Lyubarskii (**1958**)]. Although the group $G_{\mathbf{k}}$ always contains an identity representation, our algebraic procedure does not yield it in this case, because, by construction, the $O(\boldsymbol{\alpha})$ algebra contains only the *relevant* representations.

We will show in Section 8.10 that, if there are no one-dimensional representations (e.g., because of Eq. 8.9.6), most of the classes are irrelevant, that is, their Dirac characters vanish identically (in a **k** subspace). For the most complicated space groups no more than four relevant classes or Dirac characters are found. Thus the "brute-force" method of obtaining the representations from the algebra of the Dirac characters will be easy—in fact, probably the easiest of the methods available.

Another procedure likely to be easy is the method of "induced" representations, in which the representations of a group are obtained from those of a subgroup [Lomont (**1959**), Mackey (**1951**)]. It is often possible to choose the subgroup to be a symmorphic space group (Zak, **1960**), whose representations we have just shown to be obtainable from point group representations.

Döring (**1959**) has studied all the possible multiplier representations of the crystallographic point groups. Of the groups that possess "true" multiplier representations (i.e., not p-equivalent to ordinary representations), only twelve are not isomorphic. (See Problem 8.9.1.) The true multiplier representations of these groups are listed in Appendix F.

Warning: Two p-equivalent representations are listed only once in these tables even if, in the ordinary sense, these representations are inequivalent.

One of the simplest methods, then, of determining the zone boundary representations of $G_{\mathbf{k}}$ is to look up the table for the appropriate point group in Appendix F. From the totality of possible multiplier representations, those with the appropriate factor system $\lambda(\boldsymbol{\alpha},\boldsymbol{\beta})$ are selected. The factor system is described in terms of the generator algebra. For example, for the group D_{2h}, three perpendicular mirror planes can be used as generators. These can be made to obey $A^2=B^2=C^2=E$ by a factor transformation if necessary. The multiplier possibilities are then summarized by

$$CB=\alpha BC, \qquad AC=\beta CA, \qquad BA=\gamma AB \tag{8.9.7}$$

where the only choices for α, β, or γ consistent with these equations are $+1$ and -1, and at least one of the three must be negative to yield a true multiplier representation.

These methods of constructing space group representations are now less important since tables of these representations have become available. See Kovalev (**1961**), Zak, Casher, and Gluck (**1969**), Faddeev (**1964**), and Miller and Love (**1967**).

8.10. CLASS STRUCTURE AND ALGEBRAIC TREATMENT OF MULTIPLIER GROUPS

In this section we compare the group-theoretical treatment of the factor group $G_{\mathbf{k}}/T_{\mathbf{k}}$ with the multiplier representation point of view by means of an example: the group at X, that is, $\mathbf{k}=(2\pi/a)(1,0,0)$ in diamond $O_h^7=F(4_1/d)\bar{3}\,(2/m)$. From this comparison, we shall devise an algebraic procedure more efficient than either of the above methods.

Since the primitive translations for the face-centered cubic lattice are

$$\mathbf{t}_{xy}=(a/2)(1,1,0), \qquad \mathbf{t}_{yz}=(a/2)(0,1,1), \qquad \mathbf{t}_{zx}=(a/2)(1,0,1) \qquad (8.10.1)$$

the eigenvalues of these operators for $\mathbf{k}=(2\pi/a)(1,0,0)$ are

$$(\boldsymbol{\epsilon}|\mathbf{t}_{xy})=-1, \qquad (\boldsymbol{\epsilon}|\mathbf{t}_{yz})=1, \qquad (\boldsymbol{\epsilon}|\mathbf{t}_{zx})=-1 \qquad (8.10.2)$$

If $\mathbf{t}=t_1\mathbf{t}_{yz}+t_2\mathbf{t}_{zx}+t_3\mathbf{t}_{xy}$, then

$$(\boldsymbol{\epsilon}|\mathbf{t})=(-1)^{(t_2+t_3)} \qquad (8.10.3)$$

The translations with (t_2+t_3) even constitute $T_{\mathbf{k}}$. The full translation group can be decomposed into

$$T=T_{\mathbf{k}}+(\boldsymbol{\epsilon}|\mathbf{t}_{xy})T_{\mathbf{k}} \qquad (8.10.4)$$

with $(\boldsymbol{\epsilon}|0)$ and $(\boldsymbol{\epsilon}|\mathbf{t}_{xy})$ acting as representative elements of the "even" and "odd" cosets, respectively, of the factor group $T/T_{\mathbf{k}}$. The character table, Table 8.10.1, is a rearrangement of a table first computed by Herring (1942). From the group-theoretical point of view, any element $(\boldsymbol{\alpha}|\mathbf{v})$ shown in the table represents a coset $(\boldsymbol{\alpha}|\mathbf{v})T_{\mathbf{k}}$, since the $(\boldsymbol{\alpha}|\mathbf{v})$ by themselves do not constitute a group. The algebraic viewpoint is simpler in that we keep only the one element $(\boldsymbol{\alpha}|\mathbf{v})$ from each coset, and these elements obey a multiplier group.

The group $G_{\mathbf{k}}/T_{\mathbf{k}}$ has fourteen classes. Of the fourteen representations, only four obey $(\boldsymbol{\epsilon}|\mathbf{t}_{xy})=\exp(-i\mathbf{k}\cdot\mathbf{t}_{xy})=-1$. These are relevant and are shown in the table. The other ten representations, which have been computed by Elliott and Loudon (1960), are not shown because they obey $(\boldsymbol{\epsilon}|\mathbf{t}_{xy})=+1$ and are thus irrelevant.

Much of the information in the character table merely duplicates the statement $(\boldsymbol{\epsilon}|\mathbf{t}_{xy})=-1$. Thus the classes in rows 5–8 of Table 8.10.1 can be obtained from rows 1–4 by multiplying corresponding elements by $(\boldsymbol{\epsilon}|\mathbf{t}_{xy})$. The elements $(\boldsymbol{\delta}_{2x}|0)$ and $(\boldsymbol{\delta}_{2x}|\mathbf{t}_{xy})$ can serve as an example. If we restrict ourselves to considering the *relevant representations*, for which $(\boldsymbol{\epsilon}|\mathbf{t}_{xy})=-1$, the characters in rows 5–8 must be the negatives of those in corresponding rows 1–4. We need not

Table 8.10.1 Characters at $X=(2\pi/a)(1,0,0)$[a]

32	Classes[b]	X_1	X_2	X_3	X_4
1	$(\epsilon\|0)$	2	2	2	2
1	$(\delta_{2x}\|0)$	2	2	-2	-2
2	$(\delta_{2yz}\|\tau+\mathbf{t}_{xy}),\ (\delta_{2\bar{y}z}\|\tau)$	0	0	-2	2
2	$(\rho_{yz},\rho_{\bar{y}z}\|0)$	2	-2	0	0
1	$(\epsilon\|\mathbf{t}_{xy})$	-2	-2	-2	-2
1	$(\delta_{2x}\|\mathbf{t}_{xy})$	-2	-2	2	2
2	$(\delta_{2yz}\|\tau),\ (\delta_{2\bar{y}z}\|\tau+\mathbf{t}_{xy})$	0	0	2	-2
2	$(\rho_{yz},\rho_{\bar{y}z}\|\mathbf{t}_{xy})$	-2	2	0	0
2	$(\mathbf{i}\|\tau,\tau+\mathbf{t}_{xy})$	0	0	0	0
4	$(\delta_{2y},\delta_{2z}\|0,\mathbf{t}_{xy})$	0	0	0	0
4	$(\delta_{4x},\delta_{4x}^{-1}\|\tau,\tau+\mathbf{t}_{xy})$	0	0	0	0
4	$(\rho_y,\rho_z\|\tau,\tau+\mathbf{t}_{xy})$	0	0	0	0
2	$(\rho_x\|\tau,\tau+\mathbf{t}_{xy})$	0	0	0	0
4	$(\sigma_{4x},\sigma_{4x}^{-1}\|0,\mathbf{t}_{xy})$	0	0	0	0

[a]See Table 9.1.1 and Appendix B for definitions of symbols.

[b]$\tau=\frac{1}{4}(\mathbf{t}_{xy}+\mathbf{t}_{yz}+\mathbf{t}_{zx})$ is the vector from the origin atom to its nearest neighbor in the first octant.

regard $(\delta_{2x}|0)$ and $-(\delta_{2x}|0)$ as independent elements, from an algebraic viewpoint, and can discard rows 5–8.

Each of the last six lines of the table is a class that contains $R(\epsilon|\mathbf{t}_{xy})$ when it contains R. The corresponding Dirac character contains $R+(-R)$ and vanishes identically. Thus the vanishing of the characters for these classes contains *only* the translational information[3] $(\epsilon|\mathbf{t}_{xy})=-1$. These classes can be discarded.

Since the translational information $(\epsilon|\mathbf{t})=\exp(-i\mathbf{k}\cdot\mathbf{t})$ is given, once $\mathbf{k}$ is specified, we are permitted to write an abbreviated character table that contains *only* the classes needed to specify all the rotational (i.e., nontranslational) properties of the representations. These will be referred to as the *relevant classes*. With this definition we can state without proof the following theorem.

Theorem 8.10.1

The number of *relevant* representations is equal to the number of *relevant* classes.

To give the theorem content, however, we must provide a definition of "class"

[3]For the irrelevant representations, the characters of these classes are nonvanishing and are needed to distinguish among these representations.

appropriate to multiplier groups. It is easier, however, to define first the Dirac character of a class via Eq. 1.3.12:

$$\Omega_c = \frac{n_c}{g} \sum_{\beta} O(\beta)O(\alpha)[O(\beta)]^{-1} \tag{8.10.5}$$

where n_c is the number of elements in the class of α in the underlying ordinary group of order g.

It is necessary to divide by g/n_c, even for an ordinary group, because this is the number of times that each element is repeated in the sum. In the multiplier case, however, $O(\alpha)$ can be repeated in the form $\mu O(\alpha)$. But we have another theorem.

Theorem 8.10.2

If $O(\alpha)$ and $\mu O(\alpha)$ with $\mu \neq 1$ both appear in the sum 8.10.5, the Dirac character for the class containing $O(\alpha)$ vanishes identically.

Proof: Let $O(\beta)O(\alpha)[O(\beta)]^{-1} = \mu O(\alpha)$. Then

$$[O(\beta)]^n O(\alpha)[O(\beta)]^{-n} = \mu^n O(\alpha)$$

But $\beta^r = \epsilon$ for some small r. Then $[O(\beta)]^r =$ a multiple of the identity, and it follows, using $n = r$, that $\mu^r = 1$. The sum 8.10.5 then contains

$$(1 + \mu + \mu^2 + \cdots + \mu^{r-1})O(\alpha) = 0$$

which vanishes because of the roots of unity theorem 6.4.1. If $O(\gamma) = O(\chi)O(\alpha)[O(\chi)]^{-1}$, then, as n ranges from 0 to $r-1$,

$$O(\chi)[O(\beta)]^n O(\alpha)[O(\beta)]^{-n}[O(\chi)]^{-1}$$

contributes

$$(1 + \mu + \mu^2 + \cdots + \mu^{r-1})O(\gamma) = 0$$

to the Dirac character. Thus either the Dirac character vanishes identically and we can regard the class as *irrelevant*, or the class can be defined as the set of distinct terms in the sum 8.10.5. The factor n_c/g then simply cancels the identical repetition of these terms.

Furthermore, when we arrive at a multiplier group with no one-dimensional representations because of Eq. 8.9.6, $[O(\beta)]^{-1}O(\alpha)O(\beta) = \mu O(\alpha)$ and some Dirac characters must vanish identically. Thus, in the only case in which the representations cannot be obtained from those of the point group, the Dirac character algebra simplifies in that some of its elements vanish identically. In our fairly complicated example of the highly symmetric diamond structure at X, only four Dirac characters out of fourteen classes enter the algebra. In most cases, only one or two Dirac characters survive. Thus the "brute-force" evaluation of the eigenvalues of the Dirac character algebra is often the simplest method of all.

8.11. DOUBLE SPACE GROUPS

The results of the preceding sections can be generalized to include double space groups by the following theorem.

Theorem 8.11.1

The irreducible representations of all double groups of the wave vector $G'_{\mathbf{k}}$ in the interior of the Brillouin zone, and of all symmorphic (double) space groups in the interior and on the zone boundary, can be obtained from the corresponding point group $P(\mathbf{k}) = G_{\mathbf{k}}/T$ by adding a phase factor $\exp(-i\mathbf{k}\cdot\mathbf{u})$ to the matrix representation or character of $\boldsymbol{\alpha}$ in the double point group $P(\mathbf{k})'$ to obtain the corresponding double space group representation or character of $(\boldsymbol{\alpha}|\mathbf{u})$.

Proof: We first recall the following relation between a rotational operator $\boldsymbol{\alpha}$ and the corresponding barred operation $\bar{\boldsymbol{\alpha}}$:

$\boldsymbol{\alpha} = \bar{\boldsymbol{\alpha}}$ in single-valued representations

$\boldsymbol{\alpha} = -\bar{\boldsymbol{\alpha}}$ in double-valued ("spin") representations

Since a vector $\mathbf{t}$ transforms according to the representation $l=1$ of the full rotation group (a single-valued representation),

$$\boldsymbol{\alpha}\cdot\mathbf{t} = \bar{\boldsymbol{\alpha}}\cdot\mathbf{t} \tag{8.11.1}$$

Hence in space groups the double valuedness enters only through the rotational products. If the single point group $P(\mathbf{k})$ obeys

$$\boldsymbol{\alpha\beta} = \boldsymbol{\gamma} \tag{8.11.2}$$

the double point group $P(\mathbf{k})'$ will obey

$$\boldsymbol{\alpha\beta} = \mu(\boldsymbol{\alpha},\boldsymbol{\beta})\boldsymbol{\gamma} \tag{8.11.3}$$

where $\mu(\boldsymbol{\alpha},\boldsymbol{\beta}) = \pm 1$ with the sign determined by Eqs. 2.6.20–2.6.23). Similarly, the single space group product

$$(\boldsymbol{\alpha}|\mathbf{u})(\boldsymbol{\beta}|\mathbf{v}) = (\boldsymbol{\alpha\beta}|\mathbf{u}+\boldsymbol{\alpha}\cdot\mathbf{v}) \tag{8.11.4}$$

is replaced by the double space group $G'_{\mathbf{k}}$ product

$$(\boldsymbol{\alpha}|\mathbf{u})(\boldsymbol{\beta}|\mathbf{v})=\mu(\boldsymbol{\alpha},\boldsymbol{\beta})(\boldsymbol{\alpha}\boldsymbol{\beta}|\mathbf{u}+\boldsymbol{\alpha}\cdot\mathbf{v}) \tag{8.11.5}$$

with the identical $\mu(\boldsymbol{\alpha},\boldsymbol{\beta})$ as above. Correspondingly, the operators $O(\boldsymbol{\alpha}) = \exp(i\mathbf{k}\cdot\mathbf{u})(\boldsymbol{\alpha}|\mathbf{u})$ obey the single group equation

$$O(\boldsymbol{\alpha})O(\boldsymbol{\beta})=\lambda(\boldsymbol{\alpha},\boldsymbol{\beta})O(\boldsymbol{\alpha}\boldsymbol{\beta}) \tag{8.11.6}$$

and the corresponding double group equation

$$O(\boldsymbol{\alpha})O(\boldsymbol{\beta})=\mu(\boldsymbol{\alpha},\boldsymbol{\beta})\lambda(\boldsymbol{\alpha},\boldsymbol{\beta})O(\boldsymbol{\alpha}\boldsymbol{\beta}) \tag{8.11.7}$$

In Sections 8.8 and 8.9 we showed that, inside the Brillouin zone for all space groups and on the zone boundary for symmorphic space groups, $\lambda(\boldsymbol{\alpha},\boldsymbol{\beta})=1$. Thus $O(\boldsymbol{\alpha})O(\boldsymbol{\beta})=\mu(\boldsymbol{\alpha},\boldsymbol{\beta})O(\boldsymbol{\gamma})$, so that the operators $O(\boldsymbol{\alpha})$ obey the algebra of the double point group $P(\mathbf{k})'$ and have the same representations and characters. Q.E.D.

When $\lambda(\boldsymbol{\alpha},\boldsymbol{\beta})\neq 1$, Eq. 8.11.7 can be used to determine the algebra of the generators of the group. This algebra either has a one-dimensional representation and Theorem 8.9.1 can be applied to obtain the representations, or else one can use the algebra to enter the tables of multiplier representations in Appendix F.

Note: The double group at **k** may have (relevant) one-dimensional representations even if the single group has none.

Classes in Double Space Groups

To determine the classes in double space groups we consider an n-fold rotation $\boldsymbol{\delta}_n$ and its similarity transforms:

$$(\boldsymbol{\delta}_m|\mathbf{v})^{-1}(\boldsymbol{\delta}_n|\mathbf{u})(\boldsymbol{\delta}_m|\mathbf{v})=(\boldsymbol{\delta}'_n|\mathbf{u}') \tag{8.11.8}$$

$$\boldsymbol{\delta}'_n=\boldsymbol{\delta}_m^{-1}\boldsymbol{\delta}_n\boldsymbol{\delta}_m \tag{8.11.9}$$

$$\mathbf{u}'=\boldsymbol{\delta}_m^{-1}(\mathbf{u}-\mathbf{v}+\boldsymbol{\delta}_n\mathbf{v}) \tag{8.11.10}$$

Can $(\bar{\boldsymbol{\delta}}_n|\mathbf{u})$ be in the same class as $(\boldsymbol{\delta}_n|\mathbf{u})$? For this to be so, $\boldsymbol{\delta}'_n$ must have the same axis as $\boldsymbol{\delta}_n$, that is, $\boldsymbol{\delta}_m$ must be a twofold rotation perpendicular to $\boldsymbol{\delta}_n$. In this case

$$\boldsymbol{\delta}'_n=\boldsymbol{\delta}_n^{-1} \tag{8.11.11}$$

But δ_n^{-1} cannot equal $\bar{\delta}_n$ unless $n=2$. Thus, as for point groups, we have the following theorems.

Theorem 8.11.2

For $n>2$, in a double space group the class C_n of n-fold rotations (including screw axes) and the class $\bar{C}_n$ form distinct classes.

Theorem 8.11.3

For $n=2$, the classes C_n and $\bar{C}_n$ are joined in the double space group only if there exists a twofold axis $(\boldsymbol{\delta}'_2|\mathbf{v})$ perpendicular to $(\boldsymbol{\delta}_2|\mathbf{u})$ and if $\mathbf{u}'-\mathbf{u}$ belongs to $T_{\mathbf{k}}$, where

$$\mathbf{u}' = (\boldsymbol{\delta}'_2)^{-1}(\mathbf{u}-\mathbf{v}+\boldsymbol{\delta}_2\mathbf{v})$$

$$= \boldsymbol{\delta}'_2(\mathbf{u}-\mathbf{v}+\boldsymbol{\delta}_2\mathbf{v}) \qquad (8.11.12)$$

since $(\boldsymbol{\delta}'_2)^{-1}=\boldsymbol{\delta}'_2$ when acting on vectors. Stated carefully, it is $(\bar{\boldsymbol{\delta}}_2|\mathbf{u}')$ that is in the same class as $(\boldsymbol{\delta}_2|\mathbf{u})$.

We illustrate Theorem 8.11.3 by considering the point $W=(2\pi/a)(1,0,\frac{1}{2})$ in the diamond lattice, which has the twofold element $(\boldsymbol{\delta}_{2z}|0)$, and at right angles to it $(\boldsymbol{\delta}_{2xy}|\boldsymbol{\tau})$, where $\boldsymbol{\tau}=\frac{1}{4}(a,a,a)$:

$$(\boldsymbol{\delta}_{2xy}|\boldsymbol{\tau})^{-1}(\boldsymbol{\delta}_{2z}|0)(\boldsymbol{\delta}_{2xy}|\boldsymbol{\tau}) = (\boldsymbol{\delta}_{2z}^{-1}|\mathbf{t}_{xy}) \qquad (8.11.13)$$

where $\mathbf{t}_{xy}=\frac{1}{2}(a,a,0)$. Thus $(\boldsymbol{\delta}_{2z}|0)$ is in the same class as $(\bar{\boldsymbol{\delta}}_{2z}|\mathbf{t}_{xy})$, but not in the same class as $(\bar{\boldsymbol{\delta}}_{2z}|0)$. In the single group, Herring lists the class as $(\boldsymbol{\delta}_{2z}|0)$, $(\boldsymbol{\delta}_{2z}|\mathbf{t}_{xy})$. It appears as if the new class is the old one with a bar on *one* element. We could more properly regard it as

$$C = (\boldsymbol{\delta}_{2z}|0), (\boldsymbol{\delta}_{2z}^{-1}|\mathbf{t}_{xy}) \qquad (8.11.14)$$

in either the single or the double group. The class $\bar{C}$ is not distinct from the class

$$(\boldsymbol{\delta}_{2z}|\mathbf{t}_{xy}), (\boldsymbol{\delta}_{2z}^{-1}|0) \qquad (8.11.15)$$

since both $(\boldsymbol{\epsilon}|\mathbf{t}_{xy})$ and the barring operation produce a factor of -1.

8.12. SUMMARY

Symmorphic space groups are constructed by combining pure rotational elements with pure lattice translations. By the use of screw axes and glide planes among the generators, additional nonsymmorphic space groups can be constructed.

In dealing with lattice vibration problems and electronic energy bands, it is convenient to first diagonalize the pure lattice translation operators. This permits us to restrict our attention to $G_{\mathbf{k}}$, the group of the wave vector **k**, rather than considering the full space group. We show that there is a simple relation between the representations of $G_{\mathbf{k}}$ and the representations of the corresponding point group in two cases:(1) symmorphic space groups and (2) nonsymmorphic space groups for **k** inside the Brillouin zone.

Representations of nonsymmorphic space groups on the zone boundary, the only difficult case, can be divided into two categories. In the first, the group possesses a one-dimensional representation. In this case we show that, by making a factor transformation based on this one-dimensional representation, we can relate all of the space group representations to those of the corresponding point group. The remaining nontrivial problem is the construction of representations on the zone boundary for nonsymmorphic space groups that do not possess any one-dimensional relevant representations. For this case we find that many of the Dirac characters vanish identically. Thus the problem of finding such representations is reduced to that of dealing with an algebra of a very small number of elements, the nonvanishing Dirac characters. The equations obeyed by these Dirac characters can be solved algebraically for each set of eigenvalues, and these sets of eigenvalues can then be used to produce the conventional character table (see Section 1.5).

problems

8.2.1. The set of operations $(\boldsymbol{\alpha}|\mathbf{a}) \in G$ restores a crystal to itself. Show, nevertheless, that it is the operation $\boldsymbol{\alpha}\mathbf{t}$ and not $\boldsymbol{\alpha}\mathbf{t}+\mathbf{a}$ that yields a new translation vector $\mathbf{t}'$.

8.3.1. The Friedel law states that the intensity of X-ray reflection from the crystal planes $(h_1h_2h_3)$ is the same as that from the planes $(\bar{h}_1\bar{h}_2\bar{h}_3)$ (except near a resonance level). Discuss the proof of this law. How does the Friedel law limit the extent to which X-rays can be used to identify the space group of a crystal?

8.4.1. Show that a crystal with a nonsymmorphic space group *must* have more than one atom per primitive cell.

8.5.1. Show that, if even one $\mathbf{v}(\boldsymbol{\alpha}) \neq 0$, the set of operations $\{(\boldsymbol{\alpha}|\mathbf{v}(\boldsymbol{\alpha}))\}$ does not form a group.

8.6.1. Prove that $T_{\mathbf{k}}$ is an invariant subgroup of T and of $G_{\mathbf{k}}$.

8.6.2. Show that the number of points in the star of **k** is the ratio of the orders of the groups G and $G_{\mathbf{k}}$. Show that this conclusion remains true at the zone boundary provided that **k** and $\mathbf{k}+\mathbf{K}$ are counted as one point, where **K** is a reciprocal lattice vector.

8.6.3. Is the group of the wave vector $G_{\mathbf{k}}$ an invariant subgroup of G?

8.6.4. Discuss the relation

$$(G_{\mathbf{k}}/T_{\mathbf{k}})=(G_{\mathbf{k}}/T)\times(T/T_{\mathbf{k}})$$

Is it true for symmorphic groups? Is it true in general?

8.6.5. Prove the last statement of Section 8.6: that the basis vectors for the irreducible representations of $G_{\mathbf{k}}$ can be used to generate the irreducible representations of the full space group. Show that all irreducible representations can be obtained in this way.

8.8.1. Consider a simple cubic or face-centered cubic crystal along the direction $\Lambda=(k,k,k)$ inside the Brillouin zone. Show that the associated point group at Λ is C_{3v}. Show that, in contrast to the basis functions listed in Table A7, an appropriate set of basis functions consists of the following:

$$\Lambda_1 \qquad x+y+z$$

$$\Lambda_2 \qquad xy(x-y)+yz(y-z)+zx(z-x)$$

$$\Lambda_3 \qquad [(x-z),\ (y-z)]$$

8.8.2. Consider a cubic crystal at a point along $\Sigma=(k,k,0)$. Show that an appropriate character table with basis functions is

	ϵ	δ_{2xy}	ρ_z	$\rho_{\bar{x}y}$	
Σ_1	1	1	1	1	$x+y$
Σ_2	1	1	-1	-1	$z(x-y)$
Σ_3	1	-1	-1	1	z
Σ_4	1	-1	1	-1	$x-y$

See Jones (**1960**), p. 107, and Callaway (**1964**), p. 28.

8.9.1. Show that the group D_3 has no multiplier representations other than those p-equivalent to ordinary representations. *Hint*: The generators of D_3 obey $a^3=e$, $b^2=e$, $ba=a^2b$. For the multiplier representations we can require that $A^3=E$, $B^2=E$ (after a factor transformation), leaving $BA=\gamma A^2B$ to describe the factor system. Show that any choice other than $\gamma=1$ leads to a contradiction.

8.9.2. Show that the associative law

$$O(\alpha)[O(\beta)O(\gamma)]=[O(\alpha)O(\beta)]O(\gamma)$$

implies that

$$\lambda(\alpha,\beta\gamma)\lambda(\beta,\gamma)=\lambda(\alpha,\beta)\lambda(\alpha\beta,\gamma)$$

8.9.3. Show that

$$\lambda(\alpha,\epsilon)=1 \qquad (\text{all } \alpha)$$

$$\lambda(\alpha,\alpha^{-1})=\lambda(\alpha^{-1},\alpha)$$

***8.9.4.** What are the space groups for the wurtzite and the bismuth structures? (See Wyckoff, **1963**.) List a set of representative elements $(\alpha|v(\alpha))$. Give character tables for the representations at Γ and all principal symmetry points. State the compatibility relations between the representations. You may use the published literature for this question (Glasser, **1959**; Birman, **1959**; Mase, **1959**).

chapter nine

SPACE GROUP EXAMPLES

In order to illustrate space group techniques in a nontrivial but economical way,[1] we shall choose most of our examples from the nonsymmorphic "diamond" group $O_h^7 = F(4_1/d)\bar{3}(2/m)$. The diamond *structure*, shown in Figs. 9.1.1 and 6.1.11, consists of two interpenetrating face-centered cubic lattices of cube edge a. We can choose our origin at an atom in one sublattice in such a way that $\boldsymbol{\tau} = (a/4, a/4, a/4)$ connects this atom to its nearest neighbor in the first octant, and with the latter atom as starting point the second face-centered cubic lattice is generated. (See Table 6.1.4.) With this choice of origin, the 48 operations $(\boldsymbol{\alpha}|\mathbf{v}(\boldsymbol{\alpha}))$ can be divided into 24 simple operations $(\boldsymbol{\alpha}|0)$ constituting the point group T_d and 24 "compound" operations $(\boldsymbol{\alpha}|\boldsymbol{\tau})$ that can be obtained by multiplying the first 24 by $(\mathbf{i}|\boldsymbol{\tau})$, where $\mathbf{i}$ is the inversion. See Table 9.1.1. The simple operations transform each sublattice into itself, whereas the compound operations, in addition, interchange the two sublattices. For example, $I = (\mathbf{i}|\boldsymbol{\tau})$ can be regarded as inversion about a point halfway between two sublattices.

9.1. VANISHING ELECTRIC MOMENT IN DIAMOND

The strong "restrahl" line in the infrared absorption associated with lattice vibrations in an ionic crystal involves a coupling of the electromagnetic field

[1]This chapter is just the hors d'oeuvres. The main course (the principal applications) comes later!

Table 9.1.1 Symmetry Operations[a] of Factor Group of Space Group O_h^7

Class	Simple Operation $(\boldsymbol{\alpha}\|0)$		Class	Compound Operation[b] $(\boldsymbol{\alpha}\|\tau)$	
E	ϵ	XYZ	J	$\mathbf{i}$	$\bar{X}\bar{Y}\bar{Z}$
$3C_2$	δ_{2x}	$X\bar{Y}\bar{Z}$	$3JC_4^2$	ρ_x	$\bar{X}YZ$
	δ_{2y}	$\bar{X}Y\bar{Z}$		ρ_y	$X\bar{Y}Z$
	δ_{2z}	$\bar{X}\bar{Y}Z$		ρ_z	$XY\bar{Z}$
$6JC_4$	σ_{4x}	$\bar{X}Z\bar{Y}$	$6C_4$	δ_{4x}	$X\bar{Z}Y$
	$(\sigma_{4x})^{-1}$	$\bar{X}\bar{Z}Y$		$(\delta_{4x})^{-1}$	$XZ\bar{Y}$
	σ_{4y}	$Z\bar{Y}\bar{X}$		δ_{4y}	$\bar{Z}YX$
	$(\sigma_{4y})^{-1}$	$\bar{Z}\bar{Y}X$		$(\delta_{4y})^{-1}$	$ZY\bar{X}$
	σ_{4z}	$Y\bar{X}\bar{Z}$		δ_{4z}	$\bar{Y}XZ$
	$(\sigma_{4z})^{-1}$	$\bar{Y}X\bar{Z}$		$(\delta_{4z})^{-1}$	$Y\bar{X}Z$
$6JC_2$	$\rho_{\bar{y}z}$	XZY	$6C_2$	$\delta_{2\bar{y}z}$	$\bar{X}\bar{Z}\bar{Y}$
	$\rho_{\bar{z}x}$	ZYX		$\delta_{2\bar{z}x}$	$\bar{Z}\bar{Y}\bar{X}$
	$\rho_{\bar{x}y}$	YXZ		$\delta_{2\bar{x}y}$	$\bar{Y}\bar{X}\bar{Z}$
	ρ_{yz}	$X\bar{Z}\bar{Y}$		δ_{2yz}	$\bar{X}ZY$
	ρ_{zx}	$\bar{Z}Y\bar{X}$		δ_{2zx}	$Z\bar{Y}X$
	ρ_{xy}	$\bar{Y}\bar{X}Z$		δ_{2xy}	$YX\bar{Z}$
$8C_3$	δ_{3xyz}	ZXY	$8JC_3$	σ_{6xyz}	$\bar{Z}\bar{X}\bar{Y}$
	$\delta_{3x\bar{y}\bar{z}}$	$\bar{Z}\bar{X}Y$		$\sigma_{6x\bar{y}\bar{z}}$	$ZX\bar{Y}$
	$\delta_{3\bar{x}y\bar{z}}$	$Z\bar{X}\bar{Y}$		$\sigma_{6\bar{x}y\bar{z}}$	$\bar{Z}XY$
	$\delta_{3\bar{x}\bar{y}z}$	$\bar{Z}X\bar{Y}$		$\sigma_{6\bar{x}\bar{y}z}$	$Z\bar{X}Y$
	$(\delta_{3xyz})^{-1}$	YZX		$(\sigma_{6xyz})^{-1}$	$\bar{Y}\bar{Z}\bar{X}$
	$(\delta_{3x\bar{y}\bar{z}})^{-1}$	$\bar{Y}Z\bar{X}$		$(\sigma_{6x\bar{y}\bar{z}})^{-1}$	$Y\bar{Z}X$
	$(\delta_{3\bar{x}y\bar{z}})^{-1}$	$\bar{Y}\bar{Z}X$		$(\sigma_{6\bar{x}y\bar{z}})^{-1}$	$YZ\bar{X}$
	$(\delta_{3\bar{x}\bar{y}z})^{-1}$	$Y\bar{Z}\bar{X}$		$(\sigma_{6\bar{x}\bar{y}z})^{-1}$	$\bar{Y}ZX$

[a] $\sigma_4 = \bar{4} = (S_4)^{-1}$; $\sigma_6 = \bar{3} = (S_6)^{-1}$ where $\bar{4}$ and $\bar{3}$ are roto-inversion axes (International notation) and S_4 and S_6 are roto-reflections (Schoenflies notation). The notation of this table follows Herring (**1942**).

[b] Compound operations involve a displacement $\boldsymbol{\tau} = (a/4)(1,1,1)$ after the rotation.

Comments

$\boldsymbol{\delta}_{n,\mathbf{n}}$ = counterclockwise rotation of body through $\alpha = 2\pi/n$ about axis $\mathbf{n}$

$\mathbf{r}'$ $= \boldsymbol{\delta}_{n,\mathbf{n}}\mathbf{r} = [(\cos\alpha)\mathbf{r} + (1-\cos\alpha)\mathbf{n}(\mathbf{n}\cdot\mathbf{r}) + \sin\alpha(\mathbf{n}\times\mathbf{r})]$ (2.5.11)

$\boldsymbol{\delta}_{3x\bar{y}\bar{z}}$ has $\alpha = 2\pi/3$, $\mathbf{n} = (1,-1,-1)/\sqrt{3}$

$\mathbf{i}$ = inversion

$\boldsymbol{\rho}_{\mathbf{n}} = \mathbf{i}\boldsymbol{\delta}_{2\mathbf{n}}$ = reflection through plane $\perp$ to $\mathbf{n}$

$\boldsymbol{\rho}_x$ = (100), $\boldsymbol{\rho}_{xy} = (1,1,0)$, $\boldsymbol{\rho}_{\bar{x}y} = (-110)$ reflection planes

$\boldsymbol{\sigma}_{4\mathbf{n}}$ $= \mathbf{i}\boldsymbol{\delta}_{4\mathbf{n}}$; $\boldsymbol{\sigma}_{6\mathbf{n}} = \mathbf{i}\boldsymbol{\delta}_{3\mathbf{n}}$

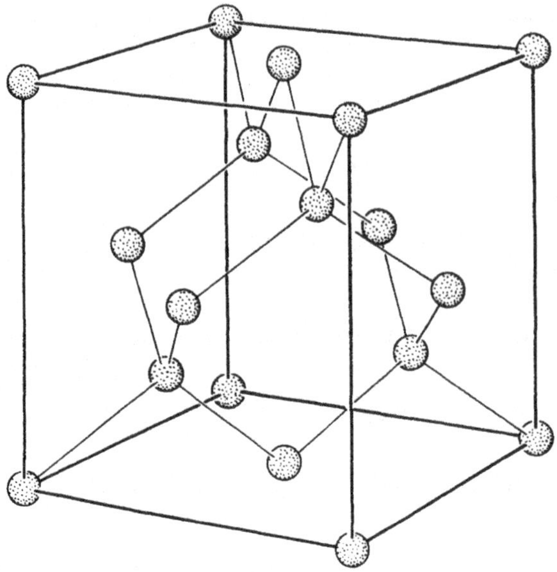

Fig. 9.1.1. The lattice structure of diamond.

with the total electric moment of the crystal induced by a lattice vibration. In this section, we shall show that to *first order* in the displacements *no total* electric moment is induced in a diamond crystal by an arbitrary displacement (Lax and Burstein, **1955**; Lax, **1965c**). Let the (linear part of the total) electric moment be given by

$$\mathbf{M} = \sum \mathbf{B}^{\alpha i} \cdot \mathbf{u}^{\alpha i} \tag{9.1.1}$$

Here $\mathbf{u}^{\alpha i}$ represents a displacement from equilibrium position $\mathbf{X}^{\alpha i}$. The single superscript m for a molecule is replaced by the double superscript αi, where i stands for a triplet of integers i_1, i_2, i_3 that specify the cell, and α is a number that specifies which site in the cell is referred to:

$$\mathbf{X}^{\alpha i} = \mathbf{X}^{\alpha} + \mathbf{X}^{i}, \qquad \mathbf{X}^{i} = i_1\mathbf{a}_1 + i_2\mathbf{a}_2 + i_3\mathbf{a}_3 \tag{9.1.2}$$

In the *primitive* cell, for diamond, there are only two atoms; these can be labeled $\alpha = -1$, $\alpha = +1$.

Under the operation $(\mathbf{S}|\mathbf{v})$, according to Eq. 5.4.11 and the following discussion, the displacement

$$(\mathbf{u}^{\alpha i})' = \mathbf{S}\mathbf{u}^{\alpha i} \tag{9.1.3}$$

is placed at the new position $(\alpha i)' = \alpha' i'$, defined by

$$\mathbf{X}^{\alpha' i'} = \mathbf{S} \cdot \mathbf{X}^{\alpha i} + \mathbf{v} \tag{9.1.4}$$

The new electric moment is therefore given by

$$\mathbf{M}' = \sum \mathbf{B}^{\alpha' i'} \cdot (\mathbf{S} \cdot \mathbf{u}^{\alpha i}) \tag{9.1.5}$$

But the total electric moment of a crystal is invariant against *arbitrary* displacements and transforms as a vector under rotations. Thus we can write

$$\mathbf{M}' = \mathbf{S} \cdot \mathbf{M} \tag{9.1.6}$$

leading to the requirement

$$\mathbf{S} \cdot \mathbf{B}^{\alpha i} = \mathbf{B}^{\alpha' i'} \cdot \mathbf{S} \tag{9.1.7}$$

The special case of a pure lattice translation $(\mathbf{S}|\mathbf{v}) = (\boldsymbol{\epsilon}|\mathbf{t})$ leads to the requirement

$$\mathbf{B}^{\alpha i} = \mathbf{B}^{\alpha, i+t} = \mathbf{B}^{\alpha} \tag{9.1.8}$$

The requirement of invariance under an arbitrary infinitesimal displacement $\Delta \mathbf{u}^{\alpha i} = \mathbf{d}$ in Eq. 9.1.1 leads to

$$\sum \mathbf{B}^{\alpha} = 0 \tag{9.1.9}$$

If we interpret $\mathbf{B}^{\alpha}$ as an effective charge, this is the statement that the primitive unit cell must be neutral.

Let us now specialize to a class of structures, including diamond, that contains two atoms per unit cell and a center of inversion halfway between them. We may label these atoms $\alpha = 1$ and $\alpha = -1$. Then the inversion operation takes α into

$$\alpha' = \bar{\alpha} = -\alpha \qquad \text{and} \qquad i' = \bar{i} = -i \tag{9.1.10}$$

Equation 9.1.7 now requires that

$$\mathbf{B}^{\alpha i} = \mathbf{B}^{\bar{\alpha} \bar{i}} \qquad \text{or} \qquad \mathbf{B}^{\alpha} = \mathbf{B}^{\bar{\alpha}} \tag{9.1.11}$$

Displacement invariance, Eq. 9.1.9, for the two-atom case reads

$$\mathbf{B}^{\alpha} + \mathbf{B}^{\bar{\alpha}} = 0 \tag{9.1.12}$$

so that $\mathbf{B}^{\alpha} = 0$ for $\alpha = \pm 1$. Thus no linear term in the total electric moment exists, and the typical resonant absorption by lattice vibrations accompanied by the emission of one phonon does not occur. The observed absorption spectrum can be explained on the basis of two phonon absorption processes (Lax and Burstein, **1955**).

The proof just given can be applied to all odd electric moments, since these reverse sign under inversion, but not to even electric moments.

9.2. INDUCED QUADRUPOLE MOMENTS IN DIAMOND

The displacement of a given atom can induce dipole moments in its neighbors. The total electric moment so induced will vanish in a diamond structure, but a more or less local total quadrupole moment will remain (Lax, **1958a**, **1965b**):

$$Q_{rs} = \sum Q^{\alpha}_{rst} u_t^{\alpha i} \tag{9.2.1}$$

where r,s,t are Cartesian indices, and we have used translational invariance to drop the i on $\mathbf{Q}^{\alpha i}$. Infinitesimal displacement invariance requires as before that

$$\sum_{\alpha} Q^{\alpha}_{rst} = 0 \tag{9.2.2}$$

Since a quadrupole moment $\mathbf{Q} \sim \mathbf{rr}$ transforms under a rotation $\mathbf{S}$ into $\mathbf{Q}' \sim (\mathbf{Sr})(\mathbf{Sr}) = \mathbf{S} \cdot \mathbf{Q} \cdot \tilde{\mathbf{S}}$, the operation $(\mathbf{S}|\mathbf{v})$ requires that

$$\mathbf{S} \cdot \mathbf{Q} \cdot \tilde{\mathbf{S}} = \sum Q^{\alpha'}_{..t} (\mathbf{S}\mathbf{u}^{\alpha i})_t \tag{9.2.3}$$

when the indices r and s are suppressed. The use of Eq. 9.2.1 in the left-hand side eventually produces the condition

$$Q^{\alpha}_{r's't'} = \sum_{rst} Q^{\alpha'}_{rst} S_{rr'} S_{ss'} S_{tt'} \tag{9.2.4}$$

Inversion produces the requirement

$$Q^{\alpha}_{rst} = -Q^{\bar{\alpha}}_{rst} \tag{9.2.5}$$

compatible with translational invariance (Eq. 9.2.2), rather than in contradiction with it, as the odd electric moments are. The remaining information is contained in the operations that do not interchange the sublattices, that is, the simple operations in Table 9.1.1, the point group T_d. It suffices to use the generators of that group. For example, using the operation $\boldsymbol{\delta}_{2z}$ in Eq. 9.2.4 reverses the signs of x and y and produces the requirement

$$Q^{\alpha}_{rst} = (-1)^{\text{no. of 1's and 2's}} Q^{\alpha}_{rst} = -(-1)^{\text{no. of 3's}} Q^{\alpha}_{rst} \tag{9.2.6}$$

where the exponents refer to the number of 1's and 2's, and the number of 3's, among the indices r,s,t. Thus we conclude that 3 must appear an odd number of times. But the cyclic permutation $\boldsymbol{\delta}_{3xyz}$ requires that the same statement be true about 1 and 2. Hence Q^{α}_{rst} vanishes unless rst is some permutation of 123. The operation $\boldsymbol{\rho}_{\bar{x}y}$, which takes XYZ into YXZ, interchanges the indices 1 and 2 with a plus sign. Similarly, $\boldsymbol{\delta}_{3xyz}$ produces a cyclic permutation of the indices. Thus Q^{α}_{rst} is independent of all permutations of the indices, and our information can be summarized in the form

$$Q^{\alpha}_{rst} = \alpha Q \,|e_{rst}| \tag{9.2.7}$$

where $\alpha = \pm 1$, e_{rst} is the Levi-Civita tensor density defined in Eq. 2.2.3, and Q is the one remaining arbitrary parameter (out of 54 in Q^{α}_{rst}!). That further use of the group T_d will not cause Q to vanish can be deduced by noticing that the right-hand side of Eq. 9.2.4 describes the triple product representation $\Gamma_p \times \Gamma_p \times \Gamma_p$, where Γ_p is the polar vector representation, and this triple product contains the identity representation just once with respect to the group T_d, *requiring one* arbitrary parameter. See Problem 9.2.1.

9.3. FORCE CONSTANTS IN CRYSTALS

With the potential energy of the crystal in the form

$$V = \frac{1}{2} \sum \mathbf{u}^{\alpha i} \cdot \mathbf{K}^{\alpha i, \beta j} \cdot \mathbf{u}^{\beta j} \tag{9.3.1}$$

invariance under lattice translations requires that

$$\mathbf{K}^{\alpha i, \beta j} = \mathbf{K}^{\alpha, i+t; \beta, j+t} = \mathbf{K}^{\alpha\beta}(\mathbf{j} - \mathbf{i}) \tag{9.3.2}$$

that is, any bond is identical to any other obtainable by a lattice translation. Thus a bond $\mathbf{K}^{\alpha\mathbf{b}}$ can be labeled by the starting atom α and the bond vector

$$\mathbf{b} = \mathbf{X}^{\beta j} - \mathbf{X}^{\alpha i} \tag{9.3.3}$$

In the space of the $\mathbf{K}^{\alpha\mathbf{b}}$ two space group operations, $(\mathbf{S}|\mathbf{v})$ and $(\mathbf{S}|\mathbf{v}+\mathbf{t})$, that differ by a lattice translation are indistinguishable. In other words, $\mathbf{K}^{\alpha\mathbf{b}}$ generates a representation (appropriate to $\mathbf{k}=0$) of the factor group G/T. Furthermore, only bond vectors $\mathbf{b}$ of a given length $|\mathbf{b}|$ are mixed by the group elements, so that one can calculate independently the number of first-neighbor constants, the number of second-neighbor constants, and so on.

As in the case of molecular force constants, the simplest procedure for determining the form and number of independent force constants is to use the group of elements that leave the bond invariant plus the elements that reverse the bond. An operation that transforms a bond to a translationally equivalent position is regarded as an element of the group of the bond. A similar extension applies to reversal elements. Equation 5.4.19, obeyed by $\mathbf{K}^{mn}$ in a molecule, is still valid with m replaced by αi and n by βj.

Since the second neighbors in a diamond lattice have the same relations to each other as do the first neighbors in a face-centered cubic (FCC) lattice, it is interesting to compare the force constants in these two cases. The operations appropriate to a simple FCC lattice are those listed in Table 9.1.1 with τ set equal to zero, that is, all operations regarded as simple point operations, yielding the point group O_h.

Example 9.3.1

Find the nearest-neighbor force constants in a simple FCC lattice.

Solution: The bond from (0,0,0) to $(\frac{1}{2},\frac{1}{2},0)$ is left invariant by the group C_{2v}: $\boldsymbol{\epsilon},\boldsymbol{\delta}_{2xy}$ (a twofold axis along the bond) and $\boldsymbol{\rho}_z,\boldsymbol{\rho}_{\bar{x}y}$ (reflection planes containing the bond). The reversal elements are these elements multiplied by the inversion: $\mathbf{i}$, $\boldsymbol{\rho}_{xy}$, $\boldsymbol{\delta}_{2z}$, and $\boldsymbol{\delta}_{2\bar{x}y}$, respectively. The number of independent constants can be calculated by means of Eq. 5.5.19:

$$N_b=\tfrac{1}{4}\left\{[\chi(\boldsymbol{\epsilon})]^2+[\chi(\boldsymbol{\delta}_{2xy})]^2+[\chi(\boldsymbol{\rho}_z)]^2+[\chi(\boldsymbol{\rho}_{\bar{x}y})]^2\right\}$$

$$=\tfrac{1}{4}\left[3^2+(-1)^2+1^2+1^2\right]=3$$

$$N'_b=\tfrac{1}{4}\left[\chi(\mathbf{i}^2)+\chi(\boldsymbol{\rho}_{xy}^2)+\chi(\boldsymbol{\delta}_{2z}^2)+\chi(\boldsymbol{\delta}_{2\bar{x}y}^2)\right] \tag{9.3.4}$$

$$=\tfrac{1}{4}\cdot 4\chi(\boldsymbol{\epsilon})=3$$

$$N_{bb}=\tfrac{1}{2}(N_b+N'_b)=3$$

where $\chi(S)$ is the character of S in the polar vector representation (x,y,z). Thus there are three independent constants, and the reversal elements do not reduce this number.

The presence of $\boldsymbol{\rho}_z$ guarantees that terms linear in z must vanish, and $\boldsymbol{\rho}_{\bar{x}y}$, which interchanges x and y, forces $\mathbf{K}$ to have the form

$$\mathbf{K}=-\begin{bmatrix}\mu & \nu & 0\\ \nu & \mu & 0\\ 0 & 0 & \lambda\end{bmatrix} \tag{9.3.5}$$

Since we are down to three constants, we stop. (The minus sign is chosen so that, when terms diagonal in the particle indices are added in, the potential energy will be positive.)

Example 9.3.2

Find the second-neighbor forces in diamond.

Solution: The bond elements are $\boldsymbol{\epsilon}$ and $\boldsymbol{\rho}_{\bar{x}y}$. The reversal elements are $\boldsymbol{\delta}_{2z}$ and $\boldsymbol{\rho}_{xy}$. [All elements $(\boldsymbol{\alpha}|\boldsymbol{\tau})$ that interchange the sublattices transfer the bonds to translationally inequivalent positions and hence are not members of the reversal group of the bond.] Here

$$N_b=\tfrac{1}{2}\left\{[\chi(\epsilon)]^2+[\chi(\rho_{\bar{x}y})]^2\right\}=\tfrac{1}{2}(3^2+1^2)=5$$

$$N'_b=\tfrac{1}{2}\left[\chi(\delta_{2z}^2)+\chi(\rho_{xy}^2)\right]=\tfrac{1}{2}(3+3)=3 \tag{9.3.6}$$

$$N_{bb}=\tfrac{1}{2}(N_b+N'_b)=4$$

Thus without a reversal symmetry requirement there are five constants; with this symmetry there are four. Symmetry of $\mathbf{K}$ under $\rho_{\bar{x}y}$ (interchange of x and y) requires

$$K_{22}=K_{11},\qquad K_{12}=K_{21}$$

$$K_{13}=K_{23},\qquad K_{31}=K_{32}$$

or

$$\mathbf{K}^{000,\frac{1}{2}\frac{1}{2}0}=\mathbf{K}^{0,\mathbf{b}}=-\begin{bmatrix}\mu & \nu & \tau\\ \nu & \mu & \tau\\ \theta & \theta & \lambda\end{bmatrix} \tag{9.3.7}$$

having five constants. [We use $\mathbf{b}$ to label the atom at $(\frac{1}{2},\frac{1}{2},0)$.] Application of $\boldsymbol{\delta}_{2z}$, which reverses the signs of x and y and sends $\mathbf{b}$ into $-\mathbf{b}$, yields $\mathbf{K}^{0,-\mathbf{b}}$ in terms of $\mathbf{K}^{0,\mathbf{b}}$ with the result

$$\mathbf{K}^{0,-\mathbf{b}}=-\begin{bmatrix}\mu & \nu & -\tau\\ \nu & \mu & -\tau\\ -\theta & -\theta & \lambda\end{bmatrix}$$

But a lattice translation of amount **b** requires that

$$\mathbf{K}^{0,-\mathbf{b}} = \mathbf{K}^{\mathbf{b},0} \tag{9.3.8}$$

Symmetry 5.4.2 yields

$$\tilde{\mathbf{K}}^{\mathbf{b},0} = \mathbf{K}^{0,\mathbf{b}} \tag{9.3.9}$$

The transpose of Eq. 9.3.8 then gives $\tilde{\mathbf{K}}^{0,-\mathbf{b}} = \mathbf{K}^{0,\mathbf{b}}$:

$$\begin{bmatrix} \mu & \nu & -\theta \\ \nu & \mu & -\theta \\ -\tau & -\tau & \lambda \end{bmatrix} = \begin{bmatrix} \mu & \nu & \tau \\ \nu & \mu & \tau \\ \theta & \theta & \lambda \end{bmatrix}$$

so that $\theta = -\tau$ and

$$\mathbf{K}^{0,\mathbf{b}} = -\begin{bmatrix} \mu & \nu & \tau \\ \nu & \mu & \tau \\ -\tau & -\tau & \lambda \end{bmatrix} \tag{9.3.10}$$

The fourth "antisymmetric" constant τ was overlooked in the classic paper of Helen Smith (**1948**) on the vibrations of diamond, apparently because of assuming that $\mathbf{K}^{mn} = \tilde{\mathbf{K}}^{mn}$ rather than obeying Eq. 5.4.2. Equation 9.3.10 appears to violate Newton's law of action and reaction, but this law is, in fact, embodied in the flip symmetry, Eq. 5.4.2, which we have used.

9.4. LOCAL ELECTRIC MOMENTS

Although we showed in Section 9.1 that the *total* electric moment induced by an arbitrary displacement vanishes in diamond, this is not necessarily true of the local moments induced in the individual atoms, for which we can write:

$$\mu^{\alpha i} = \sum_{\beta j} \mathbf{M}^{\alpha i,\beta j} \cdot \mathbf{u}^{\beta j} \tag{9.4.1}$$

Under the operation $(\mathbf{S}|\mathbf{v})$, the displacement $\mathbf{Su}^{\beta j}$ is deposited at the new position $\beta' j'$ (cf. Eq. 5.4.15), and the new moment at αi is

$$(\mu^{\alpha i})' = \sum_{\beta j} \mathbf{M}^{\alpha i,\beta' j'}(\mathbf{Su}^{\beta j}) \tag{9.4.2}$$

or

$$(\mu^{\alpha' i'})' = \sum_{\beta j} \mathbf{M}^{\alpha' i', \beta' j'} (\mathbf{S} \mathbf{u}^{\beta j}) \tag{9.4.3}$$

But $\mu^{\alpha i}$ is a vector. Thus the new moment at the new position $\alpha' i'$ is

$$(\mu^{\alpha' i'})' = \mathbf{S} \cdot \mu^{\alpha i} \tag{9.4.4}$$

Comparison with Eq. 8.4.1 now yields

$$\mathbf{M}^{\alpha i, \beta j} = \mathbf{S}^{-1} \cdot \mathbf{M}^{\alpha' i', \beta' j'} \cdot \mathbf{S} \tag{9.4.5}$$

precisely the same condition, Eqs. 5.4.17–5.4.19, as is obeyed by the force constants $\mathbf{K}^{\alpha i, \beta j}$. The only distinction is that the force constants obey the symmetry requirement 5.4.2 or

$$\mathbf{K}^{\alpha i, \beta j} = \tilde{\mathbf{K}}^{\beta j, \alpha i} \tag{9.4.6}$$

whereas no such condition is required of the moment coefficients $\mathbf{M}^{\alpha i, \beta j}$.

Example 9.4.1

Determine the nearest-neighbor force constants and the moments induced on nearest neighbors in the diamond structure.

Solution: The group of the bond [from the atom at (0,0,0) to the atom at $(a/4, a/4, a/4)$] is the group C_{3v} of point operations $\epsilon, \delta_{3xyz}, (\delta_{3xyz})^{-1}$, $\rho_{\bar{y}z}, \rho_{\bar{z}x}, \rho_{\bar{x}y}$. By Eq. 9.4.5 or 5.4.18, $\mathbf{M}$ transforms according to the direct product $\Gamma_p \times \Gamma_p$ of the polar vector representation with itself. Since $\Gamma_p \times \Gamma_p$ contains the identity representation twice for this group (see Section 4.3), there are two arbitrary constants. The cyclic permutation of indices produced by δ_{3xyz} causes $\mathbf{M}$ to have the form

$$\mathbf{M} = \begin{bmatrix} \alpha & \beta & \beta' \\ \beta' & \alpha & \beta \\ \beta & \beta' & \alpha \end{bmatrix} \tag{9.4.7}$$

The reflection plane $\rho_{\bar{x}y}$, which simply interchanges x and y (see Table 9.1.1), causes $M_{12} = M_{21}$, that is, $\beta' = \beta$ or

$$\mathbf{M} = \mathbf{K} = \begin{bmatrix} \alpha & \beta & \beta \\ \beta & \alpha & \beta \\ \beta & \beta & \alpha \end{bmatrix} \tag{9.4.8}$$

The use of the inversion element $(\mathbf{S}|\mathbf{v})=(\mathbf{i}|\tau)$ in Eq. 9.4.5 yields the relation

$$\mathbf{M}^{-10,1j}=\mathbf{M}^{10,-1\bar{j}} \tag{9.4.9}$$

In the force constant case, this could be combined with Eqs. 9.4.6 and 9.3.2 to yield

$$\mathbf{K}^{-1,1}(\mathbf{j})=\mathbf{K}^{1,-1}(-\mathbf{j})=\tilde{\mathbf{K}}^{-1,1}(\mathbf{j}) \tag{9.4.10}$$

that is, *all* the odd neighbor force constant matrices (between atoms of different sublattices) are *symmetric* in the diamond structure. For the nearest-neighbor case, then, the reversal elements add no new information, as Eq. 9.4.8 is already symmetric.

The atom at (0,0,0) has four tetrahedral bonds to atoms at $(a/4)\times[(1,1,1),(1,-1,-1),(-1,1,-1)$ and $(-1,-1,1)]$. The second atom can be obtained from the first by the twofold rotation $\boldsymbol{\delta}_{2x}$, which reverses the signs of y and z, and M (or K) can be computed via Eq. 9.4.5. However, reversal of y and z changes the signs of the xy and xz components of a tensor, but not the yz component. Thus we arrive at a complete set of matrices:

$$\text{At } (1,1,1) \quad \begin{bmatrix} \alpha & \beta & \beta \\ \beta & \alpha & \beta \\ \beta & \beta & \alpha \end{bmatrix} \qquad \text{At } (1,-1,-1) \quad \begin{bmatrix} \alpha & -\beta & -\beta \\ -\beta & \alpha & \beta \\ -\beta & \beta & \alpha \end{bmatrix}$$

$$\text{At } (-1,1,-1) \quad \begin{bmatrix} \alpha & -\beta & \beta \\ -\beta & \alpha & -\beta \\ \beta & -\beta & \alpha \end{bmatrix} \qquad \text{At } (-1,-1,1) \quad \begin{bmatrix} \alpha & \beta & -\beta \\ \beta & \alpha & -\beta \\ -\beta & -\beta & \alpha \end{bmatrix} \tag{9.4.11}$$

If the atom at the origin is displaced by the vector $\mathbf{u}$, the α terms produce an induced moment $\alpha\mathbf{u}$ on each of the neighbors, and a moment $-4\alpha\mathbf{u}$ at the origin to keep the total moment zero. The meaning of the β terms can best be illustrated by displacing the atom at the origin an amount z in the z direction. Then the four atoms acquire moments in the x-y plane that are given respectively by

$$\beta z[(1,1,0),(-1,1,0),(1,-1,0),(-1,-1,0)]$$

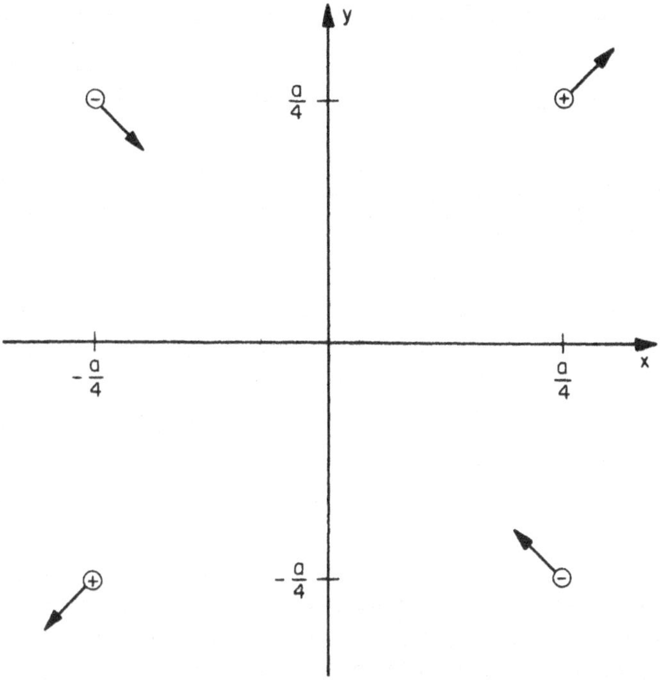

Fig. 9.4.1. Dipoles induced on the four neighbors of an atom in the diamond lattice when the central atom is displaced in the z direction. Atoms marked $\pm$ or $-$ are a distance $a/4$ above or below, respectively, the x-y plane.

and are shown in Fig. 9.4.1. The total dipole moment of this array is zero, but such an array of dipole moments produces a quadrupole tensor:

$$\mathbf{Q} = \sum_m (\boldsymbol{\mu}^m \mathbf{X}^m + \mathbf{X}^m \boldsymbol{\mu}^m) \tag{9.4.12}$$

For the above displacement in the z direction, the only nonvanishing element is

$$Q_{xy} = Q_{yx} = 2\beta a z \tag{9.4.13}$$

in agreement with Eq. 9.2.7.

9.5. SYMMETRIES OF ACOUSTIC AND OPTICAL MODES OF VIBRATION

Certain points in the Brillouin zone are labeled by symbols indicating that they are points of special symmetry (e.g., Γ, X, L) or are typical points on lines of special symmetry (e.g., Λ, Δ, Σ) or are on planes of special symmetry (e.g. A, B).

(See the Brillouin zone figures in Appendix E.) A representation of the group of the wave vector $G_{\mathbf{k}}$ is labeled by the symbol that denotes the point plus a subscript or superscript specifying a particular irreducible representation of $G_{\mathbf{k}}$. The point $\mathbf{k}=0$ is conventionally denoted as Γ.

Example 9.5.1

Show that the acoustic and optical modes of vibration at $\mathbf{k}=0$ have the symmetries $\Gamma_{15}^{-}=\Gamma_{15}$ and $\Gamma_{25}^{+}=\Gamma_{25}'$, respectively, in diamond. (See Chapters 11 and 12.)

Solution: The infinite-wavelength modes have the same motion in each cell. Thus such modes are characterized completely by the displacements $(\mathbf{u}^{1},\mathbf{u}^{-1})$ of the two atoms in the primitive cell. The actions of operators such as $(\boldsymbol{\alpha}|0)$ and $(\boldsymbol{\alpha}|\boldsymbol{\tau})$ in this six-dimensional space are represented by

$$D(\boldsymbol{\alpha}|0)=\begin{bmatrix} D^{P}(\boldsymbol{\alpha}) & 0 \\ 0 & D^{P}(\boldsymbol{\alpha}) \end{bmatrix},\qquad D(\boldsymbol{\alpha}|\boldsymbol{\tau})=\begin{bmatrix} 0 & D^{P}(\boldsymbol{\alpha}) \\ D^{P}(\boldsymbol{\alpha}) & 0 \end{bmatrix} \tag{9.5.1}$$

where $D^{P}(\boldsymbol{\alpha})$ is the ordinary 3×3 polar vector representation of $\boldsymbol{\alpha}$. The form for $D(\boldsymbol{\alpha}|\boldsymbol{\tau})$ follows from the fact that $(\boldsymbol{\alpha}|\boldsymbol{\tau})$ interchanges the two sublattices, that is, $\mathbf{u}^{1}$ and $\mathbf{u}^{-1}$. Thus we obtain the characters

$$\chi(\boldsymbol{\alpha}|0)=2\chi^{P}(\boldsymbol{\alpha}),\qquad \chi(\boldsymbol{\alpha}|\boldsymbol{\tau})=0 \tag{9.5.2}$$

The characters of the polar vector representation can be obtained from Eqs. 3.1.8 or Table 3.1.1; alternatively, one can simply remember that Γ_{15}^{-} is the polar vector representation for the group O_h. Table 9.5.1 presents the character table for the factor group G/T of the diamond group O_h^7 at $\mathbf{k}=0$; this is simply the table for the point group O_h except for the fact that the group elements are cosets, and one element from each coset ("coset representative") is shown. The last column, Γ, lists the characters obtained from Eq. 9.5.2 and clearly obeys

$$\Gamma=\Gamma_{15}^{-}+\Gamma_{25}^{+} \tag{9.5.3}$$

Thus the vibrations at $\mathbf{k}=0$ consist of two triply degenerate modes. Since

$$(\mathbf{i}|\boldsymbol{\tau})\begin{bmatrix}\mathbf{u}^{1}\\ \mathbf{u}^{-1}\end{bmatrix}=\begin{bmatrix}0 & -1\\ -1 & 0\end{bmatrix}\begin{bmatrix}\mathbf{u}^{1}\\ \mathbf{u}^{-1}\end{bmatrix}=-\begin{bmatrix}\mathbf{u}^{-1}\\ \mathbf{u}^{1}\end{bmatrix} \tag{9.5.4}$$

the fact that Γ_{15}^{-} is odd and Γ_{25}^{+} even under $(\mathbf{i}|\boldsymbol{\tau})$ permits us to deduce that the

Table 9.5.1 Character Table[a] At Γ for Factor Group O_h^7

Class	Typical Element	$\Gamma_1^{\pm}$	$\Gamma_2^{\pm}$	$\Gamma_{12}^{\pm}$	$\Gamma_{15}^{\pm}$	$\Gamma_{25}^{\pm}$	Γ
E	$(\epsilon\|0)$	1	1	2	3	3	6
$3C_4^2$	$(\delta_{2x}\|0)$	1	1	2	-1	-1	-2
$6C_4$	$(\delta_{4x}\|\tau)$	1	-1	0	1	-1	0
$6C_2$	$(\delta_{2xy}\|\tau)$	1	-1	0	-1	1	0
$8C_3$	$(\delta_{3xyz}\|0)$	1	1	-1	0	0	0
J	$(\mathbf{i}\|\tau)$	± 1	± 1	± 2	± 3	± 3	0
$3JC_4^2$	$(\rho_x\|\tau)$	± 1	± 1	± 2	∓ 1	∓ 1	0
$6JC_4$	$(\sigma_{4x}\|0)$	± 1	∓ 1	0	± 1	∓ 1	-2
$6JC_2$	$(\rho_{yz}\|0)$	± 1	∓ 1	0	∓ 1	± 1	2
$8JC_3$	$(\sigma_{6xyz}\|\tau)$	± 1	± 1	∓ 1	0	0	0

[a] $\Gamma_1^+ = \Gamma_1, \Gamma_2^+ = \Gamma_2, \Gamma_{12}^+ = \Gamma_{12}, \Gamma_{15}^+ = \Gamma'_{15}, \Gamma_{25}^+ = \Gamma'_{25}$.

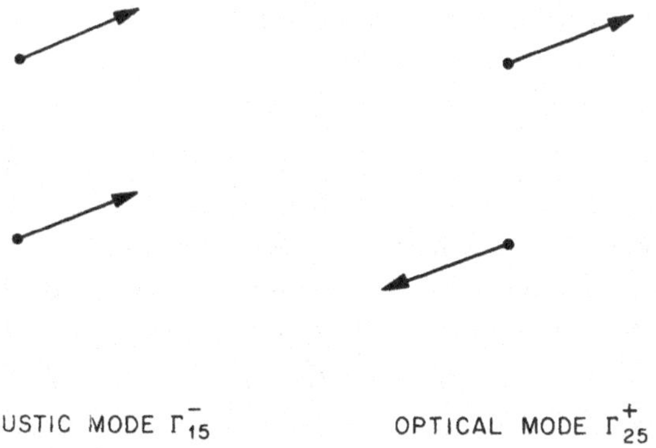

Fig. 9.5.1. Normal mode of vibrations at $\Gamma(\mathbf{k}=0)$ for the diamond structure. The modes are triply degenerate reflecting the cubic symmetry, so that displacements can be in any three orthogonal directions. The acoustic, Γ_{15}^-, mode is odd under inversion; the two sublattices move together. The optic, Γ_{25}^+, mode is even under inversion; the two sublattices move oppositely.

form of these solutions is

$$\Gamma_{15}^{-}\sim(\mathbf{u},\mathbf{u}), \qquad \Gamma_{25}^{+}\sim(\mathbf{u},-\mathbf{u}) \tag{9.5.5}$$

which are also shown pictorially in Fig. 9.5.1. The triple degeneracy consists in the fact that the displacement $\mathbf{u}$ may have arbitrary orientation. The mode Γ_{25}^{+} is referred to as "optical," since it involves a relative motion of atoms in the unit cell that, in other materials, might possibly generate an electric moment and produce optical absorption. (We have, of course, established the contrary for the diamond structure in Section 9.1.) The mode Γ_{15}^{-} is referred to as "acoustic," since all atoms in the cell have the same motion. We shall show in Chapter 11 that the frequency of this motion must vanish because of infinitesimal displacement invariance.

The results in Eq. 9.5.5 can be obtained by a procedure that is in principle more direct: Apply projection operators for the states Γ_{15}^{-} and Γ_{25}^{+} into the space $(\mathbf{u}^{1},\mathbf{u}^{-1})$. The net result is

$$\begin{aligned}\Gamma_{15}^{-}&=\tfrac{1}{2}(\mathbf{u}^{1}+\mathbf{u}^{-1},\mathbf{u}^{1}+\mathbf{u}^{-1})\\ \Gamma_{25}^{+}&=\tfrac{1}{2}(\mathbf{u}^{1}-\mathbf{u}^{-1},\mathbf{u}^{-1}-\mathbf{u}^{1})\end{aligned} \tag{9.5.6}$$

In practice, however, the construction of the projection operators, via Eq. 3.7.11, and their use require much more arithmetic. Furthermore, if we had applied projection operators for Γ_{15}^{+} or Γ_{25}^{-}, the result would have been zero. In other words, the economical use of projection operators on a space should be preceded by an analysis of the contents of the space, as in Eq. 9.5.3. But then the irreducible subspaces can often be deduced by elementary techniques, as shown here.

9.6. HOLE SCATTERING BY PHONONS

The electronic band structure of Germanium is shown in Fig. 9.6.1. Note the presence of an energy gap between the top of the filled valence band at $\Gamma_{25}^{+}=\Gamma_{25'}$ and the bottom of the conduction band[2] at L_1. At sufficiently high temperatures, or in the presence of acceptors, holes will be present near $\mathbf{k}=0$ with symmetry Γ_{25}^{+}. To determine which phonons are capable of scattering holes from one of the triply degenerate Γ_{25}^{+} states to another, we take the symmetrized product, Eq. 3.3.30, since the matrix element involves a product $\psi_{\mu}^{25}(\mathbf{r})^{*}\psi_{\nu}^{25}(\mathbf{r})=\psi_{\mu}^{25}(\mathbf{r})\psi_{\nu}^{25}(\mathbf{r})$ $=\psi_{\nu}^{25}(\mathbf{r})\psi_{\mu}^{25}(\mathbf{r})$, symmetric on the indices μ and ν. (For a discussion of the reality of wave functions see Chapter 10.) The symmetrized product is given by

$$(\Gamma_{25}^{+}\times\Gamma_{25}^{+})_{\text{sym}}=\Gamma_{25}^{+}+\Gamma_{12}^{+}+\Gamma_{1}^{+} \tag{9.6.1}$$

[2]A discussion of electronic energy bands, including symmetry properties, was presented in Chapter 7. The representations at Λ,Δ,L are discussed in Chapter 12.

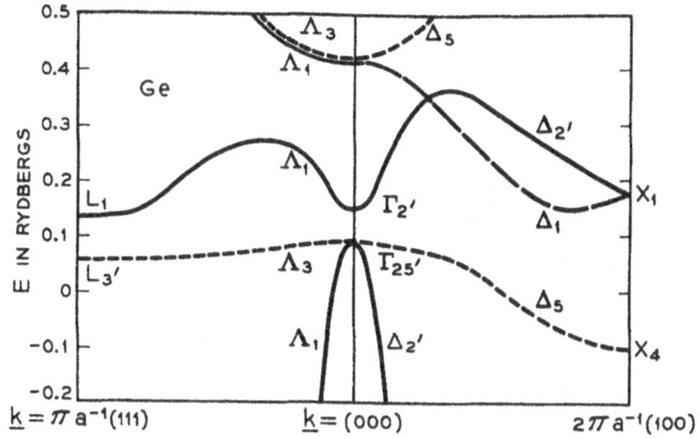

Fig. 9.6.1. Energy bands in Ge from Phillips (1958) pseudopotential calculation.

which is evaluated with the help of Table 9.5.1. Since the right-hand side contains Γ_{25}^+ once, optical phonons are allowed and only one parameter is needed to describe all possible matrix elements.

Since Γ_{15}^- does not appear, it would seem that acoustical phonons with $\mathbf{k}=0$ have vanishing matrix elements. This merely expresses the fact that pure translation will not affect the electronic states. The physically interesting question is whether or not long-wavelength acoustical phonons can scatter. The potential produced by such phonons is proportional to the strain induced by them (Bardeen and Shockley, **1950**). Since Γ_{25}^+ has basis functions (xy,yz,zx), Γ_{12}^+ has $[(x^2-y^2), (z^2-\frac{1}{2}x^2-\frac{1}{2}y^2)]$, and Γ_1^+ has $(x^2+y^2+z^2)$, we see from Eq. 9.6.1 that all possible strains are allowed. The pure compression Γ_1^+, however, because of the base vector orthogonality theorem 3.4.3, simply shifts the energy of the levels without producing scattering. Thus all scattering by (long-wave) acoustic phonons is describable in terms of two parameters, one for Γ_{25}^+ and one for Γ_{12}^+ strains.

9.7. SELECTION RULES FOR DIRECT OPTICAL ABSORPTION

The matrix element for optical absorption from a valence band state $\psi_v(\mathbf{k},\mathbf{r})$ to a conduction band state $\psi_c(\mathbf{k}',\mathbf{r})$ is governed by a matrix element of the form

$$(\psi_c(\mathbf{k}',\mathbf{r}), \exp(i\boldsymbol{\nu}\cdot\mathbf{r})\mathbf{e}\cdot\nabla\psi_v(\mathbf{k},\mathbf{r})) \tag{9.7.1}$$

where we use $\mathbf{A}(\mathbf{r})\cdot\nabla=\exp(i\boldsymbol{\nu}\cdot\mathbf{r})\mathbf{e}\cdot\nabla$ as the perturbation associated with the electric field. For any reasonable energy gap, say $\Delta E\sim 1$ eV, the photon propagation constant ν corresponding to a photon energy equal to the energy gap is

$$\nu = \frac{\Delta E}{\hbar c} \sim 10^5 \text{cm}^{-1} \tag{9.7.2}$$

and will be very small compared to $1/a \sim 10^8 \text{cm}^{-1}$, where a is a typical lattice spacing. Thus $\boldsymbol{\nu} \ll \mathbf{K}$, where $\mathbf{K}$ is any reciprocal lattice vector. Conservation of crystal momentum, Eq. 6.5.5, must take place without Umklapp processes in the form

$$\mathbf{k}' = \mathbf{k} + \boldsymbol{\nu} \approx \mathbf{k} \tag{9.7.3}$$

where the neglect of $\boldsymbol{\nu}$ is permissible because of the smallness of $\boldsymbol{\nu}$ with respect to the relevant parameter K or $1/a$. This is equivalent to the customary electric dipole approximation in which the matrix element 9.7.1 is approximated by replacing the $\exp(i\boldsymbol{\nu}\cdot\mathbf{r})$ term by unity, since the $\boldsymbol{\nu}\cdot\mathbf{r}$ corrections are of the negligible order νa.

Example 9.7.1

Is the "vertical" optical transition from the top of the valence band of Germanium at $\mathbf{k}=0$, with symmetry Γ_{25}^+, to the corresponding state Γ_2^- in the conduction band allowed? (See Fig. 9.6.1.)

Solution: In the electric dipole approximation, we are concerned with the matrix element

$$\mathbf{e} \cdot (\psi_2^-, \nabla \psi_{25}^+) \tag{9.7.4}$$

Since ∇ transforms as the polar vector representation $\Gamma_p = \Gamma_{15}^-$ in the group O_h, $\nabla \psi_{25}^+$ transforms as

$$\Gamma_{15}^- \times \Gamma_{25}^+ = \Gamma_2^- + \Gamma_{12}^- + \Gamma_{25}^- + \Gamma_{15}^- \tag{9.7.5}$$

using character table 9.5.1. Since Γ_2^- is contained, the transition is allowed. *Note*: At $\mathbf{k}=0$ we proceed with a *point* group table just as if we had an atom embedded in an O_h crystal field. Furthermore, the part of the selection rule relating to parity $(-)\times(+)=(-)$ is clearly obeyed, and the abbreviated table for O instead of O_h can be used.

Example 9.7.2

Is the "vertical" optical transition from L_3^- to the bottom of the conduction band at L_1^+ in Germanium allowed?

Solution: The point L belongs to the wave vector $\mathbf{k}=(\pi/a)(1,1,1)$ on the zone boundary for the FCC lattice. The character table appropriate to this point, for

Table 9.7.1. Group Characters at $L=(\pi/a)(1,1,1)^a$ in O_h^7

Class	Typical Element	$L_1^{\pm}$	$L_2^{\pm}$	$L_3^{\pm}$	$\Gamma_{15}^- \times L_3^-$
E	$(\epsilon\|0)$	1	1	2	6
$2C_3$	$(\delta_{3xyz}, \delta_{3xyz}^{-1}\|0)$	1	1	-1	0
$3C_2$	$(\delta_{2\bar{y}z}, \delta_{2\bar{z}x}, \delta_{2\bar{x}y}\|\tau)$	1	-1	0	0
6	$(\mathbf{i}\|\tau)(\alpha\|\mathbf{a})$	$\pm\chi(\alpha\|\mathbf{a})$			

[a] $L_1^+ = L_1, L_2^+ = L_2, L_3^+ = L_3;\ L_1^- = L_{1'}, L_2^- = L_{2'}, L_3^- = L_{3'}$.

the diamond group O_h^7, is given in Table 9.7.1. We first note that $\nabla\psi_{L_3^-}$ still has translation properties appropriate to the point L and thus couples to other states of the same wave vector. Since ∇ transforms as a polar vector Γ_{15}^-, the transformation properties of $\nabla\psi$ are given by

$$\Gamma_{15}^- \times L_3^- = L_1^+ + L_2^+ + 2L_3^+ \tag{9.7.6}$$

Hence the L_1^+ is contained, and the transition is allowed.

Since the group at L is a subgroup of the group at Γ, the characters of Γ_{15}^- for an operator in L can be found in the character table 9.5.1. One simply takes $\chi^{15-}(R)\chi^{3-}(R)$ as the appropriate character product because $\psi_\mu^{15-}(\mathbf{r})\psi_\nu^{3-}(\mathbf{r})$ constitutes a complete set of basis functions ("invariant space") with respect to the operators of L. [However, the computation of $L_3^- \times (L_1^+)^*$ is more complicated, since $\psi_\mu^{3-}(\mathbf{r})\psi_\nu^{1+}(\mathbf{r})^*$ for all μ and ν is *not* an invariant space under the operators at Γ because operators in Γ outside the group at L can generate new functions not available in the above set. This more complicated case, needed in problems such as intervalley scattering, is discussed by Lax and Hopfield (**1961**), Lax (**1962**, **1965c**), Elliott and Loudon (**1960**), Birman (**1962**, **1963**, **1966**, **1974**), Zak (**1962**), Bradley (**1966**), Birman, Lax, and Loudon (**1966**), Lax and Birman (**1972**)].

9.8. SUMMARY

In a natural extension of point group methods, space groups have been used to determine (1) the form and number of independent constants in various microscopic matrices: the force constant matrix and the electric dipole and quadrupole matrices; (2) selection rules for transitions induced by phonons or by photons; and (3) the decomposition of the space of $\mathbf{k}=0$ vibrations into its irreducible components: acoustic and optical modes.

problems

9.1.1. Show that for the zinc blende structure (like the diamond structure but with the two atoms in the unit cell distinct; see Table 6.1.4) the effective charges $\mathbf{B}^1$ and $\mathbf{B}^2$ associated with the two atoms are scalars.

9.1.2. Construct a space structure that belongs to the "diamond" group $O_h^7 = F(4_1/d)\bar{3}\,(2/m)$ but does not have the diamond structure. Is the vanishing electric moment found in diamond a property of the group or of the structure?

9.1.3. Can lattice vibrations of symmetry Γ_{15}^+ occur in a crystal whose space group is O_h^7?

9.2.1. Show that the conclusion that Q of Eq. 9.2.7 does not vanish is unaffected when the symmetry on rs of Q_{rs} is taken into account, that is, that $(\Gamma_p \times \Gamma_p)_{\text{sym}} \times \Gamma_p$ still contains the identity representation.

9.4.1. Show that the third-neighbor force constants in diamond take the form given in Eq. 12.1.1.

9.5.1. Discuss the compatibility of the representations along $\Sigma = [q,q,0]$ with those at $X = (2\pi/a)[1,0,0]$ for the space group O_h^7 appropriate to the diamond structure. You may use the character tables 12.2.1 and 12.4.1.

9.5.2. Diamond at the point X has inversion as one of the elements of the group of the wave vector (See Table 8.10.1.) But the characters $\chi(E) = 2, \chi(I) = 0$ imply that we have two-dimensional representations, each of which has one even and one odd component. Why do the representations lack a definite parity? Do any of the representations Δ at $(q,0,0)$ acquire a definite parity as $q \rightarrow 2\pi/a$, that is, the point X at which inversion is a member of the group?

chapter ten

TIME REVERSAL[1]

0.1. NATURE OF TIME-REVERSAL OPERATORS WITHOUT SPIN

Newton's third law involves second time derivatives. It is therefore even under time *inversion*: $t \rightarrow -t$. If $x(t)$ is a solution, so is $x(-t)$, and we refer to x as even, whereas the velocity $v = dx/dt$ is then odd under time inversion. In Hamiltonian language, if $x(t), p(t)$ is a solution, where $p = m\,dx/dt$ is the conjugate momentum, then $x(-t), -p(-t)$ must be a second solution of the same equations of motion. This will be true if we require the Hamiltonian $H(x,p)$ to be an even function of p:

$$H(x,-p) = H(x,p) \tag{10.1.1}$$

In quantum mechanics, the reversal of the momentum can be carried out in the Schrödinger space representation by means of the operation K_0 of complex conjugation:

$$K_0\left(-i\hbar\frac{\partial}{\partial x}\right)K_0^{-1} = i\hbar\frac{\partial}{\partial x} \tag{10.1.2}$$

[1]The basic work in time reversal was done by Wigner (**1932, 1959**).

and Eq. 10.1.1 can be written in the form

$$H\left(x, i\hbar\frac{\partial}{\partial x}\right) = H\left(x, -i\hbar\frac{\partial}{\partial x}\right) \tag{10.1.3}$$

that is, in the space representation, H is real.

Since

$$K_0\psi(x) = \psi(x)^* \tag{10.1.4}$$

if $\psi(x)$ is an eigenfunction of H with real eigenvalue E, then

$$H\psi = E\psi \qquad \text{and} \qquad K_0 H\psi = H(K_0\psi) = E(K_0\psi) \tag{10.1.5}$$

Thus $K_0\psi(x) = \psi^*(x)$ is also a solution with the same eigenvalue because H is *real* in the space representation, in other words, because H commutes with the time reversal operator K_0.

The definition of K_0 as complex conjugation, Eq. 10.1.4, is restricted to the space representation. If

$$\psi(x) = \int f(p)\exp\left(\frac{ipx}{\hbar}\right)dp \tag{10.1.6}$$

then

$$K_0\psi = \int f^*(p)\exp\left(-\frac{ipx}{\hbar}\right)dp$$

$$= \int f^*(-p)\exp\left(\frac{ipx}{\hbar}\right)dp \tag{10.1.7}$$

so that in the momentum representation it is required that

$$K_0 f(p) = f^*(-p) \tag{10.1.8}$$

We can define K_0 either by the invariant (independent of the representation) requirement

$$K_0 x K_0^{-1} = x, \qquad K_0 p K_0^{-1} = -p \tag{10.1.9}$$

that it leave the positions alone and reverse the velocities, or by the statement that K_0 can be obtained in any representation by using K_0 = complex conjugation in the space representation. The operator K_0 is referred to as Wigner's *time-reversal* operator. It is distinct from time inversion, but it uses the time-inversion symmetry of the problem to obtain a new solution.

The time-reversal operator has the property of being *antilinear*:

$$K_0(a_1\psi_1 + a_2\psi_2) = a_1^* K_0\psi_1 + a_2^* K_0\psi_2 \tag{10.1.10}$$

and *antiunitary*:

$$(K_0\Phi, K_0\Psi) = (\Psi, \Phi) \tag{10.1.11}$$

and also obeys

$$K_0^2 = 1 \qquad \text{or} \qquad K_0^{-1} = K_0$$

The relationship between these results and our classical discussion is, as usual, seen most readily in the Heisenberg representation. Thus we can write for the time-reversed variables.

$$x^T(t) = K_0 x(t) K_0^{-1} = K_0 \exp(iHt) x \exp(-iHt) K_0^{-1}$$
$$= x(-t) \tag{10.1.12}$$

since

$$K_0 \exp(iHt) K_0^{-1} = \exp(K_0 iHt K_0^{-1})$$
$$= \exp(-itK_0 H K_0^{-1}) = \exp(-itH) \tag{10.1.13}$$

where we have set $\hbar = 1$ for simplicity. Similarly, the sign change in Eq. 10.1.9 for p yields

$$p^T(t) = K_0 p(t) K_0^{-1} = -p(-t) \tag{10.1.14}$$

These results describe time reversal in the Heisenberg framework, in which the operators change with time, but the wave functions $\psi(x, t_0)$ are fixed. Corresponding statements can be made about $\psi(x, t)$ in the Schrödinger framework by transforming from one case to the other. It is easier, however, to note that, if

$$H\psi = i\hbar \frac{\partial \psi}{\partial t} \tag{10.1.15}$$

and Eq. 10.1.3 is replaced by its invariant counterpart

$$K_0 H = H K_0 \tag{10.1.16}$$

then

$$\psi^T(t) = K_0 \psi(-t) \tag{10.1.17}$$

is the corresponding "time-reversed" solution expressed in an invariant way.

10.2. TIME REVERSAL WITH SPIN

Since time reversal leaves **r** invariant and reverses **p**, we see that it also reverses $\mathbf{L} = \mathbf{r} \times \mathbf{p}$, so that **p** and **L** can be referred to as odd (or "imaginary") operators.

We anticipate, then, that in the presence of spin the Wigner time-reversal operator K will be defined by

$$K\mathbf{r}K^{-1}=\mathbf{r}, \qquad K\mathbf{p}K^{-1}=-\mathbf{p}, \qquad K\boldsymbol{\sigma}K^{-1}=-\boldsymbol{\sigma} \tag{10.2.1}$$

since the spin operators are special cases of angular momentum operators. If K is defined in this way, a Hamiltonian with spin-orbit coupling:

$$H=\frac{p^2}{2m}+\left(\frac{\hbar}{4m^2c^2}\right)(\boldsymbol{\sigma}\times\nabla V\cdot\mathbf{p})+V \tag{10.2.2}$$

will commute with time reversal:

$$KH=HK \tag{10.2.3}$$

since $\mathbf{p}$ and $\boldsymbol{\sigma}$ are simultaneously reversed, thus preserving the otherwise troublesome middle term. Thus, if $\psi(t)$ is a solution, then

$$\psi^T(t)=K\psi(-t) \tag{10.2.4}$$

is the time-reversed solution. Let us determine K by setting

$$K=UK_0 \tag{10.2.5}$$

where U is a unitary spin operator that commutes with $\mathbf{r}$ and $\mathbf{p}$. Then Eq. 10.2.1 yields

$$U\boldsymbol{\sigma}^*U^{-1}=-\boldsymbol{\sigma} \tag{10.2.6}$$

Although we have in mind particularly a particle of spin $\frac{1}{2}$, a particle of spin j would require

$$U\mathbf{J}^*U^{-1}=-\mathbf{J} \tag{10.2.7}$$

The representation chosen for $\mathbf{J}$ in Eq. 2.3.1 is such as to make J_x and J_z real and J_y purely imaginary. Thus Eq. 10.2.7 can be rewritten in the form

$$U(J_x,J_y,J_z)U^{-1}=(-J_x,J_y,-J_z) \tag{10.2.8}$$

which is simply a two fold rotation $\boldsymbol{\delta}_{2y}$ about the y axis. Aside from an irrelevant phase factor, we can choose U to be the unitary operator[2] (cf. Eq. 2.1.16):

[2]For integral $j=l$ it is possible to introduce real spherical harmonics (Fano, **1960**). In such a real basis, the matrix elements of $\mathbf{l}=-i\mathbf{r}\times\nabla$ are purely imaginary, and we can reverse $\mathbf{l}$ simply by taking $K=K_0$. (For half-integral j, the representations *cannot* be made real, and it is necessary to modify K_0 by a rotation of π.) Although K has an invariant meaning defined by the requirement that it reverse $\mathbf{p}$ and $\mathbf{J}$, we have shown that its explicit form depends on the choice of representation.

$$U = \exp(i\pi J_y) \tag{10.2.9}$$

which in our representation (Eq. 2.3.1) is real. Thus

$$K^2 = UK_0UK_0 = U^2 = \exp(i2\pi J_y) = (-1)^{2j} \tag{10.2.10}$$

That is, $K^2 = +1$ for systems of integral spin and $K^2 = -1$ for systems of half-integral (i.e., half an odd integer) spin. We now return to the special case $j = \frac{1}{2}$ and use Eq. 2.6.7 to write

$$U = \exp(\tfrac{1}{2}i\pi\sigma_y) = i\sigma_y = \begin{bmatrix} 0 & 1 \\ -1 & 0 \end{bmatrix} \tag{10.2.11}$$

for this case. We can, of course, modify U by an arbitrary phase factor; in particular, we can use $U = \sigma_y$ without affecting the conclusion:

$$K^2 = -1 \qquad \text{(spin case)} \tag{10.2.12}$$

For an n-electron system, of course, all spins must reverse and we can use

$$U = \prod_{m=1}^{n} \sigma_y^m \tag{10.2.13}$$

$$K^2 = (-1)^n \tag{10.2.14}$$

That K is an *antilinear* operator follows from Eq. 10.2.5, the linear nature of $U = i\sigma_y$, and the antilinear nature of K_0. This conclusion is independent of the representation, since it can also be derived from the commutation rule:

$$Ki\hbar K^{-1} = K[x,p]K^{-1} = [x, -p] = -i\hbar \tag{10.2.15}$$

Moreover, since U is unitary, $K = UK_0$ is *antiunitary*, that is,

$$(K\Phi, K\Psi) = (\Psi, \Phi) \tag{10.2.16}$$

Theorem 10.2.1

In the absence of spin-orbit coupling it is possible to use the operator K_0 for electrons, even though the latter have spin.

Proof: Stated one way, K_0 commutes with H. Stated another way, K_0 is not the time-reversal operator K, in this case, but a combination of this operator with U, that is, a twofold rotation of spin space. But all such rotations are permissible, since the spin motion is decoupled from the orbital motion.

10.3. TIME REVERSAL IN EXTERNAL FIELDS

The Hamiltonian in the presence of external electromagnetic fields described by the scalar and vector potentials Φ and $\mathbf{A}$ is obtained from the original Hamiltonian $H(\mathbf{p},\mathbf{r})$ by writing

$$H\left(\mathbf{p}-\frac{e\mathbf{A}}{c},\mathbf{r}\right)+e\Phi \tag{10.3.1}$$

If the original Hamiltonian was invariant under time reversal (i.e., $\mathbf{p}\to-\mathbf{p}$), the new Hamiltonian will be invariant under this operation only if one simultaneously reverses $\mathbf{A}$, but not Φ, that is, reverses the motion of the sources of the field. The currents produced by a set of charges whose motion is reversed in accord with Eqs. 10.1.12–10.1.14 will obey[3]

$$\mathbf{J}^T(\mathbf{r},t)=-\mathbf{J}(\mathbf{r},-t), \qquad \rho^T(\mathbf{r},t)=\rho(\mathbf{r},-t) \tag{10.3.2}$$

thus causing the time-reversed fields and potentials to obey

$$\begin{aligned} \mathbf{E}^T(\mathbf{r},t)&=\mathbf{E}(\mathbf{r},-t), \qquad &\mathbf{H}^T(\mathbf{r},t)&=-\mathbf{H}(\mathbf{r},-t) \\ \mathbf{A}^T(\mathbf{r},t)&=-\mathbf{A}(\mathbf{r},-t), \qquad &\Phi^T(\mathbf{r},t)&=\Phi(\mathbf{r},-t) \end{aligned} \tag{10.3.3}$$

Speaking classically, if we have a solution $x(t,\mathbf{H}), p(t,\mathbf{H})$ in the presence of a magnetic field, we know there exists another solution,

$$x^T(t,\mathbf{H})=x(-t,\mathbf{H}^T), \qquad p^T(t,\mathbf{H})=-p(-t,\mathbf{H}^T) \tag{10.3.4}$$

not of the same equations, but of the equations with the magnetic field reversed. In the quantum-mechanical case, we can write for the time-reversed Hamiltonian

$$H^T(\mathbf{H})=KH(\mathbf{H})K^{-1}=H(\mathbf{H}^T) \tag{10.3.5a}$$

or if $H=H(t,\mathbf{H})$ is explicitly time-dependent

$$H^T(t,\mathbf{H})=KH(-t,\mathbf{H})K^{-1}=H(-t,\mathbf{H}^T) \tag{10.3.5b}$$

Here K is the time-reversal operator K_0 or $\sigma_y K_0$, so that

$$\Psi^T(t,\mathbf{H}^T)=K\Psi(-t,\mathbf{H}) \tag{10.3.6}$$

is a solution of the magnetic-field-reversed problem even if, as in Eq. 10.3.5b, the Hamiltonian depends explicitly on the time.

When the kinetic energy has the form $p^2/2m$, the presence of a magnetic field leads to a term linear in p and of the form $-(e/mc)\mathbf{A}\cdot\mathbf{p}$. With $\mathbf{A}=\frac{1}{2}(\mathbf{H}\times\mathbf{r})$, for

[3] This result assumes tacitly that electric charge e is invariant under time reversal.

a uniform magnetic field, this term is

$$-\left(\frac{e}{2mc}\right)\mathbf{L}\cdot\mathbf{H} \tag{10.3.7}$$

where $\mathbf{L}=\mathbf{r}\times\mathbf{p}$ is the angular momentum of the system. Thus the sign of this term can be reversed by a twofold rotation[4] δ_2 of the system (holding the field fixed) about some axis perpendicular to $\mathbf{H}$. If such a rotation exists, which *commutes* with $H(\mathbf{p},\mathbf{r})$ in the absence of the field, then

$$\theta=\delta_2 K \tag{10.3.8}$$

will commute with the complete Hamiltonian in the presence of the field. Thus, if $\psi(t,\mathbf{H})$ is a solution,

$$\Psi^{\theta}(t,\mathbf{H})=\theta\Psi(-t,\mathbf{H}) \tag{10.3.9}$$

is also a solution with the *same* direction of the magnetic field.

In the absence of spin-orbit coupling, we can replace K by K_0. Furthermore, in this case

$$\theta^2=(\delta_2 K_0)^2=1 \tag{10.3.10}$$

so that, if Ψ is a solution, $\frac{1}{2}(\Psi+\theta\Psi)$ is a solution that obeys the constraint

$$\theta\Psi=\Psi \qquad \text{or} \qquad \delta_2\Psi^*=\Psi \tag{10.3.11}$$

and all solutions can be made to obey this constraint by appropriate choice of phase factors.[5] This result, Eq. 10.3.11, is exact, independent of the strength of the magnetic field. We have, in effect, diagonalized simultaneously the commuting operators θ and H. In the spin case, when $\theta^2=-1$, however, we shall see that, although $[\theta,H]=0$, we cannot diagonalize θ because of its antiunitary behavior.

10.4. ANTILINEAR AND ANTIUNITARY OPERATORS

We have more or less by inspection constructed an explicit form $K=UK_0$ for the time-reversal operator (with U given by Eq. 10.2.13) with the properties

$$K^2=\pm 1 \tag{10.4.1}$$

[4]A twofold rotation is the only *proper* rotation that reverses a vector. (See Corollary 2.7.2.) Since inversion has no effect on a pseudovector such as $\mathbf{H}$, the only improper reversal element is a reflection plane containing $\mathbf{H}$.

[5]Since θ is antilinear, if $\theta\Psi=-\Psi$, then $i\Psi$ obeys the desired constraint. Note that Ψ and $\theta\Psi$ may belong to different irreducible representations.

$$(K\Phi, K\Psi) = (\Psi, \Phi), \quad \text{that is, } K = \text{antiunitary} \tag{10.4.2}$$

$$K(a\Psi) = a^* K\Psi, \quad \text{that is, } K = \text{antilinear} \tag{10.4.3}$$

In this section we show that these results are, in fact, consequences of fairly fundamental quantum-mechanical arguments.

Theorem 10.4.1

If K is the time-reversal operator, then $K^2 = \pm 1$.

Proof: Two successive time reversals must restore a physical system to itself:

$$K^2\Psi = \exp(i\gamma)\Psi$$

where the superposition principle forces[6] the phase factor $\exp(i\gamma)$ to be independent of the state Ψ. However,

$$K^3\Psi = KK^2\Psi = K\exp(i\gamma)\Psi = \exp(-i\gamma)K\Psi$$

and

$$K^3\Psi = K^2(K\Psi) = \exp(i\gamma)K\Psi$$

so that

$$\exp(i\gamma) = \exp(-i\gamma) = \pm 1$$

We shall define a *symmetry* operator as effecting a transformation from one reference frame to another:

$$\Psi' = O\Psi, \qquad \Phi' = O\Phi \tag{10.4.4}$$

If

$$\hat{\Psi} = \frac{\Psi}{(\Psi,\Psi)^{1/2}}, \qquad \hat{\Phi} = \frac{\Phi}{(\Phi,\Phi)^{1/2}} \tag{10.4.5}$$

are an arbitrary pair of unit vectors, observers in two reference frames related by the symmetry operator O must see the same overlap probability:

$$|(\hat{\Phi}', \hat{\Psi}')| \equiv |(O\hat{\Phi}, O\hat{\Psi})| = |(\hat{\Phi}, \hat{\Psi})| \tag{10.4.6}$$

Although the physics of state vectors is completely describable in the subspace of unit vectors, we shall for convenience extend our description to unnormalized vectors by assuming that

$$O(a\Psi) = aO\Psi \quad \text{for real } a \tag{10.4.7}$$

This assumption permits us to rewrite Eq. 10.4.6 in a form

$$|(\Phi', \Psi')| \equiv |(O\Phi, O\Psi)| = |(\Phi, \Psi)| \tag{10.4.8}$$

[6]We have tacitly assumed Eq. 10.4.15 with $O = K^2$.

valid for unnormalized vectors. In particular, Eq. 10.4.8 implies the preservation of normalization:

$$(O\Phi, O\Phi) = (\Phi, \Phi) \tag{10.4.9}$$

We shall refer to operators O that obey Eq. 10.4.8 as Wigner mappings or more briefly as *symmetry* operators.

Theorem 10.4.2

Wigner mappings that are linear are also unitary. Wigner mappings that are antilinear are antiunitary.

Proof:

$$(O(\Phi + a\Psi), O(\Phi + a\Psi)) = (\Phi, \Phi) + |a|^2(\Psi, \Psi) + 2\,\mathrm{Re}[a(\Phi, \Psi)] \tag{10.4.10}$$

follows from Eq. 10.4.9. But we can also write

$$O(\Phi + a\Psi) = O\Phi + g(a)O\Psi, \qquad g(a) = a \text{ or } a^* \tag{10.4.11}$$

in the linear or antilinear case, respectively. Thus

$$(O\Phi + g(a)O\Psi, O\Phi + g(a)O\Psi) = (\Phi, \Phi) + |a|^2(\Psi, \Psi) + 2\,\mathrm{Re}[g(a)(O\Phi, O\Psi)] \tag{10.4.12}$$

Comparing evaluations 10.4.12 and 10.4.10 for $a = 1$ and i, respectively, we have

$$\begin{aligned} \mathrm{Re}\,(O\Phi, O\Psi) &= \mathrm{Re}\,(\Phi, \Psi) \\ \mathrm{Im}\,(O\Phi, O\Psi) &= \pm\,\mathrm{Im}\,(\Phi, \Psi) \end{aligned} \tag{10.4.13}$$

Thus the upper (lower) sign, belonging to the linear (antilinear) case, corresponds to a unitary (antiunitary) operator, where

$$(O\Phi, O\Psi) = (\Phi, \Psi) \qquad \text{(unitary)}$$

or

$$= (\Psi, \Phi) \qquad \text{(antiunitary)} \tag{10.4.14}$$

Theorem 10.4.3

Linear symmetry operators preserve the sense of time. Antilinear operators reverse the sense of time.

Proof: Let

$$\Psi(t) = \exp(-iHt)\Psi(0)$$

be the state vector in the old reference frame. In the new reference frame, the state vector will be

$$\Psi'(t) = O\Psi(\pm t) = O\exp(\mp iHt)\Psi(0)$$

where the upper sign applies when the sense of time is preserved, and the lower when it is reversed (cf., Eq. 10.2.4). But an observer in the new frame sees state vector $O\Psi(0)$ at $t=0$ and Hamiltonian $H' = OHO^{-1}$. Thus he expects the state vector at time t to be given by

$$\Psi'(t) = \exp(-iH't)\,O\Psi(0)$$

For these results to be consistent, we must have

$$O\exp(\mp iHt) = \exp(-iH't)\,O$$

or

$$\exp(\mp iHt) = O^{-1}\exp(-iH't)\,O = \exp(-O^{-1}iOHt)$$

or

$$O^{-1}iO = \pm i$$

so that if O is linear the sense of time is preserved, and if O is antilinear it is reversed.

Let us define a *projectivity* as a mapping M that carries a set of independent vectors into a set of independent vectors, and a complete set of vectors into a (possibly different) complete set of vectors. A *semilinear transformation* O is then defined by the requirements

$$O(\Phi+\Psi) = O\Phi + O\Psi \tag{10.4.15}$$

$$O(a\Phi) = g(a)\,O\Phi \tag{10.4.16}$$

where $g(a)$ is independent of Φ and constitutes an isomorphism on the number field in the sense that

$$\begin{aligned} g(a+b) &= g(a)+g(b), \qquad & g(ab) &= g(a)g(b) \\ g(1) &= 1, & g(0) &= 0 \end{aligned} \tag{10.4.17}$$

In the field of complex numbers the only solutions to Eq. 10.4.17 are $g(a) = a$ or $g(a) = a^*$, that is, a semilinear mapping is then linear or antilinear.

In Appendix G we prove (1) that all Wigner mappings are projectivities, and (2) that any projectivity M is reducible to a semilinear mapping O by a factor transformation

$$O\Psi = \alpha(\Psi)\,M\Psi \tag{10.4.18}$$

where $\alpha(\Psi)$ is a (complex) number. The results of this section can be summarized in a succint way: Symmetry operators must after a factor transformation, Eq. 10.4.18, be linear or antilinear. The linear operators preserve the sense of time and, in view of Theorem 10.4.2, are unitary, whereas the antilinear operators reverse time and are antiunitary.

We now state two theorems whose proof is trivial.

Theorem 10.4.4

The product of two unitary operators is unitary; the product of two antiunitary operators is unitary; the product of a unitary and an antiunitary operator (in either order) is antiunitary.

Theorem 10.4.5

Any antiunitary operator A can be written in the form $A = UK_0$ (or K_0U), where U is a unitary operator.

Since $U^{-1} = U^\dagger$ and $K_0^{-1} = K_0$ both exist, we can deduce two more theorems.

Theorem 10.4.6

All symmetry operators are nonsingular (i.e., possess reciprocals).

Kramers' (1930) Theorem 10.4.7

In the absence of magnetic fields, but in the presence of arbitrary electric (e.g., crystalline) fields, each energy of a system containing an odd number of electrons must be at least doubly degenerate.[7]

Proof: In the absence of a magnetic field, $[K,H]=0$, so that, if Ψ is an eigenstate, $K\Psi$ is also an eigenstate with the same energy. Furthermore, the antiunitarity of K yields, via Eq. 10.4.14b,

$$(\Psi, K\Psi) = (K^2\Psi, K\Psi) = -(\Psi, K\Psi) = 0$$

so that $K\Psi$ is independent of Ψ, establishing the theorem. All that we have used is that K is antilinear, $K^2 = -1$ for an odd number of electrons in accord with Eq. 10.2.10 or 10.2.14, and $[K,H]=0$. The proof can therefore be generalized to any antilinear operator θ with $\theta^2 = -1$, $[\theta, H]=0$.

Kramers' theorem substantiates our previous remark that it may not be possible to diagonalize antiunitary operators even when they commute with H. Clearly, if K could be diagonalized, $K\Psi$ would not be orthogonal to Ψ.

Example 10.4.1

Prove that a time-reversible Hamiltonian produces no coupling between states that are precise time reverses of one another, when $K^2 = -1$.

[7] The degeneracy must indeed be even-fold. See Problem 10.4.3.

Proof:

$$(K\Psi, \exp(-iHt)\Psi) = (K\exp(-iHt)\Psi, K^2\Psi)$$
$$= K^2(\exp(iHt)K\Psi, \Psi) = -(K\Psi, \exp(-iHt)\Psi) = 0$$

Thus $\Psi(t)$ is always orthogonal to $K\Psi$, as claimed.

Example 10.4.2

Explain the origin of the "Van Vleck cancellation" (Van Vleck, **1940**; Abrahams, **1957**) in the calculation of paramagnetic spin relaxation in Kramers salts.

Explanation: Kramers salts are salts in which the only degeneracy is the Kramers degeneracy between a pair of electronic states Ψ and $K\Psi$. In a magnetic field the degeneracy between these states is lifted, and a resonance absorption experiment can be performed, using microwaves whose frequency is given by the Zeeman splitting. The relaxation rate is the rate of transition from the upper to the lower state by phonon emission.

The perturbation for one-phonon processes has the form $V(\mathbf{r})\epsilon(\mathbf{u})$, where $\epsilon(\mathbf{u})$ is the strain associated with lattice displacement $\mathbf{u}$, and $V(\mathbf{r})$ is the orbit-lattice interaction energy per unit strain. Although the lattice does not make a transition to a time-reversed state, our matrix element contains the electronic factor $(\Phi, V(\mathbf{r})\Psi)$. The final electronic state Φ differs from $K\Psi$ slightly because of the presence of a magnetic field. Thus, after cancellation of large terms that appear in Van Vleck's and Abrahams' calculations, a net small relaxation rate is obtained that vanishes if the magnetic field goes to zero, in view of Example 10.4.1.

Example 10.4.3

Show that two or more phonons *can* induce a transition between an electronic state Ψ (via intermediate states Ψ_i) and its exact time reverse $K\Psi$, that is, show that a relaxation rate is obtained that does *not* vanish with the magnetic field (Orbach, **1961**).

Solution: With X_m a lattice state, and $X_{\bar{m}} = KX_m$ the time-reversed state, the amplitude of the transition is proportional to

$$(X_n K\Psi, \exp(-iHt)X_m\Psi) = -(X_{\bar{m}}K\Psi, \exp(-iHt)X_{\bar{n}}\Psi) \quad (10.4.19)$$

We have followed the procedure of Example 10.4.1 but do not get a vanishing because we cannot assume that $n = \bar{m}$, that is, that the full final state is the time reverse of the full initial state.

10.5. ONSAGER RELATIONS[8]

Macroscopic Derivation

The linear dissipative response of a physical system described by macroscopic parameters $\boldsymbol{\alpha} = \alpha_1, \alpha_2, \ldots, \alpha_n$ is written by Onsager in the form of a linear relation between "fluxes" $\dot{\alpha}_i$ and forces X_j:

$$\dot{\alpha}_i = \sum L_{ij} X_j \tag{10.5.1}$$

where the (thermodynamic) forces are defined by[9]

$$X_j = \frac{\partial S}{\partial \alpha_j} \tag{10.5.2}$$

where $S = S(\alpha_1, \alpha_2, \ldots, \alpha_n)$ is the entropy. Actually Eq. 10.5.1 is known to be an idealization. It is much more nearly correct if we understand that

$$\dot{\alpha}_i(t) = \frac{1}{\tau}[\alpha_i(t+\tau) - \alpha_i(t)] \tag{10.5.3}$$

where τ is small compared to all macroscopic times but is larger than the duration of the microscopic collisions that give rise to the dissipation.

Let us change j to m in Eq. 10.5.1 and multiply by α_j to obtain

$$\langle \dot{\alpha}_i \alpha_j \rangle = \sum L_{im} \langle X_m \alpha_j \rangle \tag{10.5.4}$$

where $\langle\ \rangle$ denotes an ensemble average against the distribution function $W(\boldsymbol{\alpha})$, obtained by reversing the Boltzmann definition of entropy:

$$W(\boldsymbol{\alpha}) = N \exp\left[\frac{S(\boldsymbol{\alpha})}{k}\right] \tag{10.5.5}$$

[8]Onsager (**1931a-, 1931b**), Casimir (**1945**), Wigner (**1954**), Callen (**1948**), Takahasi (**1952**), Kubo (**1957**), de Groot and Mazur (**1962**), Lax (**1958b, 1960**).

[9]The use of these forces is equivalent to choosing the entropy itself as the central thermodynamic function. See, for example, Callen (**1960**), Section 5.4.

An integration by parts permits us to write

$$\langle X_m \alpha_j \rangle = N \int \exp\left(\frac{S}{k}\right) \frac{\partial S}{\partial \alpha_m} \alpha_j \, d\alpha$$

$$= -k\left(\frac{\partial \alpha_j}{\partial \alpha_m}\right) N \int \exp\left(\frac{S}{k}\right) d\alpha$$

$$= -k\delta_{jm} \tag{10.5.6}$$

and

$$-kL_{ij} = \langle \dot{\alpha}_i \alpha_j \rangle = \frac{1}{\tau} \langle [a_i(\tau) - \alpha_i(0)]\alpha_j(0) \rangle \tag{10.5.7}$$

where we have used Eq. 10.5.3 and stationarity

$$\langle \alpha_i(t+t')\alpha_j(u+t') \rangle = \langle \alpha_i(t)\alpha_j(u) \rangle \tag{10.5.8}$$

to simplify the right-hand side of Eq. 10.5.7. Thus the symmetry of L_{ij} is determined by the symmetry of $\langle \alpha_i(t)\alpha_j(0) \rangle$.

Time-reversal invariance (in the absence of a magnetic field) implies, however, that the solutions $\alpha_i(t)$ and the time-reversed solutions $\alpha_i^T(t)$ must receive equal weight in the ensemble average.[10] Hence we can write

$$\langle \alpha_i(t)\alpha_j(0) \rangle = \langle \alpha_j^T(0)\alpha_i^T(t) \rangle$$

$$= \epsilon_i \epsilon_j \langle \alpha_j(0)\alpha_i(-t) \rangle$$

$$= \epsilon_i \epsilon_j \langle \alpha_j(t)\alpha_i(0) \rangle \tag{10.5.9}$$

In the first step, we have reversed the order of the factors, which is immaterial in a classical derivation but will be shown in Eq. 10.5.24 to be the correct order in the quantum-mechanical case. The factor $\epsilon_i = \pm 1$ according as the variable is even or odd under time inversion (cf. Eq. 10.3.4). The result in Eq. 10.5.9 immediately leads to the Onsager relations:

$$L_{ij} = L_{ji} \epsilon_j \epsilon_i \tag{10.5.10}$$

[10]This assumes thermal equilibrium between time-reversed states. A material possessing magnetic order (e.g., a ferromagnet) may have an underlying spin Hamiltonian that is even under time reversal (i.e., its energy will be unchanged by a reversal of all the spins). However, thermal equilibrium is usually maintained only between quantum states (or classical motions) that are consistent with a given direction of the magnetic order. Thus the Onsager relations to be derived relate a tensor with one direction of the magnetic order to a corresponding tensor in a crystal with the reversed order.

In the presence of a magnetic field, $\alpha_i^T(t, -\mathbf{H})$ receives the same weight (when the field is $-\mathbf{H}$) as does $\alpha_i(t,\mathbf{H})$ (when the field is $\mathbf{H}$). Thus

$$\langle \alpha_i(t,\mathbf{H})\alpha_j(0,\mathbf{H})\rangle = \epsilon_i\epsilon_j\langle \alpha_j(t,-\mathbf{H})\alpha_i(0,-\mathbf{H})\rangle \qquad (10.5.11)$$

or

$$L_{ij}(\mathbf{H}) = \epsilon_i\epsilon_j L_{ji}(-\mathbf{H}) \qquad (10.5.12)$$

The study of the "thermodynamics of irreversible processes" consists largely of the application to numerous examples of the consequences of the Onsager relations (De Groot, **1951**; Prigogine, **1967**; De Groot and Mazur, **1962**).

The power of the macroscopic method is that it places no restriction on the nature of the α's. Thus the Thomson second relation between thermoelectric power and the Peltier effect is an Onsager relation that follows when temperature differences and potential differences are the forces, and electric and heat currents are fluxes. However, the derivation is weak in that Eq. 10.5.1 implies a Markoffian assumption (Lax, **1960**) about the physical system in the parameters $\boldsymbol{\alpha}$ that is reasonable but not necessary. Also, the inability to let $\tau \to 0$ in Eq. 10.5.3 requires for explanation a detailed fundamental analysis of transport theory that it is not our intention to provide here (Zwanzig, **1961**).

Microscopic Derivation

If we restrict ourselves to processes in which the forces can be described by adding a term $V(t)$ to the Hamiltonian H, where

$$V(t) = -\sum_j F_j(t)\alpha_j \qquad (10.5.13)$$

we can give a derivation free of the assumptions of the original macroscopic Onsager derivation. We start by establishing 10.5.8, that is, stationarity.

Theorem 10.5.1

If the Hamiltonian of a system does not depend explicitly on the time, ensemble averages are stationary.

Proof: In this case we can write

$$\alpha_i(t) = \exp(iHt)\,\alpha_i \exp(-iHt) \qquad (10.5.14)$$

An equilibrium canonical ensemble is described by the density matrix (ter Haar, **1961**; Dirac, **1958**)

$$\rho(\beta) = \frac{\exp(-\beta H)}{Z} \qquad Z = \mathrm{Tr}[\exp(-\beta H)] \qquad (10.5.15)$$

where Tr = trace. The desired ensemble average can be written as

$$\langle \alpha_i(t)\alpha_j(u)\rangle = \mathrm{Tr}[\alpha_i(t)\alpha_j(u)\rho(\beta)]$$

$$= \mathrm{Tr}[\exp(iHt)\alpha_i \exp(-iHt)\exp(iHu)\alpha_j \exp(-iHu)\rho(\beta)] \quad (10.5.16)$$

Since $\exp(-iHu)$ commutes with $\rho(\beta)$, it can be moved to the extreme right. Furthermore, cyclic permutations are permitted inside traces:

$$\mathrm{Tr}\,ABC = \mathrm{Tr}\,CAB = \mathrm{Tr}\,BCA \quad (10.5.17)$$

Moving $\exp(-iHu)$ to the extreme left yields

$$\langle \alpha_i(t)\alpha_j(u)\rangle = \langle \alpha_i(t-u)\alpha_j(0)\rangle \quad (10.5.18)$$

The result is a function only of the difference $t-u$ and is stationary against $t \to t+t'$, $u \to u+t'$.

The proof of Eq. 10.5.9 requires some preliminaries.

Lemma 10.5.2

If L is any linear operator, then

$$(\Phi, L\Psi) = (K\Psi, \bar{L}K\Phi) \quad (10.5.19)$$

where

$$\bar{L} \equiv KL^{\dagger}K^{-1} = (KLK^{-1})^{\dagger} \quad (10.5.20)$$

(Adjoints of antilinear operators are defined in Problem 10.4.2.)

Proof:

$$(\Phi, L\Psi) = (KL\Psi, K\Phi) = (KLK^{-1}K\Psi, K\Phi)$$

$$= \left(K\Psi, (KLK^{-1})^{\dagger}K\Phi\right)$$

Also, by the definition of $L^{\dagger}$,

$$(\Phi, L\Psi) = (L^{\dagger}\Phi, \Psi) = (K\Psi, KL^{\dagger}\Phi) = (K\Psi, KL^{\dagger}K^{-1}K\Phi)$$

Lemma 10.5.3

Let σ be a time-symmetric space in the sense that, for every Ψ in σ, $K\Psi$ is also in σ. Let Ψ_n be a complete set of basis vectors in σ. Then $K\Psi_n$ is also a complete set of basis vectors in σ. (Arbitrary Ψ will be expandable in $K\Psi_n$, since $K\Psi$ is expandable in Ψ_n.)

If we set $\Phi = \Psi = \Psi_n$ in Eq. 10.5.19 and sum over n, we arrive at the following theorem.

Theorem 10.5.4

$$\mathrm{Tr}_\sigma L = \mathrm{Tr}_{\bar\sigma} \bar{L}$$

where $\bar\sigma = K\sigma$ spans the time-reverse states $K\Psi_n$ of the states Ψ_n in σ.

If

$$\alpha_i(t) \equiv \alpha_i(t, H) = \exp(iHt)\alpha_i(0)\exp(-iHt) \tag{10.5.21}$$

and $\bar\alpha_i(0) = \epsilon_i \alpha_i(0)$ describes the time-reversal properties of α_i without assuming that the latter is Hermitian, then

$$\begin{aligned}\overline{\alpha_i(t,H)} &= \exp(-i\bar{H}t)\bar\alpha_i(0)\exp(i\bar{H}t)\\ &= \epsilon_i \alpha_i(-t, \bar{H})\end{aligned} \tag{10.5.22}$$

Theorem 10.5.4 with $L = \alpha_i(t)\alpha_j(0)\rho(\beta)$ now yields

$$\langle \alpha_i(t,H)\alpha_j(0)\rangle_\sigma = \epsilon_i\epsilon_j \langle \alpha_j(0)\alpha_i(-t,\bar{H})\rangle_{\bar\sigma} \tag{10.5.23}$$

If H does not depend explicitly on the time, the cyclic property of traces yields the result expected from *stationarity*:

$$\langle \alpha_i(t,H)\alpha_j(0)\rangle_\sigma = \epsilon_i\epsilon_j \langle \alpha_j(t,\bar{H})\alpha_i(0)\rangle_{\bar\sigma} \tag{10.5.24}$$

If a magnetic field $\mathbf{H}$ is present, $H = H(\mathbf{H})$; then $\bar{H} = H(-\mathbf{H})$ is the Hamiltonian appropriate to the reverse magnetic field, and $\bar\sigma$ spans a set of eigenstates appropriate to the reversed field. But $\bar\sigma$ need not reduce to σ even when the magnetic field is reduced to zero.

In Appendix H, we show that the response $\dot\alpha_i$ to the forces $F_j(t)$ of Eq. 10.5.13 can be written in the form

$$\langle \dot\alpha_i(t)\rangle = \sum_j \int_{-\infty}^{t} dt' y_{ij}(t-t')F_j(t') \tag{10.5.25}$$

The admittance at frequency ω, obtained by setting $F_j(t') = F_j \exp(i\omega t')$, is given by

$$\langle \dot\alpha_i \rangle = \sum_j Y_{ij}(\omega) F_j \tag{10.5.26}$$

with

$$Y_{ij}(\omega) = \int_0^\infty y_{ij}(t)\exp(-i\omega t)\,dt \tag{10.5.27}$$

The Onsager coefficients are simply the special case

$$L_{ij} = Y_{ij}(0) \tag{10.5.28}$$

The symmetry of the admittances $Y_{ij}(\omega)$ can then be studied by examining $y_{ij}(t)$. Whereas the form of Eqs. 10.5.25–10.5.27 is simply a consequence of stationarity, the precise expression (Eq. H12):

$$y_{ij}(t) = \mathrm{Tr}\left[\dot{\alpha}_i(t)\int_0^{\beta} d\lambda \exp(-\lambda H)\dot{\alpha}_j(0)\exp(\lambda H)\rho(\beta)\right] \tag{10.5.29}$$

follows from the generalized mobility theory of Appendix H.

If we think of λ as an imaginary time, we can write

$$\exp(-\lambda H)\dot{\alpha}_j(0)\exp(\lambda H) = \dot{\alpha}_j(i\hbar\lambda) \tag{10.5.30}$$

In the classical limit, $\hbar \to 0$, we can therefore write

$$y_{ij}(t) \to \beta\langle \dot{\alpha}_i(t)\dot{\alpha}_j(0)\rangle \tag{10.5.31}$$

and the Onsager relations follow directly from Eq. 10.5.24. When $\hbar \neq 0$, the direct application of Theorem 10.5.4 to Eq. 10.5.29 yields

$$y_{ij}(t, \mathbf{H}) = \epsilon_i\epsilon_j y_{ji}(t, -\mathbf{H}) \tag{10.5.32}$$

so that the Onsager relations are valid in the more general form

$$Y_{ij}(\omega, \mathbf{H}) = \epsilon_i\epsilon_j Y_{ji}(\omega, -\mathbf{H}) \tag{10.5.33}$$

10.6. THE TIME-REVERSED REPRESENTATION

If M is a linear or antilinear operator and A an antilinear operator, then

$$AM\psi_\nu = A\sum \psi_\mu M_{\mu\nu} = \sum M^*_{\mu\nu} A\psi_\mu = \sum \psi_\lambda A_{\lambda\mu} M^*_{\mu\nu} \tag{10.6.1}$$

so that the product of the two operators is *not* represented by the product of the two matrices when the operator that acts last is antilinear. Thus, if we add time reversal K to a group, we must seek not ordinary matrix representations, but "corepresentations," that is, representations consistent with Eq. 10.6.1. Wigner (**1959**) discusses corepresentations and their reduction to irreducible forms. (See also Dimmock and Wheeler **1962**.) It is simpler, however, to avoid corepresentations and to deduce *directly* the consequences of the presence of time-reversal that interest us: added degeneracies and selection rules.

Theorem 10.6.1

The elements R of a point or space group (single or double) are "real," that is, they commute with time reversal:

$$KR = RK \tag{10.6.2}$$

Comment: If we think of R as acting on the space variables and of K as acting on the time, we would expect the operations to commute. But K is not time inversion and does affect spatial dependence. Hence a proof is required.

No-spin case. Let $\psi(\mathbf{r}) = u(\mathbf{r}) + iv(\mathbf{r})$. Then

$$K_0 R\psi(\mathbf{r}) = K_0\psi(\mathbf{rR}) = \psi(\mathbf{rR})^* = [u(\mathbf{rR}) + iv(\mathbf{rR})]^*$$

$$RK_0\psi(\mathbf{r}) = u(\mathbf{rR}) - iv(\mathbf{rR})$$

The equality of these two results is based on the fact that the matrix $\mathbf{R}$ in ordinary three space is real, that is, reality in ordinary space is equivalent to evenness under time reversal: $KRK^{-1} = R$. The same proof clearly applies if R is replaced by $(\boldsymbol{\alpha}|\mathbf{v})$, since the displacement $\mathbf{v}$ is also real.

Spin case. Using Eq. 3.5.15, we can write

$$KR\psi = (i\sigma_y K_0) D_{1/2}(R) \begin{bmatrix} \varphi_1(\mathbf{rR}) \\ \varphi_2(\mathbf{rR}) \end{bmatrix}$$

$$= i\sigma_y D_{1/2}(R)^* \begin{bmatrix} \varphi_1(\mathbf{rR}) \\ \varphi_2(\mathbf{rR}) \end{bmatrix}^*$$

$$RK\psi = R(i\sigma_y) \begin{bmatrix} \varphi_1(\mathbf{r})^* \\ \varphi_2(\mathbf{r})^* \end{bmatrix} = D_{1/2}(R)(i\sigma_y) \begin{bmatrix} \varphi_1(\mathbf{rR})^* \\ \varphi_2(\mathbf{rR})^* \end{bmatrix}$$

where $D_{1/2}(R)$ is the 2×2 matrix $\exp(-\frac{1}{2} i\alpha\boldsymbol{\sigma}\cdot\mathbf{n})$, as in Eq. 2.6.7. These two expressions will be equal provided that

$$D_{1/2}(R)^* = (i\sigma_y)^{-1} D_{1/2}(R)(i\sigma_y) \tag{10.6.3}$$

This condition, however, is immediately satisfied by any 2×2 matrix with the Cayley-Klein form (Eq. 2.6.12).

Theorem 10.6.2

The time-reversed representation is *identical* to the complex conjugate representation. Let

$$R\Psi_\nu = \sum \Psi_\mu D_{\mu\nu}(R) \tag{10.6.4}$$

$$RK\Psi_\nu = \sum K\Psi_\mu D^T_{\mu\nu}(R) \tag{10.6.5}$$

Then

$$D^T_{\mu\nu}(R) = D_{\mu\nu}(R)^* \tag{10.6.6}$$

Proof:

$$RK\Psi_\nu = KR\Psi_\nu = K\sum \Psi_\mu D_{\mu\nu}(R)$$

$$= \sum (K\Psi_\mu) D_{\mu\nu}(R)^*$$

Corollary 10.6.3

If $\theta = QK$, the basis vectors $\theta\Psi_\nu$ generate the (generalized) time-reversed representation:

$$D^\theta_{\mu\nu}(R) = D_{\mu\nu}(Q^{-1}RQ)^* \tag{10.6.7}$$

Comment: This corollary is useful when time-reversal alone (K) does not commute with the Hamiltonian but $\theta = QK$ does, where Q is some element that may reverse the magnetic field, as in Eq. 10.3.8, or the wave vector $\mathbf{k}$ of an electron. In this case θ^2 and $\theta^{-1}R\theta$(i.e., $\pm Q^2$ and $Q^{-1}RQ$) will also commute with the Hamiltonian and be elements of our set $\{R\}$.

Proof:

$$RQK\Psi_\nu = Q(Q^{-1}RQ)K\Psi_\nu = Q\sum (K\Psi_\mu) D_{\mu\nu}(Q^{-1}RQ)^*$$

uses Theorem 10.6.2 and yields the representation Eq. 10.6.7 generated by $QK\Psi_\mu$.

Theorem 10.6.4

Suppose that $D(R)$ is a set of *unitary, irreducible* matrices, containing $R = Q^2$, such that $Q^{-1}RQ$ is in the set for all R. If

$$D^\theta(R) \equiv D(Q^{-1}RQ)^* = B^{-1}D(R)B \tag{10.6.8}$$

then

$$\tilde{B} = bD(Q^{-2})B \quad \text{where } b = \pm 1 \tag{10.6.9}$$

Comment: Our theorem states that, if the (generalized) time-reversed representation is equivalent to the original, the two are related by a matrix B that for the important special case $Q=E$ must be *symmetric* or *antisymmetric*. Although our proof does not depend on any group property of the R's, in practice they will form either a group or a multiplier group.

Proof: The complex conjugate of Eq. 10. 6.8 yields

$$D(Q^{-1}RQ)=(B^{-1})^*D(R)^*B^*=(B^{-1})^*B^{-1}D(QRQ^{-1})BB^*$$

after a second use of Eq. 10.6.8. Next, we replace R by QRQ^{-1} and use the fact that products of linear operators are represented by products of matrices,

$$D(Q^2RQ^{-2})=D(Q^2)D(R)D(Q^{-2})$$

(valid for any set of operators R and Q regardless of group properties) to obtain

$$D(R)=[D(Q^{-2})BB^*]^{-1}D(R)[D(Q^{-2})BB^*]$$

Schur's lemma 1.5.2 now implies that

$$D(Q^{-2})BB^*=\frac{1}{b}=\text{a constant} \tag{10.6.10}$$

A constant such as $1/b$ is an abbreviation for $(1/b)D(E)$, a constant matrix. The unitary nature of the representations $D(R), D^{\theta}(R)$ permits us to choose B as unitary (see Problem 1.5.4) and Eq. 10.6.9 follows, but we must still prove that $b=\pm 1$. The transpose of Eq. 10.6.10 and the use of Eq. 10.6.9 to eliminate $\tilde{B}$ yield

$$B=b\tilde{B}D(Q^2)^*=b^2D(Q^{-2})BD(Q^2)^*$$

or

$$b^2D(Q^2)^*=B^{-1}D(Q^2)B \tag{10.6.11}$$

Comparison with Eq. 10.6.8 with $R=Q^2$ yields $b^2=1$, $b=\pm 1$.

Theorem 10.6.5

Under the unitary similarity transformation $D(R)=UD'(R)U^{-1}$, the matrix B transforms into

$$B'=U^{-1}B(\tilde{U})^{-1} \tag{10.6.12}$$

with

$$\tilde{B}'=bD'(Q^{-2})B'$$

preserving the "signature" b of B.

Theorem 10.6.6
The generalized time-reversed representation $D^{\theta}(R)$ can be made identical to $D(R)$ [i.e., for $Q=E$, the representation $D(R)$ can be made *real*] if and only if the matrix B is symmetric.

Proof: We seek a U for which $B'=1$. Equation 10.6.12 then requires

$$U\tilde{U}=B \tag{10.6.13}$$

which is possible if and only if B is symmetric. Moreover, we can take

$$U=B^{1/2} \tag{10.6.14}$$

provided $\tilde{U}=U$, that is, provided we choose among the square roots of the symmetric matrix B one that is symmetric. Since B is unitary and symmetric, we can write $B=\exp(iH)$, where H is Hermitian and symmetric, and we can choose $U=\exp(\frac{1}{2}iH)$. More constructive procedures for obtaining symmetric square roots are given in Problems 10.6.1 and 10.6.2.

Equation 10.6.9 implies that B is symmetric or antisymmetric only if

$$D(Q^2)=\pm b \tag{10.6.15}$$

within the irreducible representation. If this equation is not obeyed, it will often be possible to replace Q by one of the RQ, such that

$$D\left[(RQ)^2\right]=\pm b \tag{10.6.16}$$

It is necessary, in the discussion of time-reversal degeneracies in the next section, to classify the representations of a group according to the relation between the corresponding time-reversed representation $D^{\theta}(R)$ and the original representation $D(R)$.

Type 1. $D^{\theta}(R)\sim D(R)$, and $b=1$. For the case $Q=E$, $b=1$ implies that B is symmetric, and by Theorem 10.6.6 $D(R)$ can be made real. When $Q=E$, we shall call a Type 1 representation *real.*

Type 2. $D^{\theta}(R)\sim D(R)$, and $b=-1$. For the case $Q=E$, $D(R)$ cannot be made real, and type 2 can then be called pseudoreal.

Type 3. $D^{\theta}(R)$ is not equivalent to $D(R)$. For $Q=E$, we shall call such a representation complex.

Table 10.6.1 summarizes our classification of representations according to reality, that is, their behavior under complex conjugation K_0, and Table 10.6.2 classifies representations according to their behavior under $\theta=QK$. Note that D^{θ} and D are equivalent if and only if their characters are equal:

$$\chi(Q^{-1}RQ)^{*}=\chi(R) \tag{10.6.17}$$

Table 10.6.1. Representations of Groups Classified by Reality Properties

Type	Name	Equivalence	Symmetry	$D(R)\sim$Real?	$\chi(R)$	$\frac{1}{g}\Sigma\chi(R^2)$
1	Real	$D^* = B^{-1}DB$	$\tilde{B}=B$	Yes	Real	$+1$
2	Pseudoreal	$D^* = B^{-1}DB$	$\tilde{B}=-B$	No	Real	-1
3	Complex	D^* not$\sim D$	—	No	Complex	0

If Eq. 10.6.17 is satisfied, the representation is type 1 or 2. Otherwise it is type 3. For the special case $Q=E$, these remarks are equivalent to the following theorems.

Theorem 10.6.7

The characters of real and pseudoreal representations are all real. At least one character of a complex representation must be complex. The converse is also true.

Comment: A real representation (with $Q=E$) might be type 2 or 3 with respect to some $Q\neq E$.

Theorem 10.6.8

Pseudoreal representations must have even dimension.

Proof: For such representations, $\tilde{B}=-B$ and

$$\det B = \det\tilde{B} = (-1)^{l_i}\det B$$

Table 10.6.2. Representations of Groups Classified by Behavior under Generalized Time Reversal $\theta = QK$

Type	b	$\chi^\theta(R)$	Equivalence[a]	Symmetry	$D^\theta(R)=D(R)$?
1	1	$=\chi(R)$	$D^\theta(R)=B^{-1}D(R)B$	$\tilde{B}=D(Q^2)B$	If $D(Q^2)=D(E)$
2	-1	$=\chi(R)$	$D^\theta(R)=B^{-1}D(R)B$	$\tilde{B}=-D(Q^2)B$	If $D(Q^2)=-D(E)$
3	0	$\neq\chi(R)$	$D^\theta(R)$ not$\sim D(R)$	—	No

$$b\equiv\frac{1}{g}\sum_R\chi[(RQ)^2];\quad \chi^\theta(R)\equiv\chi(Q^{-1}RQ)^*;\quad D^\theta(R)\equiv D(Q^{-1}RQ)^*$$

[a]One can transform to a representation in which $D^\theta(R)=D(R)$ if $\tilde{B}=B$, that is, if $D(Q^2)=bD(E)$ as shown in last column.

Table 10.6.3. Character Table for Point Group T or 23

Typical Element	ϵ	δ_{2x}	δ_{3xyz}	$(\delta_{3xyz})^{-1}$
	E	$3C_2$	$4C_3$	$4C_3^{-1}$
Δ_1	1	1	1	1
Δ_2	1	1	ϵ	ϵ^2
Δ_3	1	1	ϵ^2	ϵ
Δ_4	3	-1	0	0

where l_i is the dimension of the representation.

Example 10.6.1

Determine the reality properties of the representations of the tetrahedral point group T.

Solution: One-dimensional representations have representatives identical to their characters. Examining Table 10.6.3, we find that Δ_1 is real and Δ_2 and Δ_3 are complex with $\Delta_2 = \Delta_3^*$. The representation Δ_4 has real characters and an odd dimension (three); hence it must be real.

A simple general method to distinguish between real and pseudoreal representations of even dimensions or between representations of types 1 and 2 when $Q \neq E$ is supplied by the following theorem.

Theorem 10.6.9

When $\chi^i(Q^{-1}RQ)^* = \chi^i(R)$, then

$$b_i = \frac{1}{g}\sum_R \chi^i\left[(RQ)^2\right] \tag{10.6.18}$$

provides a simple evaluation of the parameter b of Eq. 10.6.9 in representation i. When the representation $D^{\theta i}(R)$ is inequivalent to $D^i(R)$, Eq. 10.6.18 yields $b_i = 0$. [Thus $b_i = +1, -1, 0$ characterizes representations of types 1, 2, and 3, respectively. For $Q = E$, this is the Frobenius-Schur (1906) criterion.]

Proof:

$$b_i = \frac{1}{g}\sum_{R\mu} D^i_{\mu\mu}(RQRQ^{-1}Q^2)$$

$$= \frac{1}{g}\sum D^i_{\mu\nu}(R)D^i_{\nu\lambda}(QRQ^{-1})D^i_{\lambda\mu}(Q^2) \tag{10.6.19}$$

But

$$\sum_R D^i(R)D^i(QRQ^{-1}) = \sum_R D^i(Q^{-1}RQ)D^i(R) = \sum D^{\theta i}(R)^* D^i(R)$$

vanishes if $D^{\theta i}$ is the inequivalent to D^i. If the two are equivalent, Eq. 10.6.8 with $R \to QRQ^{-1}$ yields

$$D(QRQ^{-1}) = BD(R)^* B^{-1} \qquad (10.6.20)$$

Inserting Eq. 10.6.20 into Eq. 10.6.19 and employing the matrix orthogonality theorem 1.5.3 simplifies the right-hand side of Eq. 10.6.19 to

$$b_i = \frac{1}{l_i}\mathrm{Tr}\left[\tilde{B}B^{-1}D^i(Q^2)\right]$$

Inserting relation 10.6.9 between $\tilde{B}$ and B, we find that $b_i = b$ as desired.

By applying the Frobenius criterion, Eq. 10.6.18, with $Q = E$ to all point group representations of even dimension with real characters, we arrive at three theorems.

Theorem 10.6.10

The single crystallographic point groups have no pseudoreal representations.

Comment: If $G = H \times C_i$, the reality of representations $\Gamma_j^{\pm}$ of G is determined by the corresponding representations Γ_j of H, since $I^2 = E$. If G is a group containing improper elements (but not I), its representations (and their reality properties) are identical to those of the isomorphic proper group G_p.

Theorem 10.6.11

The double-valued representations of the crystallographic point groups are real only for one-dimensional representations. All other representations are pseudoreal or complex, according as the characters are real or complex.

Comment: These representations have dimensions one, two, and four. Only those of even dimension and real character can be pseudoreal. Application of the Frobenius criterion (Eq. 10.6.18) (with $Q = E$) to these representations shows they are *all* pseudoreal. Real double group representations are thus rare and occur only among the one-dimensional representations.

Theorem 10.6.12

The representation $D^j(R)$ of the proper rotation group $R^+(3)$ is real for j = integer and pseudoreal for j = half-integer.

Table 10.6.4. Character Table[a] for Point Group $\mathcal{C}_2$

	E	C_2	$\bar{E}$	$\bar{C}_2$
Γ_1	1	1	1	1
Γ_2	1	-1	1	-1
Γ_3	1	i	-1	$-i$
Γ_4	1	$-i$	-1	i

[a]The last two columns can be omitted as long as one remembers that $\bar{E}=E$ in the single group representation Γ_1, Γ_2, and $\bar{E}=-E$ in the double group representations Γ_3, Γ_4.

Proof: When the Hurwitz invariant volume element (Eq. 2.4.7), $[R(\alpha)]^2 = R(2\alpha)$, and the characters of Eq. 2.4.3 are used, the Frobenius criterion takes the form

$$b_j = \frac{\int_0^{\pi} [\sin(2j+1)\alpha/\sin\alpha](1-\cos\alpha)\,d\alpha}{\int_0^{\pi}(1-\cos\alpha)\,d\alpha}$$

By direct evaluation $b_0 = 1$, $b_{1/2} = -1$. From Eq. 2.4.2,

$$\chi^j(\alpha) - \chi^{j-1}(\alpha) = 2\cos j\alpha$$

so that $b_j - b_{j-1}$ involves an integral over $2\cos 2j\alpha$ which vanishes when $2j \geqslant 2$. The theorem follows by induction.

Example 10.6.2

Discuss the reality type of the single and double group representations of the group $\mathcal{C}_2$ (see Table 10.6.4) for $Q=E$ and $Q=S_4$.

Solution: First consider $Q=E$. The representations are all one dimensional, with Γ_1, Γ_2 real and Γ_3, Γ_4 complex. Now consider $Q=S_4$, which commutes with C_2. Then

$$\chi^{\theta}(C_2) = \chi(S_4^{-1}C_2S_4)^* = \chi(C_2)^*$$

Thus the θ-reversed representation is equivalent to the original for Γ_1 and Γ_2, but not for Γ_3 or Γ_4. Hence Γ_3 and Γ_4 are type 3 representations. We can verify this by noting that the square of the reversal elements are $(S_4)^2 = C_2$ and $(S_4C_2)^2 = C_2\bar{E} = \bar{C}_2$. For the double group representations, $\bar{C}_2 = -C_2$; therefore

$$b = \tfrac{1}{2}\left[\chi(C_2) + \chi(\bar{C}_2)\right] = 0$$

verifying that the representation is of type 3. For the single group representations, $\bar{C}_2 = C_2$ and $b = \chi(C_2) = \pm 1$ for representations Γ_1 and Γ_2, respectively. Thus Γ_1 is type 1, and Γ_2 is type 2. Hence type 2 representations (with $Q \neq E$) are possible for single point groups even though pseudoreal representations are not.

Example 10.6.3

Repeat Example 10.6.2 with $Q = U_2$ = a twofold axis at right angles to C_2.

Solution: $U_2^{-1}C_2U_2 = C_2^{-1}$. Thus

$$\chi(U_2^{-1}C_2U_2)^* = \chi(C_2^{-1})^* = \chi(C_2)$$

Hence all representations are equivalent to their θ reverse. Thus all representations must be of type 1 or 2. To distinguish between these, write briefly $b = \frac{1}{2}[U_2^2 + (U_2C_2)^2]$ and use $U_2^2 = C_2^2 = \bar{E}$. But $U_2C_2 = U_2C_2U_2^{-1}U_2 = C_2^{-1}U_2$, so that

$$U_2C_2U_2C_2 = U_2C_2(C_2^{-1}U_2) = (U_2)^2 = \bar{E}$$

Thus $b = \frac{1}{2}(\bar{E} + \bar{E}) = \bar{E} = +1$ for the single group representations Γ_1 and Γ_2, which are then of type 1, and $b = -1$ for the double group representations Γ_3 and Γ_4, which are of type 2.

10.7 TIME-REVERSAL DEGENERACIES

Let us consider the space σ, spanned by the functions Ψ_μ belonging to a given irreducible representation, and the corresponding space $\bar{\sigma}$, spanned by $\theta\Psi_\mu \equiv QK\Psi_\mu$. We should like to show that $\bar{\sigma}$ is either identical to σ or orthogonal to σ. In the latter case additional degeneracy due to time reversal has occurred. Our conclusions are embodied in the following theorem.

Theorem 10.7.1

For a type 3 representation, the space σ is orthogonal to the time-reversed space $\bar{\sigma}$. For representations of types 1 and 2, if $bK^2 = -1$, σ is orthogonal to $\bar{\sigma}$ and time reversal produces additional degeneracy; if $bK^2 = +1$, *we can define our states in such a way* that $\bar{\sigma} \equiv \sigma$, that is, time reversal produces no added degeneracy.

Proof: For type 3 representations, σ and $\bar{\sigma}$ are bases for conjugate but inequivalent representations. The base vector orthogonality theorem 3.4.3 then guarantees that any vector in $\bar{\sigma}$ is orthogonal to any vector in σ. This added degeneracy is referred to as a pairing, since a pair of *different* representations have the same energy.

When the two representations are equivalent as in Eq. 10.6.8, a basis in $\bar{\sigma}$, namely,

$$\Phi_\lambda = \sum (\theta \Psi_\mu)(B^{-1})_{\mu\lambda} \tag{10.7.1}$$

can be chosen that yields a representation

$$R\Phi_\nu = \sum \Phi_\mu D_{\mu\nu}(R) \tag{10.7.2}$$

identical to that generated by Ψ_ν in Eq. 10.6.4.

Let us now consider the overlap between the members of an arbitrary pair of vectors in σ and $\bar{\sigma}$:

$$(\Psi_\nu, \theta\Psi_\mu) = (\theta^2\Psi_\mu, \theta\Psi_\nu) = \sum K^2(\Psi_{\mu'} D_{\mu'\mu}(Q^2), \theta\Psi_\nu) \tag{10.7.3}$$

Let us invert Eq. 10.7.1:

$$\theta\Psi_\mu = \sum \Phi_\lambda B_{\lambda\mu} \tag{10.7.4}$$

and insert the result into the first and last terms in Eq. 10.7.3:

$$\begin{aligned}(\Psi_\nu, \Phi_\nu) B_{\nu\mu} &= K^2 \sum (\Psi_\lambda D_{\lambda\mu}(Q^2), \Phi_\lambda) B_{\lambda\nu} \\ &= K^2 \sum_\lambda (\Psi_\lambda, \Phi_\lambda) D_{\mu\lambda}(Q^{-2}) B_{\lambda\nu}\end{aligned} \tag{10.7.5}$$

where terms that vanish because of the base vector orthogonality theorem 3.4.3 have been dropped. The theorem can be used again to write

$$(\Psi_\nu, \Phi_\nu) = (\Psi_\lambda, \Phi_\lambda) = \frac{1}{l_i} \sum_{\mu=1}^{l_i} (\Psi_\mu, \Phi_\mu) \equiv (\sigma, \bar{\sigma}) \tag{10.7.6}$$

When this result is combined with $D(Q^{-2})B = b\tilde{B}$, Eq. 10.7.5 becomes

$$(1 - bK^2) B_{\nu\mu}(\sigma, \bar{\sigma}) = 0 \tag{10.7.7}$$

Since not all $B_{\mu\nu}$ vanish, if $K^2 = -b$, we have $(\sigma, \bar{\sigma}) = 0$. In this case, we can say that a *doubling* has occurred, since the two spaces that are necessarily degenerate with one another belong to the *same* representation.

We must still establish that when $K^2 = b$ we have $\bar{\sigma} \equiv \sigma$ and no additional degeneracy occurs. We apply θ to Eq. 10.7.1 with the λ and μ indices interchanged to obtain

$$\begin{aligned}\theta\Phi_\mu = \sum (\theta^2\Psi_\lambda)(B^{-1})^*_{\lambda\mu} &= K^2 \sum [\Psi_\nu D_{\nu\lambda}(Q^2)] \tilde{B}_{\lambda\mu} \\ &= bK^2 \sum \Psi_\nu B_{\nu\mu}\end{aligned} \tag{10.7.8}$$

where $D(Q^2)\tilde{B}=bB$ and $bK^2=1$ in the present case. Then we combine Eq. 10.7.8 with Eq. 10.7.4 to obtain

$$\theta(\Psi_\mu \pm \Phi_\mu) = \sum_\lambda (\Phi_\lambda \pm \Psi_\lambda) B_{\lambda\mu} \tag{10.7.9}$$

Since Φ_μ was constructed to obey the same representation (Eq. 10.7.2) as Ψ_μ, the basis vectors $\Psi_\mu \pm \Phi_\mu$ yield two spaces σ^+ and σ^-, each of which yields representation 10.7.2. Moreover, we have just established that

$$\theta\sigma^+ = \sigma^+, \qquad \theta\sigma^- = \sigma^- \tag{10.7.10}$$

so that these spaces are invariant under (generalized) time reversal, as well as under the group operations. Since no group operation, including θ, connects σ^+ and σ^-, there is no group-theoretical reason why states in one space should have the same energy as states in the other space. Thus, *if the* Ψ_μ *are understood to be eigenvectors* of a Hamiltonian with the required symmetry, then, in the absence of accidental degeneracy, σ is already identical with σ^+ (or σ^-), the other space σ^- (or σ^+) *does not exist*, and we have $\bar{\sigma} \equiv \sigma$.

These conclusions concerning degeneracies produced by time reversal are summarized in Table 10.7.1.

In the case of *accidental* degeneracy the possibilities are more complicated, and the following theorem is helpful.

Table 10.7.1. Extra Degeneracy Caused by Time Reversal $\theta = QK$

Type[a]	b	$\chi^\theta(R)$	$K^2 = +1$	$K^2 = -1$
1	1	$=\chi(R)$	None	Doubling[b]
2	−1	$=\chi(R)$	Doubling	None
3	0	$\neq\chi(R)$	Pairing[c]	Pairing

$$b = \frac{1}{g}\sum_R \chi[(RQ)^2]; \qquad \chi^\theta(R) = \chi(Q^{-1}RQ)^*$$

$$D^\theta(R) \equiv D(Q^{-1}RQ)^* = B^{-1}D(R)B, \qquad \tilde{B} = bD(Q^{-2})B$$

[a]For type 3, B does not exist.
[b]Doubling: A given representation must appear twice. The spaces are orthogonal, but the energies identical.
[c]Pairing: Two inequivalent representations are degenerate.

Theorem 10.7.2

If any vector in $\bar{\sigma}$, say $\theta\Psi_1$, is expressible in terms of the vectors of σ, for example,

$$\theta\Psi_1 = \sum c_\nu \Psi_\nu \tag{10.7.11}$$

then all vectors $\theta\Psi_\mu$ are so expressible.

Proof: An arbitrary vector in the space σ can be chosen to be one of the basis vectors, and we choose it to be Ψ_1, where $\theta\Psi_1$ is a vector common to σ and $\bar{\sigma}$. But $\theta\Psi_\mu$ can now be constructed by means of the transfer operator $P_{\mu 1}$, Eq. 3.7.4, appropriate to the representation $D^\theta(R)$ generated by $\theta\Psi_\mu$:

$$\theta\Psi_\mu = \frac{l_j}{g} \sum_R D^\theta_{\mu 1}(R)^* R(\theta\Psi_1)$$

The insertion of $\theta\Psi_1$ from Eq. 10.7.11 establishes the theorem, since $R\sigma = \sigma$.

Note: *We have established* in Theorem 10.7.2 *that σ and $\bar{\sigma}$ either are identical or have no vector in common, but not that σ and $\bar{\sigma}$ are necessarily orthogonal,* for example, the space spanned by (x,y) and the one-dimensional space $(z+x)$ are nonorthogonal and nonoverlapping. In Eq. 10.7.10, we have shown, however, how to make $\sigma \equiv \bar{\sigma}$ when $(\sigma,\bar{\sigma}) \neq 0$.

Example 10.7.1

Determine the reality type of the single and double group representations of $\mathcal{C}_{3v}$.

Solution: Representations of odd dimension are trivial, since the pseudoreal case is excluded by Theorem 10.6.8. The representations are then real or complex in accord with the characters. Thus Λ_1 and Λ_2 (see Table 10.7.2) are *real*, and Λ_4 and Λ_5 are complex.

For the two-dimensional representations Λ_3 and Λ_6, we must apply the

Table 10.7.2. Reality Properties of Representations of $\mathcal{C}_{3v}$

	E	$\bar{E}$	C_3, C_3^{-1}	$\bar{C}_3, \bar{C}_3^{-1}$	$3\sigma_v$	$3\bar{\sigma}_v$	$(1/g)\Sigma\chi(R^2)$
Λ_1	1	1	1	1	1	1	1
Λ_2	1	1	1	1	-1	-1	1
Λ_3	2	2	-1	-1	0	0	1
Λ_4	1	-1	-1	1	i	$-i$	0
Λ_5	1	-1	-1	1	$-i$	i	0
Λ_6	2	-2	1	-1	0	0	-1
R^2	E	E	$\bar{C}_3$	$\bar{C}_3$	$\bar{E}$	$\bar{E}$	

Frobenius-Schur criterion (Eq. 10.6.18). The arithmetic is simplified if we note that, when R and S are in the same class, so are R^2 and S^2. Thus only a typical element in each class need be used. *Care is needed, however, in squaring double group elements.* Thus $C_2^2 = \bar{E}$ and hence $\sigma^2 = (IC_2)^2 = \bar{E}$. Also (cf. Eq. 2.16.5) $(C_3)^2 = C_3^3 C_3^{-1} = \bar{E} C_3^{-1} = \overline{C_3^{-1}}$. Thus, using the last line of Table 10.7.2 for a typical R^2,

$$\frac{1}{12}\sum \chi(R^2) = \frac{2\chi(E) + 6\chi(\bar{E}) + 4\chi(\bar{C}_3)}{12}$$

with the result $1, 1, 1, 0, 0, -1$ for the six representations, so that Λ_3 and Λ_6 are real and pseudoreal, respectively, and the other results verify our previous remarks.

Example 10.7.2

Discuss the degeneracies produced by time reversal for the group $\mathcal{C}_{3v}$.

Solution: In the case of no spin or of an even number of electrons, we are restricted to the single group representations Λ_1, Λ_2, and Λ_3 (for which $\bar{E} = +E$). Since these representations are real and $K^2 = +1$, no added degeneracy results (cf. Table 10.7.1).

In the case of an odd number of electrons, $K^2 = -1$, and we must use the double group representations $\Lambda_4, \Lambda_5, \Lambda_6$, for which $\bar{E} = -E$. Since Λ_6 is pseudoreal, no added degeneracy results (cf. Table 10.7.1). But $\Lambda_5 = \Lambda_4^*$ are a *pair* of complex conjugate representations that are of necessity degenerate with one another. Indeed, they illustrate the following theorems.

Theorem 10.7.3

Complex (type 3) representations of a group occur in conjugate pairs.

Proof: If $D(R)$ obeys the multiplication table, so does $D(R)^*$.

Theorem 10.7.4

From Theorems 10.6.10 and 10.6.11 we deduce that for the crystallographic point groups degeneracy of the "doubling" type is absent among the single-valued representations, and occurs among the double-valued representations only for one-dimensional representations with real character.

10.8 THE HERRING CRITERION FOR SPACE GROUPS

The Frobenius criterion requires a sum over all the elements of a space group. Herring (**1937a**) has shown how to simplify this criterion, so that it can be related simply to the representative elements of the factor group of the wave vector $G_{\mathbf{k}}/T$.

We first note that, if R is a group element, its character with respect to the group of the wave vector is given by

$$\chi^{\mathbf{k}}(R)=\sum_{\mu}(\psi_{\mathbf{k}\mu}, R\psi_{\mathbf{k}\mu}) \tag{10.8.1}$$

whereas its character with respect to the entire space group requires a sum as well over the *star* of **k** (cf. Section 8.6):

$$\chi(R)=\sum_{\mathbf{k}\mu}(\psi_{\mathbf{k}\mu}, R\psi_{\mathbf{k}\mu})=\sum_{\mathbf{k}}\chi^{\mathbf{k}}(R) \tag{10.8.2}$$

The general element of a space group and its square can be written in the forms

$$R=(\alpha|\mathbf{v}(\alpha)+\mathbf{t})=(\epsilon|\mathbf{t})R_0, \qquad R_0=(\alpha|\mathbf{v}(\alpha)) \tag{10.8.3}$$

$$R^2=(\epsilon|\mathbf{t}+\alpha\cdot\mathbf{t})R_0^2=\exp[-i\mathbf{k}\cdot(\mathbf{t}+\alpha\mathbf{t})]R_0^2 \tag{10.8.4}$$

where the last form is valid only at **k**. The sum over all elements of the space group can now be written as

$$\sum_{\alpha,\mathbf{t}}\chi(R^2)=\sum_{\mathbf{k}}\sum_{\alpha}\chi^{\mathbf{k}}(R_0^2)\sum_{\mathbf{t}}\exp[-i(\mathbf{k}+\alpha^{-1}\cdot\mathbf{k})\cdot\mathbf{t}] \tag{10.8.5}$$

The sum on **t** will vanish unless $\alpha^{-1}\cdot\mathbf{k}+\mathbf{k}=0$ or **K**, that is,

$$\alpha^{-1}\mathbf{k}=-\mathbf{k}+\mathbf{K} \tag{10.8.6}$$

Thus α^{-1} (and hence α) is restricted to be an element that takes **k** into $-\mathbf{k}$ or an equivalent point. When this condition is obeyed, the sum on **t** yields $N = N_1N_2N_3$ (cf. Eq. 6.2.17). If the number of rotational elements is h, then $g=Nh$ is the order of the group, and the Frobenius criterion reads as

$$\frac{1}{g}\sum\chi(R^2)=\frac{1}{h}\sum_{\mathbf{k}}\sum_{Q}\chi^{\mathbf{k}}(Q^2) \tag{10.8.7}$$

where the sum on Q is restricted to the elements [among $(\alpha|\mathbf{v}(\alpha))$] that reverse **k**. Since all vectors of the star are similar, they contribute equally to the sum, producing a factor $M\equiv$ the number of distinct (nonequivalent) wave vectors in the star. Finally,

$$b=\frac{1}{g}\sum\chi(R^2)=\frac{1}{h_{\mathbf{k}}}\sum_{Q}\chi^{\mathbf{k}}(Q^2)=\pm1 \text{ or } 0 \tag{10.8.8}$$

where

$$h_{\mathbf{k}}=\frac{h}{M} \tag{10.8.9}$$

is the order of $G_{\mathbf{k}}/T$, that is, the number of rotational elements in the point group $P(\mathbf{k})$. Equation 10.8.8 can be used in conjunction with Table 10.7.1 to determine the existence or type of additional degeneracy produced by time reversal.

The same result can be obtained directly by considering time-reversal degeneracy as added degeneracy in the space $\sigma_{\mathbf{k}}$ of states with a given $\mathbf{k}$ vector. Since K has the effect of reversing $\mathbf{k}$, as does Q, elements such as QK leave $\sigma_{\mathbf{k}}$ invariant. For this space, the relevant group is then $G_{\mathbf{k}}+Q_0KG_{\mathbf{k}}$, the *reversal group* of the wave vector $\mathbf{k}$, where Q_0 is any one of the reversal elements Q. Since translations $(\boldsymbol{\epsilon}|\mathbf{t})$ cannot take us out of $\sigma_{\mathbf{k}}$, we can confine our attention to the algebra A of representative elements $(\boldsymbol{\alpha}|\mathbf{v}(\boldsymbol{\alpha}))$ of $G_{\mathbf{k}}$ plus the reversal elements Q_0A. *Application of Theorem* 10.6.9 *to this algebra yields a formula identical to* Eq. 10.8.8. Representations of types 1, 2, and 3 of this algebra correspond to real, pseudoreal, and complex representations, respectively, of the full space group.

The representation of $G_{\mathbf{k}}$ generated by $Q_0K\Psi_\mu^j$ (we can now drop the subscript 0 for simplicity of notation) has the character (cf. Eq. 10.6.7)

$$\chi^{Q\bar{j}}(R)=\chi^j(Q^{-1}RQ)^* \qquad (10.8.10)$$

whereas the representation generated by $K\Psi_\mu^j$ belongs to $G_{-\mathbf{k}}$. If $-\mathbf{k}$ is not in the star of $\mathbf{k}$, then Q does not exist and the Herring formula 10.8.8 yields $b=0$ or a type 3 representation which necessarily implies degeneracy. However, this is the trivial degeneracy of $\psi(\mathbf{k},\mathbf{r})$ and $K\psi(\mathbf{k},\mathbf{r})\sim\psi(-\mathbf{k},\mathbf{r})$ that always exists. The nontrivial question is whether a degeneracy arises at $\mathbf{k}$, that is, whether $QK\sigma^j$ is orthogonal to $\sigma^j=\{\Psi_\mu^j\}$ or identical to it. Since $\chi^{Q\bar{j}}(R)$ is *some* representation i at $\mathbf{k}$, we must have

$$\chi^j(Q^{-1}RQ)^*=\chi^i(R) \qquad (10.8.11)$$

If $i\neq j$, we have a nontrivial (type 3) degeneracy pairing representatives Γ^i and Γ^j. If $i=j$, representation j is equivalent to its "time reverse" QKj and is real or pseudoreal according as b in Eq. 10.8.8 equals $+1$ or -1. If the sign of b disagrees with that of K^2, degeneracy of the doubling type occurs.

To facilitate use of Eq. 10.8.10, we set $Q=(\mathbf{q}|\mathbf{v}(\mathbf{q}))$ and $R=(\boldsymbol{\alpha}|\mathbf{v}(\boldsymbol{\alpha}))$ and use Eq. 8.1.10 to write

$$(\mathbf{q}|\mathbf{v}(\mathbf{q}))^{-1}(\boldsymbol{\alpha}|\mathbf{v}(\boldsymbol{\alpha}))(\mathbf{q}|\mathbf{v}(\mathbf{q}))=\exp(-i\mathbf{k}\cdot\mathbf{t})(\mathbf{q}^{-1}\boldsymbol{\alpha}\mathbf{q}|\mathbf{v}(\mathbf{q}^{-1}\boldsymbol{\alpha}\mathbf{q})) \qquad (10.8.12)$$

where

$$\mathbf{t}=\mathbf{q}^{-1}[\mathbf{v}(\boldsymbol{\alpha})-\mathbf{v}(\mathbf{q})+\boldsymbol{\alpha}\mathbf{v}(\mathbf{q})]-\mathbf{v}(\mathbf{q}^{-1}\boldsymbol{\alpha}\mathbf{q}) \qquad (10.8.13)$$

We next define reciprocal lattice vectors $\mathbf{K}_{\mathbf{q}}$ and $\mathbf{K}_{\boldsymbol{\alpha}}$ by

$$\mathbf{k}\cdot\mathbf{q}^{-1}=-\mathbf{k}-\mathbf{K}_{\mathbf{q}}, \qquad \mathbf{k}\cdot\boldsymbol{\alpha}=\mathbf{k}-\mathbf{K}_{\boldsymbol{\alpha}} \qquad (10.8.14)$$

since α is in $G_{\mathbf{k}}$ and $(\mathbf{q}|\mathbf{v}(\mathbf{q}))$ is a reversal operation. Then

$$Q^{-1}RQ=\exp\{i\mathbf{K}_{\mathbf{q}}\cdot[\mathbf{v}(\alpha)+(\alpha-1)\mathbf{v}(\mathbf{q})]-i\mathbf{K}_{\alpha}\cdot\mathbf{v}(\mathbf{q})\}$$

$$\times\exp\{i\mathbf{k}\cdot[\mathbf{v}(\alpha)+\mathbf{v}(\mathbf{q}^{-1}\alpha\mathbf{q})]\}(\mathbf{q}^{-1}\alpha\mathbf{q}|\mathbf{v}(\mathbf{q}^{-1}\alpha\mathbf{q})) \qquad (10.8.15)$$

We may now take the matrix representation of both sides and the complex conjugate of Eq. 10.8.15 and use Eq. 10.8.10 to arrive at the following theorem.

Theorem 10.8.1

In the interior of the Brillouin zone for all space groups and on the zone boundary for symmorphic space groups, the "time-reversed" representation can be written in the succinct form

$$D^{Q\bar{j}}(O(\alpha))=D^{j}(O(\mathbf{q}^{-1}\alpha\mathbf{q}))^{*} \qquad (10.8.16)$$

and

$$\chi^{Q\bar{j}}(O(\alpha))=\chi^{j}(O(\mathbf{q}^{-1}\alpha\mathbf{q}))^{*} \qquad (10.8.17)$$

where

$$O(\alpha)=\exp[i\mathbf{k}\cdot\mathbf{v}(\alpha)](\alpha|\mathbf{v}(\alpha)) \qquad (10.8.18)$$

are the elements of our space group algebra in Section 8.7.

Proof: In the interior, Eq. 10.8.14 cannot be satisfied except with $\mathbf{K}_{\mathbf{q}}=\mathbf{K}_{\alpha}=0$. Equation 10.8.15 with the first exponential factor set equal to unity reduces immediately to 10.8.16. For symmorphic space groups the same result is valid everywhere, since $\mathbf{v}(\mathbf{q})=\mathbf{v}(\alpha)=0$.

Comment: We have already shown that the $O(\alpha)$ obey the point group multiplication table and therefore have point group characters. Nevertheless, we shall see in the examples below that space groups can have (single-valued) representations of *pseudoreal* type, whereas this possibility does not occur for (single) point groups.

Example 10.8.1

Discuss the degeneracy at a general point $\mathbf{k}$ of the zone.

Solution: The only possible reversal element is inversion $I=\iota$. If it is absent, no Q exists and we have only the trivial degeneracy between $\mathbf{k}$ and $-\mathbf{k}$. If it is present, we have $\chi(I^2)=\chi(E)=1$, so that the representation is "real." In the spinless case, then, there is no degeneracy at a general point $\mathbf{k}$. In the spin case, $K^2=-1$, and a degeneracy (of the doubling type) occurs throughout the zone. This simply means that if there is inversion symmetry, spin-orbit coupling cannot break up the trivial spin degeneracy that exists in the absence of spin-orbit coupling between the spin-up and spin-down states.

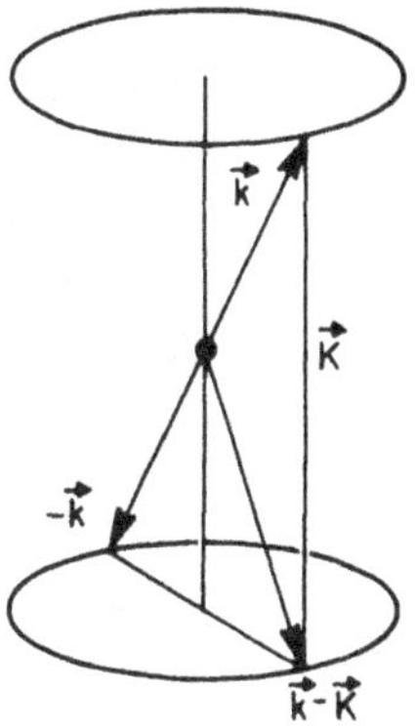

Fig. 10.8.1. A general point **k** at a zone boundary is made equivalent to $-\mathbf{k}$ by a reciprocal lattice shift **K** and a twofold rotation δ_2 or a twofold screw axis δ_2' perpendicular to the plane.

Example 10.8.2

Discuss the degeneracies at a general point of some plane in the Brillouin zone.

Solution: For most planes, the only reversal element is the inversion, and we reduce to the preceding example. If the plane passes through $\mathbf{k}=0$ or is a zone boundary plane, a twofold axis δ_2 normal to the plane reverses **k** or carries it into a point equivalent to $-\mathbf{k}$ (see Fig. 10.8.1). Thus the possible reversal elements are none, ι, or δ_2, or a screw axis δ_2', or the pairs (ι, δ_2) and (ι, δ_2').

The degeneracy results may be obtained briefly in the following way. Both δ_2' on the interior plane and δ_2 on the zone boundary have squares that reduce to $\bar{E} = \pm 1$, and no degeneracy results. However, if we possess a screw axis δ_2' with component normal to the plane, then on the zone boundary $(\delta_2')^2 = -\bar{E} = \mp 1$ when $K^2 = \pm 1$, and added degeneracy of the doubling type occurs in both cases. (Note the existence of a pseudoreal representation in the nonspin case!) The same conclusion could have been arrived at by using the Kramers' theorem 10.4.4 argument with $\theta = \delta_2' K$ and $\theta^2 = -1$. This argument does not distinguish between doubling and pairing; however, when only the identity representation is present to begin with, pairing is impossible.

If inversion ι is added to δ_2 (or δ_2' on an interior plane), $b = \frac{1}{2}(\iota^2 + \delta_2^2) = \frac{1}{2}(1 \pm 1) = 1$ or 0. Thus, in the no-spin case, we still have no degeneracy, but in the spin case there is a pairing. (Since $\mathbf{q} = \iota$ commutes with $\alpha = \epsilon$ or ρ in this case, Eq. 10.8.17 says that the paired representations are simply the two complex conjugate representations of the double point group $C_{s'}$.) However, we showed in Example 10.8.1 that inversion causes a doubling degeneracy throughout the zone. Hence no *extra* degeneracy arises in the spin case.

For ι, δ_2' at the boundary, $b = \frac{1}{2}[\iota^2 + (\delta_2')^2] = \frac{1}{2}(1 \mp 1) = 0$ or 1, leading to a pairing degeneracy in the no-spin case, and a doubling degeneracy in the spin case that simply reflects the degeneracy throughout the zone caused by ι. See Table 10.8.1 and Fig. 10.8.2 for a summary of these results.

Example 10.8.3

Discuss the degeneracies along an interior symmetry line when one of the reversal elements $(\mathbf{q}|\mathbf{v}(\mathbf{q}))$ is the inversion $\mathbf{q}=\iota$.

Solution: Whenever $\mathbf{q}$ commutes with all $\boldsymbol{\alpha}$, Eq. 10.8.16 reduces to

$$D^{Qj}(O(\boldsymbol{\alpha}))=D^{j}(O(\boldsymbol{\alpha}))^{*} \tag{10.8.19}$$

This suggests that the reality classification (and degeneracies) will be the same as for the corresponding (single or double) point group. To prove this, we note that a general Q is $(\iota|0)(\boldsymbol{\alpha}|\mathbf{v}(\boldsymbol{\alpha}))$ and that its square is given by

$$Q^{2}=(\boldsymbol{\alpha}^{2}|(\boldsymbol{\alpha}-\mathbf{1})\mathbf{v}(\boldsymbol{\alpha}))=\exp[-i\mathbf{k}\cdot(\boldsymbol{\alpha}-\mathbf{1})\cdot\mathbf{v}(\boldsymbol{\alpha})]O(\boldsymbol{\alpha}^{2})=O(\boldsymbol{\alpha}^{2})$$

since $\mathbf{k}\cdot\boldsymbol{\alpha}=\mathbf{k}$ for an interior point. Thus the Herring criterion reduces to the usual Frobenius criterion for the corresponding point group!

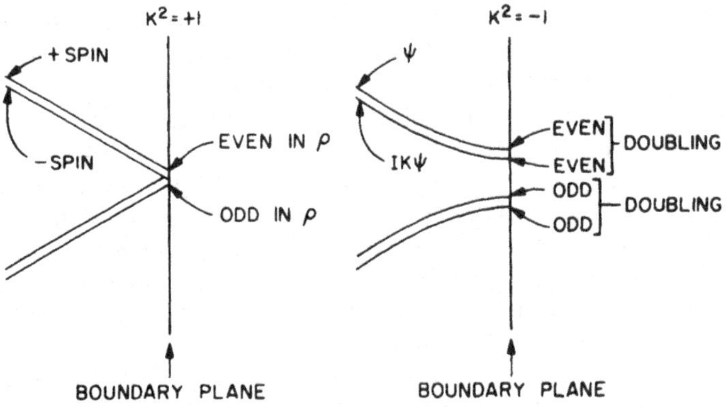

Fig. 10.8.2. Energy bands near a zone boundary plane with inversion and a twofold screw axis as the reversal elements. The case $K^2=+1$ appropriate to no spin (or, more properly, to no spin-orbit coupling) possesses the trivial degeneracy between up and down spins at a general point, and an extra (pairing) degeneracy at the zone boundary plane between representations even and odd in the reflection plane ρ. The spin case $K^2=-1$ possesses a doubling degeneracy between two representations even under ρ, and this degeneracy persists to general $\mathbf{k}$ vectors. A similar degeneracy for ρ odd also persists, but ρ even and ρ odd are not degenerate on the zone boundary plane.

Table 10.8.1. Degeneracy[a] at Zone Boundary Planes: Absence of Spin

Ordinary Elements[b]	Reversal Elements[b] Q	$\nabla E\cdot\mathbf{n}$[c]	Degeneracy	Persistence	Description of Representation
ϵ	None	$\neq 0$	No		Complex
ϵ	δ_2	0	No		Real
ϵ	ι	$\neq 0$	No		Real
ϵ	δ_2'	$\neq 0$	Doubling	No	Pseudoreal
ϵ,ρ	None	0	No		Complex
ϵ,ρ	δ_2,ι	0	No		Real
ϵ,ρ or ρ'	δ_2',ι	$\neq 0$	Pairing	No	Complex
		Presence of Spin-Orbit Coupling			
ϵ	None	$\neq 0$	No		Complex
ϵ	δ_2	0	No		Pseudoreal
ϵ	ι	$\neq 0$	Doubling	Yes	Real
ϵ	δ_2'	$\neq 0$	Doubling	No	Real
ϵ,ρ	None	0	No		Complex
ϵ,ρ	δ_2,ι	0	Pairing	Yes	Complex
ϵ,ρ or ρ'	δ_2',ι	0	Doubling	Yes	Real

[a]A degeneracy that persists throughout the zone means *no added* degeneracy at the boundary plane.
[b]δ_2' is a twofold screw axis, and ρ' a glide plane.
[c]$\nabla E\cdot\mathbf{n}$ = component of energy gradient normal to the plane.

Example 10.8.4

Discuss the degeneracies along an interior symmetry line that possesses the group (E, C_2) and the reversal elements S_4 and S_4C_2.

Solution: Since S_4 commutes with C_2, Eq. 10.8.19 remains valid and we might conclude, as in Example 10.6.2, that the reality properties are those of the point group C_2, that is, two real (single group) representations, and a pair of conjugate double group representations. But we would be *wrong* in the single group case, since $S_4^2=(S_4^{-1})^2=C_2$ and $\chi(C_2)=\pm 1$ in the two representations. Thus (using Eq. 10.8.8) the second representation, in which $C_2=-1$, is one of the rare pseudoreal representations in the spinless case and has a degeneracy of the doubling type! We can regard it as a pseudoreal representation of dimension two (including both points $\mathbf{k}$ and $-\mathbf{k}$) or as a type 2 representation of dimension one with $Q=S_4$ as a reversal element.

10.9. SELECTION RULES DUE TO TIME REVERSAL

Selection rules for matrix elements of the form

$$V_{\mu\nu}^{ij} = (\Psi_\mu^i, V\Psi_\nu^j) \tag{10.9.1}$$

are usually obtained by decomposing V into its irreducible parts,

$$V = \sum_{m\lambda} V_\lambda^m \tag{10.9.2}$$

and using the character product relation (cf. Theorem 3.4.4):

$$c^{i^*jm} = \frac{1}{g}\sum_R \chi^{i^*\times j}(R)\chi^m(R)$$

where

$$\chi^{i^*\times j}(R) = \chi^i(R)^*\chi^j(R) \tag{10.9.3}$$

and c^{i^*jm} is the number of independent parameters among the matrix elements $V_{\mu\nu\lambda}^{ijm}$ and vanishes (see Prob. 3.4.11) when *all* such matrix elements vanish.[11]

All of the information supplied by time reversal is exhausted in the use of Lemma 10.5.2 (*Eq.* 10.5.19):

$$V_{\mu\nu}^{ij} = \overline{V}_{\nu\mu}^{\bar{j}\bar{i}} \equiv \left(K\Psi_\nu^j, \overline{V}K\Psi_\mu^i\right) \tag{10.9.4}$$

where

$$\overline{V} \equiv KV^\dagger K^{-1} = FVF^{-1} \tag{10.9.5}$$

Here we introduce F as shorthand for the combined operation of Hermitian conjugation and time reversal. Since F is linear and commutes with all group elements, we can, without disturbing the decomposition 10.9.2, further decompose V into parts that are even or odd under F:

$$\overline{V} \equiv FVF^{-1} = fV, \qquad f = \pm 1 \tag{10.9.6}$$

[11] The reader should not be led to assume that matrix elements involving different V^m do not "interfere" with one another. It is well known (Rose, **1955**) that magnetic dipole and electric quadrupole matrix elements do interfere. However, for total intensities

$$\sum_{\mu\nu} |(\Psi_\mu^i, V\Psi_\nu^j)|^2$$

the different irreducible parts V^m do not interfere.

and selection rules will be obtained separately for each part. We shall see later that this decomposition procedure is valid for selection rules even if time reversal K alone is not a symmetry operation, but QK is.[12]

We refer to F as a "flipping" operator because in the absence of spin we can write Eq. 10.9.1 as[13]

$$V_{\mu\nu}^{ij} = \int \psi_\mu^i(\mathbf{r})^* V(\mathbf{r},\mathbf{r}')\psi_\nu^j(\mathbf{r}')\,d\mathbf{r}\,d\mathbf{r}' \tag{10.9.7}$$

and

$$\bar{V}(\mathbf{r},\mathbf{r}') = \left[V(\mathbf{r},\mathbf{r}')^\dagger\right]^* = V(\mathbf{r}',\mathbf{r}) \tag{10.9.8}$$

so that F interchanges $\mathbf{r}$ and $\mathbf{r}'$. This suggests that the usual product basis functions $\psi_\mu^i(\mathbf{r})^*\psi_\nu^j(\mathbf{r}')$ be replaced by the symmetric or antisymmetric combination

$$\psi_{\mu\nu}^a(\mathbf{r},\mathbf{r}') = \tfrac{1}{2}\left[\psi_\mu^i(\mathbf{r})^*\psi_\nu^j(\mathbf{r}') + a\psi_\mu^i(\mathbf{r}')^*\psi_\nu^j(\mathbf{r})\right] \tag{10.9.9}$$

with

$$a = f = \pm 1$$

In the special case

$$\psi_\mu^i = (\psi_\mu^j)^* = K_0\psi_\mu^j \tag{10.9.10}$$

the symmetry in $\mathbf{r}$ and $\mathbf{r}'$ is converted into a corresponding symmetry in $(\mu\nu)$. The argument of Eqs. 3.3.26 ff. leads immediately to the appropriately symmetrized product representation with characters

$$\chi^{j\times j}(R)_a = \tfrac{1}{2}[\chi^j(R)]^2 + \tfrac{1}{2}a\chi^j(R^2) \tag{10.9.11}$$

In this case we need not distinguish between $(j)^*\times j$ and $j\times j$. To generalize the above derivation so that it includes spin, we note that time-reversal information in the form 10.9.4 can cause a vanishing only if the *left- and right-hand sides refer*

[12] In this case, however, the number of independent parameters in V is not necessarily the sum of the corresponding numbers for the even and odd parts.

[13] A general perturbation $V = V(\mathbf{r}_{op}, \mathbf{p}_{op})$ is nonlocal. Its matrix element in the position representation has the form

$$\langle \mathbf{r}|V(\mathbf{r}_{op},\mathbf{p}_{op})|\mathbf{r}'\rangle = V(\mathbf{r},\mathbf{r}')$$

used in the matrix element 10.9.7. See Dirac (**1958**) and Messiah (**1962**).

to the same set of matrix elements. This will be the case, for example, if

$$\Psi^{i}_{\mu} = K\Psi^{j}_{\mu}, \qquad K\Psi^{i}_{\mu} = K^2\Psi^{j}_{\mu} \tag{10.9.12}$$

which provides the natural generalization of Eq. 10.9.10. Since $i = \bar{j}, \bar{i} = K^2 j$, Eq. 10.9.4 can be rewritten as

$$V^{\bar{i}\bar{j}}_{\mu\nu} = K^2 \overline{V}^{\bar{j}\bar{j}}_{\nu\mu} \tag{10.9.13}$$

or

$$V^{\bar{j}\bar{j}}_{\mu\nu} = fK^2 V^{\bar{j}\bar{j}}_{\nu\mu} \tag{10.9.14}$$

using Eq. 10.9.6. Thus we have a matrix $V_{\mu\nu}$ whose symmetry in $\mu\nu$ is given by

$$a = fK^2 \tag{10.9.15}$$

and we anticipate that the correct character product is given by Eq. 10.9.11 with $a = fK^2$, which includes a factor $K^2 = \pm 1$ for the influence of spin. (See Eq. 10.2.10.)

In Chapter 4 we illustrated the use of symmetric (antisymmetric) product representations for determining the number of independent parameters in symmetric (antisymmetric) tensors. The symmetry properties assumed in Chapter 4 were indeed a consequence of the Onsager relations established in Section 10.5. To obtain a novel illustration of symmetrized character products and the resulting selection rules, we turn to an example involving spin.

Example 10.9.1

Recent experiments on light absorption in CdS (Williams, **1960**) have displayed a shift of the band edge with strong electric field. Is this a direct effect of the field, as proposed by Franz (**1958**) and Keldysh (**1958**), or could it be that the electric field induces a piezoelectric strain that shifts the band edge (A. R. Hutson, private communication)?

Solution: The point group for CdS (the wurtzite structure) is $C_{6v} = 6mm$. (The space group is $C_{6v}^4 = P6_3mc$.) The vector space of the electric field (E_1, E_2, E_3) decomposes into $(E_1, E_2) = \Gamma_6$ and $E_3 = \Gamma_1$, where the principal axis is in the 3 or z direction. The possible strains (see Section 4.7)

$$[(x,y);z] \times [(x,y);z] = [2xy, (x^2 - y^2)] + [x^2 + y^2] + [xz, yz] + z^2 \tag{10.9.16}$$

reflect the symmetrized character product (see Problem 3.3.2):

$$(\Gamma_6 + \Gamma_1) \times (\Gamma_6 + \Gamma_1) = \Gamma_5 + \Gamma_1 + \Gamma_6 + \Gamma_1 \tag{10.9.17}$$

The basis functions for the various representations are indicated in Appendix A. The invariant energies that can be formed by combining strains and electric fields (see Section 4.5) are of the form $\Gamma_1 \times \Gamma_1$:

$$az^2E_3 + b(x^2+y^2)E_3 \tag{10.9.18}$$

and of the type $\Gamma_6 \times \Gamma_6$:

$$c(xzE_1 + yzE_2) \tag{10.9.19}$$

so that the piezoelectric tensor has three independent constants and the form just given.

In the experiments the electric field is normal to the principal axis. Thus only the last type contributes, and the question we must ask is whether a strain of $\epsilon_{13} = xz$ or $\epsilon_{23} = yz$ type can shift the energy bands.

The top of the valence band and the bottom of the conduction band have been shown to be at $\mathbf{k} = 0$ (hence our use of the point group) and to have double group representations Γ_9 and Γ_7, respectively (Thomas and Hopfield, **1960**, Fig. 5).

Without the use of time reversal, the selection rules are given by

$$\begin{aligned} \Gamma_7 \times \Gamma_7 &= \Gamma_1 + \Gamma_2 + \Gamma_6 \\ \Gamma_9 \times \Gamma_9 &= \Gamma_1 + \Gamma_2 + \Gamma_3 + \Gamma_4 \end{aligned} \tag{10.9.20}$$

so that the conduction band could be shifted by a Γ_6 strain. With time reversal, we note that the potential induced by ϵ_{13} is both Hermitian and even, under time-reversal whereas $K^2 = -1$. Thus we must apply Eq. 10.9.11 with $a = -1$ to obtain the antisymmetric products:

$$\begin{aligned} (\Gamma_7 \times \Gamma_7)_{\text{asym}} &= \Gamma_1 \\ (\Gamma_9 \times \Gamma_9)_{\text{asym}} &= \Gamma_1 \end{aligned} \tag{10.9.21}$$

so that Γ_6 type strains are now ineffective.

Similar results can be obtained by direct inspection procedures that require more detailed information but yield greater insight. For example, the representation $\Gamma_{1/2}$ of the rotation group decomposes into just Γ_7 for $\mathcal{C}_{6v}$. Thus Γ_7 is a "Kramers" doublet. The spin functions $\alpha = \Psi_+$ and $\beta = \Psi_-$ yield its matrix representations. The reflection plane σ, perpendicular to the x axis, reverses the spin (whose axis is in the z direction):

$$\sigma\Psi_+ = \lambda\Psi_-$$

Thus

$$\begin{aligned} (\Psi_+, \epsilon_{13}\Psi_+) &= (\sigma\Psi_+, \sigma\epsilon_{13}, \Psi_+) = -(\sigma\Psi_+, \epsilon_{13}\sigma\Psi_+) \\ &= -|\lambda|^2(\Psi_-, \epsilon_{13}\Psi_-) \end{aligned} \tag{10.9.22}$$

where $\sigma\epsilon_{13} = -\epsilon_{13}\sigma$ and $|\lambda|^2 = 1$. Thus the ϵ_{13} perturbation acts to split the levels. This violates Kramers' theorem, which states that the degeneracy must persist. To see this in detail, let us write the matrix elements as a 2×2 matrix

$$H = (\Psi_{\pm}, \epsilon_{13}\Psi_{\pm}) = \begin{bmatrix} a & b \\ c & d \end{bmatrix} \tag{10.9.23}$$

Time reversal now requires that

$$KHK^{-1} = i\sigma_y H^* (i\sigma_y)^{-1} = H$$

or that H has the Cayley-Klein form 2.6.12:

$$H = \begin{bmatrix} a & b \\ -b^* & a^* \end{bmatrix} \tag{10.9.24}$$

Hermiticity of H next requires that $b=0$ and that $a=a^*=\text{real}$, that is, the doublet moves as a whole and cannot be split.

Conclusion: $a=0$ as well, and all matrix elements vanish.

This proof by inspection demonstrates that the symmetrized character product procedure Eq. 10.9.11, with $a=fK^2$, efficiently takes account of the effects of *Hermiticity, time reversal,* and *spin.*

Time-Reversal Selection Rules When $[QK, H]=0$

When QK is a group element ($[QK, H]=0$), but K is not, the preceding results must be generalized to consider matrix elements of the form

$$V^{Q\bar{j}jm}_{\mu\nu\lambda} = (QK\Psi^j_\mu, V^m_\lambda \Psi^j_\nu) \tag{10.9.25}$$

By direct consideration of these matrix elements (Lax, **1965c**), we have established that the number of independent parameters among the matrix elements is given by

$$c^{(Qj)^* jm} = \frac{1}{2g} \sum_{R\epsilon G} \left\{ \chi^m(R)\chi^j(Q^{-1}RQ)\chi^j(R) + \chi^m(RQF)K^2\chi^j\left[(RQ)^2\right] \right\} \tag{10.9.26}$$

The proof used was unnecessarily general in that it also yielded all relationships among the matrix elements $V^{Qijm}_{\mu\nu\lambda}$. A shorter derivation of Eq. 10.9.26 due to E. I. Blount follows from the fact that F and QF are *linear* operators, so that addition of the symmetry operator QK can be used to enlarge our group of linear operators from G to QFG. But the operation of Hermitian conjugation contained in F cannot be applied to a single wave function but can be applied only to products of two. Let us therefore write

$$(\Phi_\mu, V\Psi_\nu) = \int \Phi^*_\mu(\mathbf{r})\, V(\mathbf{r},\mathbf{r}')\Psi_\nu(\mathbf{r}')\, d\mathbf{r}\, d\mathbf{r}'$$

$$= \mathrm{Tr}\, V(\mathbf{r},\mathbf{r}')\Psi_\nu(\mathbf{r}')\Phi^\dagger_\mu(\mathbf{r})$$

where the trace is over $\mathbf{r}$ and $\mathbf{r}'$ (and spins s and s' if they are present), and the dagger on $\Phi^\dagger_\mu$ reminds us that it comes from the left-hand ("bra") position and, by Eq. 3.4.8, generates the conjugate complex representation. Relation 10.5.19:

$$(\Phi, V\Psi) = (K\Psi, \bar{V}K\Phi)$$

where $\bar{V} = FVF^{-1} = KV^\dagger K^{-1}$, can be written as

$$\mathrm{Tr}\, V\Psi\Phi^\dagger = \mathrm{Tr}\,(FVF^{-1})K\Phi(K\Psi)^\dagger$$

so that

$$F\Psi\Phi^\dagger = K\Phi(K\Psi)^\dagger$$

is seen to take the time reverse and the Hermitian conjugate of $\Psi\Phi^\dagger$ as expected. A similar proof shows that

$$SFV(SF)^{-1} = S\bar{V}S^{-1}$$

$$SF\Psi\Phi^\dagger = SK\Phi(SK\Psi)^\dagger$$

For any R in the original group G:

$$R\Psi^j_\nu(\Phi^i_\mu)^\dagger = \sum \Psi^j_{\nu'} D^j_{\nu'\nu}(R)(\Phi^i_{\mu'})^\dagger D^i_{\mu'\mu}(R)^*$$

or

$$\chi^{i*\times j}(R) = \chi^i(R)^*\chi^j(R)$$

as expected. Any element of QGF will, however, take us from the space spanned by $\Psi^i_\nu(\Phi^i_\mu)^\dagger$ to the space spanned by $(QK\Phi^i_\mu)(QK\Psi^i_\nu)^\dagger$. If this is different from the starting space, the time-reversal operation QK will lead to relationships between different sets of matrix elements, but not to restrictive relationships among the same set of matrix elements. Therefore we can confine our attention to the case in which

$$\Phi^i_\mu = QK\Psi^j_\mu$$

that is, to matrix elements for transitions between an initial state and a final state related to the time reverse of the initial state. In this case, a typical operation RQF will restore us to the same space:

$$RQF\Psi^j_\nu(\Phi^i_\mu)^\dagger = (RQK\Phi^i_\mu)(RQK\Psi^j_\nu)^\dagger$$

$$= (RQ^2K^2\Psi^j_\mu)(RQK\Psi^j_\nu)^\dagger$$

$$= K^2\sum \Psi^j_{\mu'}D^j_{\mu'\mu}(RQ^2)\left[\sum QK\Psi^j_{\nu'}D^j_{\nu'\nu}(Q^{-1}RQ)^*\right]^\dagger$$

$$= \sum \Psi^j_{\mu'}(\Phi^i_{\nu'})^\dagger K^2 D^j_{\mu'\mu}(RQ^2)D^j_{\nu'\nu}(Q^{-1}RQ)$$

The trace is obtained by setting $\mu' = \nu$ and $\nu' = \mu$ and summing:

$$\chi^{(Q\bar{j})^*\times j}(RQF) = K^2\chi^j\left[(RQ)^2\right]$$

When the characters $\chi^m(R)$ and $\chi^m(RQF)$ are used for V^m, the coefficient c in

$$(\Gamma^{Q\bar{j}})^* \times \Gamma^j = c^{(Q\bar{j})^* jm}\Gamma^m$$

is given by averaging the character product $\chi^{(Q\bar{j})^*\times j}\chi^m$ over the $2g$ elements of $G+QGF$ and leads to the desired formula 10.9.26, where the first term (aside from the factor $\frac{1}{2}$) is the customary formula 10.9.3 with $\chi^i(R)^*$ replaced by $\chi^j(Q^{-1}RQ)$. The second term arises from a complicated symmetry of $V_{\mu\nu\lambda}$ in $(\mu\nu)$, and $\chi^m(RQF) = f\chi^m(RQ)$ if $\bar{V}^m_\lambda = fV^m_\lambda$ as in Eq. 10.9.6.

For the case $Q=E$, Eq. 10.9.26 then reduces to the use of Eq. 10.9.11 with $a = fK^2$, as required by Eq. 10.9.15.

Whether degeneracy is produced by time reversal can be learned by setting $V^m_\lambda =$ unity, since this tests the orthogonality of $QK\Psi^j_\mu$ to Ψ^j_ν. Using $f=1$, $\chi^m(R)=1$, and $\chi^j(Q^{-1}RQ)^* = \chi^i(R)$ with $i=j$ for types 1 and 2 and $i \neq j$ for type 3 yields

$$\frac{1}{g}\sum \chi^j(Q^{-1}RQ)\chi^j(R) = +1, 1, 0 = |b_j| \qquad (10.9.27)$$

for types 1, 2, and 3, respectively. Moreover, Eq. 10.6.18 is

$$\frac{1}{g}\sum \chi^j\left[(RQ)^2\right]=b_j$$

so that Eq. 10.9.26 reduce to

$$c^{(Q\bar{j})^*m}=\tfrac{1}{2}\left(|b_j|+b_jK^2\right) \tag{10.9.28}$$

When $b_j=0$ or $b_jK^2=-1$, $c^{(Q\bar{j})^*jm}=0$, and the space $QK\sigma_j$ is orthogonal to σ_j. This is the usual criterion for added degeneracy.

Note that Eq. 5.5.19 for the number of force constants in a bond is a special case of Eq. 10.9.26 with $\chi^m(R)=\chi^m(RQF)=1$, since $\mathbf{K}^{mn}$ is invariant under the group operations and even under F, that is, is symmetric on the interchange of all indices. Application of Eq. 10.9.26 to selection rules connecting different points in the Brillouin zone is given by Lax (**1962**, **1965c**).

10.10. SUMMARY

The Wigner time-reversal operator K is defined by the requirement that it reverse momentum and angular momentum but leave position alone. In the absence of spin, $K=K_0\equiv$complex conjugation in the Schrödinger representation. In the presence of spin, $K=UK_0$, where $U=\exp(i\pi J_y)$ constitutes a rotation of π about the y axis. In this case, $K^2=\pm 1$ according as the total spin of the system is integral or half-integral.

It is established that the operators of quantum mechanics, by a suitable choice of phase factors, can be made linear or antilinear. The linear operators are unitary and preserve the sense of the time. The antilinear operators are antiunitary and reverse the sense of the time.

If α_i is a variable even or odd ($\epsilon_i=\pm 1$) under time reversal, Section 10.5 leads to relations of the form

$$\langle\alpha_i(t_1)\alpha_j(t_2)\alpha_k(t_3)\rangle_{\mathbf{H}}=\epsilon_i\epsilon_j\epsilon_k\langle\alpha_k(-t_3)\alpha_j(-t_2)\alpha_i(-t_1)\rangle_{-\mathbf{H}} \tag{10.10.1}$$

where $\langle\ \rangle_{\mathbf{H}}$ denotes the ensemble average in the presence of a magnetic field $\mathbf{H}$. From these, Onsager relations can be obtained for admittances Y_{ij} such as Eq. 10.5.33:

$$Y_{ij}(\omega,\mathbf{H})=\epsilon_i\epsilon_j Y_{ji}(\omega,-\mathbf{H}) \tag{10.10.2}$$

The generalized time-reversed representation generated by $\theta\Psi_\mu$, where $\theta=QK$, is shown to be related to the representation generated by Ψ_μ by

$$D^{\theta}_{\mu\nu}(R)=D_{\mu\nu}(Q^{-1}RQ)^* \tag{10.10.3}$$

If D^{θ} is inequivalent to D (for an irreducible representation), the representation is said to be of type 3, and time reversal causes extra degeneracy between a *pair* of inequivalent representations. If

$$D^{\theta}=B^{-1}DB, \qquad \text{then } \tilde{B}=bD(Q^{-2})B$$

where $b=\pm 1$. (The representations are said to be of type 1 or type 2 according as $b=\pm 1$.) If $bK^2=-1$ added degeneracy occurs between equivalent representations (*doubling*), whereas if $bK^2=+1$ time reversal causes no added degeneracy. These results are summarized in Tables 10.6.2 and 10.7.1. A generalization of the Schur-Frobenius criterion leads to

$$b=\frac{1}{g}\sum_{R\in G}\chi\left[(RQ)^2\right]=\begin{matrix} +1 & \text{(type 1)} \\ -1 & \text{(type 2)} \\ 0 & \text{(type 3)} \end{matrix} \tag{10.10.4}$$

These results are applicable to multiplier groups as well as groups and lead immediately to the Herring criterion for space groups.

Selection rules for the matrix element $(\Phi^i_{\mu}, V^m_{\lambda}, \Psi^j_{\nu})$ are made more stringent by time reversal if the states Φ^i_{μ} are related to the time reverse of Ψ^j_{μ}. If $\Phi^i_{\mu}=QK\Psi^j_{\mu}$, the usual triple character product formula is replaced by

$$c^{Qj^*jm}=\frac{1}{2g}\sum_{R\in G}\left\{\chi^j(Q^{-1}RQ)\chi^j(R)\chi^m(R)+K^2\chi^j\left[(RQ)^2\right]\chi^m(RQF)\right\} \tag{10.10.5}$$

where $\chi^m(RQF)=\chi^m(RQ)f$ with $f=\pm 1$ if $\bar{V}\equiv KV^{\dagger}K^{-1}=fV$. This formula is also valid for the important case $Q=E$.

problems

10.2.1. Show that the (CS) Condon-Shortley (**1967**) choice of phases for the spherical harmonics [$Y_{lm}(0,0)=$ real with signs determined by Eq. 2.3.1] makes $Y_{lm}(\theta,\varphi)^*=(-1)^m Y_{l,-m}(\theta,\varphi)$. Show that the Fano-Racah (FR) choice

$$(\Psi^j_m)^{\text{FR}}=i^j(\Psi^j_m)^{\text{CS}}$$

is consistent with the choice (Eq. A22)

$$K\Psi^j_{\mu}=(-1)^{j+\mu}\Psi^j_{-\mu}$$

obeyed by the Wigner-Bargmann spinor basis functions (Eq. A10).

10.4.1. Show that $(\Phi, K\Psi) = K^2(K\Phi, \Psi)^*$, so that the adjoint of K *cannot* be defined by

$$(\Phi, K\Psi) = (K^\dagger\Phi, \Psi)$$

10.4.2. Show that a consistent definition of the adjoint for antiunitary operators is

$$(\Phi, A\Psi) = (A^\dagger\Phi, \Psi)^*$$

In particular, if $A = UK$, show that $A^\dagger = K^\dagger U^\dagger$ with $K^\dagger = K^3 = K^{-1}$.

10.4.3. Generalize Kramers' theorem 10.4.7 to show that, for an odd number of electrons, the degeneracy associated with time reversal must have an even dimension. *Hint*: If Ψ_i is in the space σ of eigenvectors, show that $K\Psi_i \neq \Psi_i$ is in σ. Can $K\Psi_i = \Psi_j$?

10.5.1. If the perturbation V is odd under barring, $(KVK^{-1})^\dagger \equiv \bar{V} = -V$, show that the first-order splitting of a degenerate set of states does not change its center of gravity.

10.5.2. Show that the spectral density of a set of states

$$n(E, \mathbf{H}) \equiv \operatorname{Tr} \delta[E - H(\mathbf{H})]$$

(with Hamiltonian H in magnetic field $\mathbf{H}$) is unaffected by reversing the direction of the magnetic field.

10.6.1. Show that any symmetric matrix $B = \tilde{B}$ can be diagonalized by an orthogonal transformation O:

$$B = O\Gamma O^{-1}, \qquad \Gamma = \text{diagonal}, \qquad \tilde{O}O = 1$$

Show that $O\Gamma^{1/2}O^{-1}$ is a symmetric square root of B. *Hint*: Paraphrase the usual proof that a Hermitian matrix can be diagonalized by a unitary matrix (Weyl, **1930**) by using $\mathbf{a}\cdot\mathbf{b}$ as the scalar product rather than $\mathbf{a}^*\cdot\mathbf{b}$, where $\mathbf{a}$ is a vector in n space when B is an $n\times n$ matrix.

10.6.2. Show that $O\Gamma^{1/2}O^{-1}$ in Problem 10.6.1 is symmetric even if O is not required to be orthogonal, provided that in $\Gamma^{1/2}$ square roots of degenerate eigenvalues are given the same sign, so as to preserve their degeneracy.

10.6.3. Show that

$$\frac{1}{g}\sum_{R\in G}\chi\left[(RQ)^2\right] = \frac{1}{g}\sum_{R\in G}\chi\left[(QR)^2\right]$$

so that Eq. 10.6.18 for b is independent of the order RQ or QR chosen. *Hint*: When does $QG = GQ$?

10.6.4. Show for a representation j of $R^+(3)$ that

$$[K, \exp(-i\varphi \mathbf{J}\cdot\mathbf{n})] = 0$$

using the behavior, $K\mathbf{J}K^{-1} = -\mathbf{J}$, of angular momentum under time reversal.

10.6.5. Show that

$$\sum_\mu (K\Phi_\mu^j)\Psi_\mu^j$$

obeys the shell theorem 4.5.6 with spin.

10.6.6. Consider an electron in a potential

$$V = A(x^2+y^2+z^2) + B(x^4+y^4+z^4)$$
$$+ iC[y^2z^2(z^2-y^2) + z^2x^2(x^2-z^2) + x^2y^2(y^2-x^2)]$$

where A, B, and C are real. Show that the point group when $C=0$ is O_h, when $C \neq 0$ is T_d. If time reversal is included, show that the appropriate group is $T_d + \delta_{2xy} K T_d$. Discuss the reality properties of the representations of T_d with this reversal operator.

10.7.1. A group G can be augmented by either of two reversal elements Q_1 and Q_2 such that (at least within a given representation j) $Q_1^2 = E, Q_2^2 = -E$. Show that for this representation $b_j = 0$, so that a type 3 degeneracy exists and the space $\theta\sigma^j$ is orthogonal to the space σ^j, where $\theta = Q_1K$. But if $\varphi = \psi + \theta\psi$ or $\varphi = (\psi - \theta\psi)/i$, we have $\theta\varphi = \varphi$. This seems to imply that we can construct our space σ^j so that $\theta\sigma^j = \sigma^j$. This contradicts the orthogonality shown above. How is this paradox resolved?

10.7.2. Consider the Abelian group of order eight generated by

$$C_2^2 = E, \qquad U^4 = E, \qquad UC_2 = C_2U$$

Show that this group has eight one-dimensional representations, four real and four complex, the latter being degenerate in pairs.

10.7.3. To the group of Problem 10.7.2 add the generator S_4 such that $S_4^2 = C_2$ and $S_4US_4^{-1} = U^{-1}$. Show that the augmented group has four real and four complex one-dimensional representations that are degenerate in pairs and two two-dimensional representations, one real and one pseudoreal.

10.7.4. How do the reality types and degeneracy differ if the group of Problem 10.7.2 is augmented by S_4K rather than by S_4?

10.7.5. Show that whether or not additional degeneracy is caused by time reversal (through $\theta = QK$) does not depend on the choice of reversal element Q among the coset QG.

10.7.6. If no extra degeneracy is produced by the operation QK, can one choose an operation $\theta = QRK$, $R \in G$, such that $D^{\theta}(R) = D(R)$? Can one find an R such that $D[(QR)^2] = D(E)$?

10.8.1. Consider a point X on the zone boundary of a crystal with tetrahedral (T) symmetry. Discuss the reality properties and degeneracies of the representations, using the reversal element S_{4x}.

10.9.1. Determine the basis functions for the representation Γ_9 of C_{6v}, and complete Example 10.9.1 by determining the selection rules for $\Gamma_9 \times \Gamma_9$ by "inspection methods."

10.9.2. Along the 111 axis in a cubic (O_h) crystal, the group of the wave vector is C_{3v} and the reversal element $Q = I =$ the inversion. Are the doubly degenerate representations Λ_3 and Λ_6 split linearly in $\Delta\mathbf{k}$ as one moves (perpendicularly) away from the 111 axis? Use $\Delta\mathbf{k} \cdot \mathbf{v}$ perturbation theory and the fact that the velocity $\mathbf{v}$ is odd under time reversal and odd under inversion to show that the appropriate character products are

$$(\Lambda_3 \times \Lambda_3)_+ = \Lambda_1^+ + \Lambda_3^+, \qquad (\Lambda_6 \times \Lambda_6)_- = \Lambda_1^+$$

(The plus superscript is the sign of the representation with respect to IF.) Since the transverse components of $\mathbf{v}$ transform as Λ_3^+, Λ_3 is linearly split but Λ_6 is not.

10.9.3. Consider the point T on an edge of the zone boundary in a cubic (O_h) space group. Show that the group of the wave vector is C_{4v} with the fourfold axis parallel to the edge in question. Show that a reflection plane σ_h is a reversal element. Discuss time-reversal degeneracies and splitting as one moves away from the edge.

10.9.4. Show that the group $G + QFG$ is isomorphic to $G + QG$.

10.9.5. Consider an atom at a site of symmetry $\mathcal{C}_{3v}$ in a crystal with a magnetic field along the threefold axis. Show that the new group is $\mathcal{C}_3 + \sigma_v K\, \mathcal{C}_3$. Show that the single group representations of $\mathcal{C}_3$ are of type 1 and the double group of type 2, so that time reversal produces no added degeneracy.

chapter eleven

LATTICE VIBRATION SPECTRA

11.1. INELASTIC NEUTRON SCATTERING

Inelastic neutron scattering has for the first time provided a reasonably reliable and direct measurement of the various branches ($t = 1, 2, \ldots$) of the vibration spectrum $\omega_t(\mathbf{q})$. (See Figs. 11.1.1 and 11.1.2 for typical results in germanium.) The energy increase of the neutron is that of the absorbed phonon:

$$E(\mathbf{k}') - E(\mathbf{k}) = \hbar\omega_t(\mathbf{q}) \tag{11.1.1}$$

Here $\mathbf{q}$ is the wave vector of the phonon and t its type, as discussed in Section 11.2. The neutron momentum change $\hbar(\mathbf{k}' - \mathbf{k})$ obeys a conservation law

$$\Delta\mathbf{k} \equiv (\mathbf{k}' - \mathbf{k}) = \mathbf{q} + \mathbf{K} \tag{11.1.2}$$

where $\mathbf{K}$ is a vector of the reciprocal lattice (cf. Sections 6.3 and 6.5). (Phonon emission can be described by replacing $\hbar\omega$ by $-\hbar\omega$ in Eq. 11.1.1 and $\mathbf{q}$ by $-\mathbf{q}$ in Eq. 11.1.2.) Thus simultaneous measurement of the energy and momentum change of the neutron permits a determination of a pair of values $\mathbf{q}$ and $\omega_t(\mathbf{q})$.

The assignment of the branch t is easy for branches well separated in energy. For branches that are near one another in energy, the distinction must be based

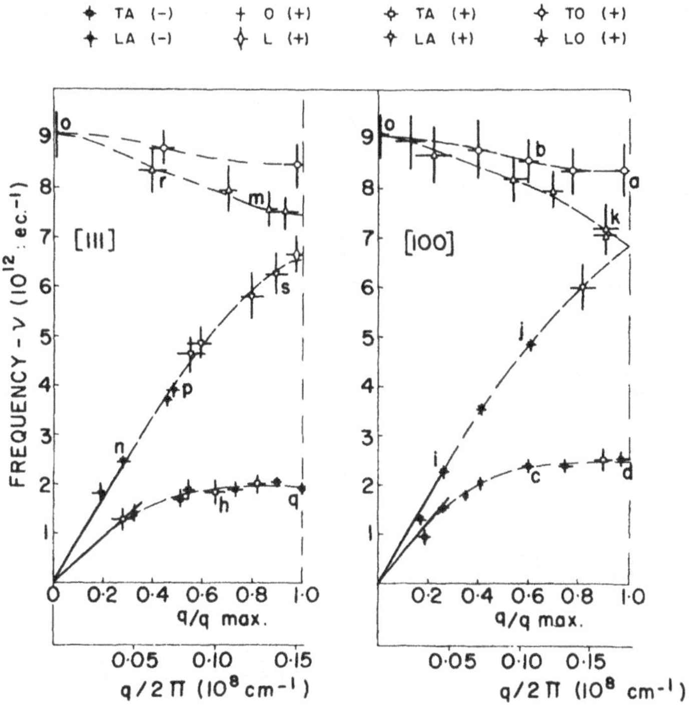

Fig. 11.1.1. Vibration spectrum of germanium in the [111] and [100] directions as measured by Brockhouse and Iyengar (**1958**).

on polarization. If a nucleus of *type* α in *cell* j has equilibrium position $\mathbf{X}^{\alpha j} = \mathbf{X}^{\alpha} + \mathbf{X}^{j}$, displacement $\mathbf{u}^{\alpha j}$, and neutron scattering amplitude f_α, the scattering amplitude for the crystal as a whole is proportional to

$$\sum_{\alpha j} f_\alpha \exp\left[i(\mathbf{k}-\mathbf{k}')\cdot(\mathbf{X}^{\alpha j}+\mathbf{u}^{\alpha j})\right]$$

This amplitude can be expanded in powers of the $\mathbf{u}^{\alpha j}$, and the one-phonon transitions are produced by the linear terms. We define phonon polarization vectors $\mathbf{b}^{\alpha}(\mathbf{q})$ by

$$\mathbf{u}^{\alpha j} = \exp(i\mathbf{q}\cdot\mathbf{X}^{j})\mathbf{b}^{\alpha}(\mathbf{q}) \tag{11.1.3}$$

for a phonon of propagation vector $\mathbf{q}$ and obtain for the one-phonon contribution to the scattered neutron intensity a result proportional to

$$\left|\sum_{j} \exp\left[i(-\Delta\mathbf{k}+\mathbf{q})\cdot\mathbf{X}^{j}\right]\right|^2 g^2 \tag{11.1.4}$$

The first factor yields the quasi-momentum-conservation law, Eq. 11.1.2, and the

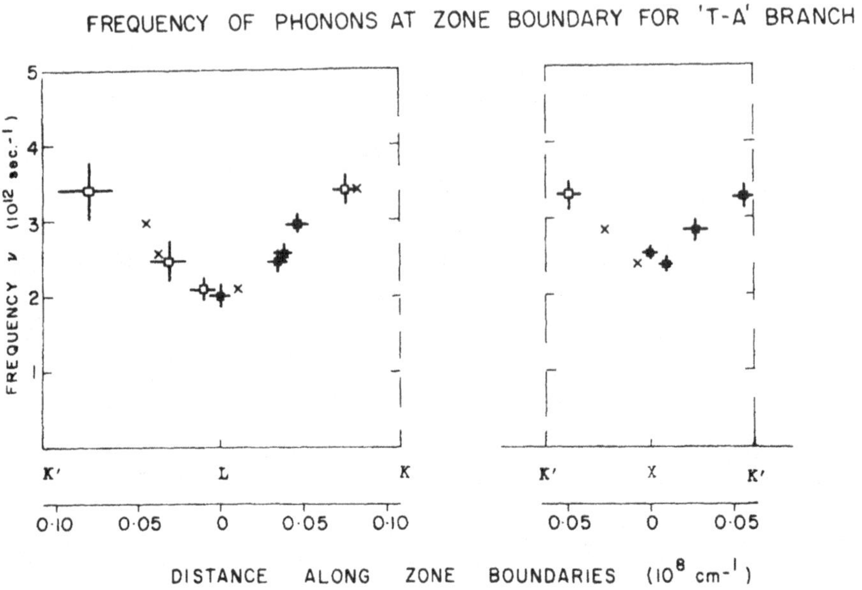

Fig. 11.1.2. The frequency of transverse acoustic phonons at the zone boundary in the 110 plane. (Brockhouse and Iyengar, **1958**). In units of $(2\pi/a)$, where a is the cube edge, the significant points are $X=[1,0,0]$, $K'=[1,\frac{1}{4},\frac{1}{4}]$, $L=[\frac{1}{2},\frac{1}{2},\frac{1}{2}]$, $K=[\frac{3}{4},\frac{3}{4},0]$. The points K' and K are equivalent.

second factor is a structure factor of the form

$$g^2=\left|\sum_{\alpha} f_\alpha \exp\left(-i\,\Delta\mathbf{k}\cdot\mathbf{X}^\alpha\right)\Delta\mathbf{k}\cdot\mathbf{b}^\alpha(\mathbf{q})\right|^2 \tag{11.1.5}$$

We see that the scattered intensity is large when the neutron momentum change $\Delta\mathbf{k}$ is parallel to the polarization vectors $\mathbf{b}^\alpha(\mathbf{q})$ and zero if it is perpendicular. If there are several atoms per unit cell, it may be necessary to make a rough model of the vibrations (e.g., using nearest-neighbor forces) to estimate the magnitudes and directions of the $\mathbf{b}^\alpha(\mathbf{q})$. In any case, the *intensity information is adequate to distinguish between branches of neighboring frequencies and differing polarizations.*

11.2. TRANSFORMATION TO NORMAL COORDINATES

We start with the kinetic and potential energies

$$T=\frac{1}{2}\sum_{\alpha i} M_\alpha(\dot{u}^{\alpha i})^2 \tag{11.2.1}$$

$$V=\frac{1}{2}\sum_{\alpha i\beta j}\mathbf{u}^{\alpha i}\cdot\mathsf{K}^{\alpha i\beta j}\cdot\mathbf{u}^{\beta j} \tag{11.2.2}$$

which lead to $3Ns$ equations of motion

$$M_\alpha\ddot{\mathbf{u}}^{\alpha i}+\sum_{\beta j}\mathsf{K}^{\alpha i\beta j}\cdot\mathbf{u}^{\beta j}=0 \tag{11.2.3}$$

where N is the number of cells and s is the number of particles in a cell ($\alpha=1,2,\ldots,s$).

The normal motions of the system are those in which all particles have a common exponential time dependence $\exp(-i\omega t)$. The symmetry of the problem with respect to lattice translations,

$$\mathsf{K}^{\alpha i\beta j}=\mathsf{K}^{\alpha\beta}(\mathbf{j}-\mathbf{i}) \tag{11.2.4}$$

forces us (cf. Section 6.2) to assume a solution of plane wave form

$$\mathbf{u}^{\alpha i}=\mathbf{b}^{\alpha}(\mathbf{q})\exp(i\mathbf{q}\cdot\mathbf{i})\exp(-i\omega t) \tag{11.2.5a}$$

or, alternatively,

$$\mathbf{u}^{\alpha i}=\mathbf{b}^{\alpha}(\mathbf{q})\exp(i\mathbf{q}\cdot\mathbf{X}^{\alpha i})\exp(-i\omega t) \tag{11.2.5b}$$

where

$$\mathbf{X}^{\alpha i}=\mathbf{X}^{\alpha}+\mathbf{X}^{i},\qquad \mathbf{X}^{i}\equiv\mathbf{i}=i_1\mathbf{a}_1+i_2\mathbf{a}_2+i_3\mathbf{a}_3 \tag{11.2.6}$$

is the position, relative to the origin, of the particle of type α in cell $\mathbf{i}$. In this chapter, equations labeled "a" or "b" refer to the choice of phase in Eq. 11.2.5a or 11.2.5b, respectively. Equations without an "a" or "b" have the same form for both choices.

The first choice (Eq. 11.2.5a) is more conventional and is followed in this section unless otherwise indicated. The second choice differs by a phase factor $\exp(i\mathbf{q}\cdot\mathbf{X}^{\alpha})$ from the first. This factor will cancel out of the final energy expressions Eqs. 11.2.16 and 11.2.17, so that these expressions are valid with either choice. The second choice, particularly for $\mathsf{R}^{\alpha\beta}(\mathbf{q})$ in Eq. 11.2.10, is almost essential for later symmetry arguments, as well as for the long-wave treatment of elasticity.

With the use of Eq. 11.2.5 and 11.2.6 the $3Ns$ equations 11.2.3 reduce to a mere $3s$ equations

$$\sum_{\beta}\mathsf{R}^{\alpha\beta}(\mathbf{q})\cdot\mathbf{b}^{\beta}(\mathbf{q})=M_\alpha\omega^2(\mathbf{q})\mathbf{b}^{\alpha}(\mathbf{q}) \tag{11.2.7}$$

for the dependence on α (and the suppressed spatial indices) of $\mathbf{b}^{\alpha}$. The absence of "*a*" or "*b*" in the designation 11.2.7 indicates that the form of this equation is valid for both choices. That the correct dependence of $\mathbf{u}^{\alpha i}$ on $\mathbf{i}$ has been assumed follows from the fact that

$$\mathsf{R}^{\alpha\beta}(\mathbf{q})=\sum_{\mathbf{j}}\mathsf{K}^{\alpha\beta}(\mathbf{j}-\mathbf{i})\exp[i\mathbf{q}\cdot(\mathbf{j}-\mathbf{i})] \tag{11.2.8a}$$

corresponding to Eq. 11.2.5a, or, alternatively,

$$\mathbf{R}^{\alpha\beta}(\mathbf{q}) = \sum_{\mathbf{j}} \mathbf{K}^{\alpha\beta}(\mathbf{j}-\mathbf{i}) \exp\left[i\mathbf{q}\cdot(\mathbf{X}^{\beta j}-\mathbf{X}^{\alpha i})\right] \qquad (11.2.8b)$$

corresponding to 11.2.5b, is independent of **i** in view of the translational invariance of **K** and the cyclic boundary conditions (Eq. 6.2.17), which we shall therefore assume. For sufficiently large crystals, Ledermann (**1944**) and Peierls (**1954**) have shown that the frequency spectrum becomes practically independent of the shape of the crystal or its boundary conditions. See also Born and Huang (**1954**), Appendix IV.

The reality of **K** yields

$$\mathbf{R}^{\alpha\beta}(\mathbf{q})^* = \mathbf{R}^{\alpha\beta}(-\mathbf{q}) \qquad (11.2.9)$$

and the symmetry of **K** (Eq. 5.4.2) with $m=\alpha i$, $n=\beta j$, yields

$$\mathbf{R}^{\alpha\beta}(\mathbf{q}) = \tilde{\mathbf{R}}^{\beta\alpha}(-\mathbf{q}) \qquad (11.2.10)$$

which combine to guarantee the Hermiticity of **R**:

$$\mathbf{R}^{\alpha\beta}(\mathbf{q})^* = \tilde{\mathbf{R}}^{\beta\alpha}(\mathbf{q}) \qquad (11.2.11)$$

Here $\tilde{\mathbf{R}}^{\beta\alpha}$ means the transpose of $\mathbf{R}^{\beta\alpha}$ with respect to the suppressed spatial indices.

Equations 11.2.7 and 11.2.8 have a set of solutions $\mathbf{b}_t^{\alpha}(\mathbf{q})$ for $t=1,2,\ldots,3s$ with frequency $\omega_t(\mathbf{q})$, where t labels the *type* of solution, for example, longitudinal acoustic. The usual orthogonality of the eigenvectors of a Hermitian matrix is modified slightly by the presence of M_α on the right-hand side of Eq. 11.2.7 and, as in the molecular case, Eq. 5.3.4, becomes

$$\sum_{\alpha} M_\alpha \mathbf{b}_t^{\alpha}(\mathbf{q})^* \cdot \mathbf{b}_{t'}^{\alpha}(\mathbf{q}) = M\,\delta(t,t') \qquad (11.2.12)$$

where M is at our disposal and can be taken to be unity or the total mass of a unit cell. See Appendix D.

Let us now make the coordinate transformation

$$\mathbf{u}^{\alpha i} = N^{-1/2} \sum_{t,\mathbf{q}} Q_t(\mathbf{q}) \mathbf{b}_t^{\alpha}(\mathbf{q}) \exp(i\mathbf{q}\cdot\mathbf{i}) \qquad (11.2.13a)$$

The kinetic energy (Eq. 11.2.1) [with $(\dot{\mathbf{u}}^{\alpha i})^2$ replaced by $(\dot{\mathbf{u}}^{\alpha i})^* \cdot \dot{\mathbf{u}}^{\alpha i}$ for convenience] now takes the form

$$T = \frac{1}{2} \sum_{tt'\mathbf{q}\mathbf{q}'} \dot{Q}_t(\mathbf{q})^* \dot{Q}_{t'}(\mathbf{q}') \sum_{\alpha} M_\alpha \mathbf{b}_t^{\alpha}(\mathbf{q})^* \cdot \mathbf{b}_{t'}^{\alpha}(\mathbf{q}')\, \Delta(\mathbf{q}'-\mathbf{q})$$

where

$$\Delta(\mathbf{q}'-\mathbf{q})=\frac{1}{N}\sum_{\mathbf{i}}\exp[i(\mathbf{q}'-\mathbf{q})\cdot\mathbf{i}]=\delta(\mathbf{q}',\mathbf{q})$$
$$T=\tfrac{1}{2}M\sum\dot{Q}_t(\mathbf{q})^*\dot{Q}_t(\mathbf{q}) \tag{11.2.14}$$

with the help of Eqs. 6.4.1 and 11.2.12.

Similarly, the potential energy becomes

$$V=\frac{1}{2}\sum_{tt'\mathbf{q}\mathbf{q}'}Q_t(\mathbf{q})^*Q_{t'}(\mathbf{q}')\frac{1}{N}\sum_{\mathbf{i}}\exp[i(\mathbf{q}'-\mathbf{q})\cdot\mathbf{i}]A$$

where

$$A=\sum_{\alpha}\mathbf{b}_t^{\alpha}(\mathbf{q})^*\cdot\sum_{j\beta}\mathbf{K}^{\alpha\beta}(\mathbf{j}-\mathbf{i})\exp[i\mathbf{q}'\cdot(\mathbf{j}-\mathbf{i})]\mathbf{b}_{t'}^{\beta}(\mathbf{q}')$$
$$=\sum_{\alpha}\mathbf{b}_t^{\alpha}(\mathbf{q})^*\cdot\sum_{\beta}\mathbf{R}^{\alpha\beta}(\mathbf{q}')\cdot\mathbf{b}_{t'}^{\beta}(\mathbf{q}')$$
$$=\sum_{\alpha}\mathbf{b}_t^{\alpha}(\mathbf{q})^*\cdot M_{\alpha}\omega_{t'}^2(\mathbf{q}')\mathbf{b}_{t'}^{\alpha}(\mathbf{q}')$$

so that in view of the orthonormality, Eq. 11.2.12:

$$V=\tfrac{1}{2}M\sum_{t\mathbf{q}}\omega_t^2(\mathbf{q})Q_t(\mathbf{q})^*Q_t(\mathbf{q}) \tag{11.2.15}$$

The coordinate transformation (Eq. 11.2.13a) expresses $3sN$ real variables $\mathbf{u}^{\alpha i}$ in terms of $3sN$ complex variables $Q_t(\mathbf{q})$. The reality of $\mathbf{u}^{\alpha i}$, however, produces the constraints

$$\mathbf{b}_t^{\alpha}(\mathbf{q})^*Q_t(\mathbf{q})^*=\mathbf{b}_t^{\alpha}(-\mathbf{q})Q_t(-\mathbf{q}) \tag{11.2.16}$$

so that, in effect, only $3sN$ independent real variables remain. It is usually convenient to pick any half-space $\mathbf{q}>0$ and to regard the $3s(N/2)$ complex variables $Q_t(\mathbf{q})$ in this half-space as the new variables. Those in the other half-space are then uniquely determined by Eq. 11.2.16.

Use of condition 11.2.9 in 11.2.7 shows that $\mathbf{b}_t(-\mathbf{q})^*$ obeys the same equation as $\mathbf{b}_t(\mathbf{q})$. Thus we can write

$$\omega_t(-\mathbf{q})=\omega_t(\mathbf{q}) \tag{11.2.17}$$

and

$$\mathbf{b}_t(-\mathbf{q})^*=\exp(i\theta)\mathbf{b}_{t'}(\mathbf{q}) \tag{11.2.18}$$

where t' is a solution of the same frequency as t

$$\omega_{t'}(\mathbf{q})=\omega_t(\mathbf{q}) \tag{11.2.19}$$

If there is no degeneracy at $\mathbf{q}$, we must have $t'=t$. As a conventional choice of phase factors, we shall take $\theta=0$ or

$$\mathbf{b}_t(-\mathbf{q})=\mathbf{b}_t(\mathbf{q})^* \tag{11.2.20}$$

The equation remains valid for all points $\mathbf{q}$ of no degeneracy, including all "general" points of the Brillouin zone. By continuity we shall extend this choice to all the symmetry points. For these, it imposes a relation between the naming of the modes at $-\mathbf{q}$ and the naming at $\mathbf{q}$.

Our normal coordinates $Q_t(\mathbf{q})$ now obey a similar requirement:

$$Q_t(-\mathbf{q})=Q_t(\mathbf{q})^* \tag{11.2.21}$$

as do the corresponding momenta:

$$P_t(-\mathbf{q})=P_t(\mathbf{q})^* \tag{11.2.22}$$

where

$$P_t(\mathbf{q})=\frac{\partial T}{\partial \dot{Q}_t(\mathbf{q})}=M\dot{Q}_t(\mathbf{q})^*=M\dot{Q}_t(-\mathbf{q}) \tag{11.2.23}$$

Dropping the indices t and $\mathbf{q}$, we find that each $Q_t(\mathbf{q})$ obeys an equation of the form

$$\ddot{Q}+\omega^2 Q=0 \tag{11.2.24}$$

or

$$Q(t)=Q_-(t)+Q_+(t), \qquad Q_-(t)=A\exp(-i\omega t), \qquad Q_+(t)=B\exp(i\omega t) \tag{11.2.25}$$

In order to obtain *traveling* waves in the direction $\mathbf{q}$, that is, waves of the form $\exp[i(\mathbf{q}\cdot\mathbf{i}-\omega t)]$, we must use solutions $\exp(i\mathbf{q}\cdot\mathbf{i})Q_-(t)$ that isolate the negative-frequency part of Q. The negative- and positive-frequency parts can be isolated by the relations

$$Q_-=\tfrac{1}{2}\left(Q-\frac{\dot{Q}}{i\omega}\right), \qquad Q_+=\tfrac{1}{2}\left(Q+\frac{\dot{Q}}{i\omega}\right) \tag{11.2.26}$$

The transformation from Q, $\dot{Q}$ to Q_-, Q_+ is a canonical (rather than coordinate) one since it involves $\dot{Q}$, that is, momenta. By restoring indices and adding a convenient normalization factor, therefore, we can define

$$a_t(\mathbf{q})=\left[\frac{M\omega_t(\mathbf{q})}{2\hbar}\right]^{1/2}\left[Q_t(\mathbf{q})-\frac{P_t(-\mathbf{q})}{iM\omega_t(\mathbf{q})}\right] \tag{11.2.27}$$

which contains only the negative-frequency part of $Q_t(\mathbf{q})$. The complex conjugate of this, using the reality requirements Eqs. 11.2.21 and 11.2.22, takes the form

$$a_t^*(\mathbf{q}) = \left[\frac{M\omega_t(\mathbf{q})}{2\hbar}\right]^{1/2}\left[Q_t(-\mathbf{q}) + \frac{P_t(\mathbf{q})}{iM\omega_t(\mathbf{q})}\right] \tag{11.2.28}$$

and contains only the positive-frequency part.

If we regard a_t^* and a_t as a set of independent variables called $a_t^\dagger$ and a_t, Eqs. 11.2.27 and 11.2.28 define a (canonical) transformation whose inverse is the equations

$$Q_t(\mathbf{q}) = \left[\frac{\hbar}{2M\omega_t(\mathbf{q})}\right]^{1/2}\left[a_t(\mathbf{q}) + a_t^\dagger(-\mathbf{q})\right] \tag{11.2.29}$$

$$\dot{Q}_t(\mathbf{q}) = \frac{P_t(-\mathbf{q})}{M} = \left[\frac{\hbar}{2M\omega_t(\mathbf{q})}\right]^{1/2}\left[-i\omega_t(\mathbf{q})\right]\left[a_t(\mathbf{q}) - a_t^\dagger(-\mathbf{q})\right] \tag{11.2.30}$$

In these new variables, the reality requirement is simply that $a_t^\dagger(\mathbf{q})$ be the complex conjugate of $a_t(\mathbf{q})$ (in quantum mechanics the adjoint operator).

11.3. QUANTIZED LATTICE OSCILLATORS: PHONONS

We can quantize our system of oscillators by introducing the commutation rules

$$[Q_t(\mathbf{q}), Q_{t'}(\mathbf{q}')] = [P_t(\mathbf{q}), P_{t'}(\mathbf{q}')] = 0 \tag{11.3.1}$$

$$[Q_t(\mathbf{q}), P_{t'}(\mathbf{q}')] = \delta(t,t')\,\delta(\mathbf{q},\mathbf{q}')\,i\hbar \tag{11.3.2}$$

The normalization constant $[\hbar/(2M\omega)]^{1/2}$ in our transformation has been chosen so as to lead to the following simple commutation rules: all a's commute, all $(a^\dagger)$'s commute, and

$$[a_t(\mathbf{q}), a_{t'}^\dagger(\mathbf{q}')] = \delta(t,t')\,\delta(\mathbf{q},\mathbf{q}') \tag{11.3.3}$$

In addition, we obtain the simple Hamiltonian

$$H = \frac{1}{2}\sum_{t\mathbf{q}} \hbar\omega_t(\mathbf{q})\left[a_t(\mathbf{q})a_t^\dagger(\mathbf{q}) + a_t^\dagger(\mathbf{q})a_t(\mathbf{q})\right] \tag{11.3.4}$$

or, in view of the commutation rules,

$$H = \sum_{t\mathbf{q}} \hbar\omega_t(\mathbf{q})\left[N_t(\mathbf{q}) + \tfrac{1}{2}\right] \tag{11.3.5}$$

where the operator

$$N_t(\mathbf{q}) \equiv a_t^\dagger(\mathbf{q}) a_t(\mathbf{q}) \qquad (11.3.6)$$

represents the number of phonons in mode $t, \mathbf{q}$.

The Heisenberg equations of motion

$$i\hbar \frac{\partial f}{\partial t} = [f, H] \qquad (11.3.7)$$

for $f = a$ and $a^\dagger$ lead to the same equations of motion:

$$\begin{aligned} \dot{a}_t(\mathbf{q}) &= -i\omega_t(\mathbf{q}) a_t(q) \\ \dot{a}_t^\dagger(\mathbf{q}) &= i\omega_t(\mathbf{q}) a_t^\dagger(\mathbf{q}) \end{aligned} \qquad (11.3.8)$$

as we anticipated in making our transformation, and the same equations as would be obtained by using the classical Hamiltonian equations of motion with a and $i\hbar a^\dagger$ as canonically conjugate variables.

We have succeeded in transforming from $3Ns$ pairs $\mathbf{u}^{\alpha l}, \mathbf{p}^{\alpha l}$ to $3Ns$ pairs $a_t(\mathbf{q}), a_t^\dagger(\mathbf{q})$. Each t and $\mathbf{q}$ corresponds to a harmonic oscillator, where $\mathbf{q}$ ranges over a full Brillouin zone, not a half zone, since there are *now* no constraints relating operators at $\mathbf{q}$ and $-\mathbf{q}$. All that remains are the constraints relating $a_t^\dagger(\mathbf{q})$ to $a_t(\mathbf{q})$, which are sufficient to give that pair 2 real rather than 2 complex degrees of freedom.

It would be possible to define *real* harmonic oscillator coordinates $x_t(\mathbf{q})$ and $p_t(\mathbf{q})$ by

$$\begin{aligned} a_t(\mathbf{q}) &= \left[\frac{M\omega_t(\mathbf{q})}{2\hbar} \right]^{1/2} \left[x_t(\mathbf{q}) - \frac{p_t(\mathbf{q})}{iM\omega_t(\mathbf{q})} \right] \\ a_t^\dagger(\mathbf{q}) &= \left[\frac{M\omega_t(\mathbf{q})}{2\hbar} \right]^{1/2} \left[x_t(\mathbf{q}) + \frac{p_t(\mathbf{q})}{iM\omega_t(\mathbf{q})} \right] \end{aligned} \qquad (11.3.9)$$

so that

$$H = \frac{1}{2} \sum_{t\mathbf{q}} \left[\frac{p_t^2(\mathbf{q})}{M} + M\omega_t^2(\mathbf{q}) x_t^2(\mathbf{q}) \right] \qquad (11.3.10)$$

[The eigenfunctions of this H are the familiar harmonic oscillator functions of elementary quantum mechanics in the real variables $x_t(\mathbf{q})$.] In practice, however, one works directly with the a's, since $a^\dagger a$ has integers n as eigenvalues, leading to energies $\hbar\omega(n+\frac{1}{2})$, and a and $a^\dagger$ are just those combinations of x and p with the simple matrix elements (see, e.g., Section 9-14 of Powell and Crasemann, **1961**):

$$a\psi_n = \sqrt{n}\,\psi_{n-1} \qquad \text{or} \qquad \langle n-1|a|n\rangle = n^{1/2} \tag{11.3.11}$$

$$a^\dagger\psi_n = \sqrt{n+1}\,\psi_{n+1} \qquad \text{or} \qquad \langle n+1|a|n\rangle = (n+1)^{1/2} \tag{11.3.12}$$

where

$$a^\dagger a\psi_n = n\psi_n \tag{11.3.13}$$

In other words, $a^\dagger$ changes ψ_n to ψ_{n+1}, increasing the energy of the system, via Eq. 11.3.5, from $(n+\frac{1}{2})\hbar\omega$ to $(n+1+\frac{1}{2})\hbar\omega$. Therefore it is referred to as a raising operator, and a is correspondingly termed a lowering operator. In precise analogy with the corresponding quantization of the electromagnetic field, a state with quantum number n is described as one containing n phonons. Thus $a_t^\dagger(\mathbf{q})$ creates a phonon of momentum $\mathbf{q}$ and type t, and $a_t(\mathbf{q})$ destroys the same phonon.

11.4. CRYSTAL MOMENTUM

The lattice translation $(\epsilon|\mathbf{t})$, which takes $\mathbf{X}^{\alpha i}$ into

$$(\mathbf{X}^{\alpha i})' = \mathbf{X}^{\alpha i} + \mathbf{t} = \mathbf{X}^{\alpha, i+t} \tag{11.4.1}$$

places at position i the displacement that was previously at $i-t$:

$$(\mathbf{u}^{\alpha i})' = (\epsilon|\mathbf{t})\mathbf{u}^{\alpha i}(\epsilon|\mathbf{t})^{-1} = \mathbf{u}^{\alpha, i-t} \tag{11.4.2}$$

(See, e.g., Eq. 2.1.8.) Thus

$$\sum_i (\mathbf{u}^{\alpha i})' \exp(-i\mathbf{q}\cdot\mathbf{i}) = \exp(-i\mathbf{q}\cdot\mathbf{t}) \sum_i \mathbf{u}^{\alpha, i-t} \exp[-i\mathbf{q}\cdot(\mathbf{i}-\mathbf{t})] \tag{11.4.3a}$$

We invert Eq. 11.2.13a and, using 11.4.3a, find

$$Q_t(\mathbf{q})' \equiv (\epsilon|\mathbf{t})Q_t(\mathbf{q})(\epsilon|\mathbf{t})^{-1} = \exp(-i\mathbf{q}\cdot\mathbf{t})Q_t(\mathbf{q}) \tag{11.4.4}$$

A similar equation holds for $\dot{Q}_t(\mathbf{q})$, so that finally

$$(\epsilon|\mathbf{t})a_t(\mathbf{q})(\epsilon|\mathbf{t})^{-1} = \exp(-i\mathbf{q}\cdot\mathbf{t})a_t(\mathbf{q})$$

or, similarly,

$$(\epsilon|\mathbf{t})a_t^\dagger(\mathbf{q})(\epsilon|\mathbf{t})^{-1} = \exp(i\mathbf{q}\cdot\mathbf{t})a_t^\dagger(\mathbf{q}) \tag{11.4.5}$$

Let $|0\rangle$ denote the ground state in which no phonons are present; $N_t(\mathbf{q})|0\rangle = 0$ for all $t, \mathbf{q}$. This state is nondegenerate. Thus

$$(\boldsymbol{\epsilon}|\mathbf{t})|0\rangle = \exp(i\theta)|0\rangle \tag{11.4.6}$$

where the arbitrary phase factor $\exp(i\theta)$ can be set equal to unity by absorbing it into the definition of a new $(\boldsymbol{\epsilon}|\mathbf{t})$. The state $a_t^\dagger(\mathbf{q})|0\rangle$ is a quantum state containing one phonon of propagation vector $\mathbf{q}$, and

$$(\boldsymbol{\epsilon}|\mathbf{t})a_t^\dagger(\mathbf{q})|0\rangle = \exp(i\mathbf{q}\cdot\mathbf{t})a_t^\dagger(\mathbf{q})|0\rangle \tag{11.4.7}$$

Similarly a state $|N\rangle$ containing $n_t(\mathbf{q})$ phonons in *each* mode $t, \mathbf{q}$ can be written as

$$|N\rangle = \prod_{t,\mathbf{q}} [a_t^\dagger(\mathbf{q})]^{n_t(\mathbf{q})}|0\rangle \tag{11.4.8}$$

$$(\boldsymbol{\epsilon}|\mathbf{t})|N\rangle = \exp\left[i\sum_{t,\mathbf{q}} n_t(\mathbf{q})\mathbf{q}\cdot\mathbf{t}\right]|N\rangle \tag{11.4.9}$$

Let us define a crystal momentum operator $\mathbf{P}$ by

$$\mathbf{P} = \sum a_t^\dagger(\mathbf{q})a_t(\mathbf{q})\hbar\mathbf{q} \tag{11.4.10}$$

Using Eqs. 11.3.6 and 11.3.13, we have

$$\mathbf{P}|N\rangle = \left[\sum N_t(\mathbf{q})\hbar\mathbf{q}\right]|N\rangle = \left[\sum n_t(\mathbf{q})\hbar\mathbf{q}\right]|N\rangle$$

$$\exp\left(\frac{i\mathbf{P}\cdot\mathbf{t}}{\hbar}\right)|N\rangle = \exp\left[i\sum_{t,\mathbf{q}} n_t(\mathbf{q})\mathbf{q}\cdot\mathbf{t}\right]|N\rangle \tag{11.4.11}$$

Comparing with Eq.11.4.9, we can equate the operators

$$\exp\left(\frac{i\mathbf{P}\cdot\mathbf{t}}{\hbar}\right) = (\boldsymbol{\epsilon}|\mathbf{t}) \tag{11.4.12}$$

since their action is the same on a complete set of states. We have thus found an explicit representation of the translation operator in terms of phonon creation and destruction operators.

Since the crystal momentum $\mathbf{q}$ can be specified only modulo a reciprocal lattice vector $\mathbf{K}$, the crystal momentum operator $\mathbf{P}$ can be specified only modulo $\hbar\mathbf{K}$. However, the more fundamental operator $\exp(i\mathbf{P}\cdot\mathbf{t}/\hbar) = (\boldsymbol{\epsilon}|\mathbf{t})$ is always unique.

The essence of the preceding discussion is that the representation in which $a_t^\dagger(\mathbf{q})a_t(\mathbf{q})$ is diagonal diagonalizes not only the total energy of the system, but also $(\boldsymbol{\epsilon}|\mathbf{t})$ and the crystal momentum.

Even in the presence of anharmonic forces, in which case the Hamiltonian and the individual operators $a_t^\dagger(\mathbf{q})a_t(\mathbf{q})$ cannot simultaneously be diagonalized, the eigenvalues of $(\boldsymbol{\epsilon}|\mathbf{t})$ and hence of the total crystal momentum $\mathbf{P}$ are good quantum numbers as long as the periodicity of the system is preserved.

In Eq. 6.5.5 we established the conservation of total crystal momentum for any collision involving several particles (phonons or phonons and electrons or phonons and neutrons or neutrons and electrons). Here "total" refers to the sum of the crystal momenta of all particles.

Although it is correct to speak of each phonon as carrying energy $\hbar\omega(\mathbf{q})$ and crystal momentum $\hbar\mathbf{q}$, the total crystal momentum is *not* identical with the total momentum of the system. In a collision with an external particle, say a neutron, the recoil momentum is surely carried by the center of mass of the crystal, that is, by the acoustic modes of propagation constant $\mathbf{q}=0$. Conservation of crystal momentum, however, describes a selection rule regarding the state of *internal* (vibrational) excitation of the crystal after collision with an external particle.

The relation between ordinary and crystal momentum can be further delineated by remarking that conservation of momentum arises because of invariance of the Hamiltonian against infinitesimal displacements $\mathbf{x}\rightarrow\mathbf{x}+\mathbf{d}$, even for an arbitrary inhomogeneous body possessing no symmetry whatever. Crystal momentum conservation, however, arises because of invariance against lattice translations $\mathbf{X}\rightarrow\mathbf{X}+\mathbf{t}$. It is, in short, a property of an infinite body possessing the material symmetry of translational invariance. For a continuum solid or liquid, the corresponding material symmetry $\mathbf{X}\rightarrow\mathbf{X}+\mathbf{d}$ would be called homogeneity.

11.5 INFINITESIMAL DISPLACEMENT AND ROTATIONAL INVARIANCE

As $\mathbf{q}\rightarrow 0$, Eq. 11.2.8 reduces to

$$\mathbf{R}^{\alpha\beta}(0)=\sum_j \mathbf{K}^{\alpha i,\beta j} \qquad (11.5.1)$$

and invariance under infinitesimal displacements 5.4.7 now takes the form

$$\sum_\beta \mathbf{R}^{\alpha\beta}(0)\cdot\mathbf{d}=0 \qquad (11.5.2)$$

where $\mathbf{d}$ is an arbitrary vector. Thus, for $\mathbf{q}=0$,

$$\mathbf{b}^{\beta}=\mathbf{d} \qquad (11.5.3)$$

corresponding to a simultaneous displacement of all particles in an *arbitrary* direction, is a solution of Eq. 11.2.7 with $\omega_t = 0$. As $\mathbf{q} \to 0$, then, three of the branches $\omega_t(\mathbf{q})$ must have frequencies that approach zero. These are known as the acoustic branches, because they correspond to the acoustic wave solutions of the equations of macroscopic elasticity.

The displacement invariance condition 5.4.7 or 11.5.2 permits us to rewrite the $\alpha\alpha$ elements in a form that does not depend on $\mathbf{K}^{\alpha i \alpha i}$:

$$\mathbf{R}^{\alpha\alpha}(\mathbf{q}) = -\sum_{j \neq i} \{1 - \exp[i\mathbf{q}\cdot(\mathbf{j}-\mathbf{i})]\}\mathbf{K}^{\alpha i, \alpha j} - \sum_{\beta \neq \alpha} \mathbf{R}^{\alpha\beta}(0) \qquad (11.5.4a)$$

Infinitesimal rotational invariance, Eq. 5.4.10, can be rewritten for the lattice case by setting $m = \alpha i$, $n = \beta j$:

$$\sum_{\beta j} \mathbf{K}^{\alpha i \beta j} \times (\mathbf{X}^{\beta j} - \mathbf{X}^{\alpha i}) = 0 \qquad (11.5.5)$$

For any point in the interior of the crystal lattice, translational invariance, Eq. 11.2.4, can be used to reduce Eq. 11.5.5 to one condition for each α (independent of i):

$$\sum_{\beta j} \mathbf{K}^{\alpha 0 \beta j}(\mathbf{j}) \times (\mathbf{X}^{\beta j} - \mathbf{X}^{\alpha}) = 0 \qquad (11.5.6)$$

Additional restrictions on the force constants in the interior may follow, however, from application of Eq. 11.5.5 to atoms at or near the boundary (Lax, **1965a**).

11.6. SYMMETRY PROPERTIES OF THE DYNAMICAL MATRIX

We take the symmetry condition obeyed by the force constant matrix 5.4.19, with $m = \alpha i$, $n = \beta j$, in the form:[1]

$$\mathbf{K}^{\alpha i \beta j} = \tilde{\mathbf{S}}\mathbf{K}^{\alpha' i' \beta' j'}\mathbf{S} \qquad (11.6.1)$$

Multiply by $\exp[i\mathbf{q}\cdot(\mathbf{X}^{\beta j} - \mathbf{X}^{\alpha i})]$, and sum on j to obtain a corresponding equation for the *second* form (11.2.8b) of the dynamical matrix:

$$\mathbf{R}^{\alpha\beta}(\mathbf{q}) = \tilde{\mathbf{S}}\mathbf{R}^{S(\alpha)S(\beta)}(\mathbf{S}\mathbf{q})\mathbf{S} \qquad (11.6.2b)$$

where $\alpha' = S(\alpha)$ is the name of the new position into which α is carried by $(\mathbf{S}|\mathbf{v})$; see Eq. 5.4.12. The $\mathbf{Sq}$ arises from writing

$$\mathbf{q}\cdot(\mathbf{X}^{\beta j} - \mathbf{X}^{\alpha i}) = \mathbf{S}\mathbf{q}\cdot\mathbf{S}(\mathbf{X}^{\beta j} - \mathbf{X}^{\alpha i}) \qquad (11.6.3)$$

[1]Relatively recent reviews of lattice dynamics from a symmetry point of view include those of Maradudin and Vosko (**1968**), Warren (**1968**), Ludwig (**1967**), and Leibfreid (**1955**).

We also use

$$\mathbf{S}(\mathbf{X}^{\beta j}-\mathbf{X}^{\alpha i})=(\mathbf{S}|\mathbf{v}(\mathbf{S}))(\mathbf{X}^{\beta j}-\mathbf{X}^{\alpha i})$$
$$=\mathbf{X}^{\beta' j'}-\mathbf{X}^{\alpha' i'}$$

to change the index in the exponential from j to j' and thus conform with the same index on $\mathbf{K}$.

In this discussion, the second form (11.2.8b) of the dynamical matrix is essential, because the first form leads to

$$\mathbf{S}(\mathbf{j}-\mathbf{i})\neq\mathbf{j}'-\mathbf{i}' \qquad (11.6.4)$$

when $\mathbf{v}(\mathbf{S})\neq 0$. If $S\in G_{\mathbf{q}}$, that is, $\mathbf{Sq}=\mathbf{q}$ or, more generally, by Eq. 8.7.6,

$$\mathbf{Sq}=\mathbf{q}-\mathbf{K}_{S^{-1}} \qquad (11.6.5)$$

then the use of

$$\mathbf{R}^{\alpha\beta}(\mathbf{q}-\mathbf{K})=\exp[i\mathbf{K}\cdot(\mathbf{X}^{\alpha}-\mathbf{X}^{\beta})]\mathbf{R}^{\alpha\beta}(\mathbf{q}) \qquad (11.6.6b)$$

together with Eq. 11.6.2, imposes conditions on $\mathbf{R}^{\alpha\beta}(\mathbf{q})$ that can serve (when combined with time reversal in Sections 11.7 and 11.8) to determine the form and number of independent constants in $\mathbf{R}^{\alpha\beta}(\mathbf{q})$. See Section 11.8.

The operations that do not leave $\mathbf{q}$ invariant do not restrict the matrix elements but provide relationships that reduce the number of independent *functions* with which it is necessary to deal. We shall illustrate this remark by considering a crystal of the diamond structure (see Table 6.1.4), with two atoms in the unit cell, $\alpha=\pm 1$.

The group O_h can be decomposed into $O_h=T_d\times C_i$. Inversion in Eq. 11.6.2 provides the relation

$$\mathbf{R}^{\alpha\beta}(\mathbf{q})=\mathbf{R}^{\overline{\alpha\beta}}(-\mathbf{q})=\mathbf{R}^{\overline{\alpha\beta}}(\mathbf{q})^{*} \qquad (11.6.7)$$

where $\bar{\alpha}=I\alpha=-\alpha$ for the diamond case. This exhausts all the information in which α and β are changed by the group operations, since the elements of T_d do not interchange the two sublattices.

The operation $\rho_{\bar{x}y}$, that is, $x\rightleftarrows y$, yields the relationships

$$\begin{aligned}R_{22}^{\alpha\beta}(q_1,q_2,q_3)&=R_{11}^{\alpha\beta}(q_2,q_1,q_3)\\ R_{21}^{\alpha\beta}(q_1,q_2,q_3)&=R_{12}^{\alpha\beta}(q_2,q_1,q_3)\end{aligned} \qquad (11.6.8b)$$

The operation $x\rightleftarrows z$ yields

$$\begin{aligned}R_{33}^{\alpha\beta}(q_1,q_2,q_3)&=R_{11}^{\alpha\beta}(q_3,q_2,q_1)\\ R_{13}^{\alpha\beta}(q_1,q_2,q_3)&=R_{31}^{\alpha\beta}(q_3,q_2,q_1)\end{aligned} \qquad (11.6.9b)$$

The transformation $x \to y \to z \to x$ yields

$$R_{23}^{\alpha\beta}(q_1, q_2, q_3) = R_{12}^{\alpha\beta}(q_2, q_3, q_1) \tag{11.6.10b}$$

$$R_{31}^{\alpha\beta}(q_1, q_2, q_3) = R_{23}^{\alpha\beta}(q_2, q_3, q_1) = R_{12}^{\alpha\beta}(q_3, q_1, q_2)$$

In this way all matrix elements are expressible in terms of the four prototype functions $R_{11}^{11}, R_{12}^{11}, R_{11}^{1-1}, R_{12}^{1-1}$.

Plane Wave Coordinates

The kinetic and potential energies are diagonalized by the transformation to normal coordinates (of the second kind):

$$\mathbf{u}^{\alpha i} = N^{-1/2} \sum_{t,\mathbf{q}} Q_t(\mathbf{q}) \mathbf{b}_t^{\alpha}(\mathbf{q}) \exp(i\mathbf{q} \cdot \mathbf{X}^{\alpha i}) \tag{11.6.11b}$$

Since this step requires knowledge of the symmetry vectors $\mathbf{b}_t^{\alpha}(\mathbf{q})$, which is not available a priori, it is convenient to introduce *plane wave coordinates*,

$$\mathbf{b}^{\alpha}(\mathbf{q}) = N^{-1/2} \sum_i \mathbf{u}^{\alpha i} \exp(-i\mathbf{q} \cdot \mathbf{X}^{\alpha i}) \tag{11.6.12b}$$

as an intermediate step that takes account of the *translational* symmetry of the lattice. The remaining decomposition

$$\mathbf{b}^{\alpha}(\mathbf{q}) = \sum_t \mathbf{b}_t^{\alpha}(\mathbf{q}) Q_t(\mathbf{q}) \tag{11.6.13b}$$

into symmetry vectors can be performed as in molecules (see Chapter 5), using the behavior of the plane wave coordinates $\mathbf{b}^{\alpha}(\mathbf{q})$ under the *rotational* transformations $(\mathbf{S}|\mathbf{v}(\mathbf{S}))$ of the space group. This behavior is governed by definition 11.6.12b with the coefficients $\exp(-i\mathbf{q} \cdot \mathbf{X}^{\alpha i})$ understood to be numbers invariant under all transformations. In this way we arrive at the following theorem.

Theorem 11.6.1

$$O(S)\mathbf{b}^{\alpha}(\mathbf{q}) = \mathbf{S}\mathbf{b}^{S^{-1}(\alpha)}(\mathbf{S}^{-1}\mathbf{q}) \tag{11.6.14b}$$

where $O(S) \equiv \exp[i\mathbf{q} \cdot \mathbf{v}(\mathbf{S})](\mathbf{S}|\mathbf{v}(\mathbf{S}))$, and $S^{-1}(\alpha)$ is the name of the point that will be carried into α by $(\mathbf{S}|\mathbf{v}(\mathbf{S}))$. The $\mathbf{S}$ on the right-hand side of Eq. 11.6.14 stands for the usual representation in three space of the rotation $\mathbf{S}$. [Strictly speaking, the left-hand side should read as $\exp[i\mathbf{q} \cdot \mathbf{v}(\mathbf{S})](\mathbf{S}|\mathbf{v}(\mathbf{S}))$ $\mathbf{b}^{\alpha}(\mathbf{q})(\mathbf{S}|\mathbf{v}(\mathbf{S}))^{-1}$ when we regard $\mathbf{b}^{\alpha}(\mathbf{q})$ as an operator; cf. Eq. 2.5.16.]

Proof: The point $S^{-1}(\alpha i)$ is carried into (αi) by S; hence the new displacement at αi is

$$(\mathbf{u}^{\alpha i})' = \mathbf{S}\mathbf{u}^{S^{-1}(\alpha i)}$$

Thus Eq. 11.6.12b implies that

$$\mathbf{b}^{\alpha}(\mathbf{q})' = N^{-1/2}\sum \exp(-i\mathbf{S}^{-1}\mathbf{q}\cdot\mathbf{S}^{-1}\mathbf{X}^{\alpha i})\mathbf{S}\mathbf{u}^{S^{-1}(\alpha i)}$$

where we have added the canceling factors $\mathbf{S}^{-1}$ in the exponent, so that with the help of

$$\mathbf{X}^{S^{-1}(\alpha i)} \equiv (\mathbf{S}|\mathbf{v}(\mathbf{S}))^{-1}\mathbf{X}^{\alpha i} = \mathbf{S}^{-1}\mathbf{X}^{\alpha i} - \mathbf{S}^{-1}\mathbf{v}(\mathbf{S})$$

we can introduce $S^{-1}(\alpha i)$ as a new summation variable, with the result

$$\mathbf{b}^{\alpha}(\mathbf{q})' \equiv (\mathbf{S}|\mathbf{v}(\mathbf{S}))\mathbf{b}^{\alpha}(\mathbf{q}) = \exp[-i\mathbf{q}\cdot\mathbf{v}(\mathbf{S})]\mathbf{S}\mathbf{b}^{S^{-1}(\alpha)}(\mathbf{S}^{-1}\mathbf{q}) \qquad (11.6.15b)$$

required by the theorem.

This is often used in regard to the *interior of the zone* with $\mathbf{S}^{-1}\mathbf{q} = \mathbf{q}$. The significance of the theorem then is that the operators $O(S)$ behave as if no translations were involved, that is, as if we could set $\mathbf{v}(\mathbf{S}) = 0$ or $\mathbf{q} = 0$. This agrees with our earlier result (Section 8.8) that $O(S)$ obeys the multiplication table and character table of the point group $P(\mathbf{q})$—or, alternatively, the space group character table with $\mathbf{q}$ set equal to zero! *In other words, provided that we act on the* $\mathbf{b}^{\alpha}(\mathbf{q})$, *we can perform all operations* $O(S)$ *as if* $\mathbf{q} = 0$, *treating* S *as a point operation. The results are then valid for* $\mathbf{q} \neq 0$ *as long as* $\mathbf{S}^{-1}\mathbf{q} = \mathbf{q}$.

For example, projection operators can be written using

$$\sum \chi(\mathbf{S}|\mathbf{v})^{*}(\mathbf{S}|\mathbf{v}) = \sum \chi[O(S)]^{*}O(S) \qquad (11.6.16)$$

with $\chi[O(S)]$ as a point group character and $O(S)$ acting as a point group operation on $\mathbf{b}^{\alpha}(\mathbf{q})$!

When $\mathbf{q}$ is at the *zone boundary*, we have additional operations such that

$$\mathbf{S}^{-1}\mathbf{q} = \mathbf{q} - \mathbf{K}_S \qquad (11.6.17)$$

and we can obtain additional symmetry information provided that we can relate $\mathbf{b}^{\alpha}(\mathbf{q}-\mathbf{K})$ to $\mathbf{b}^{\alpha}(\mathbf{q})$. Equation 11.6.12b for $\mathbf{b}^{\alpha}(\mathbf{q})$ yields directly

$$\mathbf{b}^{\alpha}(\mathbf{q}-\mathbf{K}) = \exp(\mathbf{i}\mathbf{K}\cdot\mathbf{X}^{\alpha})\mathbf{b}^{\alpha}(\mathbf{q}) \qquad (11.6.18b)$$

which is the appropriate eigenvector of the corresponding matrix $\mathbf{R}^{\alpha\beta}(\mathbf{q}-\mathbf{K})$ of Eq. 11.6.6. If we had used the *first* kind of dynamical matrix (11.2.8a), $\mathbf{R}(\mathbf{q})$ and $\mathbf{b}(\mathbf{q})$ would have been periodic functions in $\mathbf{q}$ space:

$$\mathbf{R}(\mathbf{q}-\mathbf{K}) = \mathbf{R}(\mathbf{q}), \qquad \mathbf{b}(\mathbf{q}-\mathbf{K}) = \mathbf{b}(\mathbf{q}) \qquad (11.6.19a)$$

but the price paid for this simplicity is indicated in Problem 11.6.1.

11.7. CONSEQUENCES OF TIME REVERSAL

By using translational invariance, we have reduced a problem involving a real matrix $\mathbf{K}$ of $3sN$ dimensions to a complex $3s\times 3s$ matrix $\mathbf{R}(\mathbf{q})$ (equivalent to a real $6s\times 6s$ matrix). Since the diagonalization of a matrix requires a number of operations proportional to the cube of its dimension, a factor of 8 in computing time will be saved if the matrix $\mathbf{R}(\mathbf{q})$ (and the corresponding eigenvectors) can be transformed to real form. This reduction to real form must be accomplished (when possible) by using the time-reversal invariance of the problem. In the domain of *real* displacements $\mathbf{u}^m(t)$ obeying

$$\sum \mathbf{K}^{mn}\mathbf{u}^n(t) = -M_m\frac{\partial^2\mathbf{u}^m}{\partial t^2} \tag{11.7.1}$$

the fact that the time-reversed motion $[\mathbf{u}^m(t)]^T=\mathbf{u}^m(-t)$ is also a solution is guaranteed because only *even* time derivatives occur. Even if we assume (in the usual way) a *complex* solution $\mathbf{u}^m(t)=\mathbf{u}^m\exp(-i\omega t)$, the resulting equations

$$\sum \mathbf{K}^{mn}\mathbf{u}^n = M_m\omega^2\mathbf{u}^m \tag{11.7.2}$$

for the amplitudes have only *real* coefficients. Thus, if $\mathbf{u}^m$ is a solution of Eq. 11.7.2, then so is $(\mathbf{u}^m)^*$. Hence $(\mathbf{u}^m)\exp(-i\omega t)$ and $\mathbf{u}_m^*\exp(-i\omega t)$ are both solutions of Eq. 11.7.1 with the *same sign* of the frequency. We can therefore define the time-reversed solution in the complex case to be

$$[\mathbf{u}^m(t)]^T=[\mathbf{u}^m(-t)]^* \tag{11.7.3}$$

or

$$[\mathbf{u}^m]^T=(\mathbf{u}^m)^* \tag{11.7.4}$$

in complete analogy with the corresponding definition (10.1.17) for the Schrödinger equation.

It follows from Eq. 11.7.4 and the linearity of Eq. 11.7.2 that, if $\mathbf{u}^m$ is a solution of 11.7.2, so is $\mathbf{u}^m+(\mathbf{u}^m)^*$. In other words, the reality of $\mathbf{K}^{mn}$ guarantees the possibility of real eigenvectors.

Since the matrix $\mathbf{K}^{\alpha i,\beta j}$ for the complete lattice is real, we can take a solution $\mathbf{u}^{\alpha i}$ and construct the real solution $\mathbf{u}^{\alpha i}+(\mathbf{u}^{\alpha i})^*$. Unfortunately, this procedure creates a real solution by mixing modes of propagation constant $\mathbf{q}$ and $-\mathbf{q}$. If there is *no* element of the space group connecting $\mathbf{q}$ and $-\mathbf{q}$ (i.e., $-\mathbf{q}$ is not in the star of $\mathbf{q}$), it will *not* be possible to construct a solution that is both real and of propagation constant $\mathbf{q}$. (All the representations of $G_{\mathbf{q}}$ are then of type 3, i.e., complex.) For all general vectors $\mathbf{q}$ this is indeed the state of affairs unless the space group contains inversion as an element.

If there is an element Q that reverses $\mathbf{q}$, it can be combined with K_0 = complex conjugation to produce an element QK_0 that leaves $\mathbf{q}$ invariant. In group-theoretical language, we must augment the group of the wave vector $G_{\mathbf{q}}$ by QK_0 to form the reversal group of the wave vector:

$$G_{\mathbf{q},-\mathbf{q}} = G_{\mathbf{q}} + QK_0G_{\mathbf{q}} \tag{11.7.5}$$

just as we previously introduced the group of the bond and the reversal group of the bond.

The procedure by which $\mathbf{R}(\mathbf{q})$ can be made real will now be discussed. Let us rewrite Eq. 11.6.2 in the form

$$\mathbf{R}^{\alpha\beta}(\mathbf{q}) = \tilde{\mathbf{Q}}\mathbf{R}^{Q(\alpha),Q(\beta)}(\mathbf{Qq})\mathbf{Q} \tag{11.7.6b}$$

For the special case in which $\mathbf{Qq} = -\mathbf{q}$, we can apply Eq. 11.2.9, to obtain

$$\mathbf{R}^{\alpha\beta}(\mathbf{q})^* = \tilde{\mathbf{Q}}\mathbf{R}^{Q(\alpha),Q(\beta)}(\mathbf{q})\mathbf{Q} \tag{11.7.7b}$$

or

$$\mathbf{R}^* = \Omega^{-1}\mathbf{R}\Omega$$

or

$$(R^{\alpha\beta}_{\mu\nu})^* = (\Omega^{-1})^{\alpha\alpha'}_{\mu\mu'} R^{\alpha'\beta'}_{\mu'\nu'} \Omega^{\beta'\beta}_{\nu'\nu} \tag{11.7.8}$$

where

$$\Omega^{\beta'\beta}_{\nu'\nu} = \delta[\beta', Q(\beta)]Q_{\nu'\nu} \tag{11.7.9b}$$

and $Q_{\nu'\nu}$ is the *real* 3×3 matrix representing the rotation Q.

To make R real [by the method of Theorem 10.6.6, used to make representations real], we set $R = UR'U^{-1}$. Then R' will be real if $U\tilde{U} = \Omega$, as is possible if Ω is symmetric. We can then set $U = \Omega^{1/2}$ = the symmetric square root of Ω.

Since Ω is real and unitary, $\tilde{\Omega} = \Omega^{-1}$ and the condition of symmetry $\tilde{\Omega} = \Omega$ is then equivalent to the requirement that $\Omega^2 = 1$. But Eq. 11.7.9 implies that $\Omega^2 = 1$ if $Q^2 = 1$. Thus the dynamical matrix $\mathbf{R}^{\alpha\beta}(\mathbf{q})$ can be made real if we can find a reversal operator $(\mathbf{Q}|\mathbf{v}(\mathbf{Q}))$ such that Q^2 = the identity. See, however, Problems 11.7.2 and 11.7.3.

We can avoid taking square roots of matrices by imposing the condition

$$\Omega\mathbf{b}^*(\mathbf{q}) = \mathbf{b}(\mathbf{q}) \tag{11.7.10}$$

(where all indices have been suppressed). We obtain this condition by noting that, if $\mathbf{b}$ is an eigenvector of $\mathbf{R}(\mathbf{q})$, Eq. 11.7.7 implies that $\Omega\mathbf{b}^*$ is also an eigenvector. Then $\mathbf{b}+\Omega\mathbf{b}^*$ is also an eigenvector that obeys Eq. 11.7.10. Moreover, any constraint such as 11.7.10 among $3s$ complex variables leaves only $3s$ independent real variables and automatically results in a real matrix.[2,3]

As an alternative viewpoint, we note that constraint 11.7.10 is equivalent to requiring that the eigenvectors of the real matrix $\mathbf{R}'=\Omega^{-1/2}\mathbf{R}\Omega^{1/2}$, namely, $\Omega^{-1/2}\mathbf{b}$, be real. Using $Q^{-1}=Q$, we rearrange requirement 11.7.10 in the form

$$\mathbf{b}^{\beta}(\mathbf{q})^*=\mathbf{Q}\cdot\mathbf{b}^{Q(\beta)}(\mathbf{q}) \tag{11.7.11}$$

where $\mathbf{Q}$ is a 3×3 matrix, and not the operator $(\mathbf{Q}|\mathbf{v}(\mathbf{Q}))$.

Since the only reversal element possible for a general $\mathbf{q}$ is $Q=I=$ the inversion, we shall consider this case in detail. Equation 11.7.7 becomes

$$\mathbf{R}^{\alpha\beta}(\mathbf{q})^*=\mathbf{R}^{\overline{\alpha\beta}}(\mathbf{q}) \tag{11.7.12}$$

where $\bar{\alpha}$ is an abbreviation for $I(\alpha)$, the atom inverse to α. If each atom is at a center of symmetry as in sodium chloride, $\mathbf{R}$ is already real. Constraint 11.7.11 would make all $\mathbf{b}^{\beta}$ purely imaginary. For the inversion case, we shall therefore remove a single phase factor and replace 11.7.11 by

$$\mathbf{b}^{\beta}(\mathbf{q})^*=\mathbf{b}^{\bar{\beta}}(\mathbf{q}) \tag{11.7.13}$$

Example 11.7.1

Apply these methods to the diamond structure. Make $\mathbf{R}(\mathbf{q})$ real.

Solution: If $\alpha=1, \bar{\alpha}=-1=\bar{1}$ are a pair of inverse atoms as in the diamond structure, condition 11.7.12, together with the Hermiticity of $\mathbf{R}$, guarantees that the matrix $\mathbf{R}$ has the form

$$\mathbf{R}\equiv\begin{bmatrix}\mathbf{R}^{11} & \mathbf{R}^{1\bar{1}}\\ \mathbf{R}^{\bar{1}1} & \mathbf{R}^{\bar{1}\bar{1}}\end{bmatrix}=\begin{bmatrix}\mathbf{H} & \mathbf{S}\\ \mathbf{S}^* & \mathbf{H}^*\end{bmatrix} \tag{11.7.14}$$

where $\mathbf{H}$ is Hermitian and $\mathbf{S}$ is symmetric. Our constraint 11.7.13 then takes the form

$$(\mathbf{b}^{1})^*=\mathbf{b}^{\bar{1}} \tag{11.7.15}$$

[2]If all the eigenvectors of a Hermitian matrix M are real, the matrix must be real. *Proof*: $M-M^*$ has eigenvalue zero for a complete set of eigenvectors and hence must vanish.

[3]If one first reduces $\mathbf{R}^{\alpha\beta}(\mathbf{q})$ to "irreducible" block form by introducing symmetry vectors, this procedure will not mix two blocks, provided that time reversal produces no extra degeneracy. Conversely, if extra degeneracy is produced by time reversal, two $n_i\times n_i$ complex blocks will be combined into a real $2n_i\times2n_i$ matrix, which is no saving whatsoever.

Thus $\mathbf{b}^{\bar{1}}$ can be eliminated entirely by using $\mathrm{Re}\,\mathbf{b}^1$ and $\mathrm{Im}\,\mathbf{b}^1$ as independent variables. In particular, the unitary transformation $\mathbf{b}' = \mathbf{V}\mathbf{b}$

$$\begin{bmatrix} \sqrt{2}\ \mathrm{Re}\,\mathbf{b}^1 \\ \sqrt{2}\ \mathrm{Im}\,\mathbf{b}^1 \end{bmatrix} = \frac{1}{\sqrt{2}} \begin{bmatrix} 1 & 1 \\ -i & i \end{bmatrix} \begin{bmatrix} \mathbf{b}^1 \\ \mathbf{b}^{\bar{1}} \end{bmatrix} \tag{11.7.16}$$

leads to the new matrix

$$\mathbf{R}' = \mathbf{V}\mathbf{R}\mathbf{V}^{-1} = \begin{bmatrix} \mathrm{Re}\,(\mathbf{H}+\mathbf{S}) & \mathrm{Im}\,(\mathbf{S}-\mathbf{H}) \\ \mathrm{Im}\,(\mathbf{S}+\mathbf{H}) & \mathrm{Re}\,(\mathbf{H}-\mathbf{S}) \end{bmatrix} \tag{11.7.17}$$

which is both real and symmetric!

Example 11.7.2

Make $\mathbf{R}(\mathbf{q})$ real for the calcium fluoride structure.

Solution: The CaF_2 structure is face-centered cubic with Ca at (000) and F at $(\frac{1}{4}, \frac{1}{4}, \frac{1}{4})$ and $(-\frac{1}{4}, -\frac{1}{4}, -\frac{1}{4})$, so that the calcium is at a center of symmetry, and the fluorines are inversion images of each other. The appropriate transformation matrix is then:

$$\mathbf{V} = \begin{bmatrix} 1 & 0 & 0 \\ 0 & 1/\sqrt{2} & 1/\sqrt{2} \\ 0 & -i/\sqrt{2} & i/\sqrt{2} \end{bmatrix} \tag{11.7.18}$$

where we denote Ca, F, and F by 0, 1, and $\bar{1}$ and have used $\mathbf{b}^0$, $\sqrt{2}\ \mathrm{Re}\,\mathbf{b}^1$, and $\sqrt{2}\ \mathrm{Im}\,\mathbf{b}^1$ as variables. Then the new real matrix is given by

$$\mathbf{R}' = \begin{bmatrix} \mathbf{R}^{00} & \sqrt{2}\ \mathrm{Re}\,\mathbf{R}^{01} & -\sqrt{2}\ \mathrm{Im}\,\mathbf{R}^{01} \\ \sqrt{2}\ \mathrm{Re}\,\tilde{\mathbf{R}}^{01} & \mathrm{Re}\,(\mathbf{R}^{11}+\mathbf{R}^{12}) & \mathrm{Im}\,(\mathbf{R}^{12}-\mathbf{R}^{11}) \\ -\sqrt{2}\ \mathrm{Im}\,\tilde{\mathbf{R}}^{01} & \mathrm{Im}\,(\mathbf{R}^{12}+\mathbf{R}^{11}) & \mathrm{Re}\,(\mathbf{R}^{11}-\mathbf{R}^{12}) \end{bmatrix} \tag{11.7.19}$$

Comment: Since Ca is at a center of symmetry $\bar{0}=0$, Eq. 11.7.12 tells us that $\mathbf{R}^{00}(\mathbf{q})$ is already a real matrix, and Eq. 11.7.13 requires $[\mathbf{b}^0(\mathbf{q})]^* = \mathbf{b}^0(\mathbf{q}) = \text{real}$; hence our use of $\mathbf{b}^0$ as a variable. We can avoid the matrix operations $\mathbf{V}\mathbf{R}\mathbf{V}^{-1}$ by simply imposing the constraints directly in the equations of motion. Thus we can write

$$\mathbf{R}^{00}\mathbf{b}^0+\mathbf{R}^{01}\mathbf{b}^1+\mathbf{R}^{0\bar{1}}\mathbf{b}^{\bar{1}}$$

$$=\mathbf{R}^{00}\mathbf{b}^0+\mathbf{R}^{01}(\text{Re}\,\mathbf{b}^1+i\,\text{Im}\,\mathbf{b}^1)+(\mathbf{R}^{01})^*(\text{Re}\,\mathbf{b}^1-i\,\text{Im}\,\mathbf{b}^1)$$

$$=\mathbf{R}^{00}\mathbf{b}^0+(\sqrt{2}\;\text{Re}\,\mathbf{R}^{01})(\sqrt{2}\;\text{Re}\,\mathbf{b}^1)-(\sqrt{2}\;\text{Im}\,\mathbf{R}^{01})(\sqrt{2}\;\text{Im}\,\mathbf{b}^1) \quad (11.7.20)$$

which leads immediately to the first line of matrix 11.7.19. The appearance of $\tilde{\mathbf{R}}^{01}$ in the second line of Eq. 11.7.19 can be deduced from the fact that $\mathbf{R}'$ is real and Hermitian and hence symmetric.

11.8. ▲ FORM AND NUMBER OF INDEPENDENT CONSTANTS IN THE DYNAMICAL MATRIX FOR INTERNAL AND ZONE BOUNDARY POINTS

We may combine Eqs. 11.6.2, 11.6.5, and 11.6.6b to obtain the condition

$$R_{\mu\nu}^{\alpha\beta}(\mathbf{q})=\exp\left[i\mathbf{K}_{S^{-1}}\cdot(\mathbf{X}^{S(\alpha)}-\mathbf{X}^{S(\beta)})\right]R_{\mu'\nu'}^{S(\alpha)S(\beta)}(\mathbf{q})S_{\mu'\mu}S_{\nu'\nu} \quad (11.8.1\text{b})$$

imposed on the matrix $\mathbf{R}^{\alpha\beta}(\mathbf{q})$ by an arbitrary element $(\mathbf{S}|\mathbf{v}(\mathbf{S}))$ of $G_{\mathbf{q}}$. As in the case of molecular vibrations, Section 5.5, we can rewrite Eq. 11.8.1 as a representation generated by $\mathbf{R}^{\alpha\beta}(\mathbf{q})$:

$$SR_{\mu\nu}^{\alpha\beta}(\mathbf{q})=\sum R_{\mu'\nu'}^{\alpha'\beta'}(\mathbf{q})\,\Delta_{\alpha'\beta'\mu'\nu',\alpha\beta\mu\nu}(S) \quad (11.8.2)$$

$$\Delta_{\alpha'\beta'\mu'\nu',\alpha\beta\mu\nu}(S)=\delta\left[\alpha',S(\alpha)\right]\delta\left[\beta',S(\beta)\right]S_{\mu'\mu}S_{\nu'\nu}\exp\left[i\mathbf{K}_{S^{-1}}\cdot(\mathbf{X}^{\alpha'}-\mathbf{X}^{\beta'})\right] \quad (11.8.3\text{b})$$

The trace of this representation can be written as

$$\chi_{\text{cell}}(S)=\sum_{\alpha,\beta}\delta\left[\alpha,S(\alpha)\right]\delta\left[\beta,S(\beta)\right]\left[\chi(S)\right]^2\exp\left[i\mathbf{K}_{S^{-1}}\cdot(\mathbf{X}^{\alpha}-\mathbf{X}^{\beta})\right] \quad (11.8.4)$$

where $\chi(S)$ is the character of S in the polar vector representation Γ_1^-, given by Eq. 3.1.8. We refer to the above character as the cell character, since α and β are summed over all particles in a cell.

If Q is a reversal element obeying

$$\mathbf{Q}\mathbf{q}=-\mathbf{q}-\mathbf{K}_{Q^{-1}} \quad (11.8.5)$$

then Eqs. 11.6.2b, 11.6.6b, and 11.2.10 and an interchange of dummy indices μ' and ν' can be combined to yield the condition

$$\mathbf{R}^{\alpha\beta}_{\mu\nu}(\mathbf{q}) = \exp\left[i\mathbf{K}_{Q^{-1}}\cdot(\mathbf{X}^{Q(\alpha)}-\mathbf{X}^{Q(\beta)})\right]R^{Q(\beta)Q(\alpha)}_{\mu'\nu'}(\mathbf{q})Q_{\nu'\mu}Q_{\mu'\nu} \quad (11.8.6b)$$

imposed on $\mathbf{R}(\mathbf{q})$ by each reversal element $(\mathbf{Q}|\mathbf{v}(\mathbf{Q}))$. We can rewrite Eq. 11.8.6 as a matrix representation:

$$QR^{\alpha\beta}_{\mu\nu}(\mathbf{q}) = R^{\alpha'\beta'}_{\mu'\nu'}(\mathbf{q})\Delta_{\alpha'\beta'\mu'\nu',\alpha\beta\mu\nu}(Q) \quad (11.8.7)$$

$$\Delta(Q) = \delta[\alpha', Q(\beta)]\delta[\beta', Q(\alpha)]Q_{\nu'\mu}Q_{\mu'\nu}\exp\left[i\mathbf{K}_{Q^{-1}}\cdot(\mathbf{X}^{\beta'}-\mathbf{X}^{\alpha'})\right] \quad (11.8.8b)$$

with character for the cell

$$\chi_{\text{cell}}(Q) = \sum_{\alpha\beta}\delta[\alpha, Q(\beta)]\delta[\beta, Q(\alpha)]\chi(Q^2)\exp\left[i\mathbf{K}_{Q^{-1}}\cdot(\mathbf{X}^{\beta}-\mathbf{X}^{\alpha})\right] \quad (11.8.9)$$

The number of independent parameters N_{cell} in $\mathbf{R}(\mathbf{q})$ is the number of times that Δ as a representation of the reversal group $G_{\mathbf{q},-\mathbf{q}} = G_{\mathbf{q}} + Q_0 G_{\mathbf{q}}$ contains the identity representation:

$$N_{\text{cell}} = \tfrac{1}{2}(N + N')$$

$$N = \frac{1}{g_{\mathbf{q}}}\sum_{S\in G_{\mathbf{q}}}\chi_{\text{cell}}(S) \quad (11.8.10)$$

$$N' = \frac{1}{g_{\mathbf{q}}}\sum_{Q\in Q_0 G_{\mathbf{q}}}\chi_{\text{cell}}(Q)$$

where Q_0 is any one of the reversal elements. If no Q exists, $N_{\text{cell}} = N$, since only the group $G_{\mathbf{q}}$ is then applicable.

As in the molecular case, we can obtain more specific information [namely, the number $N_{\alpha\beta}$ of independent parameters in $\mathbf{R}^{\alpha\beta}(\mathbf{q})$, for a given α and β] and have simpler computations if we restrict our attention to the reversal group of the bond $G_{\alpha\beta}$ (the subgroup of $G_{\mathbf{q}} + Q_0 G_{\mathbf{q}}$ that leaves $\alpha\beta$ invariant or interchanges α and β):

$$N_{\alpha\beta} = \frac{1}{g_{\alpha\beta}}\sum_{R}\left\{[\chi(R)]^2 J_{\alpha\beta}(R) + \chi(R^2)J'_{\alpha\beta}(R)\right\}$$

$$\times\exp\left[i\mathbf{K}_{R^{-1}}\cdot(\mathbf{X}^{R(\alpha)}-\mathbf{X}^{R(\beta)})\right] \quad (11.8.11)$$

where the factors $J(R), J'(R)$ select the point group rotations R of $(\mathbf{R}|\mathbf{v}(\mathbf{R}))$ that are relevant, that is, the J's vanish except that

$$\begin{aligned} J_{\alpha\beta}(R) &= 1 \quad \text{if} \quad R\mathbf{q} = \mathbf{q} - \mathbf{K}, \quad R(\alpha) = \alpha, \quad R(\beta) = \beta \\ J'_{\alpha\beta}(R) &= 1 \quad \text{if} \quad R\mathbf{q} = -\mathbf{q} - \mathbf{K}, \quad R(\alpha) = \beta, \quad R(\beta) = \alpha \end{aligned} \tag{11.8.12}$$

and

$$g_{\alpha\beta} = \sum_{R} [J_{\alpha\beta}(R) + J'_{\alpha\beta}(R)] \tag{11.8.13}$$

is the order of $G_{\alpha\beta}$, the reversal group of the bond $(\alpha\beta)$. This formula has been rewritten so as to remain valid when $\beta = \alpha$ and/or when no qualified reversal elements exist.

Since $(\mathbf{S}|\mathbf{v}(\mathbf{S}))$ and $(\mathbf{S}|\mathbf{v}(\mathbf{S})+\mathbf{t})$ in Eq. 11.8.2 generate the same representation and character, Eq. 11.8.10 as an *average* over the reversal group $G_{\mathbf{q},-\mathbf{q}}$ reduces (after canceling a factor equal to the order of the translation group) to an average over the reversal algebra

$$A(\mathbf{q}, -\mathbf{q}) = A(\mathbf{q}) + Q_0 A(\mathbf{q}) \tag{11.8.14}$$

where $A(\mathbf{q})$ is the set (algebra) of representative elements $(\mathbf{S}|\mathbf{v}(\mathbf{S}))$ in $G_{\mathbf{q}}$. Indeed, the algebraic viewpoint of Section 8.7 would have led us directly to Eqs. 11.8.10–11.8.13 as averages over representative elements. ▲

11.9. THE METHOD OF LONG WAVES: PRIMITIVE LATTICES

The elastic constants of a crystal can be determined by static and dynamic (sound wave) measurements. The theoretical counterpart of the static measurement consists in applying a strain to the crystal, that is, giving atom $m = (\alpha i)$ a displacement linear in its equilibrium position,

$$u_{\mu}^{m} = u_{\mu a} X_{a}^{m} \tag{11.9.1}$$

If this displacement is inserted into the potential energy (Eq. 11.2.2), the latter will be expressible as a quadratic expression in the displacement gradients $u_{\mu a}$. The elastic constants can be determined from this expression.

Unfortunately, a crystal of linear dimension L will have a number of surface atoms proportional to L^2, each given a displacement proportional to L. Thus direct use of the potential energy formula 11.2.2 endows the surface atoms with an energy proportional to L^3, that is, the volume of the crystal. In calculating the strain energy, therefore, one must not allow the crystal to become infinite in size (or use periodic boundary conditions) until the energy has been expressed in a form insensitive to the boundary. Herring has shown that this can be accomplished by using the infinitesimal rotational invariance of the potential energy (Lax, **1965a**).

The intricacies of a static calculation can be avoided by introducing sound waves into the crystal. These have displacements that do not grow linearly with position but vary sinusoidally. Thus it is possible to deal directly with an infinite crystal, as shown by Born and Huang (**1954**).The sums that appear in such a theory must, of course, show no boundary sensitivity. To see how this comes about, let us simplify our discussion by considering simple Bravais lattices. For these the polarization vectors **b** obey an equation of the form

$$M\omega^2(\mathbf{q})b_\mu = R_{\mu\nu}(\mathbf{q})b_\nu \tag{11.9.2}$$

where M is the mass per cell, and the dynamic matrix $\mathbf{R}(\mathbf{q})$ is defined by

$$R_{\mu\nu}(\mathbf{q}) = \sum_n K_{\mu\nu}^{mn} \exp\left[i\mathbf{q}\cdot(\mathbf{X}^n - \mathbf{X}^m)\right] \tag{11.9.3}$$

In general, $\mathbf{R}(\mathbf{q})^* = \mathbf{R}(-\mathbf{q})$. For a simple Bravais lattice, infinitesimal displacement invariance guarantees $\mathbf{R}(0)=0$, and inversion, which must be present in such lattices, requires $\mathbf{R}(-\mathbf{q})=\mathbf{R}(\mathbf{q})$. Thus $\mathbf{R}(\mathbf{q})$ is, in this case, a real, even function of $\mathbf{q}$.

The long-wave limit is obtained by expanding $\mathbf{R}(\mathbf{q})$ in powers of $\mathbf{q}$. The lowest nonvanishing terms are of order q^2. (And these terms are adequate for a description of sound waves, since the latter have $\omega^2 \propto q^2$.) If we divide Eq. 11.9.2 by the volume Ω_0 of the unit cell, Eq. 11.9.2 becomes

$$\rho_0\omega^2 b_\mu = \hat{C}_{\mu a\nu b}\, q_a q_b b_\nu \tag{11.9.4}$$

where summation is understood on repeated Cartesian indices and ρ_0 is the (unperturbed) density of the crystal, and

$$\hat{C}_{\mu a\nu b} = -\frac{1}{2}\frac{1}{\Omega_0}\sum_n K_{\mu\nu}^{mn}(X^n - X^m)_a (X^n - X^m)_b \tag{11.9.5}$$

This sum is boundary insensitive because it involves only differences of positions. Moreover, it possesses the symmetry

$$\hat{C}_{\mu a\nu b} = \hat{C}_{\mu b\nu a} \tag{11.9.6}$$

Since solutions of the form $\exp(i\mathbf{q}\cdot\mathbf{X} - i\omega t)$ have been assumed, the transition

$$-i\omega \rightarrow \frac{\partial}{\partial t}, \qquad iq_a \rightarrow \frac{\partial}{\partial X_a}, \qquad b_\mu = u_\mu \tag{11.9.7}$$

restores us to the macroscopic wave equation

$$\rho_0 \ddot{u}_\mu = \hat{C}_{\mu a\nu b}\left(\frac{\partial^2 u_\nu}{\partial X_a \partial X_b}\right) \tag{11.9.8}$$

In the absence of initial stress (for a more detailed discussion see Lax, **1965a**) the equation for sound waves is

$$\rho_0 \ddot{u}_\mu = C_{\mu a \nu b} \frac{\partial^2 u_\nu}{\partial X_a \, \partial X_b} \tag{11.9.9}$$

where the elasticity tensor $C_{\mu a \nu b}$ (called C^{MANB} in Section 4.6) is *not* symmetric in (ab). However, the sum over a and b in Eq. 11.9.9 uses only the part of $C_{\mu a \nu b}$ that is symmetric in (ab). Thus, in a sound wave measurement or in a long-wave calculation, we determine $\hat{C}$, and must deduce C from the relations

$$C_{\mu a \nu b} + C_{\mu b \nu a} = 2\hat{C}_{\mu a \nu b} \tag{11.9.10}$$

In general, a full matrix $C_{\mu a \nu b}$ could not be deduced from a knowledge only of the part symmetric in (ab). But we know from Section 4.6 that $C_{\mu a \nu b}$ is symmetric in (μa), in (νb), and in the double exchange (μa) into (νb). Let us interchange μ and a in Eq. 11.9.10 to obtain

$$C_{\mu a \nu b} + C_{a b \nu \mu} = 2\hat{C}_{a \mu \nu b} \tag{11.9.11}$$

The interchange of μ and ν in Eq. 11.9.11 yields

$$C_{\nu a \mu b} + C_{a b \mu \nu} = 2\hat{C}_{a \nu \mu b} \tag{11.9.12}$$

If we add Eqs. 11.9.10 and 11.9.11, subtract 11.9.12, and divide by 2, we obtain

$$C_{\mu a \nu b} = \hat{C}_{\mu a \nu b} + \hat{C}_{a \mu \nu b} - \hat{C}_{a \nu \mu b} \tag{11.9.13}$$

When combined with Eq. 11.9.5, Eq. 11.9.13 yields an expression for the elastic constant in terms of the microscopic force constants.

It is not obvious, now, that these computed elastic constants $C_{\mu a \nu b}$ have all the properties we desire for them unless the $\hat{C}_{\mu a \nu b}$ have certain as yet unknown symmetries. These can be determined directly from Eq. 11.9.10. Interchange μ and ν in 11.9.10. This merely interchanges the two left-hand terms. Hence the right-hand side must be symmetric in μ and ν:

$$\hat{C}_{\mu a \nu b} = \hat{C}_{\nu a \mu b} \tag{11.9.14}$$

a symmetry that will be immediately obvious from the microscopic expression 11.9.5 for $\hat{C}$.

A less obvious set of conditions can be obtained by interchanging μ and a and also ν and b in Eq. 11.9.10:

$$C_{a \mu b \nu} + C_{a \nu b \mu} = 2\hat{C}_{a \mu b \nu} \tag{11.9.15}$$

The left-hand side has not been altered by this interchange. Hence we must conclude that

$$\hat{C}_{\mu a \nu b} = \hat{C}_{a \mu b \nu} \tag{11.9.16}$$

This equation is the macroscopic counterpart of the Born-Huang conditions (in the absence of initial stress). Born and Huang arrived at this condition by assuming that macroscopic and microscopic theories must be in agreement. Leibfried and Ludwig (**1960**) and Hedin (**1960**) have provided a purely microscopic proof of these relations. The microscopic proof uses infinitesimal rotational invariance in precisely the same way as we have used it here. But the microscopic expressions are more complicated, and it is less easy to see how trivially these relations follow from rotational invariance, that is, from the symmetries of $C_{\mu a \nu b}$ derived in Section 4.6.

We regard the Born-Huang conditions 11.9.16, when combined with Eq. 11.9.5, as *additional constraints* imposed by infinitesimal rotational invariance on the microscopic force constants $K_{\mu\nu}^{mn}$. The need for such additional conditions is explained in Lax (**1965a**).

It is easy to verify *now* that expression 11.9.13 for the elastic constants has the desired symmetry in (μa) and in the combined exchange $(\mu\nu)$ and (ab), from which the symmetry in (νb) follows.

We should remark that Herring's static calculation (summarized by Lax, **1965a**) yields results in agreement with the method of long waves in that he finds that the potential energy per unit (undeformed) volume Σ is given (in the absence of initial stress) by

$$\Sigma = \tfrac{1}{2} C_{\mu a \nu b} u_{\mu a} u_{\nu b} \tag{11.9.17}$$

For primitive lattices, Herring's expression for $C_{\mu a \nu b}$ is related to the microscopic force constants via Eqs. 11.9.13 and 11.9.5, that is, the long-wave expression. For slowly varying strains Eq. 11.9.17 must remain valid with $u_{\mu a}$ replaced by $u_{\mu,a} = \partial u_\mu / \partial X^a =$ the local displacement gradient.

As a closing remark, we note that $C_{\mu a \nu b} q_a q_b$ in Eq. 11.9.4 or Eq. 11.9.9 is a homogeneous quadratic function of q. Thus the polarization vectors b_μ will be independent of $q = |\mathbf{q}|$ but dependent on the direction of $\mathbf{q}$:

$$b_\mu(\mathbf{q}) = b_\mu\left(\frac{\mathbf{q}}{q}\right) \tag{11.9.18}$$

Similarly, $\omega^2 \propto q^2$ or

$$\omega(\mathbf{q}) = c\left(\frac{\mathbf{q}}{q}\right) q \tag{11.9.19}$$

Therefore b_μ and ω are not analytic in $\mathbf{q}$ at $\mathbf{q} = 0$, since b_μ and the sound velocity

c depend on the direction $\mathbf{q}/q$ from which $\mathbf{q}=0$ is approached. This nonanalyticity is caused by the triple degeneracy at $\mathbf{q}=0$. The way in which a degeneracy is lifted, in degenerate perturbation theory, depends on the *form* of the perturbation matrix $V_{\mu\nu}$. In this case the *form* of

$$V_{\mu\nu}=C_{\mu a\nu b}q_a q_b=\hat{C}_{\mu a\nu b}q_a q_b \tag{11.9.20}$$

depends on the direction of $\mathbf{q}$ (but not its magnitude).

11.10. ▲ NONPRIMITIVE LATTICES AND INTERNAL SHIFTS

In this section, we wish to determine the static elastic energy associated with a given strain for a nonprimitive lattice. To this end, we impose on atom (αi) the displacement

$$u_\mu^{\alpha i}=u_\mu^\alpha+u_{\mu a}X_a^{\alpha i} \tag{11.10.1}$$

and find from Eq. 11.2.2 an energy per unit (undeformed) volume $\Sigma=V/\Omega$ of the crystal:

$$\Sigma=\frac{1}{2}\begin{bmatrix}\alpha & \beta\\ \mu & \nu\end{bmatrix}u_\mu^\alpha u_\nu^\beta+u_\mu^\alpha\begin{bmatrix}\alpha & & \\ \mu & \nu & b\end{bmatrix}u_{\nu b}+\tfrac{1}{2}\{\,\mu\nu,ab\,\}u_{\mu a}u_{\nu b} \tag{11.10.2}$$

The second term in Eq. 11.10.1 describes an imposed *external strain*. The first term u_μ^α describes an *internal shift*, that is, a motion of the αth sublattice relative to its equilibrium position. The terms in Eq. 11.10.2 describe, respectively, energies proportional to

$$(\text{internal shift})^2+\text{internal shift}\times\text{external strain}+(\text{external strain})^2 \tag{11.10.3}$$

The energy required to produce an external strain will be reduced if we allow an internal adjustment. Indeed, the internal shift will automatically adjust itself to minimize the energy (at fixed external strain):

$$\frac{\partial\Sigma}{\partial u_\mu^\alpha}=0 \tag{11.10.4}$$

If the u_μ^α so computed is reinserted into Eq. 11.10.2, Σ will be expressible as a function only of the external strain:

$$\Sigma=\tfrac{1}{2}C_{\mu a\nu b}u_{\mu a}u_{\nu b} \tag{11.10.5}$$

where

$$C_{\mu a\nu b} = \{\mu a, \nu b\} + (\mu a, \nu b) \tag{11.10.6}$$

The term $(\mu a, \nu b)$ arises from the first two terms in Eq. 11.10.2 and is associated with the internal shift. The term in $\{\mu a, \nu b\}$ is simply the last term in 11.10.2 and describes the energy associated with an external strain when the internal shifts are forced to be zero. If we write out this coefficient in detail, using Eqs. 11.2.2 and 11.10.1, we obtain

$$\{\mu a, \nu b\} = \frac{1}{\Omega} \sum_{\alpha i \beta j} K_{\mu\nu}^{\alpha i, \beta j} X_a^{\alpha i} X_b^{\beta j} \tag{11.10.7}$$

This term is, unfortunately, boundary sensitive. But it is precisely the same term as that for the primitive lattice, with the simple replacements $m \to \alpha i$, $n \to \beta j$. Thus it can be evaluated by Herring's methods, using infinitesimal rotational invariance (see Appendix B of Lax, **1965a**).

A simpler, alternative procedure is to define the part of $\{\mu a, \nu b\}$ that is symmetric in (ab) as

$$\hat{C}_{\mu a\nu b}^{\text{ext}} = \tfrac{1}{2}\{\mu a, \nu b\} + \tfrac{1}{2}\{\mu b, \nu a\} \tag{11.10.8}$$

One can show, using only *infinitesimal* displacement invariance of $\mathbf{K}^{\alpha i\beta j}$, that this expression can be rewritten in the boundary-insensitive form

$$\hat{C}_{\mu a\nu b}^{\text{ext}} = -\frac{1}{2}\frac{1}{\Omega} \sum_{\alpha i \beta j} K_{\mu\nu}^{\alpha i\beta j} (X^{\beta j} - X^{\alpha i})_a (X^{\beta j} - X^{\alpha i})_b \tag{11.10.9}$$

(For proof see Appendix A of Lax, **1965a**.) Since this sum is boundary insensitive, we can now invoke lattice translational invariance to argue that the sum over j is independent of i, so that

$$\hat{C}_{\mu a\nu b}^{\text{ext}} = -\frac{1}{2}\frac{1}{\Omega_0} \sum_{\alpha \beta j} K_{\mu\nu}^{\alpha i\beta j} (X^{\beta j} - X^{\alpha i})_a (X^{\beta j} - X^{\alpha i})_b \tag{11.10.10}$$

Moreover, as in the case of a primitive lattice, we can solve Eq. 11.10.8 for $\{\mu a, \nu b\}$ in terms of $\hat{C}^{\text{ext}}$:

$$\{\mu a, \nu b\} = \hat{C}_{\mu a\nu b}^{\text{ext}} + \hat{C}_{a\mu\nu b}^{\text{ext}} - \hat{C}_{a\nu\mu b}^{\text{ext}} \tag{11.10.11}$$

If we introduce the notation of Born and Huang (**1954**),

$$\hat{C}_{\mu a\nu b}^{\text{ext}} \equiv [\mu\nu, ab] \tag{11.10.12}$$

we can then write the elastic constants (Eq. 11.10.6) in the form

$$C_{\mu a\nu b} = [\,\mu\nu, ab\,] + [\,a\nu, \mu b\,] - [\,a\mu, \nu b\,] + (\,\mu a, \nu b\,) \tag{11.10.13}$$

where the last term is the internal contribution, which we now proceed to evaluate.

Condition 11.10.4 requires the internal shifts to obey

$$\begin{bmatrix} \alpha & \beta \\ \mu & \nu \end{bmatrix} u_\nu^\beta = -\begin{bmatrix} \alpha & & \\ \mu & \nu & b \end{bmatrix} u_{\nu b} \tag{11.10.14}$$

where Eqs. 11.10.1 and 11.10.2 yield

$$\begin{bmatrix} \alpha & \beta \\ \mu & \nu \end{bmatrix} = \frac{1}{\Omega}\sum_{ij} K_{\mu\nu}^{\alpha i\beta j} = \frac{1}{\Omega_0}\sum_{j} K_{\mu\nu}^{\alpha i\beta j} \tag{11.10.15}$$

$$\begin{bmatrix} \alpha & & \\ \mu & \nu & b \end{bmatrix} = \sum_\beta \begin{bmatrix} \alpha & \beta & \\ \mu & \nu & b \end{bmatrix} \tag{11.10.16}$$

where

$$\begin{bmatrix} \alpha & \beta & \\ \mu & \nu & b \end{bmatrix} = \frac{1}{\Omega}\sum_{ij} K_{\mu\nu}^{\alpha i\beta j} X_b^{\beta j}$$

$$= \frac{1}{\Omega_0}\sum_{j} K_{\mu\nu}^{\alpha i\beta j} X_b^{\beta j} \tag{11.10.17}$$

The internal shifts can be obtained from Eqs. 11.10.14 in the form

$$u_\lambda^\gamma = -\begin{bmatrix} \gamma & \delta \\ \lambda & \rho \end{bmatrix}^{-1} \begin{bmatrix} \delta & & \\ \rho & \mu & a \end{bmatrix} u_{\mu a} \tag{11.10.18}$$

or

$$u_\rho^\delta = -\begin{bmatrix} \delta & \gamma \\ \rho & \lambda \end{bmatrix}^{-1} \begin{bmatrix} \gamma & & \\ \lambda & \nu & b \end{bmatrix} u_{\nu b} \tag{11.10.19}$$

If we change the summation indices in the energy expression 11.10.2 so that $u_\mu^\alpha, u_\nu^\beta$ are replaced by $u_\lambda^\gamma, u_\rho^\delta$ and the above shifts are inserted, the energy takes form 11.10.6 with

$$(\mu a, \nu b) = -\begin{bmatrix} \delta & & \\ \rho & \mu & a \end{bmatrix}\begin{bmatrix} \delta & \gamma \\ \rho & \lambda \end{bmatrix}^{-1}\begin{bmatrix} \gamma & & \\ \lambda & \nu & b \end{bmatrix} \tag{11.10.20}$$

In obtaining this result, we have exploited the symmetry of $K_{\mu\nu}^{\alpha i \beta j}$ on the interchange $\alpha i \mu \rightleftarrows \beta j \nu$. It can be shown, using rotational invariance (see Problems 11.10.1–11.10.4) that $(\mu a, \nu b)$ is already symmetric in (μa) and in (νb). It is also symmetric under the double interchange $(\mu a) \rightleftarrows (\nu b)$ and is already expressed in a boundary-insensitive way.

The Method of Long Waves: Nonprimitive Lattices

The contribution of the internal shift energy to the elastic constants will be recomputed in this section, using the method of long waves. We do this (1) to illuminate the relationship between the static and dynamic methods, and (2) to facilitate the computing of the reciprocal matrix in Eq. 11.10.20 by giving this problem a simple physical interpretation.

Let us take the second form (Eq. 11.2.8b) of the dynamical matrix and expand in powers of $\mathbf{q}$, retaining terms to order q^2. The equation of motion 11.2.7, divided by Ω_0, the volume of the unit cell, reduces to

$$\rho_\alpha \omega^2(\mathbf{q}) b_\mu^\alpha = \left(\begin{bmatrix} \alpha & \beta \\ \mu & \nu \end{bmatrix} + \begin{bmatrix} \alpha & \beta & \\ \mu & \nu & b \end{bmatrix} i q_b + \begin{bmatrix} \alpha & \beta & & \\ \mu & \nu & a & b \end{bmatrix} q_a q_b \right) b_\nu^\beta \tag{11.10.21}$$

where $\rho_\alpha = M_\alpha / \Omega_0$,

$$\begin{bmatrix} \alpha & \beta & & \\ \mu & \nu & a & b \end{bmatrix} = -\frac{1}{2}\frac{1}{\Omega_0}\sum_j K_{\mu\nu}^{\alpha i \beta j}(X^{\beta j} - X^{\alpha i})_a (X^{\beta j} - X^{\alpha i})_b \tag{11.10.22b}$$

and the other symbols have been defined in Eqs. 11.10.15–11.10.17. The zeroth-order equations at $\mathbf{q} = 0$:

$$\rho_\alpha \omega^2 b_\mu^\alpha = \begin{bmatrix} \alpha & \beta \\ \mu & \nu \end{bmatrix} b_\nu^\beta \tag{11.10.23}$$

have $3s-3$ "optical" solutions in which the particles in the cell move in relation to one another (and all cells move in phase). The translational invariance condition

$$\sum_{\beta j} \mathbf{K}^{\alpha i \beta j} = 0 \tag{11.10.24}$$

guarantees that

$$\sum_{\beta} \begin{bmatrix} \alpha & \beta \\ \mu & \nu \end{bmatrix} = 0 \tag{11.10.25}$$

The acoustic solution, in which all particles in the cell are given the same motion,

$$b_\mu^\alpha = d_\mu \tag{11.10.26}$$

then has a (triply degenerate) zero eigenvalue:

$$\sum_{\beta} \begin{bmatrix} \alpha & \beta \\ \mu & \nu \end{bmatrix} d_\nu = 0 \tag{11.10.27}$$

corresponding to an absence of restoring force against an arbitrary motion of the crystal as a whole.

If we wish to calculate ω^2 accurate to order q^2, we must calculate $b_\mu^\alpha(\mathbf{q})$ accurate to order q. Thus we write

$$b_\mu^\alpha(\mathbf{q}) = d_\mu + u_\mu^\alpha \tag{11.10.28}$$

where $d_\mu = d_\mu(\mathbf{q}/q)$ is of zeroth order in q, and $u_\mu^\alpha = f_\mu^\alpha(\mathbf{q}/q)q$ is of first order in q, and both may depend on the orientation $\mathbf{q}/q$ of $\mathbf{q}$. We must choose d_μ to be a correct starting vector in the sense of degenerate perturbation theory. This starting approximation is customarily chosen to diagonalize the matrix of the perturbation taken within the degeneracy set. *If* there were no term linear in q in Eq. 11.10.21, the quadratic form to be diagonalized would be

$$\sum d_\mu \begin{bmatrix} \alpha & \beta & & \\ \mu & \nu & a & b \end{bmatrix} q_a q_b d_\nu = \sum_{\mu\nu} d_\mu [\mu\nu, ab] q_a q_b d_\nu$$

where

$$[\mu\nu, ab] = \sum_{\alpha\beta} \begin{bmatrix} \alpha & \beta & & \\ \mu & \nu & a & b \end{bmatrix} = \check{C}_{\mu a \nu b}^{\text{ext}}$$

so that, with d_1, d_2, and d_3 as the basis vectors, the matrix we would have to diagonalize would be

$$V_{\mu\nu}^{(2)} = [\,\mu\nu, ab\,] q_a q_b = \hat{C}_{\mu a \nu b}^{\text{ext}} q_a q_b \tag{11.10.29}$$

just as in Eq. 11.9.20 for the case of a primitive lattice.

The new feature in the present case is the occurrence in Eq. 11.10.21 of a term linear in q. This term seems to raise the possibility that the degeneracy could be lifted to first order in q, leading to an $\omega^2 \propto q$ (which would disagree with experiment). In this case, the form to diagonalize would be

$$\sum d_\mu \begin{bmatrix} \alpha & \beta & \\ \mu & \nu & b \end{bmatrix} i q_b d_\nu$$

However,

$$V_{\mu\nu}^{(1)} = i \sum_{\alpha\beta b} \begin{bmatrix} \alpha & \beta & \\ \mu & \nu & b \end{bmatrix} q_b \equiv 0 \tag{11.10.30}$$

since by Eq. 11.10.17 the sum over α creates a result proportional to $\Sigma_{\alpha i} \mathbf{K}^{\alpha i \beta j}$, which vanishes by displacement invariance.

Thus the degeneracy is not lifted to first order in q. Does this mean that in proceeding to second order we can use $V_{\mu\nu}^{(2)}$ and ignore the terms linear in q? No, because the term linear in $\mathbf{q}$ will couple the overall motion d_μ to some internal ("optical") motion u_μ^α. The internal motion created will be of the first order in q and will couple back to yield a correction of second order in q in the equation for d_μ. This correction should yield the internal contribution $(\mu a, \nu b)$ to the elastic constants found previously by a static argument. This procedure is essentially a generalization of degenerate perturbation theory to higher order, the formalism of which is described in Lax (**1950**) and in Problem 11.10.7.

We can proceed here, without relying on the general formalism, by inserting Eq. 11.10.28 into Eq.11.10.21 and equating the terms of a given order in q. The zeroth-order terms were obtained in Eq. 11.10.23. The first-order terms lead to

$$0 = \begin{bmatrix} \alpha & \beta \\ \mu & \nu \end{bmatrix} u_\nu^\beta + \begin{bmatrix} \alpha & & \\ \mu & \nu & b \end{bmatrix} i q_b d_\nu \tag{11.10.31}$$

with a vanishing left-hand side since ω^2 has been shown to possess no first-order term. If we regard d_ν as a continuous function of $\mathbf{X}$,

$$d_\nu = d_\nu(\mathbf{X}) = d_\nu(0) \exp(i\mathbf{q} \cdot \mathbf{X}) \tag{11.10.32}$$

then

$$iq_b d_\nu = \frac{\partial d_\nu}{\partial X^b} = u_{\nu b} \tag{11.10.33}$$

represents the external strain and u_ν^β the internal shift, and Eq. 11.10.31 reduces to our statically determined relation 11.10.14 between these two!

Our preceeding discussion of the internal shift induced by the external strain was incomplete in that it assumed that the matrix $\begin{bmatrix} \alpha & \beta \\ \mu & \nu \end{bmatrix}$ possesses a reciprocal. We see from Eq. 11.10.27, however, that in fact this matrix possesses three null vectors (vectors with eigenvalue zero) and hence is singular. This difficulty can be readily surmounted, however, by noting that

$$\sum_\alpha \begin{bmatrix} \alpha & & \\ \mu & \nu & b \end{bmatrix} = 0 \tag{11.10.34}$$

so that the inhomogeneous term in Eq. 11.10.14 or 11.10.31 (i.e., the term not involving the u_μ^β) is in fact orthogonal to the null vectors. Thus the inhomogeneous equation 11.10.31 or 11.10.14 can be "solved" using Eq. D 19. The solution then takes the form 11.10.18, provided that the reciprocal is given the form (see Eq. D 15):

$$\begin{bmatrix} \alpha & \beta \\ \mu & \nu \end{bmatrix}^{-1} = \frac{1}{\rho_0} \sum_t{}' \frac{1}{\omega_t^2} (b_\mu^{\alpha t})^* b_\nu^{\beta t} \tag{11.10.35}$$

where $\rho_0 = \Sigma_\alpha M_\alpha / \Omega_0 =$ the density in the underformed state and $b_\mu^{\alpha t}$ is the tth eigenvector of Eq. 11.10.23, that is, the tth mode at $\mathbf{q}=0$. The prime on the sum indicates that the acoustical modes, for which $\omega_t = 0$, are to be omitted. The shift u_μ^α induced by the external strain can be modified by adding an arbitrary null vector (see Eq. D 19), that is, an arbitrary uniform motion. This arbitrariness can be eliminated, and the reciprocal matrix fixed at the value 11.10.35 just given, by requiring that d_μ be the full center of mass motion:

$$d_\mu = \frac{\Sigma \rho_\alpha b_\mu^\alpha(\mathbf{q})}{\rho_0} \tag{11.10.36}$$

with none contained in u_μ^α.

We now take the appropriate linear combination of Eqs. 11.10.21 to obtain an equation for d_μ and retain all terms to second order in q with the result

$$\rho_0\omega^2 d_\mu = \begin{bmatrix} & \beta & \\ \mu & \nu & b \end{bmatrix} iq_b u_\nu^\beta + [\mu\nu, ab] q_a q_b d_\nu \tag{11.10.37}$$

where

$$\begin{bmatrix} & \beta & \\ \mu & \nu & b \end{bmatrix} \equiv \sum_\alpha \begin{bmatrix} \alpha & \beta & \\ \mu & \nu & b \end{bmatrix} = -\begin{bmatrix} \beta & & \\ \nu & \mu & b \end{bmatrix} \tag{11.10.38}$$

If the shift u_ν^β is eliminated via Eq. 11.10.31, we obtain the equation for sound waves:

$$\rho_0\omega^2 d_\mu = \{(\mu a, \nu b) + [\mu\nu, ab]\} q_a q_b d_\nu \tag{11.10.39}$$

with $(\mu a, \nu b)$ given by Eq. 11.10.20, exactly as in the static calculation.

In summary, then, the elastic constants are to be computed using Eq. 11.10.13 with $[\mu\nu, ab] = \hat{C}^{\text{ext}}_{\mu a\nu b}$ given by 11.10.10, and $(\mu a, \nu b)$ given by 11.10.20, using Eq. 11.10.16 for the coupling terms and 11.10.35 for the reciprocal of the singular matrix. ▲

11.11. SUMMARY

Translational invariance makes it possible to set up a running wave description of normal modes whose quantized excitations are referred to as phonons. In addition to carrying energy, these excitations carry a crystal momentum

$$\mathbf{P} = \sum \hbar \mathbf{q} a_t^\dagger(\mathbf{q}) a_t(\mathbf{q}) \tag{11.11.1}$$

which, added to all other crystal momenta (e.g., electron, neutron, photon), gives a total crystal momentum which is conserved (modulo a reciprocal lattice vector).

Infinitesimal displacement invariance is shown to guarantee the existence of three acoustic branches $\omega_t(\mathbf{q})$, $t = 1, 2, 3$, whose frequencies vanish at $\mathbf{q} = 0$.

Under the operation $(\mathbf{S}|\mathbf{v}(\mathbf{S}))$, which takes atom α into $S(\alpha)$, the dynamical matrix 11.2.8 is shown to obey

$$\mathbf{R}^{\alpha\beta}(\mathbf{q}) = \tilde{\mathbf{S}}\mathbf{R}^{S(\alpha)S(\beta)}(\mathbf{S}\mathbf{q})\mathbf{S} \tag{11.11.2b}$$

which relates different elements of the dynamical matrix. Restrictions on given elements can be imposed by requiring $\mathbf{S}\cdot\mathbf{q} \doteq \mathbf{q}$ (equality modulo a reciprocal lattice vector) and using, when necessary,

$$\mathbf{R}^{\alpha\beta}(\mathbf{q}-\mathbf{K}) = \exp[i\mathbf{K}\cdot(\mathbf{X}^{\alpha}-\mathbf{X}^{\beta})]\mathbf{R}^{\alpha\beta}(\mathbf{q}) \tag{11.11.3b}$$

In addition, reversal elements $(\mathbf{Q}|\mathbf{v}(\mathbf{Q}))$ such that $\mathbf{Q}\mathbf{q} \doteq -\mathbf{q}$ can be combined with $R_{\mu\nu}^{\alpha\beta}(-\mathbf{q}) = R_{\mu\nu}^{\alpha\beta}(\mathbf{q})^* = R_{\nu\mu}^{\beta\alpha}(\mathbf{q})$, using the Hermiticity of $\mathbf{R}(\mathbf{q})$, to further restrict $\mathbf{R}(\mathbf{q})$. Even more detailed information is obtained by restricting oneself to the group of the bond, that is, to S such that $S(\alpha)=\alpha$, $S(\beta)=\beta$, and to bond-reversal elements such that $Q(\alpha)=\beta$, $Q(\beta)=\alpha$. Indeed, the number of independent parameters in such a bond is given by Eq. 11.8.11.

In order to construct symmetry coordinates we need relation 11.6.14b for the action of symmetry operators on plane wave coordinates:

$$O(S)\mathbf{b}^{\alpha}(\mathbf{q}) = \mathbf{S}\mathbf{b}^{S^{-1}(\alpha)}(\mathbf{S}^{-1}\cdot\mathbf{q}) \tag{11.11.4b}$$

using $\mathbf{b}^{\alpha}(\mathbf{q}-\mathbf{K}) = \exp(i\mathbf{K}\cdot\mathbf{X}^{\alpha})\mathbf{b}^{\alpha}(\mathbf{q})$ when necessary.

Time reversal combined with a reversal element can often be used to make the dynamical matrix and its eigenvectors real.

The elastic coefficients for nonprimitive as well as primitive lattices are evaluated by the method of long waves. The relation to the static viewpoint is discussed.

problems

11.3.1. Show that the variables a and $a^{\dagger}$ of Eq. 11.3.9 can be modified by a phase transformation

$$a^{\dagger} \to \exp(i\theta)\,a^{\dagger}, \qquad a \to \exp(-i\theta)\,a$$

without affecting the nature or commutation rules of these variables.

11.3.2. Show, using the commutation rules of a and $a^{\dagger}$, that

$$\exp(ia^{\dagger}a\lambda)\,a^{\dagger}\exp(-ia^{\dagger}a\lambda) = a^{\dagger}\exp(i\lambda)$$

$$\exp(ia^{\dagger}a\lambda)\,a\exp(-ia^{\dagger}a\lambda) = a\exp(-i\lambda)$$

11.5.1. Why are there exactly three branches for which $\omega(\mathbf{q}) \to 0$ as $q \to 0$?

11.6.1. Show that the first form (11.2.8a) and the second form (11.2.8b) of the dynamical matrix are related by

$$\mathbf{R}_1^{\alpha\beta}(\mathbf{q}) = \exp[i\mathbf{q}\cdot(\mathbf{X}^{\alpha}-\mathbf{X}^{\beta})]\mathbf{R}_2^{\alpha\beta}(\mathbf{q})$$

Hence show that the first form of the dynamical matrix obeys the symmetry condition

$$\mathbf{R}_1^{\alpha\beta}(\mathbf{q}) = \exp i[\mathbf{q}\cdot(\mathbf{X}^{\alpha}-\mathbf{X}^{\beta}) - \mathbf{Sq}\cdot(\mathbf{X}^{S(\alpha)}-\mathbf{X}^{S(\beta)})]\tilde{\mathbf{S}}\mathbf{R}_1{}^{S(\alpha)S(\beta)}(\mathbf{Sq})\mathbf{S}$$

instead of the simpler condition 11.6.2b.

11.7.1. Representations of the second kind leave invariant an antisymmetric bilinear form. However, the potential energy associated with a lattice vibration is a real symmetric form left invariant by the group. Conclusion: representations of the second kind do not appear among the solutions of lattice vibration problems. Comment on these statements.

11.7.2. If $\mathbf{Qq} = -\mathbf{q} - \mathbf{K}_Q$, show that Eq. 11.7.7b is to be replaced by

$$\mathbf{R}^{\alpha\beta}(\mathbf{q})^* = \tilde{\mathbf{Q}}\mathbf{R}^{Q(\alpha),Q(\beta)}(\mathbf{q})\mathbf{Q}\exp[i\mathbf{K}_Q\cdot(\mathbf{X}^{Q(\beta)}-\mathbf{X}^{Q(\alpha)})]$$

and Eq. 11.7.9b by

$$\Omega_{\nu'\nu}^{\beta'\beta} = \exp(i\mathbf{K}_Q\cdot\mathbf{X}^{\beta'})\,\delta[\beta', Q(\beta)]Q_{\nu'\nu}$$

11.7.3. Using the results of Problem 11.7.2, determine the conditions under which $\mathbf{R}^{\alpha\beta}(\mathbf{q})$ can be made real.

11.7.4. The number of constants in the potential energy of a molecular vibration problem equals the number of *real* constants. In a lattice vibration problem, the same is true of $\mathbf{K}^{\alpha i\beta j}$, since the latter are real. In $\mathbf{R}^{\alpha\beta}(\mathbf{q})$ the constants can be complex. If one calculates the number of constants by a group-theoretic formula including the effects of time reversal, is it the number of real or complex constants that is obtained?

11.8.1. Show that the representation generated by $\mathbf{b}^{\alpha}(\mathbf{q})^*\mathbf{b}^{\beta}(\mathbf{q})$ is that given by Eq. 11.8.3.

11.10.1. Show that the space group operation $(\mathbf{S}|\mathbf{v})$ implies the relationships

$$\begin{bmatrix} \alpha & \beta \\ \mu & \nu \end{bmatrix} = \begin{bmatrix} S(\alpha) & S(\beta) \\ \mu' & \nu' \end{bmatrix} S_{\mu'\mu}S_{\nu'\nu}$$

$$\begin{bmatrix} \alpha & \beta & \\ \mu & \nu & b \end{bmatrix} = \begin{bmatrix} S(\alpha) & S(\beta) & \\ \mu' & \nu' & b' \end{bmatrix} S_{\mu'\mu}S_{\nu'\nu}S_{b'b}$$

$$\begin{bmatrix} \alpha & \beta & & \\ \mu & \nu & a & b \end{bmatrix} = \begin{bmatrix} S(\alpha) & S(\beta) & & \\ \mu' & \nu' & a' & b' \end{bmatrix} S_{\mu'\mu}S_{\nu'\nu}S_{a'a}S_{b'b}$$

11.10.2. Show that the above expressions are symmetric under the interchange $(\alpha\mu)\rightleftarrows(\beta\nu)$.

11.10.3. Use the results of Problems 11.10.1 and 11.10.2 to show that the microscopic space group symmetry of $\mathbf{K}^{\alpha i\beta j}$ implies the corresponding macroscopic point group of the elasticity tensor.

11.10.4. Use infinitesimal rotational invariance to prove that

$$\begin{bmatrix} \alpha & & \\ \mu & \nu & b \end{bmatrix} = \begin{bmatrix} \alpha & & \\ \mu & b & \nu \end{bmatrix}$$

Show that $(\mu a, \nu b) = (a\mu, \nu b) = (\nu b, \mu a)$. See Eq. 11.10.20.

11.10.5. Prove that even in the presence of internal shifts the elasticity tensor possesses the point group symmetry.

11.10.6. If each atom δ carries an effective charge per unit volume q^{δ}, show with the help of Eq. 11.10.19 that the piezoelectric tensor can be written as

$$e_{abc} = -\sum_{\delta} q^{\delta} \begin{bmatrix} \delta & \gamma \\ a & d \end{bmatrix}^{-1} \begin{bmatrix} \gamma & & \\ d & b & c \end{bmatrix}$$

$$= e_{acb}$$

11.10.7. Consider a set of states m (or n) that are nearly or exactly degenerate and a set of states a or b that are outside this set. The total matrix coupling these states, $\mathbf{H}_0+\mathbf{V}$, can be decomposed into matrix elements $\mathbf{H}_0+\mathbf{Y}$ that connect "degenerate" states and those $\mathbf{X}$ that do not:

$$\mathbf{H}_0 = \begin{bmatrix} E_m\delta_{mn} & 0 \\ 0 & E_a\delta_{ab} \end{bmatrix} \qquad \mathbf{Y} = \begin{bmatrix} V_{mn} & 0 \\ 0 & V_{ab} \end{bmatrix} = \mathbf{V}^{\text{in}}$$

$$\mathbf{X} = \begin{bmatrix} 0 & V_{mb} \\ V_{an} & 0 \end{bmatrix} = \mathbf{V}^{\text{out}}$$

Develop a perturbation theory of almost degenerate states by seeking a unitary transformation [$\exp(\mathbf{A})$ with $\mathbf{A}^{\dagger} = -\mathbf{A}$] to a new effective matrix

$$\mathbf{K} = \exp(\mathbf{A})(\mathbf{H}_0+\mathbf{Y}+\mathbf{X})\exp(-\mathbf{A})$$

such that $\mathbf{K}^{out}=0$. Use the arbitrariness that remains to choose $\mathbf{A}^{in}=0$. With $\mathbf{H}=\mathbf{H}_0+\mathbf{Y}$ show that

$$\mathbf{K}^{in}=\mathbf{H}+\frac{1}{2!}[[\mathbf{H},\mathbf{A}],\mathbf{A}]+\frac{1}{4!}[[[[\mathbf{H},\mathbf{A}],\mathbf{A}],\mathbf{A}],\mathbf{A}]+\cdots$$

$$+[\mathbf{X},\mathbf{A}]+\frac{1}{3!}[[[\mathbf{X},\mathbf{A}],\mathbf{A}],\mathbf{A}]+\cdots$$

$$\mathbf{K}^{out}=\mathbf{X}+\frac{1}{2!}[[\mathbf{X},\mathbf{A}],\mathbf{A}]+\ldots$$

$$+[\mathbf{H},\mathbf{A}]+\frac{1}{3!}[[[\mathbf{H},\mathbf{A}],\mathbf{A}],\mathbf{A}]+\ldots$$

Set $\mathbf{K}^{out}=0$ and solve for $\mathbf{A}$ by iteration.

With the notation

$$Z^H_{an}=(E_a-E_n)^{-1}Z_{an}$$

for the solution of

$$[\mathbf{H}_0,\mathbf{A}]=\mathbf{Z}, \qquad \mathbf{A}=\mathbf{Z}^H$$

show that an expansion of $\mathbf{K}$ in powers of $\mathbf{V}$ is given by

$$\mathbf{K}^{in}=\sum_{j=0}^{\infty}\mathbf{K}^{(j)}$$

$$\mathbf{K}^{(0)}=\mathbf{H}_0, \qquad \mathbf{K}^{(1)}=\mathbf{Y}$$

$$\mathbf{K}^{(2)}=\tfrac{1}{2}(\mathbf{X}^H\mathbf{X}-\mathbf{X}\mathbf{X}^H)$$

$$\mathbf{K}^{(3)}=\tfrac{1}{2}\left[\mathbf{X},[\mathbf{Y},\mathbf{X}^H]^H\right]$$

$$\mathbf{K}^{(4)}=\tfrac{1}{2}[\mathbf{X},\mathbf{A}^{(3)}]-\tfrac{1}{12}[[\mathbf{K}^{(2)},\mathbf{X}^H],\mathbf{X}^H]$$

where

$$\mathbf{A}^{(3)}=\left[[\mathbf{Y},\mathbf{X}^H]^H,\mathbf{Y}\right]^H+\tfrac{2}{3}[\mathbf{K}^{(2)},\mathbf{X}^H]^H$$

Show that

$$K_{mn}^{(2)} = \tfrac{1}{2} V_{ma} V_{an}\left[(E_m - E_a)^{-1} + (E_n - E_a)^{-1} \right]$$

$$K_{mn}^{(3)} = \frac{1}{2} \frac{V_{ma} V_{ab} V_{bn}}{(E_a - E_n)(E_b - E_n)} - \frac{1}{2} \frac{V_{ma} V_{ap} V_{pn}}{(E_a - E_p)(E_a - E_n)}$$

$$- \frac{1}{2} \frac{V_{mp} V_{pa} V_{an}}{(E_m - E_a)(E_p - E_a)} + \frac{1}{2} \frac{V_{ma} V_{ab} V_{bn}}{(E_m - E_a)(E_m - E_b)}$$

where p is in the same set as m and n. Evaluate $K_{mn}^{(4)}$ and verify that it reduces in the non-degenerate case to the fourth-order Rayleigh-Schrödinger perturbation result given by Brueckner (**1959**). For the fully degenerate case, compare with Soliverez (**1969**). A different approach is given by Roussy (**1973**). Explicit results for $\mathbf{K}^{(2)}$ have been obtained by Fawcett and McLean (**1965**). For an application of $\mathbf{K}^{(2)}$ see Eq. 7.5.7.

11.10.8. If the atoms of a lattice containing several atoms per unit cell are charged, write an expression for the piezoelectricity coefficient in terms of the internal displacements.

chapter twelve

VIBRATIONS OF LATTICES WITH THE DIAMOND STRUCTURE

12.1. FORCE CONSTANTS AND THE DYNAMICAL MATRIX

To illustrate the methods we developed in Chapter 11, we consider the vibrations of lattices with the diamond structure at the points $(q,0,0)$, that is Δ, including the point $X=(2\pi/a)(1,0,0)$ on the zone boundary; at the points $(q,q,0)$, that is, Σ; and at $\Lambda=(q,q,q)$, including $L=(2\pi/a)(\frac{1}{2},\frac{1}{2},\frac{1}{2})$. (See Figs. 11.1.1 and 11.1.2.) The diamond structure has been chosen because it illustrates all the interesting idiosyncrasies associated with a nonsymmorphic space group. We discuss first the points $(q,0,0)$ and show, using the character table, that regardless of the range of the forces the modes of vibration are pure longitudinal and pure transverse. We also discuss the relative phase of the motion of the two particles in the unit cell as we move out toward the zone boundary. (Time reversal has already been used to show that these amplitudes have the same magnitude, Eq. 11.7.15.) We learn that the relative phase for the longitudinal modes is fixed by symmetry alone! The phase for the transverse modes somewhat more conventionally depends on the force constants. Thus the symmetry of the transverse acoustic mode at the zone boundary can be X_3 or X_4, depending on the force constants.

We then discuss the $\Sigma \equiv (q,q,0)$ case to show that the symmetry of the modes can be predicted with the help of the character table even when it is not obvious intuitively: one mode will, for example, be a mixture of longitudinal acoustic and transverse optical. Previous authors[1] considering the 6×6 dynamical matrix at Σ have succeeded only in reducing it to a 2×2 problem and a 4×4 problem. We show that, in fact, it reduces to two 2×2 problems and two 1×1 problems without a priori knowledge of the force constants. The last two statements were included in the 1960-1961 version of the lecture notes which preceded this book, and the results at Σ were given in Lax (**1965b**). Since then, Maradudin and Vosko (**1968**) and others applying group-theoretic techniques have obtained the same result.

Although the form of our results will be independent of the range of the forces, for the sake of definiteness we indicate that the first-, second-, and third-neighbor forces connecting an atom at the origin (0,0,0) to atoms at $(a/4)(1,1,1)$, $(a/4)(2,2,0)$, and $(a/4)(-1,-1,-3)$ are, respectively:

$$(1,1,1)=\begin{bmatrix} \alpha & \beta & \beta \\ \beta & \alpha & \beta \\ \beta & \beta & \alpha \end{bmatrix}, \qquad (2,2,0)=\begin{bmatrix} \mu & \nu & \delta \\ \nu & \mu & \delta \\ \bar{\delta} & \bar{\delta} & \lambda \end{bmatrix},$$

$$(-1,-1,-3)=\begin{bmatrix} \mu' & \nu' & \delta' \\ \nu' & \mu' & \delta' \\ \delta' & \delta' & \lambda' \end{bmatrix} \qquad (12.1.1)$$

The form of the last matrix can be readily determined by a procedure similar to that used in Examples 9.3.2 and 9.4.1 to obtain the first two. The notation has been chosen to agree with that of Herman (**1959**), since he has made an extensive analysis out to fifth neighbors. The translational symmetry of the problem is exhausted by diagonalizing the $(\boldsymbol{\epsilon}|\mathbf{t})$, that is, by picking a $\mathbf{q}$ and constructing the dynamical matrix $\mathbf{R}(\mathbf{q})$. Equation 11.6.1 can be used to determine the force constants for all atoms in a shell from that of the prototype. Equation 11.2.8b for $\mathbf{R}(\mathbf{q})$ can then be summed shell by shell to obtain the

[1]See, for example, Smith (**1948**) and Herman (**1959**).

dynamical matrix. Typical elements are given by

$$R_{11}^{11}(\mathbf{p}) = 4\alpha + 8\mu' + 4\lambda' + 4\mu[2 - \cos\pi p_1(\cos\pi p_2 + \cos\pi p_3)]$$

$$+ 4\lambda(1 - \cos\pi p_2 \cos\pi p_3)$$

$$R_{12}^{11}(\mathbf{p}) = 4\nu \sin\pi p_1 \sin\pi p_2 - 4i\delta \sin\pi p_3(\cos\pi p_1 - \cos\pi p_2)$$

$$R_{11}^{1\bar{1}}(\mathbf{p}) = -4\alpha\left(\cos\frac{\pi}{2}p_1 \cos\frac{\pi}{2}p_2 \cos\frac{\pi}{2}p_3 - i\sin\frac{\pi}{2}p_1 \sin\frac{\pi}{2}p_2 \sin\frac{\pi}{2}p_3\right)$$

$$-4\mu'\left[\cos\frac{\pi}{2}p_1\left(\cos\frac{\pi}{2}p_2 \cos\frac{3\pi}{2}p_3 + \cos\frac{3\pi}{2}p_2 \cos\frac{\pi}{2}p_3\right)\right.$$

$$\left. + i\sin\frac{\pi}{2}p_1\left(\sin\frac{\pi}{2}p_2 \sin\frac{3\pi}{2}p_3 + \sin\frac{3\pi}{2}p_2 \sin\frac{\pi}{2}p_3\right)\right]$$

$$-4\lambda'\left(\cos\frac{3\pi}{2}p_1 \cos\frac{\pi}{2}p_2 \cos\frac{\pi}{2}p_3 + i\sin\frac{3\pi}{2}p_1 \sin\frac{\pi}{2}p_2 \sin\frac{\pi}{2}p_3\right) \qquad (12.1.2)$$

$$R_{12}^{1\bar{1}}(\mathbf{p}) = 4\beta\left(\cos\frac{\pi}{2}p_3 \sin\frac{\pi}{2}p_1 \sin\frac{\pi}{2}p_2 - i\sin\frac{\pi}{2}p_3 \cos\frac{\pi}{2}p_1 \cos\frac{\pi}{2}p_2\right)$$

$$+4\nu'\left(\cos\frac{3\pi}{2}p_3 \sin\frac{\pi}{2}p_1 \sin\frac{\pi}{2}p_2 + i\sin\frac{3\pi}{2}p_3 \cos\frac{\pi}{2}p_1 \cos\frac{\pi}{2}p_2\right)$$

$$+4\delta'\left[\cos\frac{\pi}{2}p_3\left(\sin\frac{\pi}{2}p_1 \sin\frac{3\pi}{2}p_2 + \sin\frac{3\pi}{2}p_1 \sin\frac{\pi}{2}p_2\right)\right.$$

$$\left. + i\sin\frac{\pi}{2}p_3\left(\cos\frac{\pi}{2}p_1 \cos\frac{3\pi}{2}p_2 + \cos\frac{3\pi}{2}p_1 \cos\frac{\pi}{2}p_2\right)\right]$$

where

$$\mathbf{q} = \left(\frac{2\pi}{a}\right)\mathbf{p} \qquad (12.1.3)$$

The remaining elements may be obtained from these via Eqs. 11.6.7–11.6.10 and need not be listed separately.[2]

Since the even neighbors form one face-centered cubic lattice, and the odd neighbors form another, starting at $\tau = (a/4)(1,1,1)$, we find (1) that only odd neighbor force constants enter $\mathbf{R}^{1\bar{1}}$, and (2) except for the constant terms that remain when $\mathbf{p} \to 0$, which are computed using displacement invariance (Eq. 11.5.4), only even neighbor force constants enter $\mathbf{R}^{11}$.

Since translational invariance has been exhausted, the remaining symmetry lies in the algebra of the representative elements $(\boldsymbol{\alpha}|\mathbf{v}(\boldsymbol{\alpha}))$. Moreover, having fixed $\mathbf{q}$, we can restrict ourselves to $\boldsymbol{\alpha}$ such that $\boldsymbol{\alpha}\mathbf{q} \doteq \mathbf{q}$, that is, to the elements of the point group of the wave vector $P(\mathbf{q})$. Step 1 of the procedure outlined in Chapter 5 for treating a dynamical problem has been accomplished: the relevant symmetry elements are $(\boldsymbol{\alpha}|\mathbf{v}(\boldsymbol{\alpha}))$ or $O(\boldsymbol{\alpha})$ for $\boldsymbol{\alpha} \in P(\mathbf{q})$. (To these, of course, should be added the reversal elements times time reversal.)

As step 2 we choose the space of our problem to be the six-dimensional space spanned by the vectors $\mathbf{b}^1(\mathbf{q}), \mathbf{b}^{\bar{1}}(\mathbf{q})$. As step 3, the transformation properties of $\mathbf{b}^{\alpha}(\mathbf{q})$ under the group operations $(\mathbf{S}|\mathbf{v}(\mathbf{S}))$ have already been established in Theorem 11.6.1. The choice

$$\mathbf{b}^1(\mathbf{q})^* = \mathbf{b}^{\bar{1}}(\mathbf{q}) \tag{12.1.4}$$

(see Section 11.7) at all points $\mathbf{q}$ at which ιK produces no additional degeneracy exhausts the information to be obtained from time reversal plus reversal elements.

The characters and irreducible representations spanned by $\mathbf{b}^1(\mathbf{q}), \mathbf{b}^{\bar{1}}(\mathbf{q})$ (steps 4 and 5) must now be carried out separately at each point $\mathbf{q}$ of interest.

12.2. SYMMETRY OF VIBRATIONS AT $\Delta = (q, 0, 0)$

An examination of the complete table of representative elements for the diamond space group (Table 9.1.1) permits us to select those that are within $G_{\mathbf{q}}$. The point group of the wave vector $P(\mathbf{q})$ is simply the group $\mathcal{C}_{4v}$ with the axis along the $(1,0,0)$ direction.

Since Δ is an interior point of the Brillouin zone, we can use the character table of $\mathcal{C}_{4v}$ from Appendix A, provided that we regard the $O(S)$ as the group elements; see Section 8.9. It is conventional, however, to list the elements as $(S|\mathbf{v}(S))$ and to state the characters of these elements.[3] This leads to Table 12.2.1

[2]In contrast to Herman (**1959**), we have used the second kind of dynamical matrix (11.2.8b), since the symmetry requirements that it obeys, Eq. 11.6.2b, are simpler than the corresponding requirements, Prob. 11.6.1, for the first kind of dynamical matrix.

[3]These elements do not usually form a group. In ordinary group-theoretical parlance the "character table" is the character table of the factor group $G_{\mathbf{q}}/T_{\mathbf{q}}$.

Table 12.2.1. Group Characters at $\Delta=(q,0,0)$

	Δ_1	Δ_2	Δ_2'	Δ_1'	Δ_5	Γ_{15}^-	Γ_{25}^+	X_1	X_2	X_3	X_4
$(\epsilon\|0)$	1	1	1	1	2	3	3	2	2	2	2
$(\delta_{2x}\|0)$	1	1	1	1	-2	-1	-1	2	2	-2	-2
$(\delta_{4x},\delta_{4x}^{-1}\|\tau)$	λ	$-\lambda$	$-\lambda$	λ	0	1	-1	0	0	0	0
$(\rho_y,\rho_z\|\tau)$	λ	λ	$-\lambda$	$-\lambda$	0	1	-1	0	0	0	0
$(\rho_{yz},\rho_{\bar{y}z}\|0)$	1	-1	1	-1	0	1	1	2	-2	0	0
$(\epsilon\|\mathbf{t}_{xy})(\alpha\|\mathbf{t})$	$\lambda^2\chi(\alpha\|\mathbf{t})$										

where $\lambda=\exp(-iqa/4)$, $\mathbf{t}_{xy}=(a/2)(1,1,0)$; at X, $\lambda=-i$

with the extra phase factors λ. The representations of this point group are labeled $\Delta_1,\Delta_2,\Delta_1',\Delta_2',\Delta_5$ to remind us that they are associated with the point Δ in the Brillouin zone. We have also included in this table, for future reference, some representations of Γ and X that are not necessarily irreducible with respect to the group at Δ.

By calculating characters (step 4) we can determine which of the Δ representations are contained in the six-dimensional space $[\mathbf{b}^1(\mathbf{q}),\mathbf{b}^{\bar{1}}(\mathbf{q})]$ (step 5). Theorem 11.6.1 essentially tells us that the calculation at any finite q leads to the same result as that at $(q,0,0)$ with $q\to 0$, at which point direct use can be made of the point group $\mathcal{C}_{4v}$. However, we have already investigated the problem at $\mathbf{q}=0$ and found, using characters in Section 9.5, that the six-dimensional space $(\mathbf{u}^1,\mathbf{u}^{\bar{1}})$ contains the two representations $\Gamma_{15}^-=$acoustic modes and $\Gamma_{25}^+=$optical modes. We therefore use the character table 12.2.1 at Δ and $q=0$ with characters for Γ_{15}^- and Γ_{25}^+ transferred from Table 9.5.1 to obtain the compatibility relations[4]

$$\text{acoustic:}\quad \Gamma_{15}^-\equiv\Gamma_{15}=\Delta_1+\Delta_5$$

$$\text{optical:}\quad \Gamma_{25}^+\equiv\Gamma_{25}'=\Delta_2'+\Delta_5 \tag{12.2.1}$$

Our physical space containing $\Gamma_{15}+\Gamma_{25}'$ thus contains Δ_1,Δ_2' once and Δ_5 twice. Since Δ_5 is two dimensional, we may guess that it corresponds to the transverse vibrations. It appears twice, once for transverse optical (TO) and once for transverse acoustic (TA). Since these modes have the same symmetry, they will interact, and their forms will depend on the force constants. By contrast, Δ_1 and Δ_2', which will be shown shortly to be longitudinal acoustic (LA) and longitudinal optic (LO), respectively, do not interact, and each appears once *in the*

[4]The notation $\Gamma_{25}'=\Gamma_{25}$ is evidently a mnemonic device introduced by Bouckaert, Smoluchowski, and Wigner (1936) just to remind us of these compatibility relations: $\Gamma_{25}'=\Delta_2'+\Delta_5$.

direction $(q,0,0)$. Thus the form of these modes is *uniquely determined by symmetry alone*. The Δ_5 modes appearing *twice* will involve the solution of a *pair* of simultaneous equations. The 6×6 problem is reduced to a 2×2 problem (and another identical 2×2) and two 1×1 problems (see Eqs. 12.3.8–12.3.15).

The Longitudinal Modes: $\Delta_1 = \text{LA},\ \Delta_2' = \text{LO}$

We can determine the symmetry vectors by projecting into each irreducible representation. We start with the mode Δ_1. If the mode Δ_1 is contained in the space, we shall obtain its form. If it is absent, we shall obtain zero. Since Δ_1 is one dimensional, we need not use the elaborate machinery of projection operations but may simply require that our mode be an eigenvector of each operator $(\mathbf{S}|\mathbf{v})$ with the eigenvalue $\chi^{\Delta_1}(\mathbf{S}|\mathbf{v})$, or, more simply, in accord with Theorem 11.6.1, an eigenvector of $O(S)$ regarded as a point operation S with eigenvalue $\chi^{\Delta_1}[O(S)]$ namely, the value in Table 12.2.1 with λ set equal to unity! In applying Theorem 11.6.1, it is useful to know that all operations of the form $(\mathbf{S}|0)$ are point operations about a lattice point that leave the lattices alone, whereas operations of the form $(\mathbf{S}|\boldsymbol{\tau})$ [with $\boldsymbol{\tau}=a/4(1,1,1)$] interchange the two sublattices. For these $S^{-1}(\alpha)=\bar{\alpha}$. Using the notation

$$(\boldsymbol{\delta}_{2x}|0)\left(b_x^1,b_y^1,b_z^1;\ b_x^{\bar{1}},b_y^{\bar{1}},b_z^{\bar{1}}\right)=\left(b_x^1,-b_y^1,-b_z^1;\ b_x^{\bar{1}},-b_y^{\bar{1}},-b_z^{\bar{1}}\right) \quad (12.2.2)$$

we see that the requirement that $\boldsymbol{\delta}_{2x}$ have eigenvalue $+1$ forces Δ_1 to have the longitudinal form $(b_x^1,0,0;\ b_x^{\bar{1}},0,0)$. Modes Δ_2, Δ_2', and Δ_1' also have $\boldsymbol{\delta}_{2x}=1$ and the same longitudinal form. Table 12.2.1 shows (see Problem 12.2.1) that, in representation Δ_5, $\boldsymbol{\delta}_{2x}$ is diagonal with eigenvalue -1, causing the latter mode to be pure transverse.

A longitudinal mode in the $(q,0,0)$ direction is clearly even under $(\boldsymbol{\rho}_{\bar{y}z}|0)$, that is, $y\to z$, $z\to y$, so that Δ_2 and Δ_1', which are odd, *cannot be lattice vibrations* in the diamond structure. The absence of these modes was established in Eq. 12.2.1. We will not bother to project into modes that we know are absent.

Since we can write

$$O(\boldsymbol{\rho}_y)\left(b_x^1,0,0;\ b_x^{\bar{1}},0,0\right)=\left(b_x^{\bar{1}},0,0;\ b_x^1,0,0\right) \quad (12.2.3)$$

the eigenvalue $+1$ of $O(\boldsymbol{\rho}_y)$ associated with Δ_1 makes $b_x^{\bar{1}}=b_x^1$, whereas the eigenvalue -1 of Δ_2' makes $b_x^{\bar{1}}=-b_x^1$. Thus Δ_1 is called LA and Δ_2' is called LO, and we can write

$$\begin{aligned}\Delta_1&=\text{LA}=(1,0,0;\ 1,0,0)\\ \Delta_2'&=\text{LO}=(1,0,0;\ -1,0,0)\end{aligned} \quad (12.2.4)$$

(See, however, Problem 12.2.3.) As anticipated, these modes, which are not repeated in the space, have a form that is dictated by symmetry alone. We have the unusual result that LA and LO modes are not coupled by the dynamics. For Δ_1, the $\mathbf{b}^1(\mathbf{q})$ (first three entries in Eqs. 12.2.4) and the $\mathbf{b}^{\bar{1}}(\mathbf{q})$ (last three entries in 12.2.4) are precisely in phase for all $\mathbf{q}$. In terms of the original displacements, we find, for both Δ_1 and Δ_2' that

$$\frac{u^{\bar{1}}}{u^1}=\frac{\exp\left[i\mathbf{q}\cdot\left(\mathbf{X}^{\bar{1}}-\mathbf{X}^1\right)\right]b^{\bar{1}}}{b^1}=\frac{\exp\left(i\mathbf{q}\cdot\boldsymbol{\tau}\right)b^{\bar{1}}}{b^1} \tag{12.2.5}$$

At the zone boundary point X, $\mathbf{q}=(2\pi/a)(1,0,0)$ and $\exp(i\mathbf{q}\cdot\boldsymbol{\tau})=i$, so that

$$\frac{u^{\bar{1}}}{u^1}=\frac{ib^{\bar{1}}}{b^1} \tag{12.2.6}$$

Thus the actual displacements of the particles at X in the two modes are

$$\begin{aligned}\Delta_1(X)&=(1,0,0;\ i,0,0)\\ \Delta_2'(X)&=(1,0,0;\ -i,0,0)\end{aligned} \tag{12.2.7}$$

so that at the zone boundary it is difficult to tell which of these modes is optical and which is acoustic. Indeed, the compatibility relation

$$X_1=\Delta_1+\Delta_2' \tag{12.2.8}$$

obtained with the help of Table 12.2.1 with $\lambda=-i$, shows that these modes become degenerate at this point. Figure 12.2.1 displays the experimental spectrum for germanium obtained by neutron scattering, with the symmetry classifications entered on the diagram.

The Transverse Modes: Δ_5

Only $(\epsilon|0)$ and $(\delta_{2x}|0)$ have nonvanishing characters in Δ_5. Since $(\epsilon-\delta_{2x})(x,y,z)=2(0,y,z)$, projection into the Δ_5 representation via Eq. 3.7.11 yields a mode of the form

$$\left(0,b_y^1,b_z^1;\ 0,b_y^{\bar{1}},b_z^{\bar{1}}\right) \tag{12.2.9}$$

We obtain four simultaneous equations in four unknowns, because the representation Δ_5 is repeated. Our orthogonality theorem 3.4.3 guarantees, however, that different partner functions of a given representation do not interact.

Provided that we choose corresponding partner functions to transform alike, Eq. 3.4.10 forces the 4×4 system to decompose into two *identical* 2×2 systems!

The simplest way to achieve this, as in our treatment of molecular vibrations, Section 5.2, is to force the partners of a representation to be eigenvectors of some operators of the group. (The partners are then labeled by the corresponding eigenvalues.) The simplest remaining operator is $(\rho_{\bar{y}z}|0)$, which simply interchanges y and z. For $\rho_{\bar{y}z} = -1$, we have two (odd) modes of the form

$$\left(0, b_y^{\mathrm{I}}, -b_y^{\mathrm{I}}; 0, b_y^{\bar{\mathrm{I}}}, -b_y^{\bar{\mathrm{I}}}\right) \tag{12.2.10}$$

one optical and one acoustical, depending on $b_y^{\bar{\mathrm{I}}}/b_y^{\mathrm{I}}$. By setting $b_y^{\bar{\mathrm{I}}} = 0$, we obtain one odd symmetry vector of Δ_5:

$$\mathbf{e}_1^5 = (0, 1, -1; 0, 0, 0) \tag{12.2.11}$$

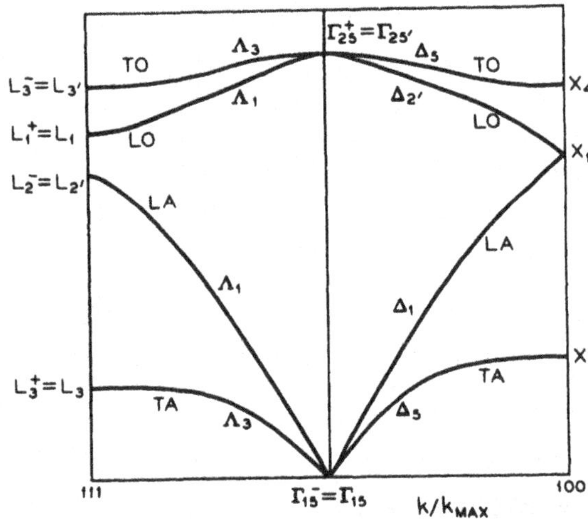

Fig. 12.2.1. The lattice vibration spectrum of germanium as determined by Brockhouse and Iyengar (**1958**) See Fig. 11.1.1. The symmetry assignment is slightly ambiguous and is made as follows: LO > LA; therefore L_1 or L_2', whichever is higher, is LO. Similarly TO > TA; therefore the higher of L_3' and L_2 (and of X_3 and X_4) is to be assigned to TO. On a nearest-neighbor model with central and noncentral forces α and β, the frequency (or rather $M\omega^2$) for each symmetry type is given by $X_4 = 4\alpha + 4\beta$, $X_1 = 4\alpha$, $X_3 = 4\alpha - 4\beta$; $L_3' = 6\alpha + 2\beta$, $L_1 = 2\alpha + 4\beta$, $L_2' = 6\alpha - 4\beta$, $L_3 = 2\alpha - 2\beta$. The choice $\beta > 0$ is required to obtain a sensible fit to the vibration spectrum, even when longer-range forces are present [Herman, (**1959**)]. This fixes all assignments except L_1 and L_2', which are close. The choice $L_2' =$ LA is determined by a comparison between the selection rules for indirect optical absorption in Germanium (Lax and Hopfield, **1961**, Fig. 2) and experiment (Haynes, Lax, and Flood, **1959**). This choice is consistent with $\beta > \frac{1}{2}\alpha$. Central forces would require $\beta = \alpha$. Lax(**1965b**) found $\beta/\alpha \sim 0.6$ fitting the spectrum, and Cochran (**1959**) obtained $\beta/\alpha \sim 0.7$.

Acting on $\mathbf{e}_1^5$ with $(\boldsymbol{\rho}_z|\boldsymbol{\tau})$, we obtain a new symmetry vector in the same irreducible space:

$$\mathbf{e}_2^5=(0,0,0;\ 0,1,1) \tag{12.2.12}$$

which is orthogonal to $\mathbf{e}_1^5$ and is even in $\boldsymbol{\rho}_{\bar{y}z}$ and hence can be used as the even partner $\mathbf{e}_2^5$ corresponding to $\mathbf{e}_1^5$.

An independent odd partner can be obtained by setting $b_y^1=0$ in Eq. 12.2.10. Forming the even partner by acting with $(\boldsymbol{\rho}_z|\boldsymbol{\tau})$ as before, we obtain a set

$$\mathbf{e}_1^{5'}=(0,0,0;\ 0,1,-1), \qquad \mathbf{e}_2^{5'}=(0,1,1;\ 0,0,0) \tag{12.2.13}$$

that transforms precisely as does our first set, Eqs. 12.2.11 and 12.2.12. The combinations $\mathbf{e}_\mu^{5''}=a\mathbf{e}_\mu^5+b\mathbf{e}_\mu^{5'}$ will transform in a manner identical to $\mathbf{e}_\mu^5$ or $\mathbf{e}_\mu^{5'}$ provided that a and b are independent of μ. In other words, if the odd solution $(\mu=1)$ is

$$(0,a,-a;\ 0,b,-b) \tag{12.2.14}$$

the even solution $(\mu=2)$ must be

$$(0,b,b;\ 0,a,a) \tag{12.2.15}$$

with the same frequency. Thus the $\mu=1$ mode is polarized in the 45° direction $(0,1,\bar{1})$. The $\mu=2$ mode, which is degenerate with it, is polarized at right angles to the first, that is, $(0,1,1)$, and the motions of the two particles in the cell have been interchanged.

The above modes are polarized at $\pm 45°$ relative to the cubic axes, that is, the two particles move along parallel directions. For point groups or symmorphic space groups, such a degeneracy between a pair of transverse modes is produced by a principal axis C_n, $n>2$. Thus, if $\mathbf{b}$ is one polarized mode, $\mathbf{b}$ and $C_n\mathbf{b}$ can be combined to produce a mode polarized in any direction in the plane defined by $\mathbf{b}$ and $C_n\mathbf{b}$. For the diamond group, however, the degeneracy between modes 12.2.14 and 12.2.15 is produced by the fourfold screw operation $(\boldsymbol{\delta}_{4x}|\boldsymbol{\tau})$, which interchanges the roles of the two particles in addition to producing a rotation. As a consequence, we arrive at the unusual result that no linear combination of modes 12.2.14 and 12.2.15 will yield plane polarized modes along any orientations except the starting ones.

The Zone Boundary Transverse Modes: X_3 and X_4

The compatibility relations

$$X_3=\Delta_5, \qquad X_4=\Delta_5 \tag{12.2.16}$$

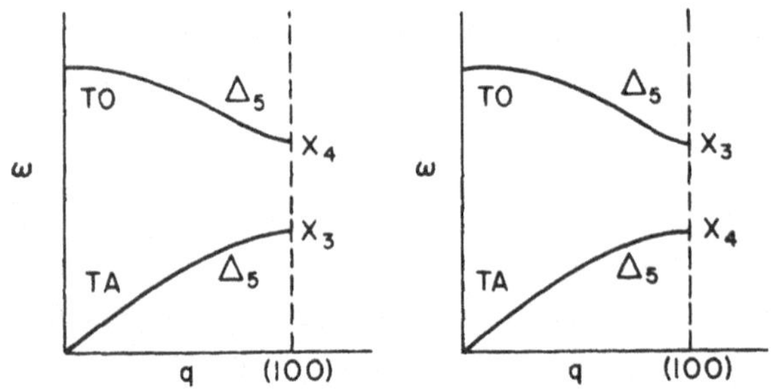

Fig. 12.2.2. Transverse vibrations in a diamond structure along $\Delta=(q,0,0)$, showing the possibilities TA$=X_3$ or TA$=X_4$ at the zero boundary. See legend for Fig. 12.2.1.

implied by Table 12.2.1 show that Δ_5 becomes either X_3 or X_4 at the zone boundary. The possibility that both sets of Δ_5 modes become degenerate at the zone boundary, giving a fourfold degeneracy there, is ruled out (barring accidental degeneracy) because the relevant modes at X, X_3, and X_4 are only twofold degenerate. There is, then, no symmetry reason for expecting a fourfold degeneracy. We are therefore left with the two possibilities shown in Fig. 12.2.2.

We can determine the nature of X_3 by projecting into it. We have already done part of this by projecting into Δ_5. The character table for X (Table 8.10.1) shows that $(\boldsymbol{\delta}_{2\bar{y}z}|\boldsymbol{\tau})$ is the only operator among the *relevant* classes that is not in the group at Δ. Indeed, it is the only relevant element that distinguishes between X_3 and X_4: its character is -2 for X_3 and $+2$ for X_4, that is, X_3 and X_4 modes are eigenvectors of $(\boldsymbol{\delta}_{2\bar{x}y}|\boldsymbol{\tau})$ with eigenvalues -1 and $+1$. See Prob. 12.2.1. We can therefore impose these requirements instead of using a projection procedure.

Equation 11.6.15 permits us to write

$$(\boldsymbol{\delta}_{2\bar{y}z}|\boldsymbol{\tau})\mathbf{b}^{\alpha}(\mathbf{q})=\exp(-i\mathbf{q}\cdot\boldsymbol{\tau})\boldsymbol{\delta}_{2\bar{y}z}\mathbf{b}^{\bar{\alpha}}(-\mathbf{q}) \qquad (12.2.17)$$

At the zone boundary point X, $-\mathbf{q}=\mathbf{q}-\mathbf{K}$ and Eq. 11.6.18 yields

$$\mathbf{b}^{\bar{\alpha}}(-\mathbf{q})\equiv\mathbf{b}^{\bar{\alpha}}(\mathbf{q}-\mathbf{K})=\exp(i\mathbf{K}\cdot\mathbf{X}^{\bar{\alpha}})\mathbf{b}^{\bar{\alpha}}(\mathbf{q}) \qquad (12.2.18)$$

It is now *essential* to stick to our *convention* that the origin of coordinates is at one atom

$$\mathbf{X}^{1}=0, \qquad \mathbf{X}^{\bar{1}}=\boldsymbol{\tau}=\left(\frac{a}{4}\right)(1,1,1) \qquad (12.2.19)$$

with the result[5]

$$(\boldsymbol{\delta}_{2\bar{y}z}|\boldsymbol{\tau})\mathbf{b}^{1}=i\left(0,-b_z^{\bar{1}},-b_y^{\bar{1}}\right) \qquad (12.2.20)$$

[5]A consistent shift of notation—origin$\to\frac{1}{2}(\mathbf{X}^{1}+\mathbf{X}^{\bar{1}})$, $(\boldsymbol{\delta}_{2\bar{y}z}|\boldsymbol{\tau})\to(\boldsymbol{\delta}_{2\bar{y}z}|0)$, $\mathbf{X}^{\bar{1}}\to\frac{1}{2}\boldsymbol{\tau}$—leads to the same physical result, Eq. 12.2.20.

For X_3, $\delta_{2\bar{y}z}=-1$ yields the condition

$$b_z^{\bar{1}}=-ib_y^1 \tag{12.2.21}$$

which leads via Eqs. 12.2.14 and 12.2.15 to the two symmetry vectors for **b**:

$$X_3:\quad \mathbf{b}_1^3=(0,1,-1;\ 0,i,-i),\qquad \mathbf{b}_2^3=(0,1,1;\ 0,-i,-i) \tag{12.2.22}$$

or, in terms of the actual displacements u_μ^α, Eqs. 12.2.21 and 12.2.5 yield $u_z^{\bar{1}}=u_y^1$, and

$$X_3:\quad (\mathbf{u}^3)_1=(0,1,-1;\ 0,-1,1),\qquad (\mathbf{u}^3)_2=(0,1,1;\ 0,1,1) \tag{12.2.23}$$

For mode X_4, the ratio of $u^{\bar{1}}$ to u^1 is opposite to that in X_3:

$$X_4:\quad (\mathbf{u}^4)_1=(0,1,-1;\ 0,1,-1),\qquad (\mathbf{u}^4)_2=(0,1,1;\ 0,-1,-1) \tag{12.2.24}$$

These motions in the y-z plane are shown in Fig. 12.2.3.

Whether mode X_4 is higher or lower in frequency than X_3 is not a group-theoretical question. We shall show in the next section, however, that for any reasonable assignment of the forces X_4 will become TO, that is, the higher-frequency mode, and X_3 will become TA. This conclusion is indeed suggested by the fact that in $(\mathbf{u}^4)_2$ the sublattices move in opposing directions, whereas in $(\mathbf{u}^3)_2$ they move together, which should require less energy.

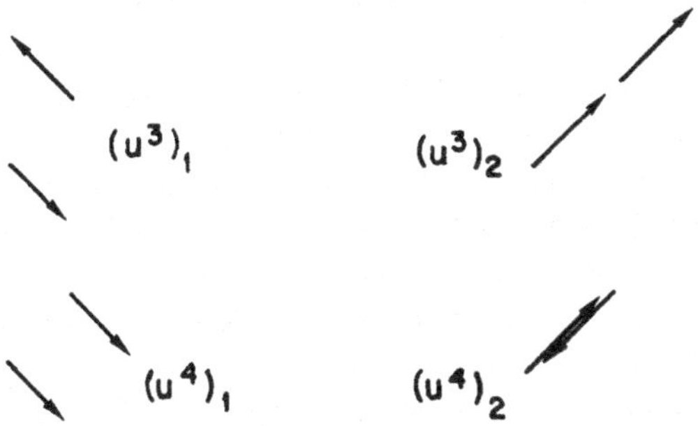

Fig. 12.2.3. The doubly degenerate transverse lattice vibrations X_3 propagate in the x direction and have displacements $(\mathbf{u}^3)_1$ [or $(\mathbf{u}^3)_2$] in the y-z plane that are odd [or even] in $\rho_{\bar{y}z}$. The corresponding odd [even] modes of X_4 are shown as $(\mathbf{u}^4)_1$ [$(\mathbf{u}^4)_2$].

12.3. R(q) AND ω(q) FOR $\mathbf{q}=(q,0,0)$

The elements of the space group that reverse $\mathbf{q}$ have been combined with time reversal to show that the $\mathbf{R}(\mathbf{q})$ matrix has the form (Eq. 11.7.14)

$$\mathbf{R}=\begin{bmatrix} \mathbf{H} & \mathbf{S} \\ \mathbf{S}^* & \mathbf{H}^* \end{bmatrix} \tag{12.3.1}$$

where $\mathbf{H}$ is Hermitian and $\mathbf{S}$ is symmetric. We therefore confine our attention to the elements of $G_{\mathbf{q}}$ and start with those that leave α and β alone, that is, do not interchange the sublattices. The element $(\delta_{2x}|0)$, that is, $y\to -y$, $z\to -z$, kills terms odd in y or z:

$$H_{12}=H_{13}=S_{12}=S_{13}=0 \tag{12.3.2}$$

The element $(\rho_{\bar{y}z}|0)$, that is, $y\to z$, $z\to y$, imposes symmetry on indices 2 and 3:

$$\begin{aligned} &H_{22}=H_{33}; && S_{22}=S_{33} \\ &H_{23}=H_{32}=\text{real}; && S_{23}=S_{32},\ \text{not new} \end{aligned} \tag{12.3.3}$$

Now consider the elements that interchange the sublattices. Element $(\rho_y|\boldsymbol{\tau})$ yields

$$\mathbf{R}^{\alpha\beta}=\rho_y\mathbf{R}^{\bar{\alpha}\bar{\beta}}\rho_y \tag{12.3.4}$$

Since ρ_y reverses y, we obtain, using Eqs. 12.3.4, 12.3.3 and 12.3.1

$$\begin{aligned} &H_{23}=-H_{23}^*=0 \\ &S_{23}=-S_{23}^*=\text{purely imaginary} \end{aligned} \tag{12.3.5}$$

Thus our matrices take the forms

$$\mathbf{H}=\begin{bmatrix} A & 0 & 0 \\ 0 & B & 0 \\ 0 & 0 & B \end{bmatrix}, \qquad \mathbf{S}=\begin{bmatrix} E & 0 & 0 \\ 0 & D & iC \\ 0 & iC & D \end{bmatrix} \tag{12.3.6}$$

leaving five independent, real parameters. See Problem 12.3.1.

Using Eqs. 11.6.7–11.6.10 and 12.1.2 up to and including second-neighbor forces, we have

$$\begin{aligned} A &= 4\alpha + 8\mu(1-\cos\pi p) \\ B &= 4\alpha + (4\mu + 4\lambda)(1-\cos\pi p) \\ C &= -4\beta \sin\frac{\pi p}{2} \\ D = E &= -4\alpha\cos\frac{\pi p}{2} \end{aligned} \tag{12.3.7}$$

but $D \neq E$ in general.

If we insert the known form of the longitudinal modes (Eq. 12.2.4) into 11.2.8 and set $M_{-1} = M_1 = M$, we obtain:

$$\begin{aligned} M\omega^2(\Delta_1) &= A + E = 4\alpha\left(1-\cos\frac{\pi p}{2}\right) + 8\mu(1-\cos\pi p) \\ M\omega^2(\Delta_2') &= A - E = 4\alpha\left(1+\cos\frac{\pi p}{2}\right) + 8\mu(1-\cos\pi p) \end{aligned} \tag{12.3.8}$$

In the first zone, Δ_1 has the lower frequency and is called LA. In the second zone, the order of frequencies is reversed and Δ_1 must be called LO. See Problem 12.2.3. At the zone boundary $\mathbf{p} = (1,0,0)$ these eigenvalues coalesce to the value

$$M\omega_{X_1}^2 = 4\alpha + 16\mu \tag{12.3.9}$$

With Eq. 12.2.14, the Δ_5 modes of type 1 ($\rho_{\bar{y}z} = -1$) provide the equations

$$\left[\begin{array}{ccc|ccc} A & 0 & 0 & E & 0 & 0 \\ 0 & B & 0 & 0 & D & iC \\ 0 & 0 & B & 0 & iC & D \\ \hline E & 0 & 0 & A & 0 & 0 \\ 0 & D & -iC & 0 & B & 0 \\ 0 & -iC & D & 0 & 0 & B \end{array}\right] \left[\begin{array}{c} 0 \\ a \\ -a \\ \hline 0 \\ b \\ -b \end{array}\right] = M\omega^2 \left[\begin{array}{c} 0 \\ a \\ -a \\ \hline 0 \\ b \\ -b \end{array}\right] \tag{12.3.10}$$

or

$$\begin{aligned} Ba+(D-iC)b&=M\omega^2 a \\ (D+iC)a+Bb&=M\omega^2 b \end{aligned} \tag{12.3.11}$$

so that

$$M\omega^2=B\pm(C^2+D^2)^{1/2} \tag{12.3.12}$$

the upper sign consistently referring to TO, the lower to TA, and

$$\frac{b}{a}=\frac{M\omega^2-B}{D-iC}=\pm\frac{(D^2+C^2)^{1/2}}{D-iC}=\pm\frac{D+iC}{(D^2+C^2)^{1/2}} \tag{12.3.13}$$

As $p\to 0$, $C\to 0$, $D\to -4\alpha$ and with $\alpha>0$, the first-partner solutions approach

$$\text{TA:}\quad \Delta_5^{(1)}\to(0,1,-1;\,0,1,-1)$$

$$\text{TO:}\quad \Delta_5^{(1)}\to(0,1,-1;\,0,-1,1) \tag{12.3.14}$$

In this case the optic and acoustic branches have the same Δ_5 symmetry. The acoustic branch can be recognized in two mutually consistent ways: (1) it is the branch whose frequency goes to zero as $q\to 0$ (hence the lowest in frequency of all modes of the same symmetry), and (2) it is the mode whose form as $q\to 0$ requires all the particles to have identical displacements.

The second-partner Δ_5 solutions have the form (Eq. 12.2.15):

$$\Delta_5^{(2)}=(0,b,b;\,0,a,a)$$

The solutions for b/a and $M\omega^2$ are the same as for the first type. As $p\to 0$,

$$\begin{aligned} \text{TA:}&\quad \Delta_5^{(2)}\to(0,1,1;\,0,1,1) \\ \text{TO:}&\quad \Delta_5^{(2)}\to(0,1,1;\,0,-1,-1) \end{aligned} \tag{12.3.15}$$

As we approach the zone boundary point X, $\mathbf{p}\to(1,0,0)$, $D\to 0$, and $C\to -4\beta$, so that

$$\frac{b}{a}\to\mp\frac{i\beta}{|\beta|}=\mp i\quad \text{if } \beta>0 \tag{12.3.16}$$

with the reverse sign if $\beta<0$. Thus, for $\beta>0$,

$$\begin{aligned} \text{TA:}&\quad \Delta_5^{(1)}\to(0,1,-1;\,0,i,-i)=X_3 \\ \text{TO:}&\quad \Delta_5^{(1)}\to(0,1,-1;\,0,-i,i)=X_4 \end{aligned} \tag{12.3.17}$$

whereas, if β were less than zero, we would have

$$\text{TA}\to X_4 \quad \text{and} \quad \text{TO}\to X_3 \tag{12.3.18}$$

In either case, mode X_3 is characterized by $b/a=i$, which, inserted into Eq. 12.3.10, yields

$$M\omega_{X_3}^2 = B + C = 4\alpha + 8(\mu+\lambda) - 4\beta \tag{12.3.19}$$

whereas X_4 has the form $b/a = -i$, and

$$M\omega_{X_4}^2 = B - C = 4\alpha + 8(\mu+\lambda) + 4\beta \tag{12.3.20}$$

so that X_3 is lower (and hence=TA) for $\beta > 0$, the choice required to obtain a sensible fit to the experimental vibration spectrum.

2.4. Σ MODES $(q,q,0)$

The character table 12.4.1 yields the compatibility relations:

$$\Gamma_{15}^- = \Sigma_1 + \Sigma_3 + \Sigma_4 = \text{acoustic}, \qquad \Gamma_{25}^+ = \Sigma_1 + \Sigma_2 + \Sigma_3 = \text{optical} \tag{12.4.1}$$

Modes Σ_1 and Σ_3 have both optical and acoustical properties! These modes appear twice in the original six-dimensional space, and therefore each will lead to a 2×2 matrix to diagonalize. Modes Σ_2 and Σ_4 appear only once and are completely determined in form by the character table. Therefore we start with these.

Since $\rho_{\bar{x}y} = -1$ for Σ_2 and Σ_4 (see Table 12.4.1) we can project into $\Sigma_2 + \Sigma_4$ by imposing $\rho_{\bar{x}y} = -1$ on a general motion which forces the specialization

$$(a,b,c;\ a',b',c') \to (a,-a,0;\ a',-a',0) \tag{12.4.2}$$

Table 12.4.1. Characters at $\Sigma = (q,q,0)$

	Σ_1	Σ_2	Σ_3	Σ_4	Γ_{15}^-	Γ_{25}^+
$(\epsilon\|0)$	1	1	1	1	3	3
$(\rho_{\bar{x}y}\|0)$	1	-1	1	-1	1	1
$(\rho_z\|\tau)$	μ	$-\mu$	$-\mu$	μ	1	-1
$(\delta_{2xy}\|\tau)$	μ	μ	$-\mu$	$-\mu$	-1	1

where $\mu = \exp(-i\mathbf{q}\cdot\boldsymbol{\tau}) = \exp(-iqa/2)$.

The condition $(\rho_z|\tau)=\pm\mu$ [or $O(\rho_z)=\pm 1$] for Σ_4 or Σ_2, respectively, forces $a'=\pm a$; thus

$$\begin{aligned}\Sigma_4&=(1,-1,0;\ 1,-1,0)=\mathrm{TA}_{1\bar{1}0}=\mathrm{TA}_{\perp z}\\ \Sigma_2&=(1,-1,0;\ -1,1,0)=\mathrm{TO}_{1\bar{1}0}=\mathrm{TO}_{\perp z}\end{aligned}\tag{12.4.3}$$

where $(1\bar{1}0)$ indicates the direction of polarization, perpendicular to z as well as to the direction of propagation.

Modes Σ_1 and Σ_3 are both restricted by $\rho_{\bar{x}y}=1$ to have the form $(a,a,c;\ a',a',c')$, but $O(\rho_z)=\pm 1$ yields $a'=\pm a$, $c'=\mp c$, or

$$\begin{aligned}\Sigma_1&=(a,a,c;\ a,a,-c)=(\mathrm{LA}+\mathrm{TO}_z)\\ \Sigma_3&=(a,a,c;\ -a,-a,c)=(\mathrm{LO}+\mathrm{TA}_z)\end{aligned}\tag{12.4.4}$$

Thus Σ_1 is a mixture of LA and TO (polarized in the z direction) motions with the ratio a/c a function of q and the force constants. If these forms are combined with the reality condition $\mathbf{b}^{\alpha}(\mathbf{q})^*=\mathbf{b}^{\bar{\alpha}}(\mathbf{q})$, we learn that a is real and c is purely imaginary. Thus Σ_1 has an LA component plus a TO component polarized in the z direction 90° out of phase with it. If we draw a rough figure (see Fig. 12.4.1) of the possible frequency spectrum in the $(q,q,0)$ direction, consistent with Eq. 12.4.1, we see that one Σ_1 mode is acoustic and one optic in nature. Thus as $q\to 0$ the Σ_1 mixtures $\mathrm{LA}+\mathrm{TO}_z$ approach

$$\Sigma_1^-=\Sigma_1^{\mathrm{ac}}\to\mathrm{LA},\qquad \Sigma_1^+=\Sigma_1^{\mathrm{op}}\to\mathrm{TO}_z$$

in the two limiting cases, although for finite q these characteristic motions are admixed. Similar remarks apply to Σ_3. Thus we see that, away from $q=0$, the labels LA, TO, TA are not fundamental; it is the classification Σ_j imposed by the symmetry of the physical system that remains generally valid. We can, however, classify our branches as Σ_1^{ac} and Σ_1^{op} in accord with their limiting behavior as $q\to 0$.

Dynamical Matrix and Frequencies at Σ

The element $(\rho_{\bar{x}y}|0)$, that is, $x\to y$, $y\to x$, yields the conditions

$$R_{11}^{\alpha\beta}=R_{22}^{\alpha\beta},\qquad R_{12}^{\alpha\beta}=R_{21}^{\alpha\beta},\qquad R_{13}^{\alpha\beta}=R_{23}^{\alpha\beta}\tag{12.4.5}$$

or

$$H_{11}=H_{22},\quad S_{11}=S_{22};\qquad H_{12}=H_{21}=\text{real};\qquad H_{13}=H_{23},\quad S_{13}=S_{23}$$

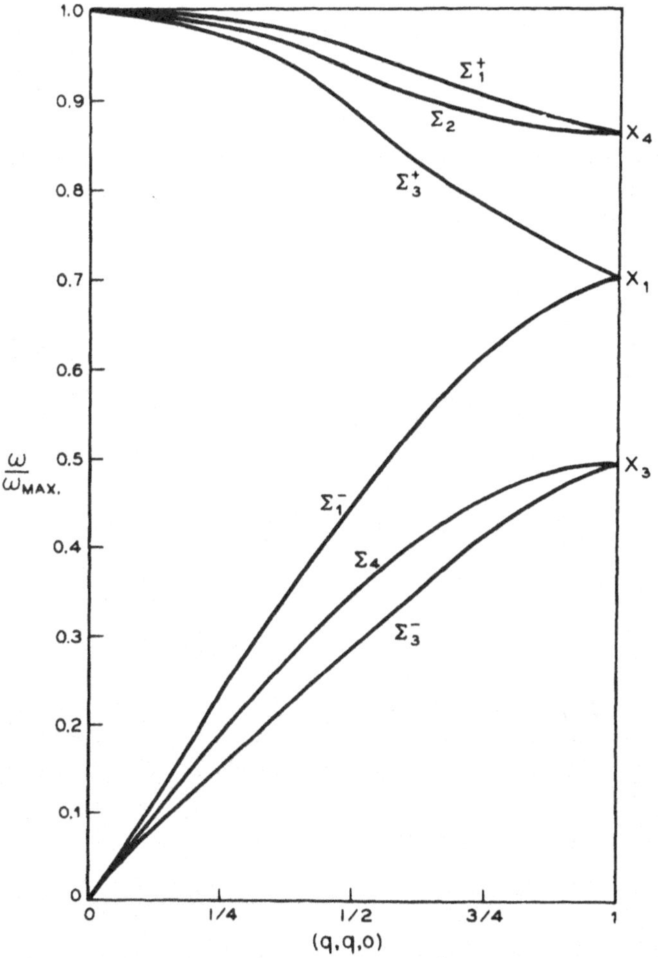

Fig. 12.4.1. *Lattice vibration spectrum* $\omega(\mathbf{q})$ *for* $\mathbf{q}$ *in the* $\Sigma=(q,q,0)$ *direction.* $\Sigma_2=\mathrm{TO}_{1\bar{1}0}=\mathrm{TO}_{\perp z}$ and $\Sigma_4=\mathrm{TA}_{1\bar{1}0}=\mathrm{TA}_{\perp z}$ are transverse optical and transverse acoustic branches polarized in the $(1\bar{1}0)$ direction perpendicular to the z direction. Σ_1 is the mixture LA + TO_z, and Σ_3 is the mixture LO + TA_z. The branches Σ_1^- and Σ_3^- are approximately LA and TA_z in character, whereas Σ_1^+ and Σ_3^+ are approximately TO_z and LO in character. Thus Σ_1^- and Σ_3^+ combine into the longitudinal mode X_1 at $(1,1,0)$, which is equivalent to $(0,0,1)$.

To nearest neighbors only, we have $\Sigma_4=4(\alpha-\beta)\sin^2(\pi p/2)$, where $q=(2\pi/a)p$, $\Sigma_2=8\alpha-\Sigma_4$, $\Sigma_1^{\pm}=4\alpha+2\beta\sin^2(\pi p/2)\pm\{[4\alpha-(4\alpha+2\beta)\sin^2(\pi p/2)]^2+8\beta^2\sin^2\pi p\}^{1/2}$, $\Sigma_3^++\Sigma_1^-=\Sigma_3^-+\Sigma_1^+=8\alpha$. What is plotted, however, is $(\Sigma/8\alpha)^{1/2}$, so that these symmetries are not apparent in the plots. We have set $\beta=\frac{1}{2}\alpha$ for simplicity.

The operation $(\rho_z|\tau)$ changes α into $\bar{\alpha}$ and reverses the z component of any vector. Thus we obtain the condition

$$R_{\mu\nu}^{\alpha\beta} = \rho_z R^{\bar{\alpha}\bar{\beta}} \rho_z = (-1)^{(\text{No. of 3's in } \mu \text{ and } \nu)} R_{\mu\nu}^{\bar{\alpha}\bar{\beta}}$$

$$= (-1)^{\text{No. of 3's}} (R_{\mu\nu}^{\alpha\beta})^* \tag{12.4.6}$$

where the last step makes use of time reversibility combined with inversion, Eq. 11.7.12. Thus elements containing 3 an even number of times (including zero) among the subscripts are purely real. The others are purely imaginary, and our matrices take the form

$$\mathbf{H} = \begin{bmatrix} A & C & iD \\ C & A & iD \\ -iD & -iD & B \end{bmatrix}, \qquad \mathbf{S} = \begin{bmatrix} E & G & iH \\ G & E & iH \\ iH & iH & F \end{bmatrix} \tag{12.4.7}$$

We can write our eigenvalue equation in the form (cf. 12.3.1):

$$\left[\begin{array}{ccc|ccc} A & C & iD & E & G & iH \\ C & A & iD & G & E & iH \\ -iD & -iD & B & iH & iH & F \\ \hline E & G & -iH & A & C & -iD \\ G & E & -iH & C & A & -iD \\ -iH & -iH & F & iD & iD & B \end{array}\right] \left[\begin{array}{c} a \\ b \\ ic \\ \hline a' \\ b' \\ ic' \end{array}\right] = M\omega^2 \left[\begin{array}{c} a \\ b \\ ic \\ \hline a' \\ b' \\ ic' \end{array}\right] \tag{12.4.8}$$

We could transform this equation to real form as in Eq. 11.7.17. We have just shown, however, that the time-reversal condition in $\mathbf{b}^\alpha(\mathbf{q})$ simply makes the z components purely imaginary. We have therefore written these components as ic and ic'. The resulting equations are, of course, entirely real after the i's are properly canceled.

The solutions of these equations might be difficult to find by inspection! If we insert the known solutions 12.4.3 for Σ_2 and Σ_4, we obtain directly the corresponding eigenvalues:

$$M\omega_2^2 = A - C - E + G = \mathrm{TO}_{\perp z}, \qquad M\omega_4^2 = A - C + E - G = \mathrm{TA}_{\perp z} \quad (12.4.9)$$

If we insert the known form of solution for Σ_1: $(a,a,ic;\ a,a,-ic)$, we obtain a pair of simultaneous equations with real coefficients:

$$\begin{aligned}(A+C+E+G)a+(H-D)c&=M\omega^2 a\\ 2(H-D)a+(B-F)c&=M\omega^2 c\end{aligned} \quad (12.4.10)$$

The remaining equations are simply repetitions of these. Thus Σ_1 has two eigenvalues:

$$2M\omega_1^2 = A+C+E+G+B-F \pm \left[(A+C+E+G-B+F)^2+8(H-D)^2\right]^{1/2} \quad (12.4.11)$$

As $q \to 0$, the higher-frequency solution must approach TO_z, whereas the lower approaches LA.

For Σ_3 we insert $(a,a,ic;\ -a,-a,ic)$ to obtain

$$\begin{aligned}(A+C-E-G)a-(H+D)c&=M\omega^2 a\\ -2(H+D)a+(B+F)c&=M\omega^2 c\end{aligned} \quad (12.4.12)$$

which differ from Σ_1 in the replacements $E,F,G,H \to -E,-F,-G,-H$. Thus

$$2M\omega_3^2 = A+C-E-G+B+F \pm \left[(A+C-E-G-B-F)^2+8(H+D)^2\right]^{1/2} \quad (12.4.13)$$

Again the higher-frequency solution must approach LO and the lower approach TA_z as $q \to 0$.

We shall now list, using Eq. 12.1.2, the values of the parameters including third-neighbor interactions:

$$A = R_{11}^{11} = 4\alpha + 8\mu' + 4\lambda' + 4\mu(2 - \cos\pi p - \cos^2\pi p) + 4\lambda(1 - \cos\pi p)$$

$$B = R_{33}^{11} = R_{11}^{11}(p_3,p_2,p_1) = 4\alpha + 8\mu' + 4\lambda' + 8\mu(1 - \cos\pi p) + 4\lambda\sin^2\pi p$$

$$C = R_{12}^{11} = 4\nu\sin^2\pi p$$

$$iD = R_{23}^{11} = R_{12}^{11}(p_2,p_3,p_1) = 4i\delta\sin\pi p(1 - \cos\pi p)$$

$$E=R_{11}^{1\bar{1}}=-4(\alpha+\mu')\cos^2\frac{\pi p}{2}-4(\mu'+\lambda')\cos\frac{\pi p}{2}\cos\frac{3\pi p}{2} \tag{12.4.14}$$

$$F=R_{33}^{1\bar{1}}=R_{11}^{1\bar{1}}(p_3,p_2,p_1)=-4(\alpha+\lambda')\cos^2\frac{\pi p}{2}-8\mu'\cos\frac{\pi p}{2}\cos\frac{3\pi p}{2}$$

$$G=R_{12}^{1\bar{1}}=4(\beta+\nu')\sin^2\frac{\pi p}{2}+8\delta'\sin\frac{\pi p}{2}\sin\frac{3\pi p}{2}$$

$$iH=R_{23}^{1\bar{1}}=R_{12}^{1\bar{1}}(p_2,p_3,p_1)$$

$$=-i2\beta\sin\pi p+i2\delta\sin 2\pi p+i2\nu'(\sin 2\pi p+\sin\pi p)$$

The zone boundary point K: $\mathbf{p}=(\frac{3}{4},\frac{3}{4},0)$ is equivalent to another zone boundary point $(-\frac{1}{4},-\frac{1}{4},-1)$, since they differ by $\mathbf{P}=(1,1,1)$, which is the body center translation of the reciprocal lattice. But no additional symmetry element seems to be available that connects these points. This can be seen also by noting that at $(\frac{3}{4},\frac{3}{4},0)$ the parameters acquire no new relationships:

$$A=4\alpha+8\mu'+4\lambda'+(6+2\sqrt{2})\mu+(4+2\sqrt{2})\lambda$$

$$B=4\alpha+8\mu'+4\lambda'+(8+4\sqrt{2})\mu+2\lambda$$

$$C=2\nu$$

$$D=(2+2\sqrt{2})\delta \tag{12.4.15}$$

$$E=-(2-\sqrt{2})\alpha+(2\sqrt{2}-2)\mu'+\sqrt{2}\lambda'$$

$$F=-(\alpha+\lambda')(2-\sqrt{2})+2\sqrt{2}\mu'$$

$$G=(2+\sqrt{2})(\beta+\nu')-2\sqrt{2}\delta'$$

$$H=-\sqrt{2}\beta-2\delta-(2-\sqrt{2})\nu'$$

If one continues, however, to the point $\mathbf{p}=(1,1,0)$, this is equivalent, via $\mathbf{P}=(-1,-1,1)$, to $\mathbf{p}=(0,0,1)$, that is, to X. The Σ_3 mode of lower frequency transforms into the TA$=X_3$ mode with eigenvalue $4\alpha-4\beta$, and the Σ_3 mode of higher frequency transforms into X_1 with eigenvalue 4α. See Fig. 12.4.1.

12.5 THE MODES $\Lambda=(q,q,q)$ AND $L=(2\pi/a)(\frac{1}{2},\frac{1}{2},\frac{1}{2})$

The compatibility relations are of the form (see Table 12.5.1)

$$\Gamma_{15}^{-}=\Lambda_1+\Lambda_3,\qquad \Gamma_{25}^{+}=\Lambda_1+\Lambda_3 \tag{12.5.1}$$

Table 12.5.1. Group Characters at $\Lambda=(q,q,q)$

	Λ_1	Λ_2	Λ_3
$(\epsilon\|0)$	1	1	2
$(\delta_{3xyz},\delta_{3xyz}^{-1}\|0)$	1	1	−1
$(\rho_{\bar{y}z},\rho_{\bar{z}x},\rho_{\bar{x}y}\|0)$	1	−1	0

The invariance of Λ_1 under the threefold operation δ_{3xyz} guarantees that Λ_1 is longitudinal, with the form $(a,a,a;\ b,b,b)$. The solution of a pair of simultaneous equations for b/a determines the LO and LA modes and frequencies.

Projection into Λ_3 shows that it is transverse, with all transverse directions degenerate. We can take $(1,-1,0)$ and $(1,1,-2)$ as orthogonal to $(1,1,1)$ and to each other. Thus we can assume modes of the form

$$\Lambda_3^{(1)}=(a,-a,0;\ b,-b,0),\qquad \Lambda_3^{(2)}=(a,a,-2a;\ b,b,-2b) \tag{12.5.2}$$

and solve for the ratio b/a.

The dynamical matrix has the form of Eq. 12.3.1 with

$$\mathbf{R}^{11}(\mathbf{p})=\mathbf{H}=\begin{bmatrix} A & B & B\\ B & A & B\\ B & B & A\end{bmatrix},\qquad \mathbf{R}^{12}(\mathbf{p})=\mathbf{S}=\begin{bmatrix} C & D & D\\ D & C & D\\ D & D & C\end{bmatrix} \tag{12.5.3}$$

Table 12.5.2. Group Characters at $L=(\pi/a)(1,1,1)$

	$L_1^{\pm}$	$L_2^{\pm}$	$L_3^{\pm}$
$(\epsilon\|0)$	1	1	2
$(\delta_{3xyz},\delta_{3xyz}^{-1}\|0)$	1	1	−1
$(\delta_{2\bar{y}z},\delta_{2\bar{z}x},\delta_{2\bar{x}y}\|\tau)$	1	−1	0
$(\mathbf{i}\|\tau)(\alpha\|\tau)$	$\pm\chi(\alpha\|\tau)$		

where $L_j^+=L_j$; $L_j^-=L_j'$

where, to second-neighbor forces:

$$
\begin{aligned}
A &= 4\alpha + (8\mu + 4\lambda)\sin^2 \pi p \\
B &= 4\nu \sin^2 \pi p \\
C &= -4\alpha\left(\cos^3 \frac{\pi p}{2} - i\sin^3 \frac{\pi p}{2}\right) \\
D &= 2\beta \sin \pi p \exp\left(-\frac{i\pi p}{2}\right)
\end{aligned}
\tag{12.5.4}
$$

At the zone boundary, the compatibility relations obtained by comparing Table 12.5.2 with Table 12.5.1 are as follows:

$$L_1^+ = \Lambda_1, \qquad L_2^- = \Lambda_1, \qquad L_3^+ = \Lambda_3, \qquad L_3^- = \Lambda_3 \tag{12.5.5}$$

so that one longitudinal mode approaches L_1^+ and the other L_2^-. Whichever has the lower frequency will be called LA. Although LA is odd at $\mathbf{q}=0$, it *is not necessarily odd at* the zone boundary. Similar considerations apply to the transverse modes. See Fig. 12.2.1.

The representations at the zone boundary are distinguished from those in the interior by being even or odd under inversion! Since

$$(\iota|\tau)\mathbf{u}^{\alpha i} = -\mathbf{u}^{\bar{\alpha}\bar{i}} = -\mathbf{u}^{\bar{\alpha}i} = \pm\mathbf{u}^{\alpha i} \tag{12.5.6}$$

we have

$$\frac{u^{\bar{1}}}{u^1} = \mp 1 \qquad \text{or} \qquad \frac{b}{a} = \pm\exp\left(\frac{i\pi}{4}\right) \tag{12.5.7}$$

where the upper sign refers to even and the lower to odd modes. The form of the modes is now completely determined, and we find for the eigenvalues:

$$
\begin{aligned}
&L_3^+:\quad M\omega^2 = 2\alpha - 2\beta + 8\mu + 4\lambda - 4\nu, &&L_1^+:\quad M\omega^2 = 2\alpha + 4\beta + 8\mu + 8\nu + 4\lambda \\
&L_3^-:\quad M\omega^2 = 6\alpha + 2\beta + 8\mu + 4\lambda - 4\nu, &&L_2^-:\quad M\omega^2 = 6\alpha - 4\beta + 8\mu + 8\nu + 4\lambda
\end{aligned}
\tag{12.5.8}
$$

Thus L_3^+ is surely TA, and L_3^- is surely TO with parity opposite to that at $\mathbf{q}=0$. The assignment of L_1^+ and L_2^- to LO and LA is harder. Central forces, $\beta=\alpha$, or even moderately noncentral forces, $\beta > \frac{1}{2}\alpha > 0$, lead to the choice $L_1^+ = \text{LO}$, $L_2^- = \text{LA}$. See the legend for Fig. 12.2.1 for further discussion.

12.6. ELASTIC PROPERTIES OF THE DIAMOND STRUCTURE

In Sections 11.9 and 11.10 we provided explicit expressions for the elastic constants in terms of sums over the force constants. Much of the work in performing these sums has already been done in setting up the dynamical matrix $\mathbf{R}(\mathbf{q})$ in Eqs. 12.1.2. We can take advantage of this work by making a direct comparison between the equation for sound waves (11.9.9) and the long-wave limit of the dynamical equations. If we multiply Eq. 11.9.9 by $a^3/8$, the volume per particle, and write $M=\rho a^3/8=$ the mass of one atom, the sound wave equations take the form

$$M\omega^2\mathbf{u}=\left(\frac{\pi^2}{2}\right)\mathbf{D}\cdot\mathbf{u} \tag{12.6.1}$$

where

$$D_{\mu\nu}=aC_{\mu a\nu b}p_a p_b \tag{12.6.2}$$

and $\mathbf{q}=(2\pi/a)\mathbf{p}$, as defined in Eq. 12.1.3.

Our dynamical equations of motion in real form, 11.7.17, can be written as

$$\begin{aligned}\mathrm{Re}\,(\mathbf{H}+\mathbf{S})\mathbf{u}+\mathrm{Im}\,(\mathbf{S}-\mathbf{H})\mathbf{v}&=M\omega^2\mathbf{u}\\ \mathrm{Im}\,(\mathbf{H}+\mathbf{S})\mathbf{u}+\mathrm{Re}\,(\mathbf{H}-\mathbf{S})\mathbf{v}&=M\omega^2\mathbf{v}\end{aligned} \tag{12.6.3}$$

where

$$\mathbf{H}=\mathbf{R}^{11}(\mathbf{p}),\qquad \mathbf{S}=\mathbf{R}^{1\bar{1}}(\mathbf{p}) \tag{12.6.4}$$

are given in Eqs. 12.1.2, $\mathbf{u}=(\mathbf{b}^1+\mathbf{b}^{\bar{1}})/\sqrt{2}=\sqrt{2}\,\mathrm{Re}\,\mathbf{b}^1$ represents the center of mass motion, and $\mathbf{v}=i(\mathbf{b}^{\bar{1}}-\mathbf{b}^1)/\sqrt{2}=\sqrt{2}\,\mathrm{Im}\,\mathbf{b}^1$ represents the relative motion. By eliminating the relative motion $\mathbf{v}$, we obtain an effective equation for the center of mass motion of the same form as 12.6.1 with $\mathbf{D}$ given by

$$\left(\frac{\pi^2}{2}\right)\mathbf{D}=\mathrm{Re}(\mathbf{H}+\mathbf{S})-\mathrm{Im}(\mathbf{S}-\mathbf{H})[\mathrm{Re}(\mathbf{H}-\mathbf{S}-M\omega^2)]^{-1}\mathrm{Im}(\mathbf{S}+\mathbf{H}) \tag{12.6.5}$$

From our macroscopic equation 12.6.2, $\mathbf{D}$ for a cubic crystal has the typical elements

$$\begin{aligned}D_{11}&=aC_{11}p_1^2+aC_{44}(p_2^2+p_3^2)\\ D_{12}&=a(C_{12}+C_{44})p_1p_2\end{aligned} \tag{12.6.6}$$

where we use the abbreviation of Eq. 4.7.5, that is,

$$C_{11,11}=C_{11},\qquad C_{11,22}=C_{12},\qquad C_{12,12}=C_{44} \tag{12.6.7}$$

Thus the elastic constants can be determined by evaluating Eq. 12.6.5 in the limit of small **p** accurate to order p^2. Since $\mathrm{Im}(\mathbf{S}-\mathbf{H})$ and $\mathrm{Im}(\mathbf{S}+\mathbf{H})$ are already of order p, it is permissible to drop $M\omega^2 \propto p^2$ and evaluate $\mathrm{Re}(\mathbf{H}-\mathbf{S})$ to the zeroth order of accuracy. Moreover, the Hermiticity of **H** can be combined with the tetrahedral point group T_d to guarantee that, to first order in p, $\mathrm{Im}\,\mathbf{H}=0$. Thus

$$\left(\frac{\pi^2}{2}\right)\mathbf{D}=\mathrm{Re}(\mathbf{H}+\mathbf{S})-\mathrm{Im}\,\mathbf{S}[\mathrm{Re}(\mathbf{H}-\mathbf{S})]^{-1}\mathrm{Im}\,\mathbf{S} \qquad (12.6.8)$$

Evaluating $\mathrm{Re}(\mathbf{H}-\mathbf{S})$, $\mathrm{Im}\,\mathbf{S}$, and $\mathrm{Re}(\mathbf{H}+\mathbf{S})$ to zeroth, first, and second order, respectively, in **p**, we obtain typical matrix elements from Eqs. 12.1.2:

$$\begin{aligned}
\mathrm{Re}(H+S)_{11}&=\frac{\pi^2}{2}[A_1p_1^2+A_3(p_2^2+p_3^2)]\\
\mathrm{Re}(H+S)_{12}&=\left(\frac{\pi^2}{2}\right)A_2p_1p_2\\
\mathrm{Re}(H-S)_{11}&=8A_5\\
\mathrm{Re}(H-S)_{12}&=0
\end{aligned} \qquad (12.6.9)$$

$$\begin{aligned}
\mathrm{Im}\,S_{11}&=0\\
\mathrm{Im}\,S_{12}&=-2\pi A_4p_3\\
\mathrm{Im}\,H_{11}&=\mathrm{Im}\,H_{12}=0
\end{aligned}$$

where, out to third neighbors, we have

$$\begin{aligned}
A_1&=\alpha+8\mu+(2\mu'+9\lambda')\\
A_2&=2\beta+8\nu+2(\nu'+6\delta')\\
A_3&=\alpha+4(\mu+\lambda)+(10\mu'+\lambda')\\
A_4&=\beta-(3\nu'+2\delta')\\
A_5&=\alpha+(2\mu'+\lambda')
\end{aligned} \qquad (12.6.10)$$

Inserting Eq. 12.6.9 and cyclic equivalents such as $\operatorname{Im} S_{23} = -2\pi A_4 p_1$ into 12.6.8, we find that

$$D_{11} = A_1 p_1^2 + \left(A_3 - \frac{A_4^2}{A_5} \right)(p_2^2 + p_3^2)$$

$$D_{12} = \left(A_2 - \frac{A_4^2}{A_5} \right) p_1 p_2 \tag{12.6.11}$$

Comparison of Eq. 12.6.11 with 12.6.6 now yields

$$aC_{11} = A_1, \qquad aC_{12} = A_2 - A_3, \qquad aC_{44} = A_3 - \frac{(A_4)^2}{A_5} \tag{12.6.12}$$

and, via Eq. 12.6.10, expresses the elastic constants in terms of the force constants.

Equation 12.1.2 leads to matrix elements of the form 12.6.9 when carried to third neighbors. We leave it as an exercise for the reader to show that this form is correct to all neighbors. *Hint*: Use Eq. 11.6.2b and Problem 12.6.1.

In spite of our dicussion in Section 11.10 of singular matrices, the matrix $\operatorname{Re}(\mathbf{H} - \mathbf{S})$ whose reciprocal we needed in Eq. 12.6.8 is nonsingular, since the center of mass motion has been explicitly removed. Moreover, $\operatorname{Re}(\mathbf{H} - \mathbf{S})$ is a constant times the unit matrix because of cubic symmetry. This simplicity also enters our general expression 11.10.35 for this reciprocal matrix because of the triple degeneracy of the optical branch Γ'_{25} at $\mathbf{q} = 0$. Indeed, the frequency of this $\mathbf{q} = 0$ optical mode (often referred to as the Raman frequency, since that is how it is observed in diamond) is given by

$$M\omega_{\mathrm{RA}}^2 = (H + S)_{11}\big|_{\mathbf{q}=0} = 8A_5 \tag{12.6.13}$$

12.7. COMPARISON WITH EXPERIMENT

Figure 12.2.1 displays a very large ratio of ω_{LA} to ω_{TA} at the zone boundary for germanium. The same is true for silicon (both germanium and silicon possess the diamond structure). Attempts to fit these data plus the elastic constants fail unless forces relating at least fifth neighbors are introduced (Herman, **1959**). We saw, however, in Section 9.4 that displacements of the nuclei can induce dipole moments on neighboring atoms. Although the total amount so induced is zero, as shown in Section 9.2, a net quadrupole moment is induced. Thus at great distances a net quadrupole-quadrupole interaction is induced, which varies as $1/R^5$. Lax (**1958a, 1965b**) has shown that these electrostatic forces tend to raise ω_{LA} and depress ω_{TA}, thus improving agreement with experiment.

In discussing ionic crystals, Dick and Overhauser (**1958**) introduced a shell model in which the polarization of an atom is represented by a massless, charged shell that can be displaced in relation to the core by core-shell interactions. In addition, as Cochran (**1959**, **1963**) has shown, in applying the shell model to germanium and silicon (as well as to ionic crystals), short-range shell-shell interactions can be introduced that simulate the stiffness of the valence electron gas. By using both these effects (stiffness and the long-range electrostatic interactions between the induced dipoles), an excellent fit is obtained to the experimental data. For a summary see Lax (**1965b**) and Cochran (**1965**).

12.8. SUMMARY

In this chapter we carry through a full-scale space group example. We use group-theoretical techniques to find (1) the symmetry vectors, (2) the force constants in each bond, and (3) the dynamical matrix.

We show that at $\Gamma(\mathbf{q}=0)$, $\Delta(q,0,0)$, $\Lambda(q,q,q)$, $\Sigma(q,q,0)$, $L(2\pi/a)(\frac{1}{2},\frac{1}{2},\frac{1}{2})$, and $X(2\pi/a)(1,0,0)$ no representation is repeated more than twice. The dynamical problem is therefore solved at each of these points without encountering anything worse than a 2×2 system.

The relation between the elastic constants and the microscopic force constants is discussed, as well as a comparison with experiment.

problems

***12.1.1.** Make an analysis of the vibrations of some other crystal structure, such as the hexagonal close-packed structure or the wurtzite structure (see Table 6.1.4), applying the methods developed in this chapter for the diamond structure.

12.2.1. Show that an $n\times n$ unitary matrix whose trace is n must be an identity matrix.

12.2.2. Show that Δ_2 (and Δ_1') is not a lattice vibration in the diamond structure. Is this true for all crystals belonging to the same space group, O_h^7?

12.2.3. Show that the representation Δ_1 and Δ_2' are related by

$$\Delta_1\left(q+\frac{4\pi}{a}\right)=\Delta_2'(q)$$

so that Δ_1 in the second Brillouin zone is related to Δ_2' in the first. Thus Δ_1 should be referred to as LA in the first zone but as LO in the second, a result compatible with the frequencies found in Eq. 12.3.8 (Lax and Birman, **1972**; Streitwolf, **1970**).

12.2.4. Construct the symmetry coordinates for the representation Δ_5 in the diamond structure that transform as $y=(0,1,0)$ and $z=(0,0,1)$ by using the projection procedure 3.7.14. *Hint*: Set $(\psi_\nu^i)^\dagger=(0,1,0)$:

$$\psi=(a,b,c;\ a',b',c')$$

Show that the right-hand side of Eq. 3.7.14 takes the form

$$(0,1,0)(0,A,B;\ 0,A,-B)+(0,0,1)(0,B,A;\ 0,-B,A)$$

where $A=\frac{1}{2}(b+b'), B=\frac{1}{2}(c-c')$. How are the resulting symmetry coordinates related to those of Eqs. 12.2.11–12.2.14?

12.2.5. Show that for vibrations of the diamond structure the space at $X=(2\pi/a)(1,0,0)$ contains three representations X_1, X_3, and X_4, each exactly once.

12.3.1. Devise a character formula for the number of independent, real parameters in $\mathbf{R}^{\alpha\beta}(\mathbf{q})$. Verify the number for diamond at $(q,0,0)$, shown in Eq. 12.3.6.

12.6.1. Show that the forms 12.6.9 of the dynamic matrix for small wave vectors derived using interactions to third neighbors are valid for interactions to all neighbors.

(*a*) Show, using displacement invariance (Eq. 11.5.2), that $\mathbf{H}+\mathbf{S}$ has no terms independent of wave vector $\mathbf{q}$ or $\mathbf{p}$.

(*b*) Show that terms linear in $\mathbf{q}$ are imaginary. (See Eq. 11.10.21.)

(*c*) Show with the help of Problem 11.10.1 that the coefficients of the linear terms in $\mathbf{q}$ obey

$$\begin{bmatrix} \alpha & \beta & \\ \mu & \nu & b \end{bmatrix} = \begin{bmatrix} \alpha & \beta & \\ \mu' & \nu' & b' \end{bmatrix} S_{\mu'\mu} S_{\nu'\nu} S_{b'b}$$

for the elements S of the subgroup T_d which do not interchange the sublattices. Show that this condition implies that

$$\begin{bmatrix} \alpha & \beta & \\ \mu & \nu & b \end{bmatrix} = iC^{\alpha\beta}|e_{\mu\nu b}|$$

where $|e_{\mu\nu b}|=1$ for any permutation of the indices 1,2,3 but vanishes if the members of any pair of indices are equal.

(*d*) Show that Hermiticity implies that Im $\mathbf{R}^{11}=0$ to first order in $\mathbf{q}$.

(*e*) Problem 11.10.1 shows that the terms second order in $\mathbf{q}$ transform like a fourth-rank tensor under T_d. Use the direct inspection methods of Chapter 4 to determine the form of such a fourth-rank tensor. Combine the above results to confirm Eq. 12.6.9.

chapter thirteen

SYMMETRY OF MOLECULAR WAVE FUNCTIONS

13.1. MOLECULAR ORBITAL THEORY

The molecular orbital (MO) method consists in representing the many-body wave function of the molecule as a *single* determinant of one-electron "molecular orbitals" each of which belongs to an irreducible representation of the full symmetry group of the molecule. Bloch orbitals, for example, are molecular orbitals that possess the translational symmetry of a crystal. The use of the full symmetry group generally forces a molecular orbital to extend over many atoms in the molecule. Valence orbitals, by contrast, sit on one or two atoms in the molecule.

The use of a single determinant, the Hartree-Fock (HF) method (Section 13.4), leads to an effective one-body Hamiltonian h that depends on the occupied molecular orbitals. Thus an initial choice must be made for the orbitals, h must be computed and its eigenfunctions obtained, and the procedure must be iterated until a self-consistent (SC) solution is reached. Partly because the SCHF method is difficult, and partly because even the best single determinant does not yield a sufficiently accurate binding energy in molecules, a simpler approach has been adopted: the use of linear combinations of atomic orbitals (LCAO's) as molecular orbitals (MO's). There is no point in finding the optimum MO's if it is necessary to carry out a second-order perturbation treatment or to take a linear combination of determinants ("configuration interaction"). It may be better to use simpler orbitals for which all the matrix elements are easily computed. With only a modest amount of additional labor, the numerical coefficients in the LCAO expansion can be varied to give the best Hartree-Fock energy. This is the LCAO-SCF (self-consistent field) method of Roothaan (**1951**). Daudel, Lefebvre, and Moser (**1960**) and Daudel (**1966**, **1968**) provide an account of calculations based on this method.

The labor in the LCAO-SCF method is reasonable, because group theory can be used to construct symmetry orbitals from the LCAO's, so that the matrix to be diagonalized will contain only interactions between identical partners of repeated representations, as in Eq. 3.4.10. Indeed, the main business of this section is to illustrate the construction of such symmetry orbitals. This symmetry orbital procedure, however, is based on the assumption that the SCF Hamiltonian, H_{SCF}, possesses the symmetry of the molecule.

We show in Section 13.4 that this will be the case provided that the molecular or symmetry orbitals used to construct H_{SCF} span a space invariant under the operations of the molecular symmetry group, that is, if any partner function of an irreducible representation is occupied, all partner functions must be occupied. This is the analog of a closed shell in the atomic case.

The Hückel Approximation

The Hückel (**1931**, **1932**) theory is an extremely simple, widely used approach to the molecular orbital problem that is used to make qualitative comparisons with experiment. The Hückel theory is an LCAO theory in which the AO's are assumed to be orthogonal. [In principle, orthogonal atomic orbitals (OAO's) could always be constructed from the original AO's.] A self-consistent one-body Hamiltonian is assumed to exist but is not actually computed. Matrix elements of H_{SCF} are assumed to exist only between an atom and itself, and between nearest-neighbor atoms. These matrix elements are not computed but are regarded as parameters chosen to fit experiment.

Example 13.1.1

Discuss benzene, C_6H_6, by the Hückel molecular orbital approximation.

Solution: The benzene molecule is a planar molecule with the symmetry D_6 of a hexagon. The subgroup $\mathcal{C}_6$ is a cyclic group precisely analogous to the translation group of a crystal. The symmetry of our molecular orbital will be the analog of that required by the Bloch theorem for energy band wave functions. If we have one available orbital $\varphi(\mathbf{r})$ to place at each carbon site j [so that $\varphi_j = \varphi(\mathbf{r}-\mathbf{R}_j)$], our wave function will have the form

$$\psi(\mathbf{r}) = \sum a_j \varphi_j = \sum a_j \varphi(\mathbf{r}-\mathbf{R}_j) \tag{13.1.1}$$

These orbitals differ from one another by a 60° rotation:

$$C_6\varphi_j = \varphi_{j-1} \tag{13.1.2}$$

and are assumed to be orthonormalized in the Hückel approximation. As in our proof of Bloch's theorem (Section 6.2), we diagonalize the generator C_6 of our cyclic subgroup by setting

$$C_6\psi = \lambda\psi, \quad \text{where } \lambda^6 = 1 \tag{13.1.3}$$

to obtain

$$C_6\psi = \sum a_j C_6\varphi_j = \sum a_j\varphi_{j-1} = \sum a_{j+1}\varphi_j = \lambda \sum a_j\varphi_j$$

or

$$a_{j+1} = \lambda a_j, \qquad a_j = \lambda^j a_0 \tag{13.1.4}$$

or by setting $a_0 = 6^{-1/2}$ to obtain normalization,

$$\psi = 6^{-1/2} \sum \lambda^j \varphi_j \tag{13.1.5}$$

where $\lambda = \exp(2\pi i m/6)$, $m = 0, 1, 2, \ldots 5$, is one of the 6th roots of unity. In the Hückel approximation, we assume that the only nonvanishing matrix elements are

$$(\varphi_j, H\varphi_j) = \alpha, \qquad (\varphi_j, H\varphi_{j\pm 1}) = \beta \tag{13.1.6}$$

so that the energies of the molecular orbitals are given by

$$E_m = \frac{1}{6}\sum_{i,j} (\lambda^i\varphi_i, H\lambda^j\varphi_j)$$

$$= \sum_j \lambda^j(\varphi_0, H\varphi_j) = \alpha + \lambda\beta + \lambda^{-1}\beta \tag{13.1.7}$$

$$E_m = \alpha + 2\beta\cos\frac{2\pi m}{6}$$

The states associated with $m = 1$ and -1 (or 5) are degenerate, and ψ_1 and $\psi_{-1} = \psi_1^*$ can be combined to yield real wave functions. Similar remarks apply to ψ_2 and $\psi_{-2} = \psi_4 = \psi_2^*$. The remaining states ψ_0 and ψ_3 are nondegenerate and already have real wave functions.

The orbitals φ_j in the above discussion are p-like orbitals oriented normal to the plane of the hexagon. This is what permitted the simple relation 13.1.2. These orbitals are known as π orbitals and are discussed in more detail in the next section.

The LCAO method with orbitals of a given symmetry at each atomic site is treated by group-theoretical procedures similar to those used in the molecular vibration problem of Chapter 5. Symmetry orbitals are constructed by inspection, by projection, or by diagonalizing certain operators or Dirac characters. If only one symmetry orbital exists for a given partner of an irreducible representation, the form of this orbital is already determined. If several sets of identically transforming symmetry orbitals exist, linear combinations are taken of corresponding partners with coefficients that are determined by the Hamiltonian rather than group theory.

The oversimplified Hückel theory has the advantage of concentrating on the symmetry properties. It is not adequate as an a priori calculation, but if the parameters (e.g., α and β) are fitted to experiment, it provides a good framework for understanding the experimental data.

13.2. VALENCE BOND ORBITALS

In this section we show how group theory can be used to construct *directed valence bond orbitals*. Let us start with the by now classic example of tetrahedrally bonded methane, CH_4. Our problem is to construct four linear combinations of carbon orbitals (i.e., hybrid orbitals) that will be directed toward each of the four hydrogen atoms, that is, in the directions

$$\mathbf{n}_1 = \frac{(1,1,1)}{\sqrt{3}}, \qquad \mathbf{n}_2 = \frac{(1,-1,-1)}{\sqrt{3}}$$

$$\mathbf{n}_3 = \frac{(-1,1,-1)}{\sqrt{3}}, \qquad \mathbf{n}_4 = \frac{(-1,-1,1)}{\sqrt{3}}$$

Thus each orbital can be pictured as one of the vectors in Fig. 13.2.1.

Character Arguments

To decide which atomic orbitals must be used to construct the desired hybrid orbitals, let us determine the character of the four-dimensional space spanned by the four "vector-like" orbitals $\varphi_1, \varphi_2, \varphi_3, \varphi_4$ with respect to the elements of the symmetry group T_d of the molecule. The individual group operations simply permute the functions φ_μ. Thus

$$\sigma_{xy}\varphi_1 = \varphi_1, \qquad \sigma_{xy}\varphi_2 = \varphi_3, \qquad \sigma_{xy}\varphi_3 = \varphi_2, \qquad \sigma_{xy}\varphi_4 = \varphi_4 \tag{13.2.1}$$

Thus the character $\chi(\sigma_{xy}) = 2$, and, in general, the character of any element is equal to the number of vectors unshifted by the group operation. In this way we

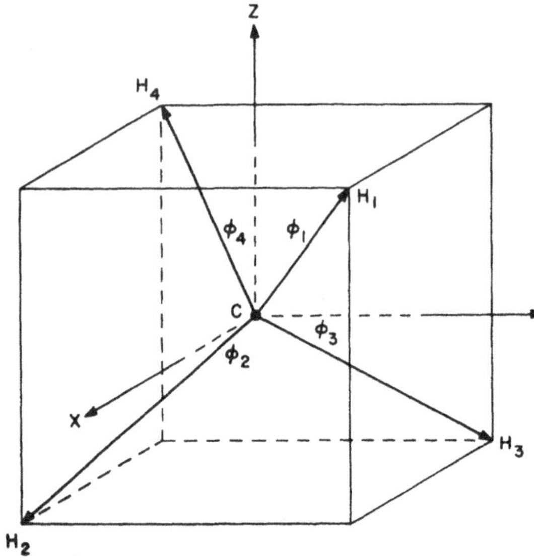

Fig. 13.2.1. The four valence bond orbitals of methane, CH_4, are represented by four tetrahedrally oriented vectors.

arrive at the characters:

$$\begin{array}{c|ccccc} & E & 8C_{3xyz} & 3C_{2x} & 6S_{4x} & 6\sigma_{xy} \\ \hline \Gamma^{\sigma}_{\text{tetra}} & 4 & 1 & 0 & 0 & 2 \end{array} \tag{13.2.2}$$

Decomposing this representation with respect to T_d, we find

$$\Gamma^{\sigma}_{\text{tetra}} = A_1 + T_2 \tag{13.2.3}$$

Table A11 indicates that the possible orbitals for A_1 are $(x^2+y^2+z^2)$, $(x^4+y^4+z^4)$, $x^2y^2z^2$, $x^8y^8z^8$, and xyz, and so forth. The carbon atom has the ground state $1s^22s^22p^2$, and the lowest-lying d state has a much higher energy than these. Thus only the $(x^2+y^2+z^2)$ state has low enough energy to enter the hybrid orbital with any appreciable probability. Table A11 lists the orbitals for representation T_2 (with respect to T_d) as Γ_5^+ and Γ_4^- (with respect to O). The T_2 orbitals of lowest energy are, therefore,

$$T_2: \qquad (x,y,z),(yz,zx,xy) \tag{13.2.4}$$

and in carbon only the p_x,p_y,p_z orbitals are likely to enter.

Linear Combination of Atomic Orbitals

We therefore seek four orbitals of the form

$$as + bp_x + cp_y + dp_z \tag{13.2.5}$$

with four sets of numerical coefficients a, b, c, d to be chosen to yield four sets of orbitals, each of which transforms as a vector in Fig. 13.2.1.

Let us consider the orbital pointing in the (1, 1, 1) direction. It must be invariant with respect to the subgroup $E, C_{3xyz}, C_{3xyz}^{-1}$ of T_d. Such an invariant is immediately constructed from the scalar product $[p_x, p_y, p_z]\cdot\mathbf{n}_1$, so that

$$\varphi_1 = as + b(p_x + p_y + p_z) \tag{13.2.6a}$$

Similarly,

$$\varphi_2 = as + b(p_x - p_y - p_z) \tag{13.2.6b}$$

$$\varphi_3 = as + b(-p_x + p_y - p_z) \tag{13.2.6c}$$

$$\varphi_4 = as + b(-p_x - p_y + p_z) \tag{13.2.6d}$$

We could also have constructed φ_2, φ_3, and φ_4, using group operations:

$$\varphi_2 = C_{2x}\varphi_1, \qquad \varphi_3 = C_{2y}\varphi_1, \qquad \varphi_4 = C_{2z}\varphi_1 \tag{13.2.7}$$

Sets of orbitals in this way by group operations are known as *equivalent orbitals* (Lennard-Jones, **1949**; Hall, **1951**). If we define s, p_x, p_y, p_z to be normalized orbitals (they are automatically orthogonal), normalization of the φ_μ yields the relationship

$$|a|^2 + 3|b|^2 = 1 \tag{13.2.8}$$

Bond Strengths

With an orbital written in factored form $R(r)Y(\theta,\varphi)$, such that $\int |Y(\theta,\varphi)|^2 d\Omega/4\pi = 1$, Pauling (**1960**, Chapter 4) defined the *strength of a bond* as the maximum numerical value of $|Y(\theta,\varphi)|$. For our directed orbital φ_1, the angular dependence is

$$\varphi_1 \sim a + \sqrt{3}\, b\left(\frac{x}{r} + \frac{y}{r} + \frac{z}{r}\right) \tag{13.2.9}$$

since $\sqrt{3}\, x/r$ is a normalized p function, that is, the average of $3x^2/r^2$ over a sphere is unity. The maximum occurs when $x=y=z=r/\sqrt{3}$, leading to the bond strength

$$B=a+3b \tag{13.2.10}$$

Pauling suggests that the ratio b/a be determined by maximizing the *bond strength B*. (This will make the carbon orbital point more sharply toward the appropriate hydrogen atom.) Since a can be chosen real and positive, we must therefore require $b=[(1-a^2)/3]^{1/2}$ to be real and positive and choose a to maximize

$$B(a)=a+[3(1-a^2)]^{1/2} \tag{13.2.11}$$

which leads to $a=b=\frac{1}{2}$ and a bond strength $B=2$. A recent example of the use of such tetrahedrally directed bonds in solids is given by Harrison (**1973**). A comprehensive discussion of bonds in solids is given by Phillips (**1973**).

Orthogonality Requirements

Pauling's procedure is certainly qualitatively correct; in particular, b/a should be taken as real. But the quantitative determination of a seems slightly arbitrary. It is interesting to note that, if we require our orbitals to be *orthogonal*, then

$$(\varphi_1,\varphi_2)=|a|^2-|b|^2=0 \tag{13.2.12}$$

leads to $a=b=\frac{1}{2}$ if a,b are assumed to be real—the same result as Pauling's!

Conversely, if we construct two hybrid orbitals directed along arbitrary directions $\mathbf{n}$ and $\mathbf{n}'$, and require maximum bond strength for each orbital *and* orthogonality between the orbitals, we find that $\mathbf{n}\cdot\mathbf{n}'=-\frac{1}{3}$, that is, the orbitals must make the tetrahedral angle 109°28′ with respect to one another.

Promotion Energy

Since the ground state of atomic carbon is $1s^2 2s^2 2p^2$, we must "promote" one $2s$ orbital to a $2p$ state in order to have the one s and three p orbitals needed to construct our four tetrahedral bonds. The increase of the total binding energy associated with the formation of these bonds is more than enough to compensate for the required *promotion energy*.

Dynamic Considerations

The foregoing symmetry arguments are equally valid with p_x, p_y, p_z replaced by d_{yz}, d_{zx}, d_{xy}, respectively, or by any (identical) linear combination of corresponding basis functions. The appropriate ratio of d to p is not given by symmetry considerations but can be determined by a variational procedure that minimizes the total energy of the molecule. In CH_4 the d/p ratio should be small, because of the large energy required to promote electrons into d states. The ion MnO_4^-, however, has $3d$ and $4p$ orbitals available for bonding. In this case, the d orbitals would be favored by energetic considerations.

Σ and π Orbitals

The directed bonds described above, which transform as vectors along the bond direction, are referred to as σ bonds. In addition, there are orbitals that transform under the molecular group as vectors perpendicular to the bond direction. These are known as π bonds, or orbitals. The classic example is ethylene, C_2H_4, shown in Fig. 13.2.2. Each C atom has three σ bonds, $\varphi_1, \varphi_2, \varphi_3$, in the x-y plane separated by angles of 120°. The pair of σ orbitals φ_1 and φ_1' directed from each carbon to the other constitutes the chemist's *single bond.* In addition, the π orbitals φ_4 and φ_4' are both in the $+z$ direction and constitute a second bond, which, together with the σ bond, constitutes the chemist's *double bond.* Thus the quantum description of ethylene corresponds to the semiempiri-

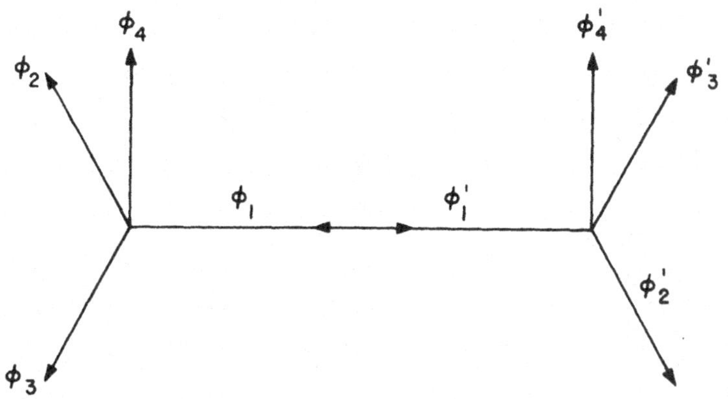

Fig. 13.2.2. The σ bonds $\varphi_1, \varphi_2, \varphi_3$ and $\varphi_1', \varphi_2', \varphi_3'$ of ethylene, C_2H_4, are shown by vectors in the plane of the molecule making 120° angles. The π bonds φ_4 and φ_4' are oriented perpendicular to this plane.

cal chemical formula

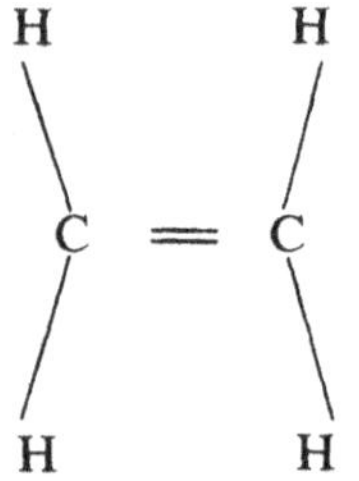

$$\begin{array}{c}HH\\ \diagdown\diagup\\ C=C\\ \diagup\diagdown\\ HH\end{array} \qquad (13.2.13)$$

Triple bonds may also be present; for example, in acetylene there is a C-C triple bond consisting of one σ bond and two π bonds. The binding energies associated with C−C, C=C, and C≡C are nearly independent of whatever else is in the molecule and are found to be roughly 2.6, 4.3, and 5.3 eV, respectively. We see that the π orbitals, with less overlap than σ orbitals that are directed toward one another, produce less bond strength.

The names σ and π derive from a classification of orbitals with respect to their properties under rotations about the bond axis. More specifically, the group of all rotations that have a bond direction invariant is the group $\mathcal{C}_{\infty v}$ containing rotations C_φ through all possible angles φ about the symmetry axis and reflections σ_v in any plane containing the axis. This is the group appropriate to a heteronuclear diatomic molecule. Its character table is given in Table 13.2.1.

We see that $\Sigma, \pi, \Delta \ldots$ orbitals transform as z or $(x+iy)^0$, $[(x+iy),(x-iy)]$, $[(x+iy)^2, (x-iy)^2], \ldots$, that is, as orbitals $\exp(im\varphi)$ with $m = 0, \pm 1, \pm 2, \ldots$. The $\Sigma, \pi, \Delta \ldots$ notation for this two-dimensional rotation group is the Greek transliteration of the $S, P, D \ldots$ notation of the full rotation group. When referring to one-electron orbitals, lower-case letters $\sigma, \pi, \delta \ldots$ are used analogously to $s, p, d \ldots$ in $R(3)$.

Table 13.2.1. $\mathcal{C}_{\infty v} \times \mathcal{C}_i = D_{\infty h}$ (∞/mm)

$\mathcal{C}_{\infty v}$ (∞m)		E	$C_\varphi, C_\varphi^{-1}$	σ_v	$\mathcal{C}_{\infty v}^+$	$\mathcal{C}_{\infty v}^-$
Σ^+	A_1	1	1	1	$[x^2+y^2, z^{2m}]$	z^{2m+1}
Σ^-	A_2	1	1	-1	$x_1y_2 - y_1x_2$	$z_1(x_1y_2 - y_1x_2)$
Π	E_1	2	$2\cos\varphi$	0	$[zx, zy]$	$[x, y]$
Δ	E_2	2	$2\cos 2\varphi$	0	$[(x+iy)^2, (x-iy)^2]$	$z[(x+iy)^2, (x-iy)^2]$
	E_n	2	$2\cos n\varphi$	0	$z^m[(x+iy)^n, (x-iy)^n]$ $m+n=$ even	$z^m[(x+iy)^n, (x-iy)^n]$ $m+n=$ odd

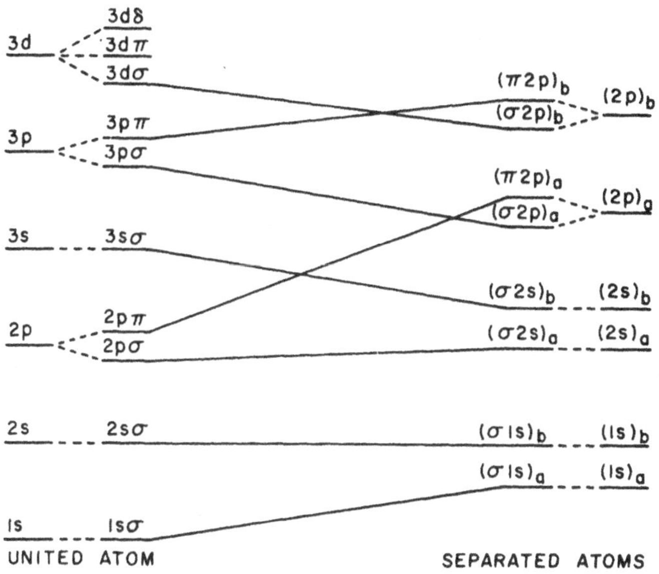

Fig. 13.2.3. A correlation diagram for heteronuclear diatomic molecules (taken from Eyring, Walter, and Kimball, 1944) that relates the symmetry of the wavefunctions in the separate *a* and *b* atom case, through intermediate cases, to the united atom case.

The reader may note that there is no one-electron orbital that transforms as Σ^-, just as there are no odd parity one-electron states of even l for $R(3)$.

It should be noted also that δ orbitals with respect to the group $\mathcal{C}_{\infty v}$ of *all* possible rotations about the bond axis might decompose into orbitals that transform as vectors parallel to or perpendicular to the bond axis ("σ or π orbitals") with respect to the rotations about this axis that are available in the group of the molecule.

Diatomic Molecules

The states of a heteronuclear diatomic molecule can be adequately classified using $R(3)$ when the two atoms are very far apart ("separated atoms") or when the nuclei are at the same point ("united atom"). The compatibility relations between $R(3)$ and $\mathcal{C}_{\infty v}$, and the requirement that levels of like symmetry cannot cross, essentially determine the correlation diagram (Fig. 13.2.3) that relates states in the diatomic molecule to states in the separated or united atom molecule.

For homonuclear diatomic molecules, the appropriate group $D_{\infty h}$ is obtained

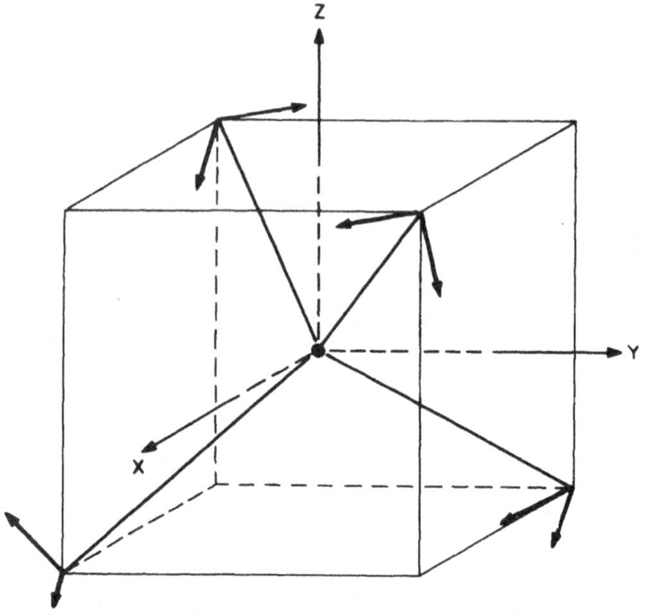

Fig. 13.2.4. The π orbitals for tetrahedral bonding are represented by vectors.

by adding inversion to $C_{\infty v}$. The reader is encouraged to draw a correlation diagram for the homonuclear case and to compare his results with Fig. 11.5 of Eyring, Walter, and Kimball (**1944**).

Example 13.2.1

Determine the symmetry character of the orbitals required to yield a set of four pairs of tetrahedrally related π orbitals. See Fig. 13.2.4.

Solution: These pairs transform as vectors perpendicular to the bond direction. The character of the space spanned by these eight vectors is precisely the same as in the corresponding molecular vibration problem, and Eq. 5.1.7 for that case can be transcribed to read as

$$\chi^{\pi}_{\text{tetra}}(R) = N(R)\chi^{\pi}(R) \tag{13.2.14}$$

where $N(R)$ is the number of atoms left invariant by R, $\chi^{\pi}(R)$ is the character of R in the two-dimensional space of a pair of π orbitals, and $\chi^{\pi}_{\text{tetra}}(R)$ is the character of R for the full tetrahedral array. Since $N(E)=4$, $N(C_{3xyz})=1$, $N(C_{2x})=0$, $N(S_{4x})=0$, $N(\sigma_{xy})=1$, we need only evaluate $\chi^{\pi}(R)$ for $R=E$, C_{3xyz}, and σ_{xy} (or other elements in the same classes). It is convenient, moreover, to use these particular elements, since they belong to the *group of the bond*

(1, 1, 1), that is, they leave the (1, 1, 1) bond invariant. This group is simply the group $\mathcal{C}_{3v}$. And from representation $\Gamma_3(x,y)$ of $\mathcal{C}_{3v}$ in Table A7 we can read the characters 2, -1, 0 for E, C_3, and σ, respectively. The complete set of characters is now given by

	E	$8C_{3xyz}$	$3C_{2x}$	$6S_{4x}$	$6\sigma_{xy}$
$\Gamma^{\pi}_{\text{tetra}}$	8	-1	0	0	0

(13.2.15)

Comparison with Table A11 for T_d yields

$$\Gamma^{\pi}_{\text{tetra}} = (T_2 + E) + T_1 \tag{13.2.16}$$

The simplest basis functions for T_2 are the p and d_{yz}, d_{xy}, d_{zx} states cited in Eq. 13.2.4. The simplest T_1 states from Table A11 are

$$T_1 = [x(y^2 - z^2), y(z^2 - x^2), z(x^2 - y^2)], [yz(y^2 - z^2), zx(z^2 - x^2), xy(x^2 - y^2)] \tag{13.2.17}$$

and the simplest basis functions for E are

$$E: \qquad [u, v] = \left[z^2 - \tfrac{1}{2}(x^2 + y^2), \tfrac{1}{2}\sqrt{3}\,(x^2 - y^2)\right] \tag{13.2.18}$$

Example 13.2.2

Construct the eight tetrahedral π orbitals of Example 13.2.1.

Solution: If we use $[X_1, Y_1]$ as a brief labeling of the pair of π orbitals that transform as a pair of vectors perpendicular to the (1, 1, 1) axis, the *equivalent orbitals* appropriate to the other three bonds are given by

$$[X_2, Y_2] = C_{2x}[X_1, Y_1]; \qquad [X_3, Y_3] = C_{2y}[X_1, Y_1]; \qquad [X_4, Y_4] = C_{2z}[X_1, Y_1] \tag{13.2.19}$$

as in Eq. 13.2.7. Our problem is thus reduced to that of finding in the space $E + T_2 + T_1$ the basis functions that transform as $[X_1, Y_1]$ or, more briefly, as $[X, Y]$. These transformation properties can be described by

$$C_3[X, Y] = [X, Y]\begin{bmatrix} -\tfrac{1}{2}, & -\tfrac{1}{2}\sqrt{3} \\ \tfrac{1}{2}\sqrt{3}, & -\tfrac{1}{2} \end{bmatrix} = \left[-\tfrac{1}{2}X + \tfrac{1}{2}\sqrt{3}\,Y, -\tfrac{1}{2}\sqrt{3}\,X - \tfrac{1}{2}Y\right] \tag{13.2.20}$$

$$\sigma_{xy}[X, Y] = [-X, Y] \tag{13.2.21}$$

See Example 1.4.1 for comparison. The fact that $(1,-1,0)/\sqrt{2}$ is a vector orthogonal to $(1,1,1)/\sqrt{3}$ suggests that $(x-y)/\sqrt{2}$, which is odd under σ_{xy} transforms as X. Using

$$C_{3xyz}[x,y,z]=[y,z,x] \tag{13.2.22}$$

(see Table 9.1.1), we note that

$$C_{3xyz}\frac{x-y}{\sqrt{2}}=\frac{y-z}{\sqrt{2}}=-\frac{1}{2}\frac{x-y}{\sqrt{2}}+\frac{1}{2}\sqrt{3}\,\frac{x+y-2z}{\sqrt{6}} \tag{13.2.23}$$

from which we can conclude that

$$[X,Y]\sim\left[\frac{x-y}{\sqrt{2}},\frac{x+y-2z}{\sqrt{6}}\right] \tag{13.2.24}$$

constitute a pair of π orbitals obtained from the p_x,p_y,p_z states, that is, the odd parity T_2 states of T_d. Since the even parity T_2 states transform in the same way, we immediately obtain another set for T_2:

$$[X,Y]\sim\left[\frac{yz-zx}{\sqrt{2}},\frac{yz+zx-2xy}{\sqrt{6}}\right] \tag{13.2.25}$$

We can construct a set for E in a similar manner or by using the projection formula 3.7.14. If we note, however, that x^2-y^2 is already odd under σ_{xy}, we need not use the full group of the bond $\mathcal{C}_{3v}$, but only the subgroup $\mathcal{C}_3$:

$$[E+C_3+C_3^{-1}]X(x^2-y^2)=X(x^2-y^2)+\left[-\tfrac{1}{2}X+\tfrac{1}{2}\sqrt{3}\,Y\right](y^2-z^2)$$

$$+\left[-\tfrac{1}{2}X-\tfrac{1}{2}\sqrt{3}\,Y\right][z^2-x^2] \tag{13.2.26}$$

By reading off the coefficients of X and Y, we find that

$$[X,Y]\sim\left[\frac{x^2-y^2}{\sqrt{2}},\frac{x^2+y^2-2z^2}{\sqrt{6}}\right]$$

In a similar manner, the T_1 states yield:

$$[X,Y]\sim\left[\frac{x(y^2-z^2)+y(z^2-x^2)-2z(x^2-y^2)}{\sqrt{12}},\frac{y(z^2-x^2)-x(y^2-z^2)}{2}\right]$$

Thus, if only $d=E+T_2$ and $f=(T_1+\cdots)$ states are available, we write

$$X=\frac{a(x^2-y^2)}{\sqrt{2}}+\frac{b(yz-zx)}{\sqrt{2}}+\frac{c[x(y^2-z^2)+y(z^2-x^2)-2z(x^2-y^2)]}{\sqrt{12}}$$

$$Y=\frac{a(x^2+y^2-2z^2)}{\sqrt{6}}+\frac{b(yz+zx-2xy)}{\sqrt{6}}+c\tfrac{1}{2}[y(z^2-x^2)-x(y^2-z^2)]$$

(13.2.27)

where yz, for example, stands for a normalized d_{yz} state, so that $(yz-zx)/\sqrt{2}$ is also normalized. Thus our normalization condition takes the form

$$|a|^2+|b|^2+|c|^2=1 \tag{13.2.28}$$

The orthogonality conditions yield

$$(X,C_{2x}X)=|a|^2-(\tfrac{2}{3})|c|^2=0$$

$$(X,C_{2y}X)=|a|^2-(\tfrac{2}{3})|c|^2=0 \tag{13.2.29}$$

$$(X,C_{2z}X)=|a|^2-|b|^2+(\tfrac{1}{3})|c|^2=0$$

Therefore, we shall set

$$a=\tfrac{1}{2},\qquad b=(\tfrac{3}{8})^{1/2},\qquad c=-(\tfrac{3}{8})^{1/2} \tag{13.2.30}$$

where the relative phases have been so chosen that, for $[x,y,z]$ in the $[1,-1,0]$ or $[1,1,-2]$ lobe direction, the three contributions in Eq. 13.2.27 all have the same sign. This choice maximizes the Pauling bond strengths in the bond directions.

Sharing of Orbitals: Weak Bonds

If both p_x,p_y,p_z and d_{yz},d_{zx},d_{xy} orbitals are available, the p orbitals can be used in the σ bonds and the d orbitals in the π bonds, or vice versa. The correct assumption is a mixture of p and d for the σ bonds, and the orthogonal mixture for the π bonds. The amount of admixture is determined by dynamical rather than symmetry considerations. (An example of such mixing in a trigonal bipyramidal AB_5 molecule is given by Cotton, **1963**, p. 115.)

If only p orbitals are available, they will be partly used up in the σ orbitals and will lead to *weak* π orbitals. If the central atom possesses s, p, and d but no $f(T_1)$ electrons, it is possible to form the four σ orbitals but only five of the eight π orbitals. In this case, the five π orbitals will be *shared* by the four neighboring atoms. The nature of this sharing will be elucidated in the discussion of *resonance* in Section 13.3.

13.3. MANY-BODY WAVE FUNCTIONS AND CHEMICAL STRUCTURES

The valence theory of the chemical bond starts with the classic Heitler-London (**1927**) theory of the H_2 molecule. We shall not repeat this theory since it is reviewed in many other books (see, e.g., Tinkham, **1964** or Eyring, Walter, and Kimball, **1944**). If $a(1)$ stands for electron 1 in an orbital centered about atom a, then

$$\psi(^1\Sigma_g^+) = [a(1)b(2) + a(2)b(1)][\alpha(1)\beta(2) - \alpha(2)\beta(1)] \quad (13.3.1)$$

is a state symmetric in space and antisymmetric in spin (the α and β refer to up- and down-spin states respectively) and represents the singlet ground state of H_2. Its energy is lower than that of the corresponding triplet state:

$$\psi(^3\Sigma_u^+) = [a(1)b(2) - a(2)b(1)] \times \begin{bmatrix} \alpha(1)\alpha(2) \\ (\alpha(1)\beta(2) + \alpha(2)\beta(1))/\sqrt{2} \\ \beta(1)\beta(2) \end{bmatrix} \quad (13.3.2)$$

because it is symmetric in space and concentrates the electronic charge in the region between the two nuclei where it can be attracted to *both*, thus enhancing the potential energy. The Pauli principle, which requires antisymmetry on the interchange of both space and spin coordinates, permits a symmetric spatial configuration only with an antisymmetric spin configuration. It is not surprising, therefore, that the ground state of most molecules is a singlet state.

The simplest many-electron wave function that obeys the Pauli principle of complete antisymmetry is a Slater determinant (See Slater, **1928**, **1929**):

$$\varphi = \frac{1}{(n!)^{1/2}} \begin{vmatrix} (a\alpha)_1 & (b\alpha)_1 & (c\beta)_1 & \cdots & (n\alpha)_1 \\ (a\alpha)_2 & (b\alpha)_2 & (c\beta)_2 & \cdots & (n\alpha)_2 \\ \cdots & & & & \\ (a\alpha)_n & (b\alpha)_n & (c\beta)_n & \cdots & (n\alpha)_n \end{vmatrix} \quad (13.3.3)$$

formed of n orbitals $a(1)b(2)\ldots n(n)$ and one of the 2^n possible assignments of spins to these orbitals. For ground states of molecules, we can confine ourselves to spin assignments with total $S_z=0$. Thus, for a four-orbital problem, there are six possible assignments:

$$\begin{array}{lcccc} & a & b & c & d \\ \varphi_1 & \alpha & \alpha & \beta & \beta \\ \varphi_2 & \alpha & \beta & \alpha & \beta \\ \varphi_3 & \beta & \alpha & \alpha & \beta \\ \varphi_4 & \alpha & \beta & \beta & \alpha \\ \varphi_5 & \beta & \alpha & \beta & \alpha \\ \varphi_6 & \beta & \beta & \alpha & \alpha \end{array} \tag{13.3.4}$$

A state corresponding to the chemist's bond diagram

$$\begin{array}{c} a\text{——}b \\ d\text{——}c \end{array}$$

can be formed by taking linear combinations of the six φ's in such a way as to antisymmetrize the spin states between a and b, and to antisymmetrize them between c and d. (Antisymmetrization of spin permits spatial symmetrization or piling up of charge in the bond directions.) A bond eigenfunction or *structure function* with this symmetry is

$$\psi_{ab,cd}=\varphi_2-\varphi_3-\varphi_4+\varphi_5 \tag{13.3.5a}$$

Similarly

$$\psi_{ad,bc}=\varphi_1-\varphi_2-\varphi_5+\varphi_6 \tag{13.3.5b}$$

$$\psi_{ac,bd}=\varphi_1-\varphi_3-\varphi_4+\varphi_6 \tag{13.3.5c}$$

In general,

$$\psi_{ij,kl}=\sum_{\mu=1}^{6}\delta_{ij}(\mu)\,\delta_{kl}(\mu)\,\varphi_\mu \tag{13.3.5d}$$

where $\delta_{ij}(\mu)=1$ if i has spin α and j has spin β, $\delta_{ij}(\mu)=-1$ if i has β and j has α, $\delta_{ij}(\mu)=0$ if i and j have the same spin. This expression is readily generalized to the case of more than four orbitals. These structure functions are necessarily symmetric in the spatial coordinates of any pair in a bond, thus improving the bonding. Moreover, the bonding is *enhanced by using directed bond orbitals rather than atomic orbitals*. The result is similar to the simple Heitler-London case: a pair of orbitals from two sites is placed in a singlet spin state. The composition

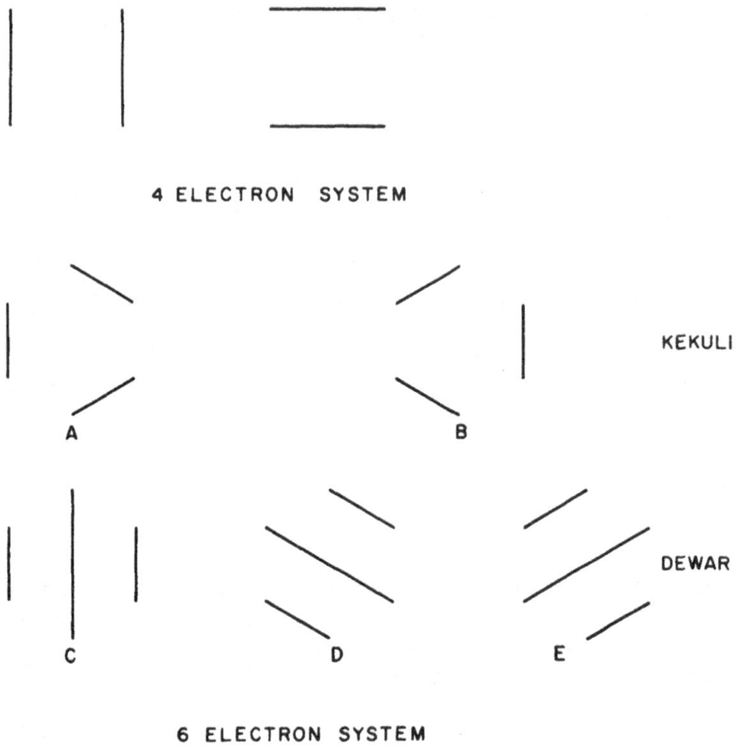

Fig. 13.3.1. Rumer diagrams that describe molecular structures (and the corresponding wave functions) for the four- and six-electron cases. (The six-electron structures were introduced in the study of benzene, C_6H_6.)

of all these singlet pair states necessarily leads to a singlet state for the molecule as a whole (Eyring, Walter, and Kimball, **1944**; Simpson, **1962**).

The above procedure produces a simple relation between an organic chemist's formula and the many-electron wave function needed to describe it in the valence bond theory. The structure functions are not all independent, however:

$$\psi_{ac,bd} = \psi_{ab,cd} + \psi_{ad,bc} \tag{13.3.6}$$

Rumer (**1932**) provided a simple means of reducing all possible structures to a "canonical" set of independent structures. The canonical diagrams are obtained by (1) drawing a circle, (2) placing one point on the circle for each nucleus in any order (not necessarily that found in the molecule), and (3) connecting the points in pairs, making sure that no lines cross.

Figure 13.3.1 shows the Rumer diagrams for a four-electron system and a six-electron system. The number of such diagrams for an $n = 2N$ electron system can be shown by combinatorial analysis (Motzkin, **1948**) to be

$$N_{\text{Rumer}}=\frac{(2N)!}{N!(N+1)!} \tag{13.3.7}$$

The bond eigenfunctions corresponding to Rumer diagrams can readily be seen to be singlet states, that is, eigenfunctions of S^2 with eigenvalue zero, where S is the total spin operator for the system (Eyring, Walter, and Kimball, **1944**; Simpson, **1962**). What is not entirely clear is that all possible singlet states are included. But the number N_S of such singlet states using vector coupling $\mathbf{S}=\mathbf{S}_1+\mathbf{S}_2+\ldots+\mathbf{S}_{2N}$ of the spins is the number of times that $\Gamma_{1/2}\times\Gamma_{1/2}\times\ldots\times\Gamma_{1/2}$ (with $2N$ factors) contains the state Γ_0. With $\chi^{1/2}(\varphi)=\exp\,(i\varphi/2)+\exp\times(-i\varphi/2)=2\cos(\varphi/2)$, we have

$$N_S=\frac{\int_0^{2\pi}(2\cos\tfrac{1}{2}\varphi)^{2N}(1-\cos\varphi)\,d\varphi}{\int_0^{2\pi}(1-\cos\varphi)\,d\varphi} \tag{13.3.8}$$

since $1-\cos\varphi$ is the Hurwitz invariant volume element appropriate to the rotation group. An evaluation of Eq. 13.3.8 yields $N_S=N_{\text{Rumer}}$, so that the bond eigenfunction method of Rumer does indeed produce all singlet states.

Resonance

The binding energy $(\psi,H\psi)$ can be calculated for each structure function or chemical diagram. The structure function with the largest binding energy is assumed to describe the bound state of the molecule. If we exploit the rotational symmetry of a molecule, we may find, however, that several structures have the same energy. For the benzene molecule, for example, the two Kekuli structures ψ_A and ψ_B of Fig. 13.3.1 differ by a rotation of 60° and are clearly degenerate. The molecule is said to resonate between these two structures. The increase in binding energy produced by the best linear combination of structure functions, as compared to the best binding energy of a single structure function, is called the *resonance energy*.

The determination of the best linear combination of structure functions can be facilitated by first constructing symmetry orbitals. Let us illustrate the procedure for benzene and use the group D_6 (since inversion yields no new information in this example).

Each group operation simply permutes the structures within an invariant set (e.g., the Kekuli set or the Dewar set of Fig. 13.1.1). Thus the character of an operation R in the space of a set of structures is the number of structures left invariant by R. These characters for the Kekuli and Dewar sets are as follows:

	E	$2C_{3z}$	$3C_{2y}$	C_{2z}	$2C_{6z}$	$3C_{2x}$
Γ_{Kekuli}	2	2	0	0	0	2
Γ_{Dewar}	3	0	1	3	0	1

(13.3.9)

Comparison with Table A9 for D_6 shows that

$$\Gamma_{\text{Kekuli}} = \Gamma_1 + \Gamma_4, \qquad \Gamma_{\text{Dewar}} = \Gamma_1 + \Gamma_6 \tag{13.3.10}$$

Since no representation is repeated, the decomposition into symmetry orbitals is uniquely determined. By projection or, more easily, by inspection, we find that in terms of the structure functions $\psi_A, \ldots, \psi_E$ of Fig. 13.3.1

$$\Gamma_1(\text{Kekuli}) = \psi_A + \psi_B$$

$$\Gamma_4 = \psi_A - \psi_B$$

$$\Gamma_1(\text{Dewar}) = \psi_C + \psi_D + \psi_E$$

$$\Gamma_6 = [\psi_C + \omega\psi_D + \omega^2\psi_E, \psi_C + \omega^2\psi_D + \omega\psi_E]$$

where $\omega = \exp(2\pi i/3)$, and we could have chosen any linear combination of ψ_C, ψ_D, and ψ_E orthogonal to $\psi_C + \psi_D + \psi_E$. Our choice diagonalizes C_{3z}.

Of course, the Hamiltonian couples together the two Γ_1 wave functions. The unnormalized ground-state wave function of benzene is found to be

$$\psi = \psi_A + \psi_B + 0.434(\psi_C + \psi_D + \psi_E)$$

The resonance energy associated with this wave function is found to be 1.11α (where α is an overlap integral), whereas, if only Kekuli structures are used, it is 0.90α, so that most of the resonance energy is supplied by the Kekuli structures. For a more detailed discussion of benzene see Eyring, Walter, and Kimball (**1944**) or Simpson (**1962**).

In a molecule for which the number of available orbitals (of low energy) is less than the number of bonds needed, we can assign the orbitals to the bonds in all possible ways and then allow a *sharing* of the orbitals by permitting the individual structures to "resonate" with one another.

13.4. HARTREE-FOCK WAVE FUNCTIONS AND BROKEN SYMMETRY

The possibility of broken symmetry—that the one-electron orbitals can have a symmetry less than is anticipated from that of the many-electron Hamiltonian—is explained in this section, using the Hartree-Fock wave function.

The Hartree-Fock method (for atoms, molecules, or solids) consists in using a single determinant of one-electron (space-spin) orbitals (as in Eq. 13.3.3) as a trial function in a variational method. Variation of the individual orbitals to find the lowest total binding energy leads to a set of equations for the one-electron orbitals. These equations are referred to as self-consistent field (SCF) equations, because the equation for each orbital depends on the (as yet unknown) wave functions of the remaining orbitals. The solution of these equations requires an iteration procedure until self-consistency is obtained.

With $x=\mathbf{r},m$ as an abbreviation for space *and* spin coordinates, the Hamiltonian of our many-electron system takes the form

$$H=\sum_i\left[\frac{(\mathbf{p}_i)^2}{2m}+V(x_i)\right]+\sum_{i<j}w(x_i,x_j) \tag{13.4.1}$$

where $V(x)$ is the potential at $\mathbf{r}$ due to all the nuclei [including an electrostatic part $V(\mathbf{r})$ and spin-dependent hyperfine interactions if desired], and $w(x,x')$ is the interaction between a pair of electrons $w(x,x')=e^2/|\mathbf{r}-\mathbf{r}'|+$spin-spin and spin-orbit terms (see Bethe and Salpeter, (**1957**), for the spin interactions that are omitted in most binding energy calculations).

With a Hamiltonian (13.4.1) restricted to one- and two-body interactions, the HF-SCF equations for the one-electron orbitals $\psi_\mu(x)$ take the form (Messiah, **1962**, Chapter 18, §9)

$$\left[-\frac{\hbar^2}{2m}\nabla^2+V(x)\right]\psi_\lambda(x)+W(x)\psi_\lambda(x)$$

$$-\int W_{\text{exc}}(x,x')\,dx'\psi_\lambda(x')=e_\lambda\psi_\lambda(x) \tag{13.4.2}$$

where

$$W_{\text{exc}}(x,x')=\rho(x;x')w(x,x') \tag{13.4.3}$$

$$W(x)=\int\rho(x')w(x,x')\,dx' \tag{13.4.4}$$

$$\rho(x;x')\equiv\sum_\mu\psi_\mu(x)\psi_\mu^\dagger(x') \tag{13.4.5}$$

$$\rho(x)\equiv\rho(x;x)=\sum_\mu|\psi_\mu(x)|^2 \tag{13.4.6}$$

and an integral over x' means an integral over $\mathbf{r}'$ and a sum over m'. Note that $\psi_\mu(x)=\psi_\mu(\mathbf{r},m)$ is not necessarily of the product form $a_\mu(\mathbf{r})\alpha(m)$.

The interaction $w(x,x')=w(x',x)$ is invariant under an arbitrary simultaneous rotation of both particles. *If* the $\psi_\mu(x)$ span a space invariant under the elements of a group G, then, by the shell theorem including spin (Theorem 4.5.6), $\rho(x;x')$ is invariant under rotations

$$\rho(x;x')=O_R\rho(x;x')O_R^{-1} \tag{13.4.7}$$

for $R\in G$. In this case, $\psi_\lambda(x)$ in Eq. 13.4.2 is an eigenfunction of an effective one-body Hamiltonian that commutes with the elements of G. For an atom, this reduces to the usual statement that, in the case of closed shells, the electrons move in a spherically symmetric potential and form bases for the irreducible representations of $R(3)$. In a molecule with a *complete* set of *occupied* orbitals, G will be the symmetry group G_m of the molecule, and our orbitals will be molecular orbitals.

In general, however, we have an open-shell case and the group G under which $\{\psi_\mu(x)\}$ is an invariant space will be some subgroup of the symmetry group G_m of the original many-body Hamiltonian; the one-body Hamiltonian and wave functions will then have lower symmetry than might have been anticipated. Barring accidental exact or near degeneracy, the many-electron wave function Ψ must belong to an irreducible representation of the full group G_m under which the many-electron Hamiltonian is invariant. However, Ψ may belong to a one-dimensional or a multidimensional representation of G_m. We shall see that in the former case the set of one-electron orbitals is invariant under G_m, and the corresponding one-electron Hamiltonian commutes with G_m. For the latter case, the same remarks will apply to that (largest) subgroup of G_m under which Ψ is a one-dimensional basis.

Theorem 13.4.1

If the determinantal wave function Ψ is composed of one-electron orbitals φ_μ, and these orbitals form an invariant space under the elements of some group G, then Ψ must be a one-dimensional basis function for G.

Proof: Under any unitary change of basis

$$\varphi_{\mu'}=\sum\varphi_\mu S_{\mu\mu'} \tag{13.4.8}$$

the determinantal wave function Ψ is simply multiplied by the Jacobian of the transformation

$$\Psi'=(\det S)\Psi \tag{13.4.9}$$

that is, by a phase factor.

But any $R\in G$ generates such a unitary transformation $S=D(R)$, and this establishes our theorem.

Thus, for G to equal G_m (i.e., for our orbitals φ_μ to form an invariant space under G_m), Ψ must be one-dimensional under G_m. In general, of course, Ψ will be one partner in a basis for an irreducible representation of G_m; for example, it could transform as z under the point group O_h. Then G is the subgroup that takes Ψ (e.g., z) into a multiple of itself (e.g., C_{4h}).

The converse of these ideas can be stated as follows.

Theorem 13.4.2

If the determinantal wave function $\Psi = \det[\psi_\mu(x)]$ is a basis for a one-dimensional representation of a group G, the set of one-electron orbitals $\{\psi_\mu(x)\}$ is invariant under G.

Proof: If the space $\sigma = \{\psi_\mu(x)\}$ is not invariant under $R \in G$, then $\{R\psi_\mu(x)\} = R\sigma \neq \sigma$ and $R\Psi$ must differ from Ψ by more than a phase factor.

Theorem 13.4.2 can be restated in a more general way.

Theorem 13.4.3

If $\Psi = \Psi(x_1, x_2, \ldots, x_N) \equiv \Psi(1,2,\ldots,N)$ is any many-body wave function (not necessarily a determinant) that is a basis for a one-dimensional representation of a group G, the split charge density $\rho(x;x')$ is invariant under G.

Proof: The appropriate generalization of Eq. 13.4.5 to a general Ψ is defined by

$$\rho(x;x') = \frac{1}{N}\sum_{j=1}^{N} \int \Psi(1,2,\ldots,j-1,x,j+1,\ldots,N)$$

$$\times \Psi^*(1,2,\ldots,j-1,x',j+1,\ldots,N)\, d1\, d2 \ldots dN/dj \qquad (13.4.10)$$

where $d3$ means $\Sigma_{m_3} \int d\mathbf{r}_3$. The average over all electrons j is unnecessary if Ψ is completely antisymmetric. Since $\Psi(1,2,\ldots,N)\Psi^*(1',2',\ldots,N')$ is by hypothesis invariant under any $R \in G$, and this property is not lost by integrating over $N-1$ variables, $\rho(x;x')$ is invariant:

$$\rho(x;x') = O_R \rho(x;x') O_R^{-1} \qquad (13.4.11)$$

or

$$\rho(\mathbf{r}m;\mathbf{r}'m') = \sum_{n,n'} D^{1/2}_{mn}(R)\rho(R^{-1}\mathbf{r},n;R^{-1}\mathbf{r}',n')D^{1/2}_{n'm'}(R^{-1}) \qquad (13.4.12)$$

Higher-order correlation functions obey the same invariance requirements.

The nature of the one-electron orbitals contained in an arbitrary many-electron Ψ can be studied by expanding $\rho(x;x')$ in any complete set $\psi_\mu(x)$ of basis functions:

$$\rho(x;x') = \sum \psi_\mu(x)\psi_\nu(x')^*\rho_{\mu\nu} \qquad (13.4.13)$$

The matrix $\rho_{\mu\nu}$ is Hermitian and can be diagonalized by a change of basis. The orbitals that diagonalize $\rho_{\mu\nu}$ are the *natural spin orbitals* of Löwdin (**1956a, 1956b**). For a determinantal Ψ, $\rho_{\mu\nu}$ in diagonal form has 1's for all occupied orbitals and 0's for all empty orbitals. For a more general Ψ (e.g., a linear combination of determinants), the diagonalized elements can be between 0 and 1, indicating a finite probability that a given orbital is occupied.

Theorem 13.4.4

If a many-body Ψ is a one-dimensional representation of G, all of the basis functions $\varphi_\mu^{ia}(x)$ of a given subspace σ^{ia} appear with equal probability in Ψ.

Proof: Equation 13.4.13 yields

$$\rho_{\mu\nu} = \int \psi_\mu(x)^\dagger \rho(x;x')\psi_\nu(x')\,dx\,dx'$$

$$= (\psi_\mu, \rho_{\text{op}}(x;x')\psi_\nu)$$

But $\rho(x;x')$ regarded as an operator commutes with all $R \in G$. By Schur's lemma 1.5.2 or by the base vector orthogonality theorem 3.4.3, ρ is a constant within each irreducible representation or, more precisely,

$$(\psi_\mu^{ia}, \rho_{\text{op}}(x;x')\psi_\nu^{jb}) = \delta_{ij}\delta_{\mu\nu}\rho^{ia,ib}$$

that is,

$$\rho(x;x') = \sum_{i,\mu,a,b} \psi_\mu^{ia}(x)\psi_\mu^{ib}(x')^*\rho^{ia,ib}$$

A change of basis that mixes only basis functions of the same i and μ can be used to reduce $\rho^{ia,ib}$ to diagonal form $\rho^{ia,ia}$. This matrix is independent of μ, yielding the equal contributions of all partner functions (of a given ia) to Ψ.

The preceding theorems can be immediately generalized to include time reversal.

Theorem 13.4.5

If $\sigma \equiv \{\psi_\mu(x)\}$ is invariant under group G and time reversal K, then $\Psi = \det[\psi_\mu(x)]$ is a basis for a one-dimensional irreducible representation of G, and a phase factor can be chosen so that $K\Psi = \Psi$, that is, no added degeneracy is produced by time reversal.

Theorem 13.4.6

If $\Psi = \det[\psi_\mu(x)]$ is nondegenerate under the elements of G and under time reversal, then the one-electron orbitals $\psi_\mu(x)$ are eigenfunctions of a one-electron Hamiltonian that is invariant under G and time reversal.

Comment: In a ferromagnetic or antiferromagnetic material, $K\Psi \neq \Psi$. Thus the one-electron Hamiltonian no longer possesses time-reversal symmetry, although some operations of the form RK may still commute with the Hamiltonian.

Theorem 13.4.7

If $K\Psi=\Psi$ and $\psi_\mu(x)$ participates in Ψ, then $K\psi_\mu(x)$ participates with equal probability.

Proof: If K is applied to the definition of $\rho(x;x')$, we learn that

$$\rho(x;x')=K\rho(x;x')K^{-1}$$

or

$$\sum \psi_\mu(x)\psi_\nu(x')^\dagger\rho_{\mu\nu}=\sum K\psi_\mu(x)[K\psi_\nu(x')]^\dagger\rho^*_{\mu\nu}$$

or

$$\rho_{\mu\nu}=\rho^*_{\bar{\mu}\bar{\nu}}$$

Choosing a diagonal representation, we have $\rho_{\bar{\mu}\bar{\mu}}=\rho_{\mu\mu}$ as desired.

Comment: In the Hartree-Fock case, if $K\Psi=\Psi$, the effective one-body Hamiltonian will commute with K.

The usual application of the Hartree-Fock method (Seitz, **1940**; Slater, **1960**) places two electrons, one with spin up and one with spin down, into each orbital state. The unrestricted Hartree-Fock (UHF) method permits the spin-up and spin-down orbitals to differ. The UHF method abandons $K\Psi=\Psi$ and thus leads to one-body Hamiltonians that are not invariant under time reversal. If Ψ contains an odd number of electrons, Kramers' theorem requires $(K\Psi,\Psi)=0$, so that the UHF method is then necessary. Moreover, the UHF method may lead to a lower ground-state energy than the HF method, even for an even number of electrons.

The UHF method is a special case of the generalized UHF (GUHF) method, in which an arbitrary set of symmetry elements of the many-body Hamiltonian (possibly including K) is ignored. The GUHF method is particularly important in treating "open-shell" configurations, for which a reduction in symmetry is expected.

13.5. THE JAHN-TELLER EFFECT

In Chapter 5 we considered molecular vibrations, and in this chapter electronic states. The electronic Hamiltonian H can be expanded, however, in terms of the vibrational symmetry coordinates q^m_λ belonging to all available representations among the molecular vibrations:

$$H=H_0+\sum_{m\lambda} H^m_\lambda q^m_\lambda+\sum_{mn\lambda\rho} H^{mn}_{\lambda\rho}q^m_\lambda q^n_\rho+\ldots \tag{13.5.1}$$

If the molecule is in a degenerate state, Ψ_ν^j belongs to representation j (and partner ν); then $K\Psi_\mu^j$ is necessarily a state with the same energy. If

$$V_{\mu\nu\lambda}^{\bar{j}jm} = (K\Psi_\mu^j, H_\lambda^m \Psi_\nu^j) \neq 0 \tag{13.5.2}$$

there will be a coupling between the electronic states that is linear in the symmetry coordinates. Under such circumstances, the minimum energy cannot occur at $q_\lambda^m = 0$, and the molecule will, in general, distort.

If m represents the identity representation, this distortion preserves the molecular shape. But then the equilibrium configuration will have been chosen to make this energy a minimum.

Jahn and Teller (**1937**) have stated a remarkable theorem: Except for linear molecules, all *degenerate* states of polyatomic molecules (except for simple Kramers degeneracy) will, in fact, be split (and distorted) by some symmetry coordinate q_λ^m that is not totally symmetric. In view of the fact that H_λ^m can be taken as Hermitian in the electron coordinates, as well as even in time reversal, the Jahn-Teller theorem is equivalent, using Eq. 10.9.15 or Eq. 10.9.26, to a statement that the character product

$$\tfrac{1}{2}[X^j(R)]^2 + \tfrac{1}{2}K^2 X^j(R^2) \tag{13.5.3}$$

contains some representation m, other than the identity representation, which can be associated with a molecular vibration. For an even number of electrons, $K^2 = 1$ and the symmetric product is taken. For an odd number of electrons, $K^2 = -1$ and the antisymmetric product is taken. The proof by Jahn and Teller (**1937**) and Jahn (**1938**), an exhaustive evaluation of all the possibilities, will not be repeated here. General, but difficult, proofs have been given by Blount (**1971**) and by Ruch and Schonhofer (**1965**). The dynamic Jahn-Teller effect has been discussed by Longuet-Higgins et al. (**1958**), Moffitt and Thorsen (**1957**), and Liehr (**1963**). This effect occurs when the H_λ^m of Eq. 13.5.1 do not commute, so that different symmetry coordinates compete in attempting to distort the molecule.

13.6. SUMMARY

The relation between Hartree-Fock many-body wave functions and molecular orbitals, especially as linear combinations of atomic orbitals, is clarified. These results are contrasted with the description of chemical structure and resonance by means of many-body wave functions involving valence bond orbitals.

It is proved that the symmetry of the one-electron orbitals is less than the full symmetry of the many-body Hamiltonian if and only if the ground-state wave function is degenerate because of group symmetry and/or time reversal.

The Jahn-Teller effect is described as the (inevitable) breaking of electronic degeneracies by coupling between electronic and vibrational motion.

problems

13.1.1. Discuss cyclobutodiene, C_4H_4, from the LCAO-MO (Hückel) point of view (see Cotton, **1963**).

13.2.1. Show that three equivalent coplanar σ bonds separated by an angle of 120° have the symmetry D_{3h}. What irreducible representations are present in these orbitals? Construct such equivalent orbitals from any of the electron configurations sp^2, sd^2, dp^2, and d^3.

13.2.2. Repeat the discussion of Problem 13.2.1 for a trigonal arrangement of π bonds.

13.2.3. Use the configuration dsp^2 to construct a tetragonal plane arrangement of bonds.

13.2.4. Show that $2s+2p_x$, $2s+2p_y$, $2s-2p_x$, $2s-2p_y$ constitute four directed valence orbitals at 90° to each other.

13.3.1. Compare the resonance energy of the benzene molecule calculated by the resonance bond method of Section 13.3 with the value obtained using the Hückel MO method of Section 13.1.

13.4.1. Prove Theorems 13.4.5, 13.4.6, and 13.4.7, relating time reversal, degeneracy, and effective one-body Hamiltonians.

13.5.1. Show that for a two-dimensional representation (Jahn, **1938**)

$$\tfrac{1}{2}\left\{[\chi(R)]^2-\chi(R^2)\right\}=\det D(R)$$

Jahn alleges that for a proper rotation $\det D(R)=+1$, and for inversion in two dimensions $\det D(I)=1$; hence for all two-dimensional representations $(\Gamma\times\Gamma)_{\text{asym}}=\Gamma_1$, so that no Jahn-Teller effect occurs. Show that, for the group T, $(\Gamma_6\times\Gamma_6)_{\text{asym}}=\Gamma_3$, so that a Γ_3 distortion can induce a Jahn-Teller splitting in the two-dimensional representation Γ_6. What is wrong with Jahn's argument?

13.5.2. The Raman effect: For a molecule or a crystal an applied electric field, $\mathbf{E}\exp(i\omega_L t)$ (due to light) induces a polarization $\mathbf{p}$ via a polarizability tensor $\boldsymbol{\alpha}$. If $\boldsymbol{\alpha}$ is modulated by a vibrational coordinate, $\mathbf{q}=\mathbf{q}(0)\exp(i\omega_V t)+\text{c.c.}$, the resulting polarization $\mathbf{p}$ will contain the frequencies $\omega_L\pm\omega_V$, which are reradiated as Stokes $(\omega_L-\omega_V)$ or anti-Stokes $(\omega_L+\omega_V)$ radiation. With $p_i=(\partial\alpha_{ij}/\partial q^\mu)q^\mu E_j$, show that the selection rule for the existence of a Raman scattering appropriate to a given vibration q^μ is that the symmetric tensor $\boldsymbol{\alpha}$, which transforms as the symmetric product of a vector $\mathbf{r}$ with itself, contain the representation Γ^μ associated with the vibration q^μ. Apply this selection rule to determine which of the vibrations of the ammonia molecule discussed in Chapter 5 are Raman active.

appendix A

CHARACTER TABLES AND BASIS FUNCTIONS FOR THE SINGLE AND DOUBLE POINT GROUPS[1]

This appendix contains eleven tables, one for each of the proper point groups. Groups isomorphic to a proper point group are included in the same table.

The decomposition of direct product representations for all representations of dimensions greater than one is shown beneath each character table. Thus for the group D_4 we have $5 \times 5 = 1 + [2] + 3 + 4$, where [2] tells us that the antisymmetric product $(\Gamma_5 \times \Gamma_5)_{\text{asym}} = \Gamma_2$ and $(\Gamma_5 \times \Gamma_5)_{\text{sym}} = \Gamma_1 + \Gamma_3 + \Gamma_4$.

NOTATION FOR THE REPRESENTATIONS

The representations $\Gamma_1, \Gamma_2, \Gamma_3, \ldots$, listed immediately adjacent to the characters, follow the sequence[2] used by Koster, Dimmock, Wheeler, and Statz (**1963**) (hereafter referred to as KDWS). Each representation is also labeled to the left by the chemical (Mulliken, **1933**) notation. Further to the left, other alternatives (such as the Bouckaert, Smoluchowski, and Wigner, **1936** notation) are listed when available.

The chemical notation uses A, B for one-dimensional representations (B if the representation is odd under the smallest rotation of the principal axis), E for two-dimensional representations, and T, U, V, W for three-, four-, five-, and six-dimensional representations. (Some authors use F, G, H, I instead of

[1] Please notify me of any errors in Tables A1–A11.

[2] The Koster sequence interchanges Γ_5 and Γ_6 for the group D_6 in relation to the Bethe (**1929**) notation. The Bethe notation has been followed by Herring (**1942**), Schiff (**1968**), Glasser (**1959**), and Rashba (**1959**); the Koster notation, by Birman (**1959**), Casella (**1959**), and Hopfield (**1959**).

T, U, V, W, but this choice can be confused with spectroscopic notation.) Double group representations are indicated by Griffith (**1961**) and Landau and Lifschitz (**1958**) by the addition of a prime. We have shifted the prime to the left to avoid confusion with single group representations such as A' and A'' in $\mathcal{C}_{3h}$. Thus $'E$ stands for the Kramers doublet with spinor basis functions $[\xi, \eta]$, discussed later in this appendix. We use $'A_j$ for a one-dimensional double group representation with character ω^j and basis function ξ^j, and $'A_{-j}$ for one with character ω^{-j} and basis function η^j; $'B_j$ differs from $'A_j$ in that (1) it is odd under the smallest rotation of the principal axis, and (2) it can be a homogeneous polynomial of degree j in ξ and η. The symbols $'E_j$, $'T_j$, and so on designate higher-dimensional representations whose lowest-degree polynomial basis functions are of degree j, that is, the subscript j denotes the lowest-dimensional D_j of the full rotation group that contains the representation.

Chemists use a different notation for the representation of $\mathcal{C}_{3h}$ than for the isomorphic group D_6. We omit the chemical names of the representations of the improper groups and use the same Γ_j name for two representations of isomorphic groups with the same character.

If inversion is added to a group, such as D_6, chemists add a subscript g (= *gerade*) or u (= *ungerade*) for even or odd parity representations. Thus, for D_{6h}, $B_{1g} = \Gamma_3^+$ and $B_{1u} = \Gamma_3^-$.

Chemists also use one symbol, E, to denote a pair of complex conjugate representations such as Γ_3 and Γ_4 of $\mathcal{C}_4$. We have assigned to these representations individual chemical names, A_2 and A_{-2}, since the character of C_4 is ω^2 and ω^{-2} for these representations. Similarly, $\Gamma_2 = B$ has the alternative designation A_4.

The notation for the rotations and reflections is explained in detail in Appendix B. A notation $3C_{2y} \equiv 3\delta_{2y}$ for the group D_3 indicates a class containing three (counterclockwise) rotations through π, one of which is about the positive y axis. (Since the threefold operation is C_{3z}, the remaining twofold operations are about axes in the x-y plane, making 120° angles with respect to each other and the y axis.) The group $\mathcal{C}_{3v}$ is obtained from D_3 by replacing C_{2y} (as a generator) by the reflection plane $\boldsymbol{\sigma}_y$.

A rotation of angle $2\pi/n$ about axis $\mathbf{n}$ is described by the operator

$$C_{n,\mathbf{n}} = R\left(\frac{2\pi}{n}, \mathbf{n}\right) = \exp\left[-i\left(\frac{2\pi}{n}\right)\mathbf{J}\cdot\mathbf{n}\right] \tag{A1}$$

acting on a wave function. Thus a rotation of α about the z axis acts on the basis function $x+iy$ to yield

$$\exp(-i\alpha l_z)(x+iy) = \exp(-i\alpha)(x+iy) \tag{A2}$$

since $l_z = L_z/\hbar$ has eigenvalue $+1$ in the state $(x+iy)$, as discussed in Section

2.5. This convention is opposite to that used by Koster, so that for the same basis functions (labeled by the same Γ_j) our characters are the complex conjugates of those in the KDWS (**1963**) tables.

Representations above and below the solid horizontal line are single- and double-valued representations, respectively, and obey $\chi(\bar{R})=\pm\chi(R)$, respectively. Thus the characters of barred elements have been omitted from the tables as redundant information.

ARRANGEMENT OF THE TABLES

If $G=H\times\mathcal{C}_i$ is the direct product of the group H with the group $\mathcal{C}_i=(E,I)$, the representations Γ_j of H lead to representations $\Gamma_j^{\pm}$ of G with characters $\chi^{j\pm}(R)$ related to those of H by

$$\chi^{j\pm}(R)=\chi^j(R), \qquad \chi^{j\pm}(IR)=\pm\chi^j(R) \tag{A3}$$

Thus, for example, we need not supply a separate table for $D_{6h}=D_6\times\mathcal{C}_i$. Instead, in the table for D_6, we give both even and odd parity basis functions for the representations of D_6. These are indicated under the headings D_6^+ and D_6^-, respectively. Groups D_{3h} and C_{6v}, which contain improper rotations, are isomorphic to D_6 and have the same character table provided that the classes are relabeled as shown in the top lines of Table A9. The even parity basis functions of C_{6v} (or D_{3h}) are identical to those of D_6 and hence are not shown explicitly. (Such basis functions cannot distinguish between an improper rotation and the corresponding proper rotation.) An odd parity basis function such as iz is a basis for Γ_2^- of D_{6h} or Γ_2 of D_6. But it is also a basis for Γ_1 for C_{6v} and for Γ_4^- of D_{3h}, as indicated by the notation Γ_4^- in the second row of D_{3h}.

In order to have a concise notation for basis functions, we use, in addition to $\mathbf{r}=(x,y,z)$, the spin or angular momentum vector $\mathbf{S}=(S_x,S_y,S_z)$, which is even under inversion and odd under time reversal, and the vector $\mathbf{R}\sim\mathbf{r}_1\times\mathbf{r}_2$, which is even under both inversion and time reversal. It is always possible, however, to find basis functions for the single group representations that are functions only of x, y, and z. Enough additional basis functions for each representation Γ_j are indicated below the table to make a complete set l_j in the sense of Theorem 3.8.1. These extra basis functions, as well as the extra basis functions for the identity basis, are useful in finding (by inspection) the coefficients $U_{\mu\nu,\lambda}^{ij,m}$ in

$$\varphi_\mu^i=\sum V_\lambda^m\psi_\nu^j U_{\lambda\nu,\mu}^{mj,i} \tag{A4}$$

that reduce the product representation $\Gamma_m\times\Gamma_j$ into irreducible components Γ_i. These coefficients are known as vector coupling, Wigner, or Clebsch-Gordan coefficients.

For the full rotation group and for all noncubic point groups, $\Gamma_i\times\Gamma_j$ contains any Γ_m no more than once, so that the Clebsch-Gordan coefficients are uniquely determined once the form of the basis functions (including phase) is specified. For the cubic point groups some products $\Gamma_j \times \Gamma_m$ contain some Γ_i twice. In such a case, two matrices $U^{mj,i}_{\lambda\nu,\mu}$ exist and lead to two independent sets of φ^i_μ.

In principle, we could have omitted from the tables all negative parity basis functions, except for Γ_1^-, since it is always possible to write

$$\Gamma_j^- = \Gamma_j^+ \times \Gamma_1^- \tag{A5}$$

Single group basis functions for Γ_j^- are listed, however, since it always proved easy to find bases for Γ_j^- that avoided the use of **S** and **R**, or that were simpler than the direct use of Eq. A5. The spin functions used for the double group representations are always even under inversion, however, and Eq. A5 usually represents the simplest procedure for obtaining Γ_j^-. It is essential, however, to use the Γ_1^- appropriate to the group in question. Thus, if explicitly written out, Table A9 would have the following extra entries in line 7:

$$\begin{array}{cccc} & D_6^- & \mathcal{C}_{6v}^- & D_{3h}^- \\ \Gamma_7 & izS_z[\xi,\eta] & iz[\xi,\eta] & (y^3-3yx^2)[\xi,\eta] \end{array} \tag{A6}$$

We omit from our tables, however, such trivially obtainable entries. Usually an entry is made only if an additional simpler form is found.

TRANSFORMATIONS OF THE SPINORS ξ,η

In these tables ξ and η represent complex variables that transform as the fundamental spinors $\psi^{1/2}_{1/2}$ and $\psi^{1/2}_{-1/2}$, respectively, that is, as in Eq. B6. The set of monomials $\xi^{j+\mu}\eta^{j-\mu}$, $\mu=-j,-j+1,\ldots,j$, constitutes an invariant space under the unitary transformations $\exp(-\frac{1}{2}i\alpha\boldsymbol{\sigma}\cdot\mathbf{n})$ of $[\xi,\eta]$, since these homogeneous transformations will not change the degree of a polynomial. The character of any $R(\alpha,\mathbf{n})$ in this space can be computed using a typical element in the class, a rotation about the z axis:

$$R(\alpha,\hat{z})[\xi,\eta] = [\xi,\eta]\exp(-\tfrac{1}{2}i\alpha\sigma_z)$$

$$= \left[\exp\left(\frac{-i\alpha}{2}\right)\xi, \exp\left(\frac{i\alpha}{2}\right)\eta\right] \tag{A7}$$

$$R(\alpha,\hat{z})\xi^{j+\mu}\eta^{j-\mu} = \exp(-i\mu\alpha)\xi^{j+\mu}\eta^{j-\mu} \tag{A8}$$

and

$$\chi^j[R(\alpha,\mathbf{n})] = \sum_{\mu=-j}^{j} \exp(-i\mu\alpha) \tag{A9}$$

is just the character of the irreducible representation j. Thus the single- and double-valued representations of $R^+(3)$ are in one-to-one correspondence with the single-valued representations of the special unitary group $SU(2)$, whose elements are $\exp(-\frac{1}{2}i\alpha\boldsymbol{\sigma}\cdot\mathbf{n})$. The basis functions for these representations can be written as

$$\varphi_\mu^j = |j,\mu\rangle = \frac{\xi^{j+\mu}\eta^{j-\mu}}{[(j+\mu)!(j-\mu)!]^{1/2}} \tag{A10}$$

where the numerical factor has been chosen to make the representation unitary (Wigner, **1959**; Bargmann, **1962**). Since ξ and η are even under inversion, we have identical transformation properties for

$$\begin{aligned}
|1,1\rangle &\sim \frac{\xi^2}{\sqrt{2}} \sim \frac{-i(R_x+iR_y)}{\sqrt{2}} \equiv R^+ \\
|1,0\rangle &\sim \xi\eta \sim iR_z \equiv R^0 \\
|1,-1\rangle &\sim \frac{\eta^2}{\sqrt{2}} \sim \frac{i(R_x-iR_y)}{\sqrt{2}} \equiv R^-
\end{aligned} \tag{A11}$$

(compare with Eqs. 2.5.5 and A14). Under proper rotations, R_x, R_y, R_z can be replaced by x,y,z. (In general, they differ by the factor Γ_1^-.)

The influence of parity on selection rules and Clebsch-Gordan coefficients is trivial. The basic problem is to determine the Clebsch-Gordan relations among even parity wave functions. For this purpose, any spinor basis function (see Eq. B5) can be expressed as

$$\begin{bmatrix} \psi_1(\mathbf{r}) \\ \psi_2(\mathbf{r}) \end{bmatrix} = \xi\psi_1(\mathbf{r}) + \eta\psi_2(\mathbf{r})$$

Ignoring the difference between $\mathbf{r}$ and $\mathbf{R}$ in the even parity case, we can express x,y,z in terms of ξ and η by inverting Eq. A11:

$$x = \tfrac{1}{2}i(\xi^2-\eta^2), \qquad y = \tfrac{1}{2}(\xi^2+\eta^2), \qquad z = -i\xi\eta \tag{A12}$$

whereupon

$$\begin{bmatrix} \psi_1(\mathbf{r}) \\ \psi_2(\mathbf{r}) \end{bmatrix} = \xi\psi_1(\mathbf{r}) + \eta\psi_2(\mathbf{r}) = F(\xi,\eta) \tag{A13}$$

is reduced to a function of ξ and η alone. The same replacement (A12) can be made in single-valued basis functions $\varphi(\mathbf{r}) = G(\xi,\eta)$. In this way, all basis functions are reduced to polynomials in ξ and η, and the Wigner coefficients to relations between these polynomials that can be written by inspection.

CHOICE OF PHASE AND BASIS FUNCTIONS

The representation Γ_4^- of O_h can have real basis functions $\mathbf{r} = [x,y,z]$, or

$$r^+ = \frac{-i(x+iy)}{\sqrt{2}}, \qquad r^0 = iz, \qquad r^- = \frac{i(x-iy)}{\sqrt{2}} \tag{A14}$$

the Fano-Racah (**1959**) contrastandard choice, or a factor i can be removed to obtain the Wigner (**1959**, Eq. 15.34) choice.

We shall fix the *form* and *phase* of all our basis functions by requiring that for any one group all basis functions obey

$$\theta\psi_\mu = \psi_\mu, \qquad \theta = QK \tag{A15}$$

where K = time reversal, and Q = some rotation or reflection operator, preferably, but not necessarily, within the group. The advantage of this choice is that

$$(\varphi, V\psi) = (\theta\varphi, \theta V\theta^{-1}\theta\psi)^* = (\varphi, V\psi)^* \tag{A16}$$

when $\theta V\theta^{-1} = V$. All matrix elements and Wigner coefficients will then be *real*. The time-reversal operator K is discussed in Chapter 10.

For the single-valued representations of O_h, one could choose $Q = \epsilon$, that is, make all single-valued basis functions real. This choice is useful, and some real basis functions are listed at the bottom of Table A11. But such a choice is not possible for the double-valued representations for which $K^2 = -1$. This minus sign can be compensated for, however, by choosing Q = some twofold operation.

The Fano-Racah choice of phase for the rotation group $R^+(3)$ is equivalent to the choice

$$QK\psi_\mu^j = \psi_\mu^j, \qquad Q = \delta_{2y} \tag{A17}$$

Application of K to Eq. A13 leads to the requirement

$$K\xi = -\eta, \qquad K\eta = \xi, \qquad KF(\xi,\eta) = [F(-\eta^*,\xi^*)]^* \tag{A18}$$

where the last form uses the antilinearity of K to require that any numerical coefficients be replaced by their complex conjugates. Since

$$\delta_{2y}\xi = \eta, \qquad \delta_{2y}\eta = -\xi \tag{A19}$$

$$\delta_{2y}K\xi = \xi, \qquad \delta_{2y}K\eta = \eta \tag{A20}$$

our fundamental spinors obey Eq. A17. Since

$$\delta_{2y} KF(\xi,\eta) = [F(\xi^*,\eta^*)]^* \tag{A21}$$

Eq. A17 is equivalent to the requirement that ψ_μ^j as a function of ξ and η contain only real coefficients, that is, ψ_μ^j is defined to within a plus or minus sign. That all the signs be alike in $\psi_\mu^j \sim \xi^{j+\mu}\eta^{j-\mu}$ is required by the conventional choice (Eq. 2.3.1) for the matrix elements of J_+ and J_- (see Problem A1). Moreover, with the choice A10 we can write

$$K\psi_\mu^j = (-1)^{j+\mu}\psi_{-\mu}^j \tag{A22}$$

This result and all our phase conventions and matrix elements for the group $R(3)$ agree with Bargmann (**1962**).[3] For construction of the Wigner coefficients (and $3j$ symbols) in this ξ,η notation see Sharp (**1960**) and Bargmann (**1962**).

Because it is often desirable to decompose ψ_μ^j into representations of the point groups, we have usually chosen $Q = \delta_{2y}$ for the latter. [Note that the choice of basis functions

$$S^+ = -(S_x + iS_y), \qquad S^0 = S_z, \qquad S^- = S_x - iS_y \tag{A23}$$

in Table A5 and elsewhere is consistent with $Q = \delta_{2y}$, since **S** is odd under time reversal.] For point groups containing complex representations the choice $Q = \delta_{2y}$ is sometimes impossible (see D_3, T). The choice of Q is indicated at the head of each table. Our choice of phase still leaves an ambiguity of sign. If all signs in one basis set are chosen, the *relative* signs are determined in any other set yielding the identical representation.

REALITY PROPERTIES OF THE REPRESENTATIONS

We have not explicitly indicated whether the representations are real, pseudoreal, or complex, since this can be read directly from the character table if we remember two general rules:

1. If the characters are complex, the representation is complex.
2. If the characters are real and the representation is of odd dimension, the representation is real.

In addition, the special rule appropriate to the crystallographic point groups must be kept in mind:

[3]See, for example, his Eq. 2.16c, except that *his* symbol K denotes *not* time reversal but the operation described in Eq. A21.

3. If the characters are real and the representation is of even dimension, no extra time-reversal degeneracy is produced, that is, the representation is real if single valued and pseudoreal if double valued.

The only cases of doubling degeneracy among the crystallographic point groups (Γ_2 in $\mathcal{C}_1$ and Γ_6 in $\mathcal{C}_3$) are real one-dimensional double group representations. In such cases ψ^i and $K\psi^i$ are listed separated by the word "or," (e.g., [ξ^3 or η^3]) to remind us that these basis functions are in orthogonal spaces.

problems

A1. Show that the operators

$$M_+ = \xi\frac{\partial}{\partial\eta}, \qquad M_- = \eta\frac{\partial}{\partial\xi}, \qquad M_z = \tfrac{1}{2}\left(\xi\frac{\partial}{\partial\xi} - \eta\frac{\partial}{\partial\eta}\right)$$

obey the same commutation rules as the angular momenta J_+, J_-, and J_z.

A2. With $M_\pm = M_x \pm iM_y$, show that M_x, M_y, M_z are the infinitesimal generators of the group $SU(2)$ by examining

$$\exp(-i\varphi\mathbf{n}\cdot\mathbf{M})f(\xi,\eta) = f([\xi,\eta]\cdot\exp(-\tfrac{1}{2}i\varphi\boldsymbol{\sigma}\cdot\mathbf{n}))$$

for $\varphi \to 0$.

A3. Show that the complete set of basis vectors ψ^i_μ in Eq. A10 can be deduced from $\psi^j_j = \xi^{2j}/[(2j)!]^{1/2}$ by applying the lowering operator M_- in agreement with the conventional choice (Eq. 2.3.1) for the matrix elements.

A4. Show that the transformation in the spinors

$$[\xi',\eta'] = [\xi,\eta]\exp(-i\alpha\boldsymbol{\sigma}\cdot\mathbf{n})$$

generates in

$$x = \tfrac{1}{2}[\xi^2 - \eta^2], \qquad y = \tfrac{1}{2}(\xi^2 + \eta^2), \qquad z = -i\xi\eta$$

the transformation

$$[x', y', z'] = [x, y, z]\mathbf{R}$$

where

$$\mathbf{R} = \exp[-i\alpha\mathbf{n}\cdot\mathbf{D}(1)]$$

is explicitly stated in Problem 2.5.2.

Table A1. $\mathcal{C}_1 \times \mathcal{C}_i = \mathcal{C}_i \quad (\bar{1})$ $\qquad Q = \boldsymbol{\delta}_{2y}$

$\mathcal{C}_1$	(1)	E	$\mathcal{C}_1^+$	$\mathcal{C}_1^-$
A	Γ_1	1	$R_y, S_x, S_z, x^2, y^2, z^2$	y, ix, iz
$'A$	Γ_2	1	$[\xi \text{ or } \eta]^a$	$\Gamma_2^+ \times \Gamma_1^- \sim [y\xi \text{ or } y\eta]$

[a] $[\xi$ or $\eta]$ indicates a *doubling* degeneracy: the time-reversed states in the pair ξ, η (the fundamental spin-up and spin-down states) are both basis functions for Γ_2 and are necessarily degenerate: $\mathbf{R} \sim \mathbf{r}_1 \times \mathbf{r}_2$, $\mathbf{S} \sim \mathbf{r} \times \mathbf{p} \sim i\mathbf{R}$.

Table A2. $\mathcal{C}_2 \times \mathcal{C}_i = \mathcal{C}_{2h} \quad (2/m)$ $\qquad Q = \boldsymbol{\delta}_{2y}$

$\mathcal{C}_s$	(m)	E	σ_z			$\mathcal{C}_s^-$
$\mathcal{C}_2$	(2)	E	C_{2z}	$\mathcal{C}_2^+$	$\mathcal{C}_2^-$	
A	Γ_1	1	1	S_z, x^2, y^2, z^2, xy	z, xyz	$\Gamma_2^- \sim x$
B	Γ_2	1	-1	R_x, R_y, xz, yz	x, y	$\Gamma_1^- \sim z$
$'A_1$	Γ_3	1	ω	ξ	$z\xi$	$\Gamma_3^+ \times \Gamma_2^- \sim x\xi$
$'A_{-1}$	Γ_4	1	ω^*	η	$z\eta$	$\Gamma_4^+ \times \Gamma_2^- \sim x\eta$

$$\omega = \exp\left(-\tfrac{1}{2}\pi i\right) = -i$$

Table A3. $D_2 \times \mathcal{C}_i = D_{2h} \quad (mmm)$ $\qquad Q = \boldsymbol{\delta}_{2y}$

BSW	$\mathcal{C}_{2v}$	$(2mm)$	ϵ	δ_{2xy}	ρ_z	$\rho_{\bar{x}y}$			
	$\mathcal{C}_{2v}$	$(2mm)$	E	C_{2z}	σ_y	σ_x			$\mathcal{C}_{2v}^-$
	D_2	(222)	E	C_{2z}	C_{2y}	C_{2x}	D_2^+	D_2^-	
Σ_1	A	Γ_1	1	1	1	1	x^2, y^2, z^2	xyz	$\Gamma_3^- \sim iz$
Σ_4	B_2	Γ_2	1	-1	1	-1	R_y, xz	y	$\Gamma_4^- \sim ix$
Σ_2	B_1	Γ_3	1	1	-1	-1	S_z, ixy	iz	$\Gamma_1^- \sim xyz$
Σ_3	B_3	Γ_4	1	-1	-1	1	S_x, iyz	ix	$\Gamma_2^- \sim y$
	$'E$	Γ_5	2	0	0	0	$[\xi, \eta]$	$xyz[\xi, \eta]$	$iz[\xi, \eta]$

$$\Gamma_5 \times \Gamma_5 = [\Gamma_1] + \Gamma_2 + \Gamma_3 + \Gamma_4; \ [\quad] = \text{antisymmetric part}$$

$$\Gamma_5: [\xi, \eta] \sim [\eta^3, -\xi^3] \sim [\xi^2\eta, -\eta^2\xi]$$

BSW: $\Sigma = (k, k, 0)$. Use top-line operators and different basis functions.

Table A4. $\mathcal{C}_4 \times \mathcal{C}_i = \mathcal{C}_{4h}$ $(4/m)$ $Q = \delta_{2y}$

$\mathcal{S}_4$	$(\bar{4})$	E	S_4^{-1}	C_2	S_4			$\mathcal{S}_4^-$
$\mathcal{C}_4$	(4)	E	C_4	C_2	C_4^{-1}	$\mathcal{C}_4^+$	$\mathcal{C}_4^-$	
A	Γ_1	1	1	1	1	x^2+y^2	iz	$\Gamma_2^- \sim xyz$
$A_4 = B$	Γ_2	1	-1	1	-1	ixy	xyz	$\Gamma_1^- \sim iz$
A_2 E	Γ_3	1	ϵ	-1	ϵ^{-1}	$-(S_x+iS_y)$	$-i(x+iy)$	$\Gamma_4^- \sim i(x-iy)$
A_{-2} E	Γ_4	1	ϵ^{-1}	-1	ϵ	S_x-iS_y	$i(x-iy)$	$\Gamma_3^- \sim -i(x+iy)$
$'A_1$	Γ_5	1	ω	ω^2	ω^{-1}	ξ	$iz\xi$	$xyz\xi$
$'A_{-1}$	Γ_6	1	ω^{-1}	ω^{-2}	ω	η	$iz\eta$	$xyz\eta$
$'A_{-3}$	Γ_7	1	ω^{-3}	ω^{-6}	ω^3	η^3	$iz\eta^3$	$xyz\eta^3$
$'A_3$	Γ_8	1	ω^3	ω^6	ω^{-3}	ξ^3	$iz\xi^3$	$xyz\xi^3$

$$\omega = \exp(-i\pi/4),\ \epsilon = \omega^2 = \exp(-i\pi/2) = -i$$

$$\Gamma_1^+ \sim z^2,\ (x \pm iy)^4$$

$$\Gamma_2^+ \sim (x^2-y^2)$$

$$\Gamma_3^+ \sim z(x+iy);\ \Gamma_3^- \sim [i(x-iy)]^3$$

$$\Gamma_4^+ \sim z(x-iy);\ \Gamma_4^- \sim [-i(x+iy)]^3$$

Table A5. $D_4 \times \mathcal{C}_i = D_{4h}$ $(4/mmm)$ $Q = \delta_{2y}$

BSW		D_{2d}	$(\bar{4}2m)$	ϵ	$2\sigma_{4z}$	δ_{2z}	δ_{2x} δ_{2y}	ρ_{xy} $\rho_{\bar{x}y}$				D_{2d}^-
	$\mathcal{C}_{4v}$	$\mathcal{C}_{4v}$	$(4mm)$	E	$2C_{4z}$	C_{2z}	σ_x σ_y	σ_{xy} $\sigma_{\bar{x}y}$			$\mathcal{C}_{4v}^-$	
D_{4h}		D_4	(422)	E	$2C_{4z}$	C_{2z}	C_{2x} C_{2y}	C_{2xy} $C_{2\bar{x}y}$	D_4^+	D_4^-		
X_1	Δ_1	A_1	Γ_1	1	1	1	1	1	x^2+y^2	$(x^2-y^2)xyz$	Γ_2^-	Γ_3^-
X_4	Δ_1'	A_2	Γ_2	1	1	1	-1	-1	S_z	iz	Γ_1^-	Γ_4^-
X_2	Δ_2	B_1	Γ_3	1	-1	1	1	-1	x^2-y^2	xyz	Γ_4^-	Γ_1^-
X_3	Δ_2'	B_2	Γ_4	1	-1	1	-1	1	ixy	$(x^2-y^2)iz$	Γ_3^-	Γ_2^-
X_5	Δ_5	E	Γ_5	2	0	-2	0	0	$[S^+, S^-]$	$[r^+, r^-]$	Γ_5^-	Γ_5^-
	$D_{1/2}$	$'E$	Γ_6	2	$\sqrt{2}$	0	0	0	$[\xi, \eta]$	$\Gamma_6^+ \times \Gamma_1^-$	$\Gamma_6^+ \times \Gamma_2^-$	$\Gamma_6^+ \times \Gamma_3^-$
		$'E_3$	Γ_7	2	$-\sqrt{2}$	0	0	0	$[\eta^3, -\xi^3]$	$\Gamma_6^+ \times \Gamma_3^-$	$\Gamma_6^+ \times \Gamma_4^-$	$\Gamma_6^+ \times \Gamma_1^-$

Table A5. (*Continued*)

Product Decomposition Table for D_4

5	6	7	
1+[2]+3+4	6+7	6+7	5
	[1]+2+5	3+4+5	6
		[1]+2+5	7

[ ]=anti-symmetric part

Γ_1^+: $z^2, x^2+y^2, x^2y^2, x^4+y^4, x^6+y^6$

Γ_2^+: $ixy(x^2-y^2) \equiv \Gamma_3^+\Gamma_4^+$

Γ_5^+: $[S^+, S^-] \sim iz[r^+, r^-]; \Gamma_5^- \sim [(r^-)^3, (r^+)^3]$

Γ_6^+: $[\xi, \eta] \sim [\xi^2\eta, -\eta^2\xi]$

Γ_7^+: $[\eta^3, -\xi^3] \sim \Gamma_6^+ \times \Gamma_3^+ \sim ixy[\xi, -\eta]$

$r^+ \equiv -i(x+iy)/\sqrt{2}$, $r^- \equiv i(x-iy)/\sqrt{2}$, $S^+ \equiv -(S_x+iS_y)/\sqrt{2}$, $S^- \equiv (S_x-iS_y)/\sqrt{2}$

BSW: $\Delta=(k,0,0)$, group $\mathcal{C}_{4v}$; $X=(\pi/a)(1,0,0)$, group D_{4h}. For both these, rotate operators and basis functions $z\to x, x\to y, y\to z$. Use $X_j^+ = X_j, X_j^- = X_j'$. The compatibility $X_1'\to\Delta_1', X_2'\to\Delta_2'$ forces the naming of these Δ's. For $M=(\pi/a)(1,1,0)$, the group is D_{4h} with the operators as shown, and $X_j\to M_j$.

Table A6. $\mathcal{C}_3 \times \mathcal{C}_i = \mathcal{C}_{3i} = \mathcal{S}_6$ $(\bar{3})$ $Q=\delta_{2y}$

	$\mathcal{C}_3$	(3)	E	C_3	C_3^{-1}	$\mathcal{C}_3^+$	$\mathcal{C}_3^-$
	A	Γ_1	1	1	1	S_z	iz
E	A_2	Γ_2	1	ϵ	ϵ^2	$-(S_x+iS_y)$	$-i(x+iy)$
E	A_{-2}	Γ_3	1	ϵ^2	ϵ	(S_x-iS_y)	$i(x-iy)$
	$'A_1$	Γ_4	1	ω	ω^{-1}	ξ	$iz\xi$
	$'A_{-1}$	Γ_5	1	ω^{-1}	ω	η	$iz\eta$
	$'B_3$	Γ_6	1	-1	-1	$[\xi^3$ or $\eta^3]$	$iz[\xi^3$ or $\eta^3]$

$\omega=\exp(-\pi i/3)$, $\epsilon=\omega^2=\exp(-2\pi i/3)$

Γ_1^+: z^2, x^2+y^2; Γ_1^-: $(x\pm iy)^3$

Γ_2^+: $z(x+iy), (x-iy)^2$

Γ_3^+: $z(x-iy), (x+iy)^2$

Table A7. $D_3 \times \mathcal{C}_i = D_{3d}$ $(\bar{3}m)$ $Q = \delta_{2z}$

BSW	$\mathcal{C}_{3v}$	$(3m)$	ϵ	$2\delta_{3xyz}$	$3\rho_{\bar{x}y}$			
	$\mathcal{C}_{3v}$	$(3m)$	E	$2C_{3z}$	$3\sigma_y$			$\mathcal{C}_{3v}$
	D_3	(32)	E	$2C_{3z}$	$3C_{2y}$	D_3^+	D_3^-	
Λ_1	A_1	Γ_1	1	1	1	z^2	$i(3x^2y-y^3)$	Γ_2^-
Λ_2	A_2	Γ_2	1	1	-1	R_z	z	Γ_1^-
Λ_3	E	Γ_3	2	-1	0	$[-S_y, S_x]$	$[ix, iy]$	Γ_3^-
	$'E$	Γ_4	2	1	0	$[\eta - i\xi, \xi - i\eta]$	$\Gamma_j^- = \Gamma_j^+ \times \Gamma_1^-$	$\Gamma_4^+ \times \Gamma_2^-$
	$'B_3$	Γ_5	1	-1	$-i$	$\xi^3 + i\eta^3$		$z(\xi^3 + i\eta^3)$
	$'B_3^*$	Γ_6	1	-1	i	$\eta^3 + i\xi^3$		$z(\eta^3 + i\xi^3)$

Product Decomposition Table for D_3

3	4	
$1+[2]+3$	$4+5+6$	3
	$[1]+2+3$	4

$[\ \ ]$ = antisymmetric part

Γ_1^+: $z^2, (x^2+y^2), iz(x^3-3xy^2), (x+iy)^6+(x-iy)^6$

Γ_2^+: $iz(3x^2y-y^3)$; Γ_2^-: $i(x^3-3xy^2)$

Γ_3^+: $[S_x, S_y] \sim iz[y, -x] \sim [2xy, x^2-y^2]$

Γ_4^+: $[\xi, \eta] \sim [\xi^2\eta, -\eta^2\xi]$

BSW: $\Lambda = (k,k,k)$. Use operators in top line and modify basis functions. At $L = (\pi/a)(1,1,1)$, group is D_{3h}. BSW call $L_j^+ = L_j$ and $L_j^- = L_j'$. See Tables 12.5.1 and 12.5.2.

Table A8. $\mathcal{C}_6 \times \mathcal{C}_i = \mathcal{C}_{6h}$ $(6/m)$ $Q = \boldsymbol{\sigma}_y$

	$\mathcal{C}_{3h}$	$(\bar{6})$	E	C_3	C_3^{-1}	σ_h	S_3	S_3^{-1}			$\mathcal{C}_{3h}^-$
	$\mathcal{C}_6$	(6)	E	C_3	C_3^{-1}	C_2	C_6^{-1}	C_6	$\mathcal{C}_6^+$	$\mathcal{C}_6^-$	
	A	Γ_1	1	1	1	1	1	1	S_z, z^2	z	Γ_4^-
A_{-4}	E''	Γ_2	1	ϵ	ϵ^2	1	ϵ	ϵ^2	$(x-iy)^2$	$z(x-iy)^2$	Γ_5^-
A_4		Γ_3	1	ϵ^2	ϵ	1	ϵ^2	ϵ	$(x+iy)^2$	$z(x+iy)^2$	Γ_6^-
A_6	B	Γ_4	1	1	1	-1	-1	-1	$z(x\pm iy)^3$	$(x\pm iy)^3$	Γ_1^-
A_2	E'	Γ_5	1	ϵ	ϵ^2	-1	$-\epsilon$	$-\epsilon^2$	$-(S_x+iS_y)$	$-(x+iy)$	Γ_2^-
A_{-2}		Γ_6	1	ϵ^2	ϵ	-1	$-\epsilon^2$	$-\epsilon$	(S_x-iS_y)	$(x-iy)$	Γ_3^-
	$'A_1$	Γ_7	1	ω^2	ω^{-2}	ω^3	ω^{-1}	ω	ξ	$z\xi$	$\Gamma_j^- = \Gamma_j^+ \times (x\pm iy)^3$
	$'A_{-1}$	Γ_8	1	ω^{-2}	ω^2	ω^{-3}	ω	ω^{-1}	η	$z\eta$	
	$'A_{-5}$	Γ_9	1	ω^2	ω^{-2}	ω^{-3}	ω^5	ω^{-5}	η^5	$z\eta^5$	
	$'A_5$	Γ_{10}	1	ω^{-2}	ω^2	ω^3	ω^{-5}	ω^5	ξ^5	$z\xi^5$	
	$'A_{-3}$	Γ_{11}	1	-1	-1	ω^3	ω^3	ω^{-3}	η^3	$z\eta^3$	
	$'A_3$	Γ_{12}	1	-1	-1	ω^{-3}	ω^{-3}	ω^3	ξ^3	$z\xi^3$	

$$\omega = \exp(-\pi i/6), \epsilon = \exp(-2\pi i/3) = \omega^4$$

$$\Gamma_1^+ \sim (x\pm iy)^6$$

$$\Gamma_5^+ \sim -z(x+iy); \qquad \Gamma_5^- \sim -(x-iy)^5$$

$$\Gamma_6^+ \sim z(x-iy); \qquad \Gamma_6^- \sim (x+iy)^5$$

Table A9. $D_6 \times \mathcal{C}_i = D_{6h}$ (6/*mmm*) $Q = \delta_{2y}$

	D_{3h}	$(\bar{6}m2)$	E	$2C_3$	$3C_{2y}$	σ_h	$2S_3$	$3\sigma_x$				D_{3h}^-
	$\mathcal{C}_{6v}$	$(6mm)$	E	$2C_3$	$3\sigma_y$	C_2	$2C_6$	$3\sigma_x$			$\mathcal{C}_{6v}^-$	
H	D_6	(622)	E	$2C_3$	$3C_{2y}$	C_2	$2C_6$	$3C_{2x}$	D_6^+	D_6^-		
Γ_1	A_1	Γ_1	1	1	1	1	1	1	z^2	izS_z	Γ_2^-	Γ_3^-
Γ_2	A_2	Γ_2	1	1	−1	1	1	−1	S_z	iz	Γ_1^-	Γ_4^-
Γ_3	B_1	Γ_3	1	1	1	−1	−1	−1	$z(x^3-3xy^2)$	y^3-3yx^2	Γ_4^-	Γ_1^-
Γ_4	B_2	Γ_4	1	1	−1	−1	−1	1	$iz(y^3-3yx^2)$	$i(x^3-3xy^2)$	Γ_3^-	Γ_2^-
Γ_6	E_1	Γ_5	2	−1	0	−2	1	0	$[S^+,S^-]$	$[r^+,r^-]$	Γ_5^-	Γ_6^-
Γ_5	E_2	Γ_6	2	−1	0	2	−1	0	$\Gamma_5^+ \times \Gamma_3^+$	$\Gamma_5^- \times \Gamma_3^+$	Γ_6^-	Γ_5^-
	$'E$	Γ_7	2	1	0	0	$\sqrt{3}$	0	$[\xi,\eta]$			
	$'E_5$	Γ_8	2	1	0	0	$-\sqrt{3}$	0	$[\eta^5,-\xi^5]$			
	$'E_3$	Γ_9	2	−2	0	0	0	0	$[\xi^3,\eta^3]$	$[r^+\xi,r^-\eta]$		

Product Decomposition Table for D_6

5	6	7	8	9	
1+[2]+6	3+4+5	7+9	8+9	7+8	5
	1+[2]+6	8+9	7+9	7+8	6
		[1]+2+5	3+4+6	5+6	7
			[1]+2+5	5+6	8
				[1]+2+3+4	9

Γ_1^+: $(x^2+y^2), \mathrm{Re}(x+iy)^6, z(yX-xY)$; Γ_1^-: $izS_z \sim z\,\mathrm{Im}(x+iy)^6$

Γ_2^+: $\Gamma_3^- \times \Gamma_4^- \equiv \frac{i}{4}[(x-iy)^6-(x+iy)^6]$

Γ_5^+: $[S^+,S^-] \sim z[r^+,-r^-]$; Γ_5^-: $z^2[r^+,r^-] \sim [(r^-)^5,(r^+)^5]$

Γ_6^+: $\Gamma_3^+[S^+,S^-] \sim \Gamma_3^- \times \Gamma_5^- \sim [(r^-)^2,(r^+)^2] \sim [(r^+)^4,(r^-)^4]$

$\sim \Gamma_4^+[-S^+,S^-] \sim \Gamma_4^-[-r^+,r^-]$;

Γ_6^-: $\Gamma_3^+ \times \Gamma_5^- \equiv z(x^3-3xy^2)[r^+,r^-] \sim iz[-(r^-)^2,(r^+)^2]$

Γ_7^+: $[\xi,\eta] \sim [\xi^2\eta, -\eta^2\xi]$

Γ_8^+: $\Gamma_7 \times \Gamma_3 \sim [\eta^5, -\xi^5]$

Γ_9^+: $[\xi^3,\eta^3] \sim [\xi^4\eta, -\eta^4\xi]$

H = Herring (**1942**) notation for hexagonal close-packed lattice oriented so that $C_6 = C_{6z}$ and one nearest neighbor atom is in the y direction. Note interchange of Γ_5 and Γ_6 between Herring (Bethe) notation and the KDWS notation of third column. See footnote on p. 416.

Table A10. $T \times \mathcal{C}_i = T_h$ $(m3)$ $Q = \rho_{\bar{x}y}$

T	(23)	E	$4C_{3xyz}$	$4C_{3xyz}^{-1}$	$3C_{2x}$	T^+	T^-
A	Γ_1	1	1	1	1	$x^4+y^4+z^4$	xyz
E	Γ_2	1	ω	ω^2	1	$u+iv$	$xyz(u+iv)$
	Γ_3	1	ω^2	ω	1	$u-iv$	$xyz(u-iv)$
T	Γ_4	3	0	0	-1	$[R_x+R_y, i(R_x-R_y), R_z]$	$[x+y, i(x-y), z]$
$'E$	Γ_5	2	1	1	0	$[\xi_1, \eta_1]$	$xyz\Gamma_j^+$
$'E_3$	Γ_6	2	ω	ω^2	0	$[\sqrt{3}\,\xi_1^2\eta_1+\eta_1^3, \sqrt{3}\,\xi_1\eta_1^2-\xi_1^3]$	
$'E_3^*$	Γ_7	2	ω^2	ω	0	$[\sqrt{3}\,\xi_1\eta_1^2+\xi_1^3, -\sqrt{3}\,\xi_1^2\eta_1+\eta_1^3]$	

The basis functions for Γ_6 and Γ_7 are interchanged relative to those of KDWS.

$$\omega = \exp(-2\pi i/3),\ u = \tfrac{1}{2}(3z^2-r^2),\ v = \tfrac{1}{2}\sqrt{3}\,(x^2-y^2)$$

$$\xi_1 = \xi \exp(-i\pi/8),\ \eta_1 = \eta \exp(i\pi/8)$$

Product Decomposition Table for T

4	5	6	7	
1+2+3+4+[4]	5+6+7	5+6+7	5+6+7	4
	[1]+4	2+4	3+4	5
		[3]+4	1+4	6
			[2]+4	7

[ ] = antisymmetric part

Γ_1^+: $x^6+y^6+z^6$

Γ_2^+: $u+iv \equiv \omega^2x^2+\omega y^2+z^2 \sim \omega^2x^4+\omega y^4+z^4$

$\Gamma_3 = K\Gamma_2$

Γ_4^+: $[R_x, R_y, R_z] \sim [yz, zx, xy] \sim [yzx^2, zxy^2, xyz^2], \mathbf{R} \sim \mathbf{r}_1 \times \mathbf{r}_2$;

Γ_4^-: $[x, y, z] \sim [x^3, y^3, z^3] \sim [xz^2, yx^2, zy^2]$

Γ_5^+: $[\xi, \eta] \sim [\xi^5 - 5\xi\eta^4, \eta^5 - 5\xi^4\eta] \sim [\sqrt{2}\,R^+\eta - R^0\xi, -\sqrt{2}\,R^-\xi + R^0\eta]$

or $[\xi_1, \eta_1] \sim [\xi_1^5 + 5\xi_1\eta_1^4, -(\eta_1^5 + 5\xi_1^4\eta_1)]$

$\sim [R_z\xi_1 + (R_ye^{i\pi/4} + R_xe^{-i\pi/4})\eta_1, -R_z\eta_1 + (R_xe^{i\pi/4} + R_ye^{-i\pi/4})\xi_1]$;

Γ_5^-: $xyz\Gamma_5^+$ or replace **R** by **r** in Γ_5^+

Γ_6^+: $\Gamma_2^{\pm}\Gamma_5^{\pm} \sim [R_z\xi_1 + i(R_x - R_y)\eta_1\cos(\pi/12) + (R_x + R_y)\eta_1\sin(\pi/12)$,

$-R_z\eta_1 + i(R_x - R_y)\xi_1\sin(\pi/12) - (R_x + R_y)\xi_1\cos(\pi/12)$

Γ_6^-: $xyz\Gamma_6^+, \Gamma_2^{\pm}\Gamma_5^{\mp}$ or $\mathbf{R} \to \mathbf{r}$ in Γ_6^+

$\Gamma_7^+ \equiv K\Gamma_6^+ = \Gamma_3^+(K\Gamma_5^+)$

Note: The basis functions for Γ_4^+ and Γ_4^- listed below the table must be rearranged to parallel those listed in the table if we wish them to be invariant under $\rho_{\bar{x}y}K$.

Table A11. $O \times \mathcal{C}_i = O_h\ (m3m)$ $\qquad Q = \delta_{2y}$

BSW	T_d / O	$(\bar{4}3m)$ / (432)	E / E	$8C_3$ / $8C_3$	$3C_{2x}$ / $3C_{2x}$	$6\sigma_{4x}$ / $6C_{4x}$	$6\sigma_{xy}$ / $6C_{2xy}$	O^+	O^-	T_d^-
Γ_1	A_1	Γ_1	1	1	1	1	1	$x^4+y^4+z^4$	$\Gamma_2^+ \times \Gamma_2^-$	Γ_2^-
Γ_2	A_2	Γ_2	1	1	1	-1	-1	$R_xR_yR_z$	xyz	Γ_1^-
Γ_{12}	E	Γ_3	2	-1	2	0	0	$[u,v]$	$xyz[v,-u]$	Γ_3^-
Γ_{15}	T_1	Γ_4	3	0	-1	1	-1	$[R^+,R^0,R^-]$	$[r^+,r^0,r^-]$	Γ_5^-
Γ_{25}	T_2	Γ_5	3	0	-1	-1	1	$[iyz,zx,ixy]$	$\Gamma_5^+ \times \Gamma_1^-$	Γ_4^-
$D_{1/2}$	$'E$	Γ_6	2	1	0	$\sqrt{2}$	0	$[\xi,\eta]$		$xyz[\xi,\eta]$
	$'E_5$	Γ_7	2	1	0	$-\sqrt{2}$	0	$\Gamma_6^+ \times \Gamma_2^+$	$xyz[\xi,\eta]$	
$D_{3/2}$	$'U_3$	Γ_8	4	-1	0	0	0	$[\lvert\frac{3}{2},\frac{3}{2}\rangle,\lvert\frac{3}{2},\frac{1}{2}\rangle,$ $\lvert\frac{3}{2},-\frac{1}{2}\rangle,\lvert\frac{3}{2},-\frac{3}{2}\rangle]$		

$$u=z^2-\tfrac{1}{2}(x^2+y^2),\ v=\tfrac{1}{2}\sqrt{3}\,(x^2-y^2),$$

$$r^+=-i(x+iy)\sqrt{2}\ ,\ r^0=iz,\ r^-=i(x-iy)/\sqrt{2}$$

$$R^+=-i(R_x+iR_y)/\sqrt{2}\ ,\ R^0=iR_z,\ R^-=-i(R_x-iR_y)/\sqrt{2}$$

Product Decomposition Table for Group O

3	4	5	6	7	8	
1+[2]+3	4+5	4+5	8	8	6+7+8	3
	1+3+[4]+5	2+3+4+5	6+8	7+8	6+7+8+8	4
		1+3+[4]+5	7+8	6+8	6+7+8+8	5
			[1]+4	2+5	3+4+5	6
				[1]+4	3+4+5	7
					[1+3+5] +2+4+4+5	8

For example, $(\Gamma_3 \times \Gamma_3)_{\text{sym}} = \Gamma_1 + \Gamma_3$
$(\Gamma_3 \times \Gamma_3)_{\text{asym}} = \Gamma_2$

Table A11. *(Continued)*

Γ_1^+: $x^2y^2z^2, x^8+y^8+z^8$; Γ_1^-: $xyz[x^4(y^2-z^2)+y^4(z^2-x^2)+z^4(x^2-y^2)]$

Γ_2^+: $x^4(y^2-z^2)+y^4(z^2-x^2)+z^4(x^2-y^2)$

Γ_3^+: $\left[z^4-\frac{1}{2}(x^4+y^4), \frac{1}{2}\sqrt{3}\,(x^4-y^4)\right] \sim \left[z^6-\frac{1}{2}(x^6+y^6), \frac{1}{2}\sqrt{3}\,(x^6-y^6)\right]$

Γ_4^+: $[R_x, R_y, R_z] \sim [x^n yz(y^2-z^2), y^n zx(z^2-x^2), z^n xy(x^2-y^2)]$, $n=0,2,4,\ldots$;

Γ_4^-: $[x,y,z] \sim [x^3,y^3,z^3] \sim [x^5,y^5,z^5] \sim [x(y^2+z^2), y(z^2+x^2), z(x^2+y^2)]$

Γ_5^+: $[yz,zx,xy] \sim [yzx^2, zxy^2, xyz^2] \sim [yzx^4, zxy^4, xyz^4]$;

Γ_5^-: $\Gamma_1^-[yz,zx,xy] \sim [x^n(y^2-z^2), y^n(z^2-x^2), z^n(x^2-y^2)]$, $n=1,3,5,\ldots$

See Von der Lage and Bethe (**1947**) for the orthonormalized form of these "Kubic Harmonics." (In the Bethe expression for α_8, replace the term $\frac{1}{6}\rho^8$ by $\frac{1}{3}\rho^8$.)

Γ_6^+: $[\xi,\eta] \sim [\sqrt{2}\,R^+\eta - R^0\xi, -\sqrt{2}\,R^-\xi + R^0\eta]/\sqrt{6}$;

Γ_6^-: $\Gamma_1^- \cdot \Gamma_6^+ \sim [\sqrt{2}\,r^+\eta - r^0\xi, -\sqrt{2}\,r^-\xi + r^0\eta]/\sqrt{6}$

Γ_7^+: $R_xR_yR_z\,[\xi,\eta] \sim [z(x-iy)\eta - ixy\xi, -z(x+iy)\xi + ixy\eta]$;

Γ_7^-: $\Gamma_2^- \cdot \Gamma_6^+$

Γ_8^+: $D_{3/2} \equiv [\xi^3/\sqrt{6}, \xi^2\eta/\sqrt{2}, \xi\eta^2/\sqrt{2}, \eta^3/\sqrt{6}]$

$\sim [(R_y-iR_x)\xi/\sqrt{2}, \{(R_y-iR_x)\eta+2iR_z\xi\}/\sqrt{6},$

$\{(R_y+iR_x)\xi+2iR_z\eta\}/\sqrt{6}, (R_y+iR_x)\eta/\sqrt{2}]$

$\sim [(x+iy)z\xi/\sqrt{6} + 2ixy\eta/\sqrt{6}, -(x+iy)z\eta/\sqrt{2},$

$-(x-iy)z\xi/\sqrt{2}, (x-iy)z\eta/\sqrt{6} + 2ixy\xi/\sqrt{6}]$

$\sim [-v\eta, u\xi, -u\eta, v\xi]$;

Γ_8^-: $[(y-ix)\xi/\sqrt{2}, \{(y-ix)\eta+2iz\xi\}/\sqrt{6}, \{(y+ix)\xi+2iz\eta\}/\sqrt{6}, (y+ix)\eta/\sqrt{2}]$

appendix B

SCHOENFLIES, INTERNATIONAL, AND HERRING NOTATIONS

POINT GROUP OPERATIONS

The action of $R(\alpha,\mathbf{n})$, a counterclockwise rotation of the system through angle α about axis $\mathbf{n}$, has by Eq. 2.5.17, the following effect on a wave function:

$$R(\alpha,\mathbf{n})\psi(\mathbf{r})=\psi[R(\alpha,\mathbf{n})\mathbf{r}]=\psi(\mathbf{rR}) \tag{B1}$$

where

$$\begin{aligned}R(\alpha,\mathbf{n})\mathbf{r}&=\exp(-i\alpha\mathbf{l}\cdot\mathbf{n})[x,y,z]=[x,y,z]\exp[-i\alpha\mathbf{n}\cdot\mathbf{D}(\mathbf{l})]\\&=\mathbf{r}\cdot\mathbf{R}=\mathbf{r}\cos\alpha+\mathbf{r}\cdot\mathbf{nn}(1-\cos\alpha)+\mathbf{r}\times\mathbf{n}\sin\alpha\end{aligned} \tag{B2}$$

Here $\mathbf{l}=-i\mathbf{r}\times\nabla$, and $\mathbf{D}(\mathbf{l})$ is the 3×3 matrix representative of $\mathbf{l}$ (or $\mathbf{J}$) in the space of $3p$ states that transform as x,y,z. (See Problem 2.5.1.)

The action on spinors is given by Eqs. 3.5.14 and 3.5.15:

$$R(\alpha,\mathbf{n})\begin{bmatrix}\psi_1(\mathbf{r})\\\psi_2(\mathbf{r})\end{bmatrix}=\mathbf{D}^{1/2}(R)\begin{bmatrix}\psi_1(\mathbf{rR})\\\psi_2(\mathbf{rR})\end{bmatrix} \tag{B3}$$

with

$$\mathbf{D}^{1/2}(R)=\exp(-\tfrac{1}{2}i\alpha\boldsymbol{\sigma}\cdot\mathbf{n}) \tag{B4}$$

where $\boldsymbol{\sigma}$ are the Pauli spin matrices 2.6.3. For improper rotations see Eq. 3.5.18.

A two-component wave function can also be written in the form

$$\Psi = \xi\psi_1(\mathbf{r}) + \eta\psi_2(\mathbf{r}) = [\xi,\eta]\cdot\begin{bmatrix} \psi_1(\mathbf{r}) \\ \psi_2(\mathbf{r}) \end{bmatrix} \tag{B5}$$

provided that ξ,η transform as the fundamental spin states $\psi_{1/2}^{1/2}$, $\psi_{-1/2}^{1/2}$, respectively. Thus

$$R(\alpha,\mathbf{n})[\xi,\eta] = [\xi,\eta]\cdot\mathbf{D}^{1/2}(R) \tag{B6}$$

$$R(\alpha,\mathbf{n})\Psi = [\xi,\eta]\cdot\mathbf{D}^{1/2}(R)\begin{bmatrix} \psi_1(\mathbf{rR}) \\ \psi_2(\mathbf{rR}) \end{bmatrix} \tag{B7}$$

which is consistent with Eq. B3.

We present the point group notations in the following order: Schoenflies = International = Herring, plus comments.

$E = 1 \quad = \epsilon \quad = R(4\pi) \quad =$ identity
$\bar{E} = ? \quad = \bar{\epsilon} \quad = R(2\pi) = E$ in single group, $-E$ in double group, representations
$C_n = n \quad = \boldsymbol{\delta}_n \quad = R(2\pi/n)$. Axis understood to be z (i.e., $C_{nz} = n_z = \boldsymbol{\delta}_{nz}$)
$C_{n\mathbf{n}} = n_{\mathbf{n}} \quad = \boldsymbol{\delta}_{n\mathbf{n}} \quad = R(2\pi/n,\mathbf{n})$. Axis $\parallel$ to $\mathbf{n}$; $(C_{n\mathbf{n}})^n = \bar{E}$
$C_{3x\bar{y}\bar{z}} = 3_{x\bar{y}\bar{z}} = \boldsymbol{\delta}_{3x\bar{y}\bar{z}}$; $n = 3, \mathbf{n} = (1,-1,-1)/\sqrt{3}$
$U_2 = C'_2 = \quad = R(\pi)$ about $\mathbf{n}\perp$ to z axis
$C_{2d} = 2_d \quad = \boldsymbol{\delta}_{2d} \quad = R(\pi)$ about a diagonal axis (e.g., $\boldsymbol{\delta}_{2xy}$ or $\boldsymbol{\delta}_{2\bar{x}y}$)
$I = \bar{1} \quad = \iota \quad =$ inversion, $I^2 = E$
$\boldsymbol{\sigma}_{\mathbf{n}} = m_{\mathbf{n}} = \boldsymbol{\rho}_{\mathbf{n}} \quad = IC_{2\mathbf{n}} = \iota\boldsymbol{\delta}_{2\mathbf{n}} =$ mirror $\perp$ to $\mathbf{n}$
$\boldsymbol{\sigma}_h = m_h = \boldsymbol{\rho}_h \quad =$ Horizontal reflection plane ($\perp$ to principal axis)
$\boldsymbol{\sigma}_v = m_v = \boldsymbol{\rho}_v \quad =$ Vertical reflection plane (contains principal axis)
$\boldsymbol{\sigma}_d = m_d = \boldsymbol{\rho}_d \quad =$ diagonal reflection plane (e.g., $\boldsymbol{\rho}_{xy}$ or $\boldsymbol{\rho}_{\bar{x}y}$)
$IC_n = \bar{n} \quad =$ rotoinversion
$S_4^{-1} = \bar{4} = \boldsymbol{\sigma}_4$
$S_6^{-1} = \bar{3} = \boldsymbol{\sigma}_6$
$S_3^{-1} = \bar{6} = \boldsymbol{\sigma}_3$
$S_n = \tilde{n} = \quad =$ rotoreflection $= \boldsymbol{\sigma}_h C_n \bar{E}$

For specific examples of operations see Table 9.1.1.

NAMES OF GROUPS

The International name contains the generators but is often redundant:

n/m = an n-fold axis C_n with a mirror plane ($\boldsymbol{\sigma}_h$) perpendicular to it.
nm = an n-fold axis C_n with a mirror plane ($\boldsymbol{\sigma}_v$) containing it.
$n2$ = an n-fold axis C_n with a twofold axis (U_2) perpendicular to it.

The correspondence between the abbreviated international and Schoenflies symbols for the groups is shown by the following

Triclinic		Monoclinic		Trigonal		Tetragonal		Hexagonal		Cubic	
1	$\mathcal{C}_1$	2	$\mathcal{C}_2$	3	$\mathcal{C}_3$	4	$\mathcal{C}_4$	6	$\mathcal{C}_6$	23	T
$\bar{1}$	$\mathcal{C}_i$	m	$\mathcal{C}_s$	$\bar{3}$	$\mathcal{S}_6$	$\bar{4}$	$\mathcal{S}_4$	$\bar{6}$	$\mathcal{C}_{3h}$	$m3$	T_h
		$2/m$	$\mathcal{C}_{2h}$			$4/m$	$\mathcal{C}_{4h}$	$6/m$	$\mathcal{C}_{6h}$	$\bar{4}3m$	T_d
		Orthorhombic		$3m$	$\mathcal{C}_{3v}$	$4mm$	$\mathcal{C}_{4v}$	$6mm$	$\mathcal{C}_{6v}$		
		$2mm$	$\mathcal{C}_{2v}$	$\bar{3}m$	D_{3d}	$\bar{4}2m$	D_{2d}	$\bar{6}m2$	D_{3h}	432	O
		222	D_2	32	D_3	422	D_4	622	D_6	$m3m$	O_h
		mmm	D_{2h}			$4/mmm$	D_{4h}	$6/mmm$	D_{6h}		

Some international symbols are abbreviated. To obey the above rules, they should be listed by the corresponding full symmetry symbol:

Short Symbol	Full Symbol	Schoenflies	Short Symbol	Full Symbol	Schoenflies
mmm	$\left(\frac{2}{m}\right)\left(\frac{2}{m}\right)\left(\frac{2}{m}\right)$	D_{2h}	$6/mmm$	$\left(\frac{6}{m}\right)\left(\frac{2}{m}\right)\left(\frac{2}{m}\right)$	D_{6h}
$4/mmm$	$\left(\frac{4}{m}\right)\left(\frac{2}{m}\right)\left(\frac{2}{m}\right)$	D_{4h}	$m3$	$\left(\frac{2}{m}\right)\bar{3}$	T_h
$\bar{3}m$	$\bar{3}\left(\frac{2}{m}\right)$	D_{3d}	$m3m$	$\left(\frac{4}{m}\right)\bar{3}\left(\frac{2}{m}\right)$	O_h

appendix C

DECOMPOSITION OF $D_J^{\pm}$ OF FULL ROTATION GROUP INTO POINT GROUP REPRESENTATIONS

The decompositions are listed only for proper point groups, but the results for the corresponding improper groups can be readily deduced with the help of the last two columns in the character tables of Appendix A. Thus $D_2 = \Gamma_3 + \Gamma_5$ for group O implies

$$D_2^{\pm} = \Gamma_3^{\pm} + \Gamma_5^{\pm} \qquad \text{for } O_h$$

$$D_2^{+} = \Gamma_3^{+} + \Gamma_5^{+} = \Gamma_3 + \Gamma_5 \qquad \text{for } O \text{ and } T_d$$

$$D_2^{-} = \Gamma_3^{-} + \Gamma_5^{-} = \Gamma_3 + \Gamma_5 \qquad \text{for } O \text{ but} = \Gamma_3 + \Gamma_4 \text{ for } T_d$$

since, from Table AII, Γ_5^- for group O transforms as Γ_4 under T_d. In the following tables, l is an integer and j is half an odd integer.

D_J	$\mathcal{C}_2$	D_2
D_0	Γ_1	Γ_1
D_1	$\Gamma_1 + 2\Gamma_2$	$\Gamma_2 + \Gamma_3 + \Gamma_4$
D_2	$3\Gamma_1 + 2\Gamma_2$	$2\Gamma_1 + \Gamma_2 + \Gamma_3 + \Gamma_4$
D_3	$3\Gamma_1 + 4\Gamma_2$	$\Gamma_1 + 2\Gamma_2 + 2\Gamma_3 + 2\Gamma_4$
D_4	$5\Gamma_1 + 4\Gamma_2$	$3\Gamma_1 + 2\Gamma_2 + 2\Gamma_3 + 2\Gamma_4$
D_5	$5\Gamma_1 + 6\Gamma_2$	$2\Gamma_1 + 3\Gamma_2 + 3\Gamma_3 + 3\Gamma_4$
D_6	$7\Gamma_1 + 6\Gamma_2$	$4\Gamma_1 + 3\Gamma_2 + 3\Gamma_3 + 3\Gamma_4$
D_{l+2}	$D_l + 2(\Gamma_1 + \Gamma_2)$	$D_l + (\Gamma_1 + \Gamma_2 + \Gamma_3 + \Gamma_4)$
D_j	$(j+\frac{1}{2})(\Gamma_3 + \Gamma_4)$	$(j+\frac{1}{2})\Gamma_5$

D_J	$\mathcal{C}_3$	D_3
D_0	Γ_1	Γ_1
D_1	$\Gamma_1+\Gamma_2+\Gamma_3$	$\Gamma_2+\Gamma_3$
D_2	$\Gamma_1+2\Gamma_2+2\Gamma_3$	$\Gamma_1+\qquad+2\Gamma_3$
D_3	$3\Gamma_1+2\Gamma_2+2\Gamma_3$	$\Gamma_1+2\Gamma_2+2\Gamma_3$
D_4	$3\Gamma_1+3\Gamma_2+3\Gamma_3$	$2\Gamma_1+\Gamma_2+3\Gamma_3$
D_5	$3\Gamma_1+4\Gamma_2+4\Gamma_3$	$\Gamma_1+2\Gamma_2+4\Gamma_3$
D_6	$5\Gamma_1+4\Gamma_2+4\Gamma_3$	$3\Gamma_1+2\Gamma_2+4\Gamma_3$
D_{l+6}	$D_l+4(\Gamma_1+\Gamma_2+\Gamma_3)$	$D_l+2(\Gamma_1+\Gamma_2+2\Gamma_3)$
$D_{1/2}$	$\Gamma_4+\Gamma_5$	Γ_4
$D_{3/2}$	$\Gamma_4+\Gamma_5+2\Gamma_6$	$\Gamma_4+\Gamma_5+\Gamma_6$
$D_{5/2}$	$2\Gamma_4+2\Gamma_5+2\Gamma_6$	$2\Gamma_4+\Gamma_5+\Gamma_6$
$D_{7/2}$	$3\Gamma_4+3\Gamma_5+2\Gamma_6$	$3\Gamma_4+\Gamma_5+\Gamma_6$
$D_{9/2}$	$3\Gamma_4+3\Gamma_5+4\Gamma_6$	$3\Gamma_4+2\Gamma_5+2\Gamma_6$
$D_{11/2}$	$4\Gamma_4+4\Gamma_5+4\Gamma_6$	$4\Gamma_4+2\Gamma_5+2\Gamma_6$
$D_{13/2}$	$5\Gamma_4+5\Gamma_5+4\Gamma_6$	$5\Gamma_4+2\Gamma_5+2\Gamma_6$
D_{j+6}	$D_j+4(\Gamma_4+\Gamma_5+\Gamma_6)$	$D_j+2(2\Gamma_4+\Gamma_5+\Gamma_6)$

D_J	$\mathcal{C}_4$	D_4
D_0	Γ_1	Γ_1
D_1	$\Gamma_1+\qquad\Gamma_3+\Gamma_4$	$\Gamma_2\qquad\qquad+\Gamma_5$
D_2	$\Gamma_1+2\Gamma_2+\Gamma_3+\Gamma_4$	$\Gamma_1\qquad+\Gamma_3+\Gamma_4+\Gamma_5$
D_3	$\Gamma_1+2\Gamma_2+2\Gamma_3+2\Gamma_4$	$+\Gamma_2+\Gamma_3+\Gamma_4+2\Gamma_5$
D_4	$3\Gamma_1+2\Gamma_2+2\Gamma_3+2\Gamma_4$	$2\Gamma_1+\Gamma_2+\Gamma_3+\Gamma_4+2\Gamma_5$
D_5	$3\Gamma_1+2\Gamma_2+3\Gamma_3+3\Gamma_4$	$\Gamma_1+2\Gamma_2+\Gamma_3+\Gamma_4+3\Gamma_5$
D_6	$3\Gamma_1+4\Gamma_2+3\Gamma_3+3\Gamma_4$	$2\Gamma_1+\Gamma_2+2\Gamma_3+2\Gamma_4+3\Gamma_5$
D_{l+4}	$D_l+2(\Gamma_1+\Gamma_2+\Gamma_3+\Gamma_4)$	$D_l+(\Gamma_1+\Gamma_2+\Gamma_3+\Gamma_4+2\Gamma_5)$
$D_{1/2}$	$\Gamma_5+\Gamma_6$	Γ_6
$D_{3/2}$	$\Gamma_5+\Gamma_6+\Gamma_7+\Gamma_8$	$\Gamma_6+\Gamma_7$
$D_{5/2}$	$\Gamma_5+\Gamma_6+2\Gamma_7+2\Gamma_8$	$\Gamma_6+2\Gamma_7$
$D_{7/2}$	$2\Gamma_5+2\Gamma_6+2\Gamma_7+2\Gamma_8$	$2\Gamma_6+2\Gamma_7$
$D_{9/2}$	$3\Gamma_5+3\Gamma_6+2\Gamma_7+2\Gamma_8$	$3\Gamma_6+2\Gamma_7$
$D_{11/2}$	$3\Gamma_5+3\Gamma_6+3\Gamma_7+3\Gamma_8$	$3\Gamma_6+3\Gamma_7$
$D_{13/2}$	$3\Gamma_5+3\Gamma_6+4\Gamma_7+4\Gamma_8$	$3\Gamma_6+4\Gamma_7$
D_{j+4}	$D_j+2(\Gamma_5+\Gamma_6+\Gamma_7+\Gamma_8)$	$D_j+2(\Gamma_6+\Gamma_7)$

D_J	$\mathcal{C}_6$	D_6
D_0	Γ_1	Γ_1
D_1	$\Gamma_1 + \Gamma_5 + \Gamma_6$	$\Gamma_2 + \Gamma_5$
D_2	$\Gamma_1 + \Gamma_2 + \Gamma_3 + \Gamma_5 + \Gamma_6$	$\Gamma_1 + \Gamma_5 + \Gamma_6$
D_3	$\Gamma_1 + \Gamma_2 + \Gamma_3 + 2\Gamma_4 + \Gamma_5 + \Gamma_6$	$\Gamma_2 + \Gamma_3 + \Gamma_4 + \Gamma_5 + \Gamma_6$
D_4	$\Gamma_1 + 2\Gamma_2 + 2\Gamma_3 + 2\Gamma_4 + \Gamma_5 + \Gamma_6$	$\Gamma_1 \quad \Gamma_2 + \Gamma_3 + \Gamma_4 + \Gamma_5 + 2\Gamma_6$
D_5	$\Gamma_1 + 2\Gamma_2 + 2\Gamma_3 + 2\Gamma_4 + 2\Gamma_5 + 2\Gamma_6$	$\Gamma_2 + \Gamma_3 + \Gamma_4 + 2\Gamma_5 + 2\Gamma_6$
D_6	$3\Gamma_1 + 2\Gamma_2 + 2\Gamma_3 + 2\Gamma_4 + 2\Gamma_5 + 2\Gamma_6$	$2\Gamma_1 + \Gamma_2 + \Gamma_3 + \Gamma_4 + 2\Gamma_5 + 2\Gamma_6$
D_{l+6}	$D_l + 2(\Gamma_1 + \Gamma_2 + \Gamma_3 + \Gamma_4 + \Gamma_5 + \Gamma_6)$	$D_l + (\Gamma_1 + \Gamma_2 + \Gamma_3 + \Gamma_4 + 2\Gamma_5 + 2\Gamma_6)$
$D_{1/2}$	$\Gamma_7 + \Gamma_8$	Γ_7
$D_{3/2}$	$\Gamma_7 + \Gamma_8 + \Gamma_{11} + \Gamma_{12}$	$\Gamma_7 + \Gamma_9$
$D_{5/2}$	$\Gamma_7 + \Gamma_8 + \Gamma_9 + \Gamma_{10} + \Gamma_{11} + \Gamma_{12}$	$\Gamma_7 + \Gamma_8 + \Gamma_9$
$D_{7/2}$	$\Gamma_7 + \Gamma_8 + 2\Gamma_9 + 2\Gamma_{10} + \Gamma_{11} + \Gamma_{12}$	$\Gamma_7 + 2\Gamma_8 + \Gamma_9$
$D_{9/2}$	$\Gamma_7 + \Gamma_8 + 2\Gamma_9 + 2\Gamma_{10} + 2\Gamma_{11} + 2\Gamma_{12}$	$\Gamma_7 + 2\Gamma_8 + 2\Gamma_9$
$D_{11/2}$	$2\Gamma_7 + 2\Gamma_8 + 2\Gamma_9 + 2\Gamma_{10} + 2\Gamma_{11} + 2\Gamma_{12}$	$2\Gamma_7 + 2\Gamma_8 + 2\Gamma_9$
$D_{13/2}$	$3\Gamma_7 + 3\Gamma_8 + 2\Gamma_9 + 2\Gamma_{10} + 2\Gamma_{11} + 2\Gamma_{12}$	$3\Gamma_7 + 2\Gamma_8 + 2\Gamma_9$
D_{j+6}	$D_j + 2(\Gamma_7 + \Gamma_8 + \Gamma_9 + \Gamma_{10} + \Gamma_{11} + \Gamma_{12})$	$D_j + 2(\Gamma_7 + \Gamma_8 + \Gamma_9)$

D_J	T	O
D_0	Γ_1	Γ_1
D_1	Γ_4	Γ_4
D_2	$\Gamma_2 + \Gamma_3 + \Gamma_4$	$\Gamma_3 + \Gamma_5$
D_3	$\Gamma_1 + 2\Gamma_4$	$\Gamma_2 + \Gamma_4 + \Gamma_5$
D_4	$\Gamma_1 + \Gamma_2 + \Gamma_3 + 2\Gamma_4$	$\Gamma_1 + \Gamma_3 + \Gamma_4 + \Gamma_5$
D_5	$\Gamma_2 + \Gamma_3 + 3\Gamma_4$	$\Gamma_3 + 2\Gamma_4 + \Gamma_5$
D_6	$2\Gamma_1 + \Gamma_2 + \Gamma_3 + 3\Gamma_4$	$\Gamma_1 + \Gamma_2 + \Gamma_3 + \Gamma_4 + 2\Gamma_5$
D_{l+6}	$D_l + (\Gamma_1 + \Gamma_2 + \Gamma_3 + 3\Gamma_4)$	
D_{l+5}		$\Gamma_1 + \Gamma_2 + 2\Gamma_3 + 3\Gamma_4 + 3\Gamma_5 - D_{6-l}$
D_{l+12}		$D_l + (\Gamma_1 + \Gamma_2 + 2\Gamma_3 + 3\Gamma_4 + 3\Gamma_5)$
$D_{1/2}$	Γ_5	Γ_6
$D_{3/2}$	$\Gamma_6 + \Gamma_7$	Γ_8
$D_{5/2}$	$\Gamma_5 + \Gamma_6 + \Gamma_7$	$\Gamma_7 + \Gamma_8$
$D_{7/2}$	$2\Gamma_5 + \Gamma_6 + \Gamma_7$	$\Gamma_6 + \Gamma_7 + \Gamma_8$
$D_{9/2}$	$\Gamma_5 + 2\Gamma_6 + 2\Gamma_7$	$\Gamma_6 + 2\Gamma_8$
$D_{11/2}$	$2\Gamma_5 + 2\Gamma_6 + 2\Gamma_7$	$\Gamma_6 + \Gamma_7 + 2\Gamma_8$
$D_{13/2}$	$3\Gamma_5 + 2\Gamma_6 + 2\Gamma_7$	$\Gamma_6 + 2\Gamma_7 + 2\Gamma_8$
D_{j+6}	$D_j + 2(\Gamma_5 + \Gamma_6 + \Gamma_7)$	$D_{j'} + (\Gamma_6 + \Gamma_7 + 2\Gamma_8)$
D_{j+12}		$D_j + (\Gamma_6 + \Gamma_7 + 2\Gamma_8)$

$D_{j'} = D_j$ with Γ_6 and Γ_7 interchanged

appendix D

ORTHOGONALITY PROPERTIES OF EIGENVECTORS OF THE EQUATION $A\Psi=\lambda B\Psi$; RECIPROCALS OF SINGULAR MATRICES

Let Ψ_t, Ψ_s be two eigenvectors obeying

$$A\Psi_t=\lambda_t B\Psi_t \tag{D1}$$

$$A\Psi_s=\lambda_s B\Psi_s \tag{D2}$$

where A and B are Hermitian operators. Multiply Eq. D1 on the left by Ψ_s, and D2 on the right by Ψ_t, taking the scalar products

$$(\Psi_s, A\Psi_t)-(A\Psi_s,\Psi_t)=\lambda_t(\Psi_s, B\Psi_t)-\lambda_s(B\Psi_s,\Psi_t)$$

The Hermiticity of A and B causes the left-hand side to vanish and the right-hand side to yield

$$(\lambda_t-\lambda_s)(\Psi_s, B\Psi_t)=0 \tag{D3}$$

Thus

$$(\Psi_s, B\Psi_t)=0 \quad \text{if } \lambda_s\neq\lambda_t \tag{D4}$$

In the degenerate case, we can choose to maintain this orthogonality, just as in the usual case in which B is the identity operator. Thus we can write

$$(\Psi_s, B\Psi_t)=B_{tt}\delta_{st} \tag{D5}$$

where B_{tt} is a normalization constant at our disposal and is often set equal to unity.

In vibration problems, the eigenvalue equation takes the form

$$\sum_{n\nu} K_{\mu\nu}^{mn} u_\nu^n = \omega_t^2 M_m u_\mu^m \tag{D6}$$

with the correspondence $A \to K$, $\lambda_t \to \omega_t^2$, and $B \to M$ or, more precisely,

$$B_{\mu\nu}^{mn} = M_m \delta_{mn} \delta_{\mu\nu} \tag{D7}$$

so that the orthonormality condition between modes s and t can be taken in the form

$$\sum_n M_n (\mathbf{b}^{ns})^* \cdot \mathbf{b}^{nt} = M\delta(s,t) \tag{D8}$$

or

$$\sum_r M_r (b_r^s)^* b_r^t = M\delta(s,t) \tag{D9}$$

where r is a single index that combines the name of the particle n and the three Cartesian components. With $s = ia\mu$, $t = jb\nu$, we obtain the condition 5.3.4 given in the text.

RECIPROCALS OF MATRICES

With the help of the orthonormalization relation D5, an arbitrary vector Φ can be expanded in either of the forms

$$\Phi = \sum_t \frac{\Psi_t (B\Psi_t, \Phi)}{B_{tt}} \tag{D10}$$

$$\Phi = \sum_t \frac{B\Psi_t (\Psi_t, \Phi)}{B_{tt}} \tag{D11}$$

Form D11 is useful in solving the inhomogeneous equation

$$A\Psi = \Phi \tag{D12}$$

with the result

$$A^{-1}\Phi = \sum_t \frac{\Psi_t (\Psi_t, \Phi)}{[\lambda_t B_{tt}]} \tag{D13}$$

$$= \sum_t \frac{\Psi_t (\Psi_t, \Phi)}{(\Psi_t, A\Psi_t)} \tag{D14}$$

so that in Dirac notation the reciprocal is given by

$$A^{-1}=\sum_t \frac{|\Psi_t\rangle\langle\Psi_t|}{(\Psi_t,A\Psi_t)}=\sum_t \frac{|\Psi_t\rangle\langle\Psi_t|}{\lambda_t(\Psi_t,B\Psi_t)} \tag{D15}$$

in a form that does not depend on the choice of normalization.

Equation D15 is a strange formula in that A^{-1} seems to depend on B, whereas, in fact, it must be independent of the choice of B. The requirement on Ψ_t may be relaxed from the condition that it be an eigenvector in Eq. D1 to the simpler requirement that

$$(\Psi_s,A\Psi_t)=(\Psi_t,A\Psi_t)\,\delta(s,t) \tag{D16}$$

We can therefore proceed as follows. Choose Ψ_1 arbitrarily. Compute $A\Psi_1$. Choose Ψ_2 as any vector orthogonal to $A\Psi_1$. Compute $A\Psi_2$. Choose Ψ_3 as any vector orthogonal to $A\Psi_1$ and $A\Psi_2$, and so on. The reciprocal can then be computed using the first form of Eq. D15 with these nearly arbitrarily chosen Ψ_t. This procedure is indeed equivalent to choosing $B=A$, $\lambda_t=1$. Try Problem D1.

RECIPROCAL OF SINGULAR MATRICES

If A possesses a null space, that is, a space spanned by the set of vectors obeying

$$A\Psi_t=0, \qquad t=1,2,\ldots,f \tag{D17}$$

then A^{-1} fails to exist. The inhomogeneous equation $A\Psi=\Phi$ can still be solved, however, provided that the inhomogeneous term Φ is orthogonal to the null space σ:

$$(\Psi_t,\Phi)=0, \qquad t=1,2,\ldots,f \tag{D18}$$

The general solution expressed in terms of the eigenvectors Ψ_t of Eq. D1 is then

$$A^{-1}\Phi=\sum_{t=1}^{f} C_t\Psi_t+\sum_{t=f+1}^{n}\frac{\Psi_t(\Psi_t,\Phi)}{(\Psi_t,A\Psi_t)} \tag{D19}$$

The first term is the general solution of the homogeneous equation $A\Psi=0$, that is, an arbitrary vector in the null space σ_0. The second term is the particular solution Ψ of the inhomogeneous equation that is orthogonal to the space σ_0 with weight B:

$$(\Psi_t,B\Psi)=0, \qquad t=1,2,\ldots,f \tag{D20}$$

This particular solution appears to depend on the choice of B, although the complete expression for $A^{-1}\Phi$ is independent of B. If Φ' is also orthogonal to σ_0 in the same sense as Φ, then

$$(\Phi', A^{-1}\Phi) = \sum_{t=f+1}^{n} \frac{(\Phi', \Psi_t)(\Psi_t, \Phi)}{(\Psi_t, A\Psi_t)} \tag{D21}$$

is an answer that does not depend on the choice of B.

problems

D1. Show that

$$\begin{bmatrix} 1 & 1 & 1 \\ 1 & 2 & 1 \\ 1 & 1 & 3 \end{bmatrix}^{-1} = \frac{1}{2}\begin{bmatrix} 5 & -2 & -1 \\ -2 & -2 & 0 \\ -1 & 0 & 1 \end{bmatrix}$$

using Eq. D15 without finding the eigenvectors of the original matrix.

D2. Show that procedure D21, used in determining the reciprocal of a singular matrix, is equivalent to choosing a basis in which matrix A and its "reciprocal" R are expressed in the forms

$$A = \begin{bmatrix} 0 & 0 \\ 0 & S \end{bmatrix}, \qquad R = \begin{bmatrix} 0 & 0 \\ 0 & S^{-1} \end{bmatrix}$$

where S is a nonsingular submatrix in the space complementary to the space σ_0 spanned by the null vectors. Show that the "reciprocal" R of A in this singular case obeys

$$ARA = A, \qquad RAR = R$$

but not $AR = RA = 1$.

appendix E

THE BRILLOUIN ZONES

Coordinates of points in the Brillouin zone in the table below, are expressed in fractions of conventional rather than primitive reciprocal lattice vectors. The point Γ is always 000 and is not listed in the tables. In Figs. E1–E4, when more than one shape of the Brillouin zone exists, the coordinates apply to figure (a) and are chosen to agree with Miller and Love (**1967**). The points in figure (b) are equivalent to points of the same name in figure (a), but their coordinates may differ by a reciprocal lattice vector. General points on planes of symmetry are not shown.

Monoclinic lattices have z as the symmetry axis (or as a normal to the symmetry plane).

A-centered monoclinic lattice: figure (a) has $c>b$; figure (b) has $c<b$. See Fig. E4.

Face-centered orthorhombic and cubic lattices have body-centered reciprocal lattices, and vice versa. See Fig. E1.

Body-centered tetragonal lattice has a face-centered reciprocal lattice equivalent to the original body-centered lattice rotated 45° about the principal (z) axis. Figure (a) has $c<\sqrt{2}\,a$; figure (b) has $c>\sqrt{2}\,a$. See Fig. E2.

Face-centered orthorhombic lattice: figure (a) has $c^2>a^2+b^2$; figure (b) has $c^2<a^2+b^2$. See Fig. E3.

Rhombohedral lattices use hexagonal coordinates with $z=3$ as principal axis and 1 and 2 axes perpendicular to z and 120° apart. Figure (a) has $c^2>a^2+b^2$; figure (b) has $c^2<a^2+b^2$. See Fig. E4.

Hexagonal primitive reciprocal lattice vectors, 60° apart and perpendicular to the z axis, are used as the basis vectors for the hexagonal Brillouin zone.

Table E1 Coordinates of Symmetry Points in the 14 Brillouin Zones

Point	Triclinic	Simple Monoclinic	Centered Monoclinic	Simple Orthorhombic	Base Centered Orthorhombic	Face Centered Orthorhombic	Body Centered Orthorhombic
Δ				$0\alpha 0$	$0\alpha 0$	$0\alpha 0$	$0\alpha 0$
Λ		$0 \quad 0\alpha$	00α	00α	00α	00α	00α
Σ				$\alpha 00$	$\alpha 00$	$\alpha 00$	$\alpha 00$
A		$\frac{1}{2} - \frac{1}{2} 0$	$\frac{1}{2}00$	$\alpha 0 \frac{1}{2}$	$\alpha 0 \frac{1}{2}$	$\alpha 0 \frac{1}{2}$	
B		$\frac{1}{2} \quad 00$		$0\alpha \frac{1}{2}$	$0\alpha \frac{1}{2}$	$0\alpha \frac{1}{2}$	
C		$0 \quad \frac{1}{2}\frac{1}{2}$		$\alpha \frac{1}{2} 0$	$\alpha \frac{1}{2} 0$	$\alpha \frac{1}{2} 0$	
D		$\frac{1}{2} \quad 0\frac{1}{2}$		$\frac{1}{2}\alpha\theta$	$\frac{1}{4}\frac{1}{4}\alpha$		$\alpha \frac{1}{4}\frac{1}{4}$
E		$\frac{1}{2} - \frac{1}{2}\frac{1}{2}$		$\alpha \frac{1}{2}\frac{1}{2}$	$\alpha \frac{1}{2}\frac{1}{2}$		
F							
G				$\frac{1}{2}0\alpha$			$\frac{1}{2}0\alpha$
H				$0\frac{1}{2}\alpha$	$0\frac{1}{2}\alpha$	$0\frac{1}{2}\alpha$	
L			$\frac{1}{2}\frac{1}{4}\frac{1}{4}$			$\frac{1}{4}\frac{1}{4}\frac{1}{4}$	
M			$\frac{1}{2}0\frac{1}{2}$				

N							
P				$\frac{1}{2}\alpha\frac{1}{2}$			$\frac{1}{4}\frac{1}{4}\alpha$
Q				$\frac{1}{2}\frac{1}{2}\alpha$			$\frac{1}{4}\alpha\frac{1}{4}$
R	$\frac{1}{2}\frac{1}{2}\frac{1}{2}$			$\frac{1}{2}\frac{1}{2}\frac{1}{2}$	$\frac{1}{4}\frac{1}{4}\frac{1}{2}$		$\frac{1}{4}0\frac{1}{4}$
S				$\frac{1}{2}\frac{1}{2}0$	$\frac{1}{4}\frac{1}{4}0$		$0\frac{1}{4}\frac{1}{4}$
T				$0\frac{1}{2}\frac{1}{2}$	$0\frac{1}{2}\frac{1}{2}$	$0\frac{1}{2}\frac{1}{2}$	$\frac{1}{4}\frac{1}{4}0$
U	$\frac{1}{2}0\frac{1}{2}$	$\frac{1}{2}\ -\frac{1}{2}\alpha$	$\frac{1}{2}0\alpha$	$\frac{1}{2}0\frac{1}{2}$		$\alpha\frac{1}{2}\frac{1}{2}$	$\frac{1}{2}\alpha 0$
V	$\frac{1}{2}\frac{1}{2}0$	$\frac{1}{2}\ \ 0\alpha$	$0\frac{1}{4}\frac{1}{4}$				
W		$0\ \ \frac{1}{2}\alpha$					$\frac{1}{4}\frac{1}{4}\frac{1}{4}$
X	$\frac{1}{2}00$			$\frac{1}{2}00$			$\frac{1}{2}00$
Y		$0\ \ \frac{1}{2}0$	$00\frac{1}{2}$	$0\frac{1}{2}0$	$0\frac{1}{2}0$	$0\frac{1}{2}0$	
Z	$00\frac{1}{2}$	$0\ \ 0\frac{1}{2}$		$00\frac{1}{2}$	$00\frac{1}{2}$	$00\frac{1}{2}$	

Table E1 (*Continued*)

	Simple Tetragonal	BC Tetragonal	SC	FCC	BCC	Rhombohedral	Hexagonal
Γ							
Δ	$0\alpha0$	$0\ \alpha\ 0$	$0\alpha0$	$0\ \alpha\ 0$	$0\ \alpha\ 0$		$0\ 0\ \alpha$
Λ	00α	$0\ 0\ \alpha$	$\alpha\alpha\alpha$	$\alpha\ \alpha\ \alpha$	$\alpha\ \alpha\ \alpha$	00α	$\alpha\ \alpha\ 0$
Σ	$\alpha\alpha0$	$\alpha\ \alpha\ 0$	$\alpha\alpha0$	$\alpha\ \alpha\ 0$	$\alpha\ \alpha\ 0$	$\alpha00$	$\alpha\ 0\ 0$
A	$\frac{1}{2}\frac{1}{2}\frac{1}{2}$						$0\ 0\ \frac{1}{2}$
B							
C							
D					$\frac{1}{4}\ \frac{1}{4}\ \alpha$		
E							
F					$\alpha\frac{1}{2}-\alpha\alpha$	$\frac{1}{2}00$	
G					$\alpha\frac{1}{2}-\alpha0$		
H					$0\ \frac{1}{2}\ 0$		$\frac{1}{3}\ \frac{1}{3}\ \frac{1}{2}$
K							$\frac{1}{3}\ \frac{1}{3}\ 0$
L				$\frac{1}{4}\ \frac{1}{4}\ \frac{1}{4}$		$\frac{1}{6}\frac{1}{3}\frac{1}{6}$	$\frac{1}{2}\ 0\ \frac{1}{2}$
M	$\frac{1}{2}\frac{1}{2}0$	$\frac{1}{2}\ \frac{1}{2}\ 0$	$\frac{1}{2}\frac{1}{2}0$				$\frac{1}{2}\ 0\ 0$

N		$\frac{1}{4}$ $\frac{1}{4}$ $\frac{1}{4}$			$\frac{1}{4}$ $\frac{1}{4}$ 0		
P		0 $\frac{1}{2}$ $\frac{1}{4}$			$\frac{1}{4}$ $\frac{1}{4}$ $\frac{1}{4}$	$\frac{2}{3}\frac{1}{3}\alpha$	$\frac{1}{3}$ $\frac{1}{3}$ α
Q		α $\frac{1}{2}-\alpha$ α		$\frac{1}{4}$ $\frac{1}{2}-\alpha$ α			α α $\frac{1}{2}$
R	$0\frac{1}{2}\frac{1}{2}$		$\frac{1}{2}\frac{1}{2}\frac{1}{2}$				α 0 $\frac{1}{2}$
S	$\alpha\alpha\frac{1}{2}$		$\alpha\frac{1}{2}\alpha$	α $\frac{1}{2}$ α			$\frac{1}{2}-\alpha$ 2α $\frac{1}{2}$
T	$\alpha\frac{1}{2}\frac{1}{2}$		$\frac{1}{2}\frac{1}{2}\alpha$			$\frac{2}{3}\frac{1}{3}\frac{1}{6}$	$\frac{1}{2}-\alpha$ 2α 0
U	$0\alpha\frac{1}{2}$						$\frac{1}{2}$ 0 α
V	$\frac{1}{2}\frac{1}{2}\alpha$	$\frac{1}{2}$ $\frac{1}{2}$ α		α $\frac{1}{2}$ 0			
W	$0\frac{1}{2}\alpha$	0 $\frac{1}{2}$ α		$\frac{1}{4}$ $\frac{1}{2}$ 0			
X	$0\frac{1}{2}0$	0 $\frac{1}{2}$ 0	$0\frac{1}{2}0$	0 $\frac{1}{2}$ 0			
Y	$\alpha\frac{1}{2}0$	α $\frac{1}{2}$ 0				$\alpha\frac{1}{3}\frac{1}{6}$	
Z	$00\frac{1}{2}$		$\alpha\frac{1}{2}0$				

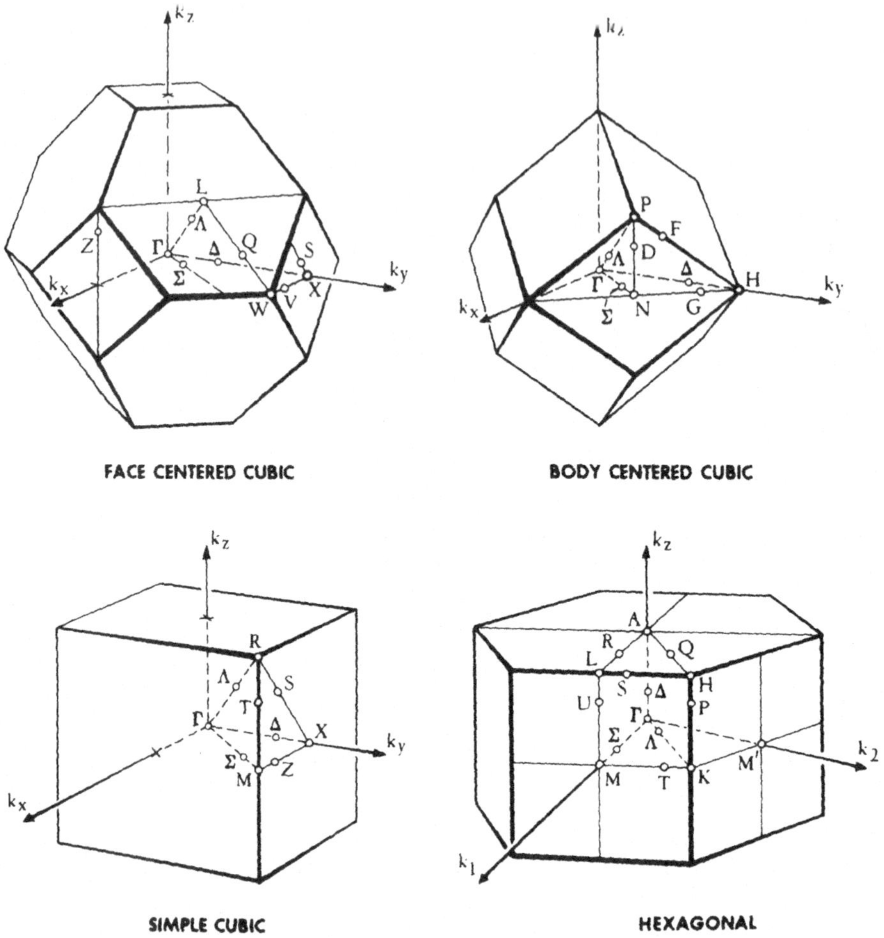

Fig. E1. The Brillouin zones.

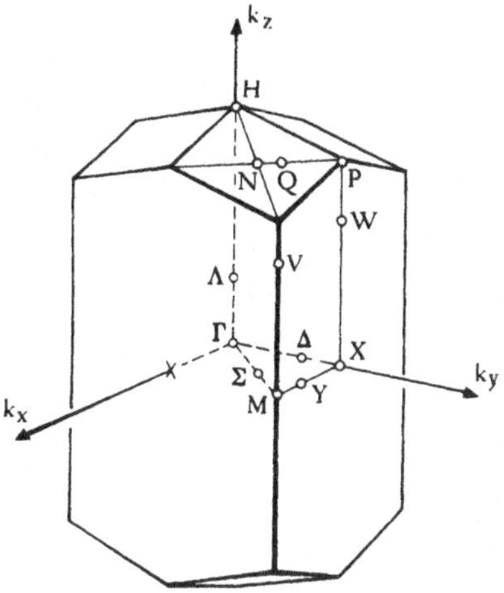

BODY CENTERED TETRAGONAL (a)

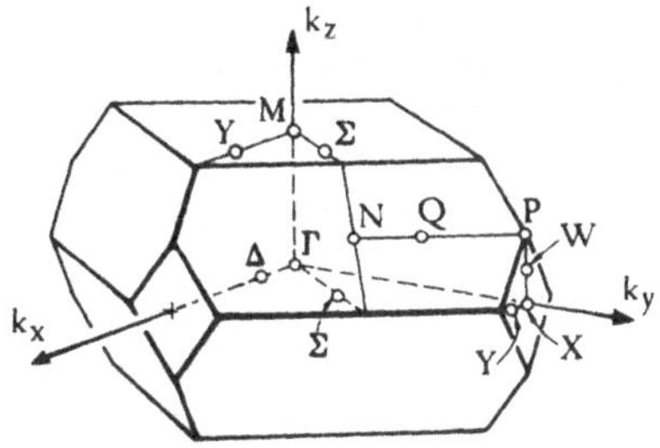

BODY CENTERED TETRAGONAL (b)

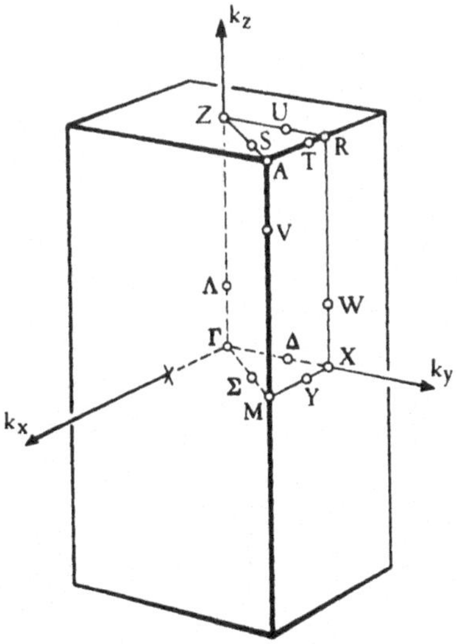

SIMPLE TETRAGONAL

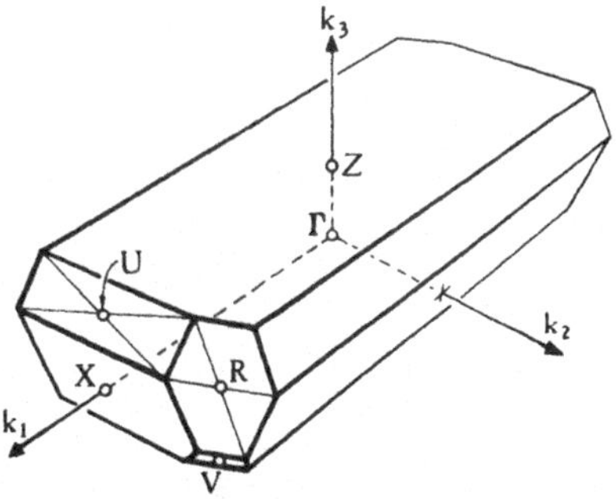

TRICLINIC

Fig. E2. The Brillouin zones.

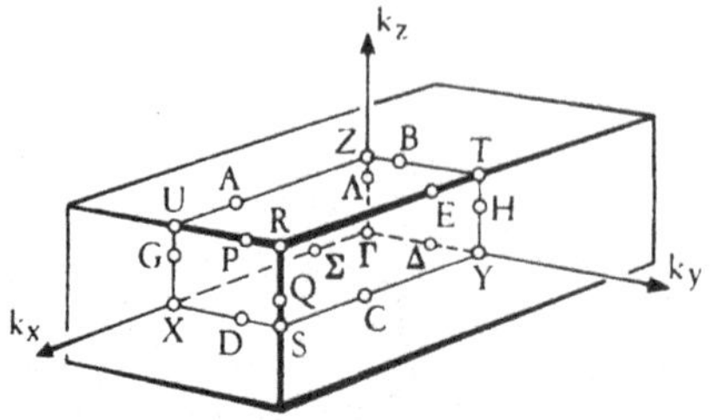

SIMPLE ORTHORHOMBIC

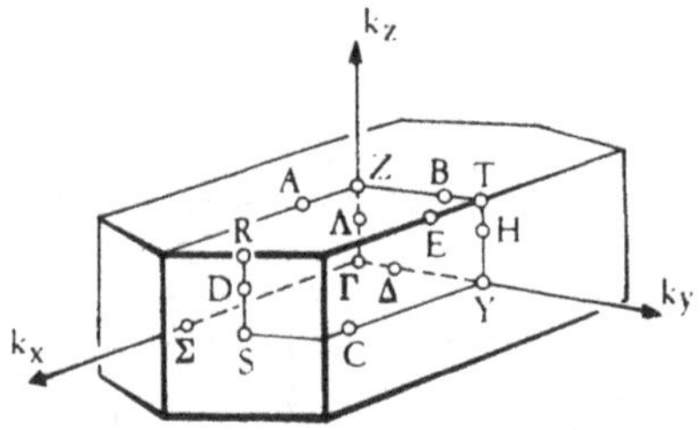

BASE CENTERED ORTHORHOMBIC

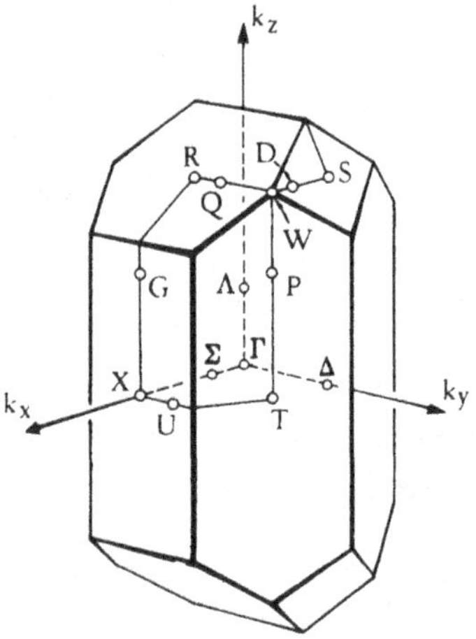

BODY CENTERED ORTHORHOMBIC

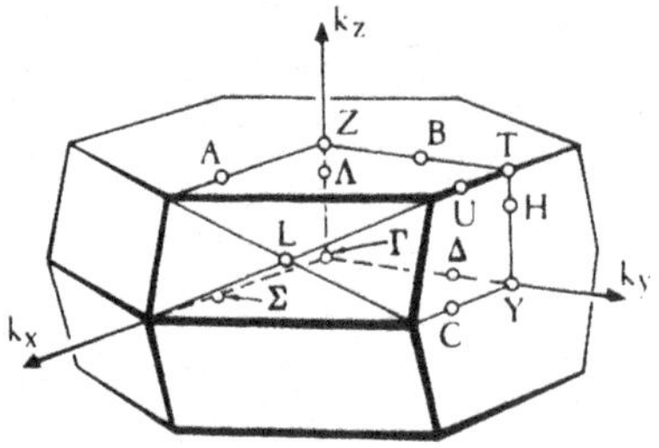

ALL FACE CENTERED ORTHORHOMBIC (a)

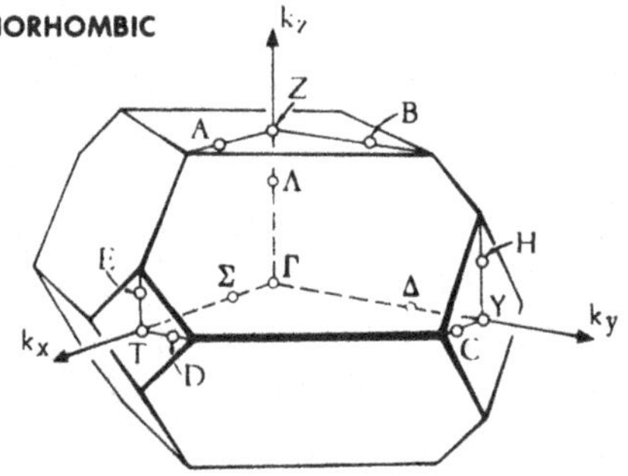

ALL FACE CENTERED ORTHORHOMBIC (b)

Fig. E3. The Brillouin zones.

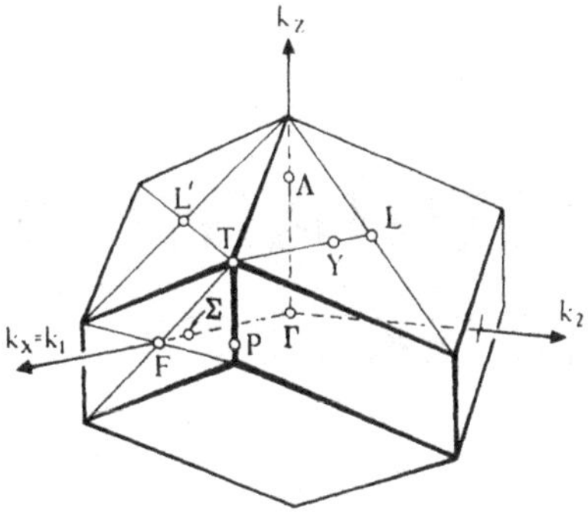

RHOMBOHEDRAL (a)

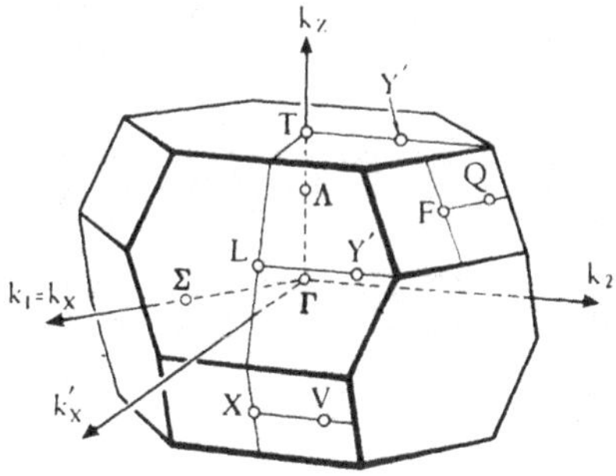

RHOMBOHEDRAL (b)

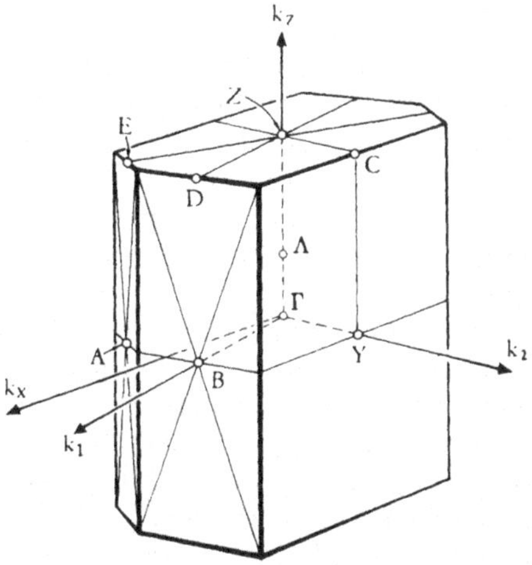

SIMPLE MONOCLINIC

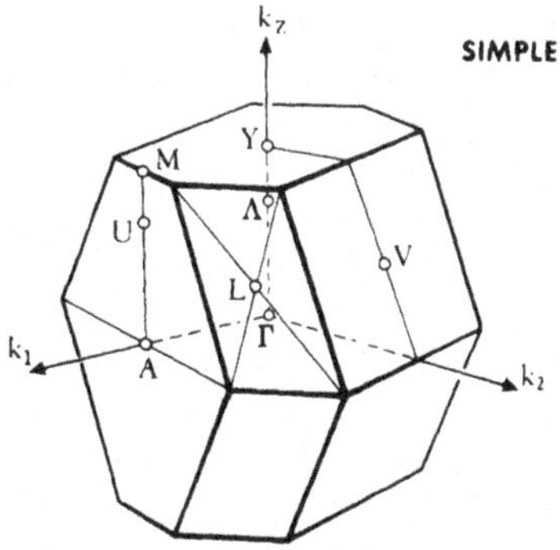

ONE FACE CENTERED MONOCLINIC (a)

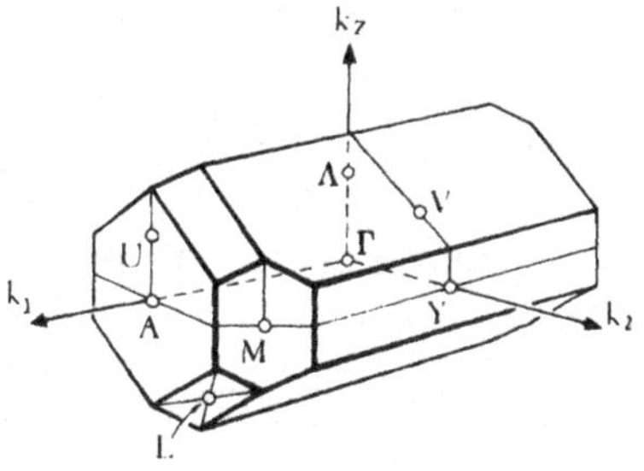

ONE FACE CENTERED MONOCLINIC (b)

Fig. E4. The Brillouin zones.

appendix F

MULTIPLIER REPRESENTATIONS FOR THE POINT GROUPS

In this appendix we present a set of tables supplying the characters of all possible multiplier (ray) representations associated with the 32 crystallographic point groups that are not projective equivalent to the ordinary (vector) representations. Such tables were originally prepared by Döring (**1959**). The tables presented here have been checked by an independent method (the algebra of the Dirac characters), permitting the elimination of (possibly typographical) errors in the Döring tables.

A multiplier group is described by the algebra of its generators, for example, $BA = \alpha AB$. If the numerical multipliers α, β, γ are set equal to unity, the ordinary group with which the table is labeled is obtained. For the case of isomorphic groups, the generator algebra will be unchanged, but the meaning of the generators will be altered as indicated. The ordinary group generators are indicated by lower-case letters, with their meanings.

The Döring tables have been greatly compressed by listing in each row all equivalent elements. For example, in Table F8 for D_{4h}, the elements AC and γA^3C are equivalent. The Dirac character associated with any element X, such as AC, is defined by Eq. 1.3.12 as

$$\Omega(X) = \frac{n_c}{g} \sum_{R \in G} RXR^{-1} \tag{F1}$$

where n_c is the number of elements in the class of X in the ordinary group. For this example, we find

$$\Omega(AC) = (AC + \gamma A^3C)\frac{1+\beta}{2} \tag{F2}$$

Thus AC and βAC are in the same class and necessarily have vanishing character if $\beta = -1$. Our table lists AC, γA^3C for the equivalent elements and also gives the multiplier $M=(1+\beta)/2$, which explains the vanishing of the characters for $\beta=-1$. For each line in each table, the Dirac character can be formed by summing the listed elements and multiplying by the multiplier M shown at the right.

The contents of the table are the usual characters (traces) related to the eigenvalues Ω^i of the Dirac characters in the conventional way (See Eq. 1.5.2):

$$\chi^i(X)=\frac{l_i\Omega^i(X)}{n_c} \tag{F3}$$

when l_i is the dimension of representation i, and n_c is the number of elements in the class containing X.

If two or more representations are projective equivalent, only one is listed. The remaining representations can be obtained by replacing the generators A, B, C, and so forth by their original values times certain roots of unity. The permissible roots of unity are listed as ϵ_n, where $(\epsilon_n)^n=1$.

Table F1

C_{2v} (2mm)		C_{2h} (2/m)	D_2 (222)
$a=(2)$ $b=m\parallel a$		$a=(2)$ $b=m\perp a$	$a=(2)$ $b=(2)\perp a$
$A^2=B^2=E$,		$BA=\alpha AB$,	$\alpha=\pm 1$
α	-1	M	
E	2	1	
A	0	$(1+\alpha)/2$	
B	0	$(1+\alpha)/2$	
AB	0	$(1+\alpha)/2$	

Arbitrary factor: ϵ_2 for A or B.

Table F2

C_{4v} (4mm)	D_4 (422)	D_{2d} ($\bar{4}2m$)
$a=(4)$ $b=m\parallel a$	$a=(4)$ $b=(2)\perp a$	$a=(\bar{4})$ $b=(2)\perp a$

$A^4=\alpha E \qquad B^2=E \qquad BA=A^3B, \qquad \alpha=\pm 1$

α	-1	M
E	2	1
A,A^3	$i\sqrt{2}$	1
A^2	0	$(1+\alpha)/2$
B,A^2B	0	$(1+\alpha)/2$
AB,A^3B	0	$(1+\alpha)/2$

Arbitrary factor: ϵ_2 for A or B.

Table F3

C_{4h} (4/m)	$a=(4)$	$b=m\perp a$
$A^4=B^2=E$	$BA=\alpha AB$	$\alpha=\pm 1$

α	-1	M
E	2	1
A^2	2	1
A	0	$(1+\alpha)/2$
A^3	0	$(1+\alpha)/2$
B	0	$(1+\alpha)/2$
AB	0	$(1+\alpha)/2$
A^2B	0	$(1+\alpha)/2$
A^3B	0	$(1+\alpha)/2$

Arbitrary factor: ϵ_4 for A; ϵ_2 for B

Table F4

D_{2h} (mmm), $a=m$, $b=m\perp a$, $c=m\perp a, \perp b$								
$A^2=B^2=C^2=E$, $CB=\alpha BC$, $AC=\beta CA$, $BA=\gamma AB$ $\alpha=\pm1, \beta=\pm1, \gamma=\pm1$								
α	-1	1	1	1	-1	-1	-1	
β	1	-1	1	-1	-1	-1	-1	M
γ	1	1	-1	-1	-1	1	-1	
E	2	2	2	2	2	2	2	1
A	2	0	0	0	0	0	0	$(1+\beta)(1+\gamma)/4$
B	0	2	0	0	0	0	0	$(1+\alpha)(1+\gamma)/4$
C	0	0	2	0	0	0	0	$(1+\alpha)(1+\beta)/4$
AB	0	0	0	0	0	2	0	$(1+\gamma)(1+\alpha\beta)/4$
BC	0	0	0	2	0	0	0	$(1+\alpha)(1+\beta\gamma)/4$
AC	0	0	0	0	2	0	0	$(1+\beta)(1+\alpha\gamma)/4$
ABC	0	0	0	0	0	0	2	$(1+\alpha\gamma)(1+\alpha\beta)/4$

Arbitrary factor: ϵ_2 for A, B or C

Table F5

C_{6v} $(6mm)$	D_6 (622)	D_{3d} $(\bar{3}2/m)$	D_{3h} $(\bar{6}m2)$
$a=(6)$ $b=m\parallel a$	$a=(6)$ $b=(2)\perp a$	$a=(\bar{3})$ $b=m\parallel a$	$a=(\bar{6})$ $b=m\parallel a$
$A^6=B^2=E$	$BA=\alpha A^5B$	$\alpha=\pm1$	
α	-1	-1	M
E	2	2	1
$A, \alpha A^5$	0	$i\sqrt{3}$	1
A^2, A^4	2	-1	1
A^3	0	0	$(1+\alpha)/2$
B, A^2B, A^4B	0	0	$(1+\alpha)/2$
AB, A^3B, A^5B	0	0	$(1+\alpha)/2$

Arbitrary factor: ϵ_2 for A or B

Table F6

C_{6h} (6/m)	$a=(6)$	$b=m\perp a$
$A^6=B^2=\alpha E$	$BA=\alpha AB$	$\alpha=\pm 1$
α	-1	M
E	2	1
A^2	2	1
A^4	2	1
A	0	$(1+\alpha)/2$
A^3	0	$(1+\alpha)/2$
A^5	0	$(1+\alpha)/2$
B	0	$(1+\alpha)/2$
AB	0	$(1+\alpha)/2$
A^2B	0	$(1+\alpha)/2$
A^3B	0	$(1+\alpha)/2$
A^4B	0	$(1+\alpha)/2$
A^5B	0	$(1+\alpha)/2$

Arbitrary factors: ϵ_6 for A; ϵ_2 for B

Table F7

T (23) and T_h ($m3$)

$a=(2)$ $b=(2)\perp a$, $c=(3)$ at equal angles to a, b, and ab; $i=$inversion (only in T_h)

$A^2=B^2=\alpha E,\quad C^3=E,\quad CA=BC,\quad CB=ABC,\quad \alpha=\pm 1$

$BA=\alpha AB,\quad I^2=E,\quad IA=AI,\quad IB=BI,\quad IC=CI$

α	-1	M
E	2	1
A,B,AB	0	$(1+\alpha)/2$
$AC,BC,ABC,\alpha C$	1	1
C^2,AC^2,BC^2,ABC^2	-1	1

Arbitrary factor: ϵ_3 for C.

For T_h, $\chi^{\pm}(IR)=\pm\chi(R)$ for the even and odd representations respectively.

Table F8

D_{4h} (4/*mmm*), $a=(4)$, $b=m\|a$, $c=m\perp a$
$A^4=\alpha E$, $B^2=C^2=E$, $BA=A^3B$, $\alpha=\pm 1$
$CA=\beta AC$, $CB=\gamma BC$, $\beta=\pm 1$, $\gamma=\pm 1$

α	1	1	1	1	1	1	−1	−1	−1	−1	
β	−1	−1	1	1	−1	−1	1	1	−1	−1	.M
γ	1	1	−1	−1	−1	−1	1	−1	1	−1	
E	2	2	2	2	2	2	2	2	4	4	1
A,A^3	0	0	2	0	0	0	$i\sqrt{2}$	$i\sqrt{2}$	0	0	$(1+\beta)/2$
A^2	2	−2	2	−2	2	−2	0	0	0	0	$(1+\alpha)/2$
B,A^2B	2	0	0	0	0	0	0	0	0	0	$(1+\alpha)(1+\gamma)/4$
AB,A^3B	0	0	0	0	2	0	0	0	0	0	$(1+\alpha)(1+\beta\gamma)/4$
C	0	0	0	0	0	0	2	0	0	0	$(1+\beta)(1+\gamma)/4$
$AC,\gamma A^3C$	0	0	0	$2i$	0	0	$i\sqrt{2}$	$\sqrt{2}$	0	0	$(1+\beta)/2$
A^2C	0	0	0	0	0	0	0	$2i$	0	0	$(1+\beta)(1+\alpha\gamma)/4$
$BC,\beta A^2BC$	0	2	0	0	0	0	0	0	0	0	$(1+\alpha)(1+\gamma)/4$
$ABC,\beta A^3BC$	0	0	0	0	0	2	0	0	0	0	$(1+\alpha)(1+\beta\gamma)/4$

Arbitrary factor: ϵ_2 for A, B, or C

Table F9

D_{6h} (6/*mmm*) $a=(6)$, $b=m\|a$ $c=m\perp a$

$A^6=B^2=C^2=E$; $BA=\alpha A^5B$, $CB=\beta BC$, $CA=\gamma AC$; $\alpha=\pm1$, $\beta=\pm1$, $\gamma=\pm1$

α	−1	−1	1	1	−1	−1	1	1	−1	−1	1	1	−1	−1	M
β	1	1	−1	−1	−1	−1	1	1	1	1	−1	−1	−1	−1	
γ	1	1	1	1	1	1	−1	−1	−1	−1	−1	−1	−1	−1	
E	2	2	2	2	2	2	2	4	2	4	2	4	2	4	1
$A,\alpha A^5$	0	w	2	1	0	w	0	0	0	0	0	0	0	0	$(1+\gamma)/2$
A^2,A^4	2	−1	2	−1	2	−1	2	−2	2	−2	2	−2	2	−2	1
A^3	0	0	2	−2	0	0	0	0	0	0	0	0	0	0	$(1+\alpha)(1+\gamma)/4$
B,A^2B,A^4B	0	0	0	0	0	0	2	0	0	0	0	0	0	0	$(1+\alpha)(1+\beta)/4$
AB,A^3B,A^5B	0	0	0	0	0	0	0	0	0	0	2	0	0	0	$(1+\alpha)(1+\beta\gamma)/4$
C	2	2	0	0	0	0	0	0	0	0	0	0	0	0	$(1+\beta)(1+\gamma)/4$
$AC,\alpha\beta A^5C$	0	w	0	w	2	1	0	0	0	0	0	0	0	0	$(1+\gamma)/2$
$A^2C,\beta A^4C$	2	−1	0	w	0	w	0	0	0	0	0	0	0	0	$(1+\gamma)/2$
A^3C	0	0	0	0	2	−2	0	0	0	0	0	0	0	0	$(1+\gamma)(1+\alpha\beta)/4$
BC,A^2BC,A^4BC	0	0	0	0	0	0	0	0	2	0	0	0	0	0	$(1+\alpha\gamma)(1+\beta)/4$
ABC,A^3BC,A^5BC	0	0	0	0	0	0	0	0	0	0	0	0	$2i$	0	$(1+\alpha\gamma)(1+\alpha\beta)/4$

$w=i\sqrt{3}$ Arbitrary factor: ϵ_2 for A, B, or C

Table F10

O (432)	T_d ($\bar{4}3m$)
$d=\boldsymbol{\delta}_{2xy}$	$d=\boldsymbol{\sigma}_{xy}=i\boldsymbol{\delta}_{2xy}$

$a=\boldsymbol{\delta}_{2x}$, $b=\boldsymbol{\delta}_{2y}$, $c=\boldsymbol{\delta}_{xyz}$ for both

$A^2=B^2=\alpha E, \quad C^3=D^2=E, \quad DC=C^2D$

$CA=BC, \quad BA=\alpha AB, \quad DA=\alpha BD, \quad DB=\alpha AD$

$CB=ABC \quad \alpha=\pm 1$

α	-1	-1	M
E	4	2	1
A,B,AB	0	0	$(1+\alpha)/2$
$C,C^2,AC^2,BC^2,ABC^2,\alpha AC,\alpha BC,\alpha ABC$	1	-1	1
$D,CD,C^2D,ABD,BCD,AC^2D,$	0	0	$(1+\alpha)/2$
$AD,ABCD,BC^2D,\alpha BD,\alpha ACD,\alpha ABC^2D$	0	$i\sqrt{2}$	1

Arbitrary factor: ϵ_2 for D

Table F11

O_h $(m3m)$, $a=\boldsymbol{\delta}_{2x}$, $b=\boldsymbol{\delta}_{2y}$, $c=\boldsymbol{\delta}_{3xyz}$, $d=\boldsymbol{\delta}_{2xy}$, $i=$ inversion

$A^2=B^2=\alpha E$, $C^3=D^2=E$, $CA=BC$, $DC=C^2D$

$BA=\alpha AB$, $DA=\alpha BD$, $DB=\alpha AD$, $CB=ABC$; $\alpha=\pm 1$

$I^2=E$, $IA=AI$, $IB=BI$, $IC=CI$, $ID=\beta DI$; $\beta=\pm 1$

α	1	1	1	-1	-1	-1	-1	
β	-1	-1	-1	1	1	-1	-1	M
E	2	2	6	4	2	4	4	1
A, B, AB	2	2	-2	0	0	0	0	$(1+\alpha)/2$
$C, C^2, AC^2, BC^2, ABC^2, \alpha AC, \alpha BC, \alpha ABC$	2	-1	0	1	-1	1	-2	1
$D, CD, C^2D, ABD, BCD, AC^2D$	0	0	0	0	0	0	0	$(1+\alpha)(1+\beta)/4$
$AD, ABCD, BC^2D, \alpha BD, \alpha ACD, \alpha ABC^2D$	0	0	0	0	$i\sqrt{2}$	0	0	$(1+\beta)/2$
I	0	0	0	4	2	0	0	$(1+\beta)/2$
AI, BI, ABI	0	0	0	0	0	0	0	$(1+\alpha)(1+\beta)/4$
$CI, \alpha ACI, \alpha BCI, \alpha ABCI, \beta C^2I, \beta AC^2I, \beta BC^2I, \beta ABC^2I$	0	w	0	1	-1	$-w$	0	1
$DI, CDI, C^2DI, BCDI, AC^2DI, ABDI$	0	0	0	0	0	0	0	$(1+\alpha)(1+\beta)/4$
$ADI, ABCDI, BC^2DI, \alpha BDI, \alpha ACDI, \alpha ABC^2DI$	0	0	0	0	$i\sqrt{2}$	0	0	$(1+\beta)/2$

appendix G

WIGNER MAPPINGS AND THE FUNDAMENTAL THEOREM OF PROJECTIVE GEOMETRY[1]

The proof that the symmetry operators of quantum mechanics (i.e., Wigner mappings) are unitary or antiunitary has been broken down into the following chain of steps:

Wigner mapping = projectivity = semilinearity = unitarity or antiunitarity

The last step was given in Section 10.4, and the first step will be presented now, by conventional methods. The equivalence of projectivity to semilinearity is presented separately later in this appendix, because these concepts and the proof of their equivalence do not involve the existence of a metric. Indeed this equivalence is the fundamental theorem of projective geometry (Baer, **1952**).

PROPERTIES OF WIGNER MAPPINGS

Theorem G1

A Wigner mapping carries an orthonormal set of vectors $\Psi_1, \Psi_2, \ldots, \Psi_n$ that spans S into an orthonormal set $M\Psi_1, M\Psi_2, \ldots, M\Psi_n$ that spans $S' = MS$.

[1]The subject of Section 10.4 and this appendix was initiated by Wigner (**1932, 1959**) and treated by Hagedorn (**1959, 1961**), Uhlhorn (**1963**), and others. Uhlhorn describes in detail the shortcomings of all previous proofs. The proof given here was constructed for an earlier draft of this chapter when Uhlhorn's paper was unavailable to the author. The key steps, by a not so remarkable coincidence, are identical to those of Uhlhorn. The reader is referred to Uhlhorn, however, for a mathematically rigorous discussion.

Proof: The orthonormality is an immediate consequence of Eq. 10.4.8. To prove completeness we define

$$I \equiv (\Phi,\Phi) - \sum_{i=1}^{n} |(\Psi_i,\Phi)|^2$$

$$I' \equiv (M\Phi,M\Phi) - \sum_{i=1}^{n} |(M\Psi_i,M\Phi)|^2$$

Completeness of the initial set of vectors implies that $I=0$ for any Φ in S. Using Eq. 10.4.8, we have $I'=I=0$. Since any vector in S' can be written in the form $M\Phi$ or is a linear combination of such terms, the mapped set of vectors $\{M\Psi_i\}$ is complete.

Theorem G2

A Wigner mapping takes independent vectors into independent vectors.

Proof: Let $\Phi_1,\Phi_2,\ldots,\Phi_n$ be a set of independent vectors, and $\Psi_1,\Psi_2,\ldots,\Psi_n$ a set of orthonormal vectors that span the same space S. If Φ is an additional vector independent of $\Phi_1,\ldots,\Phi_n$, it must also be independent of $\Psi_1,\ldots,\Psi_n$. Hence $I>0$ and $I'>0$. Thus $M\Phi$ must be independent of $M\Psi_1,\ldots,M\Psi_n$ and hence also of $M\Phi_1,\ldots,M\Phi_n$, since the latter are expressible in terms of the former by Theorem G1.

Note: Theorems G1 and G2 are easily proved without assumption 10.4.7 by using $\hat{\Phi}$ instead of Φ.

Theorem G3

A Wigner mapping is a projectivity, that is, a set of vectors $\Phi_1,\ldots,\Phi_n$ independent and complete in S, is carried by a Wigner mapping into a set of vectors $M\Phi_1,\ldots,M\Phi_n$, independent and complete in $S'=MS$.

Proof: By Theorem G2 the $M\Phi_1,\ldots,M\Phi_n$ constitute a set of n independent vectors in S'. Hence they must span S'. Thus completeness and, with Theorem G2, independence are preserved by Wigner mappings.

FUNDAMENTAL THEOREM OF PROJECTIVE GEOMETRY

Theorem G4

A projectivity on a linear manifold of three or more dimensions can be reduced by a (never vanishing) factor transformation

$$O\Psi = \alpha(\Psi)M\Psi$$

to a semilinear transformation, that is, one obeying Eqs. 10.4.15–10.4.17.

Proof: If Φ and Ψ are independent vectors, Theorem G3 implies that

$$M(\Phi+\Psi)=\lambda(\Phi,\Psi)M\Phi+\mu(\Phi,\Psi)M\Psi \tag{G1}$$

where

$$\mu(\Phi,\Psi)=\lambda(\Psi,\Phi)$$

are complex numerical coefficients.We select three *independent* vectors Φ,Ψ,Γ and use Eq. G1 to write

$$\begin{aligned}M[(\Phi+\Psi)+\Gamma]&=\lambda(\Phi+\Psi,\Gamma)M(\Phi+\Psi)+\lambda(\Gamma,\Phi+\Psi)M(\Gamma)\\&=\lambda(\Phi+\Psi,\Gamma)[\lambda(\Phi,\Psi)M(\Phi)+\lambda(\Psi,\Phi)M(\Psi)]+\lambda(\Gamma,\Phi+\Psi)M(\Gamma)\end{aligned} \tag{G2}$$

Since the vectors of a linear manifold constitute an additive, Abelian group,

$$M[(\Phi+\Psi)+\Gamma]=M[(\Phi+\Gamma)+\Psi] \tag{G3}$$

We may therefore interchange Ψ and Γ in Eq. G2 and compare the coefficients of the independent vectors $M\Psi$, $M\Gamma$, and $M\Phi$. The first yields

$$\lambda(\Phi+\Psi,\Gamma)\lambda(\Psi,\Phi)=\lambda(\Psi,\Phi+\Gamma) \tag{G4}$$

and the remaining two yield conditions derivable from Eq. G4.

We now make the key ansatz:

$$\lambda(\Phi,\Psi)=\nu(\Phi+\Psi,\Phi) \tag{G5}$$

to obtain

$$\nu(\Phi+\Psi+\Gamma,\Phi+\Psi)\nu(\Phi+\Psi,\Psi)=\nu(\Phi+\Psi+\Gamma,\Psi)$$

Next we introduce the new independent vectors:

$$\Phi'=\Phi+\Psi+\Gamma, \qquad \Gamma'=\Phi+\Psi, \qquad \Psi'=\Psi \tag{G6}$$

and (after dropping the primes) obtain the functional equation

$$\nu(\Phi,\Psi)=\nu(\Phi,\Gamma)\nu(\Gamma,\Psi) \tag{G7}$$

Lemma G5

The function $\nu(\Phi,\Psi)$ does not vanish anywhere.

Proof: If it did, it would vanish everywhere since

$$\nu(\Lambda,\Gamma)=\nu(\Lambda,\Phi)\nu(\Phi,\Psi)\nu(\Psi,\Gamma) \tag{G8}$$

Construction of $\alpha(\Psi)$

Since $\lambda(\Phi,0)=\nu(\Phi,\Phi)=1$ (see Eq. G1), we have

$$\nu(\Phi,\Gamma)\nu(\Gamma,\Phi)=1 \qquad \text{or} \qquad \nu(\Phi,\Gamma)=[\nu(\Gamma,\Phi)]^{-1} \tag{G9}$$

Hence

$$\nu(\Phi,\Psi)=[\nu(\Gamma,\Phi)]^{-1}\nu(\Gamma,\Psi) \tag{G10}$$

and this result is clearly independent of the choice of Γ. Moreover, it can be used to define $\nu(\Phi,\Psi)$ when Φ and Ψ are dependent vectors. If we make the particular choice

$$\alpha(\Psi)=\nu(\Gamma_0,\Psi) \tag{G11}$$

then

$$\nu(\Phi,\Psi)=[\alpha(\Phi)]^{-1}\alpha(\Psi), \qquad \lambda(\Phi,\Psi)=[\alpha(\Phi+\Psi)]^{-1}\alpha(\Phi) \tag{G12}$$

Thus Eq. G1 can be written in the form

$$\alpha(\Phi+\Psi)M(\Phi+\Psi)=\alpha(\Phi)M\Phi+\alpha(\Psi)M\Psi \tag{G13}$$

Hence the mapping $O\Psi=\alpha(\Psi)M\Psi$ yields the desired isomorphism:

$$O(\Phi+\Psi)=O\Phi+O\Psi \tag{G14}$$

If we now define

$$O(a\Psi)=g(a,\Psi)O\Psi \tag{G15}$$

we can write

$$\begin{aligned} g(a,\Phi+\Psi)O(\Phi)+g(a,\Phi+\Psi)O(\Psi) &= g(a,\Phi+\Psi)O(\Phi+\Psi) \\ &= O[a(\Phi+\Psi)]=O(a\Phi)+O(a\Psi) \\ &= g(a,\Phi)O(\Phi)+g(a,\Psi)O(\Psi) \end{aligned}$$

from which we conclude that

$$g(a,\Phi+\Psi)=g(a,\Phi)=g(a,\Psi)=g(a) \tag{G16}$$

is independent of Ψ. Now we have

$$O(ab\Psi)=g(a)O(b\Psi)=g(a)g(b)O(\Psi)$$

Hence

$$g(ab)=g(a)g(b) \tag{G17}$$

Also

$$O[(a+b)\Psi]=O(a\Psi)+O(b\Psi)$$

or

$$g(a+b)=g(a)+g(b) \tag{G18}$$

Conditions G16–G18 establish that $g(a,\Psi)=g(a)$ is independent of Ψ and constitutes an isomorphism on the number field.

Theorem G6

The semilinear transformation associated with a projectivity is unique (modulo a scale factor).

Proof: If a second semilinear mapping $O'\Psi=\alpha'(\Psi)M\Psi$ existed, $O'(\Phi+\Psi)=O'\Phi+O'\Psi$ would require that

$$\lambda(\Phi,\Psi)=\frac{1}{\alpha'(\Phi+\Psi)}\alpha'(\Phi)=\frac{1}{\alpha(\Phi+\Psi)}\alpha(\Phi) \tag{G19}$$

Since Φ and Ψ are independent, we may conclude that $\alpha'(\Phi)=a\alpha(\Phi)$, where a is independent of Φ and is simply a scale factor.

Moreover, if we start with a different vector mapping

$$M'\Psi=h(\Psi)M\Psi \tag{G20}$$

consistent with the same projectivity, λ is changed to

$$\lambda'(\Phi,\Psi)=h(\Phi+\Psi)\lambda(\Phi,\Psi)[h(\Phi)]^{-1} \tag{G21}$$

so that

$$\alpha'(\Phi)=a\alpha(\Phi)[h(\Phi)]^{-1}$$

or

$$O'\Phi=\alpha'(\Phi)M'(\Phi)=aO\Phi \tag{G22}$$

Thus our mapping is unique if the arbitrary scale factor a is fixed by prescribing the normalization and phase of $O\Psi$ for one Ψ. This completes our proof of the fundamental theorem.

appendix H

GENERALIZED MOBILITY THEORY[1]

We start with the equation for the density matrix (with $\hbar$ set equal to unity):

$$i\frac{\partial\rho}{\partial t}+[\rho,H+V(t)]+\frac{i[\rho-\rho(\beta)]}{\tau}=0 \tag{H1}$$

where H is the Hamiltonian of a (possibly) many-body system including interactions (e.g., electron energy + phonon energy + electron-electron interactions + phonon-phonon interactions + electron-phonon interactions),

$$V(t)=-\sum F_j(t)\alpha_j \tag{H2}$$

is the potential associated with the external forces, and the last term represents a weak interaction with the surroundings that causes the system in the absence of forces to return to the equilibrium state:

$$\rho(\beta)=\frac{\exp(-\beta H)}{\mathrm{Tr}[\exp(-\beta H)]} \tag{H3}$$

Since we are concerned here only with the part of the response linear in the forces F_j, heating effects, which are second order, do not enter. The term in $1/\tau$ needed to prevent such effects can therefore be allowed to approach zero, *after* the linear part of the response is taken. In the resulting integrals, it can be retained as a mild convergence factor that will kill oscillatory transients bearing no relation to the correct steady-state response. (The unimportance, for this calculation, of the nature of the interactions with the thermal reservoir is what permits us to treat these interactions as a simple relaxation.)

[1]Fundamental papers on this subject include those of Callen and Welton (**1951**), Callen, Barasch, and Jackson (**1952**), Jackson (**1952**), Kaplan (**1956**), Kubo (**1956, 1957**), Lax (**1955, 1958b, 1964**), Mori (**1956**), and Nakano (**1956**). See also Bernard and Callen (**1959**).

Let us regard the term in $[\rho, V(t)]$ as a known inhomogeneous term. The transformation

$$\rho = \exp(-iHt)\rho' \exp(iHt) \tag{H4}$$

eliminates the terms in $[\rho, H]$. The transformation

$$\rho' - \rho(\beta) = \exp\left(\frac{-t}{\tau}\right)g$$

eliminates the term in $(1/\tau)[\rho' - \rho(\beta)]$, and we are left with an equation of the form $\partial g/\partial t =$ known function of t. Integrating and restoring the original variables, we find that

$$\rho(t) = \rho(\beta) + i\int_{-\infty}^{t} e^{(t'-t)/\tau} e^{iH(t'-t)} [\rho(t'), V(t')]\, e^{-iH(t'-t)} dt' \tag{H5}$$

is an exact solution of Eq. H1 subject to the initial condition $\rho(-\infty) = \rho(\beta)$. Since V is regarded as small, we may solve by iteration:

$$\rho = \rho(\beta) + \rho_1 + \rho_2 + \cdots$$

The first-order response $\rho_1(t)$, which is all that concerns us, is obtained by replacing ρ by $\rho(\beta)$ on the right-hand side. The flux is then given to first order by

$$\langle \Delta\dot{\alpha}_i(t)\rangle = \mathrm{Tr}[\dot{\alpha}_i \rho_1(t)] = \sum_j \int_{-\infty}^{t} dt' \exp\left(\frac{t'-t}{\tau}\right) y_{ij}(t,t') F_j(t') \tag{H6}$$

where

$$y_{ij}(t,t') = -i\mathrm{Tr}\{\dot{\alpha}_i(t)[\rho(\beta), \alpha_j(t')]\} \tag{H7}$$

$$= y_{ij}(t-t', 0) \tag{H8}$$

and the last step, expressing stationarity, is achieved by shifting the $\exp(-iHt')$ factors until they can be incorporated with the $\exp(iHt)$ factors. Let

$$[\exp(-\beta H), \alpha_j] \equiv R(\beta)\exp(-\beta H) \tag{H9}$$

Then

$$R(\beta) = \exp(-\beta H)\alpha_j \exp(\beta H) - \alpha_j$$

$$\frac{dR}{d\beta} = \exp(-\beta H)[\alpha_j, H]\exp(\beta H)$$

Integrating this expression, using $R(0)=0$, yields

$$R(\beta)=\int_0^{\beta} d\lambda \exp(-\lambda H)[\alpha_j, H]\exp(\lambda H) \tag{H10}$$

Using the Heisenberg equation

$$i\dot{\alpha}_j=[\alpha_j, H] \tag{H11}$$

we can combine the preceding results to obtain

$$y_{ij}(t)\equiv y_{ij}(t,0)=\mathrm{Tr}\left[\dot{\alpha}_i(t)\int_0^{\beta} d\lambda \exp(-\lambda H)\,\dot{\alpha}_j(0)\exp(\lambda H)\,\rho(\beta)\right] \tag{H12}$$

that is, Eq. 10.5.29.

AUTHOR INDEX AND BIBLIOGRAPHY*

Abragam, A. and Pryce, M. H. L. (1951) "Theory of the Nuclear Hyperfine Structure of Paramagnetic Resonance Spectra in Crystals." *Proc. Roy. Soc.* **A205** (135-153) *98*

Abrahams, E. (1957) "Donor Electron Spin Relaxation in Silicon." *Phys. Rev.* **107** (491-496) *286*

Adams, E. N., II (1952) "Motion of an Electron in a Perturbed Periodic Potential" *Phys. Rev.* **85** (41-50) *214*

Adams, E. N., II (1953) "The Crystal Momentum as a Quantum Mechanical Operator." *J. Chem. Phys.* **21** (2013-2017) *214*

Aero, E. L. and Kuvshinskii, E. V. (1960) "Fundamental Equations of the Theory of Elastic Media with Rotationally Interacting Particles." *Soviet Phys. Solid State Phys.* **2** (1272-1281) *126*

Anderson, Orson (1963) "A Simplified Method for Calculating the Debye Temperature from Elastic Constants." *J. Phys. Chem. Solids* **24** (909-917) *132*

Anderson, P. W. (1963) *Concepts in Solids.* W. A. Benjamin.#

Baer, R. (1952) *Linear Algebra and Projective Geometry.* Academic *461*

Barasch, M. (1952) see Callen, H. B., and Jackson, J., *466*

Bardeen, J. and Shockley, W. (1950) "Deformation Potentials and Mobilities in Non-Polar Crystals." *Phys. Rev.* **80** (72-80) *271*

Bargmann, V. (1962) "On the Representations of the Rotation Group." *Rev. Mod. Phys.* **34** (829-845) *420; 422*

Bauer, E. (1962) see Meijer, P. H. E.,#

Bernard, W. and Callen, H. B. (1959) "Irreversible Thermodynamics of Nonlinear Processes and Noise in Driven Systems." *Rev. Mod. Phys.* **31** (1017-1044) *466*

Bethe, H. (1929) "Term Splitting in Crystals." *Ann. Physik* **3** (133-208) *23; 62; 416*

Bethe, H. A. (1947) see Von der Lage, F. C., *101; 110; 432.*

---

*References with a # constitute additional bibliography not explicitly referred to in the text. Numbers in italics represent pages in this book on which the given reference has been referred to.

This index and the ones which follow were printed on a Phototypesetter operated by Bell Laboratories. The help of J. F. Ossanna in writing the code for this typesetter is gratefully acknowledged.

Bethe, H. and Salpeter, E. E. (1957) *Quantum Mechanics of One and Two Electron Atoms.* Academic *409*

Bir, G. I. and Pikus G. Ye., (1972) *Symmetry and Deformational Effects in Semiconductors,* "Nauka" Press, Moscow#

Birman, J. L. (1959) "Simplified LCAO Method for Zincblende, Wurtzite and Mixed Crystal Structures." *Phys. Rev.* **115** (1493-1505) *255; 416*

Birman, J. L. (1962) "Space Group Selection Rules: Diamond and Zinc Blende." *Phys. Rev.* **127** (1093-1106) *273*

Birman, J. L. (1963) "Theory of Infrared and Raman Processes in Crystals: Selection Rules in Diamond and Zinc Blende." *Phys. Rev.* **131** (1489-1496) *273*

Birman, J. L. (1966) "Full Group and Subgroup Methods in Crystal Physics." *Phys. Rev.* **150** (771-782) *273*

Birman, J. L., Lax, M. and Loudon, R. (1966) "Intervalley Scattering Selection Rules in III-V Semiconductors." *Phys. Rev.* **145** (620-622) *273*

Birman, J. L. (1972) see Lax, M., *120; 273; 388.*

Birman, J. L. (1974) "Theory of Crystal Space Groups and Infrared and Raman Lattice Processes of Insulating Crystals." *Handbuch der Physik* Vol. XXV/26, Springer-Verlag *242; 273*

Bleaney, B. and Stevens, K. W. H. (1953) "Paramagnetic Resonance." *Rep. Prog. Phys.* **16** (108-159) *98*

Blount, E. I. Private Communication *317*

Blount E. I. (1962a) "Formalisms of Band Theory" in Vol. 13 *Solid State Physics* Seitz, F. and Turnbull, D., eds., Academic *182; 214; 217; 230*

Blount E. I. (1962b) "Bloch Electrons in a Magnetic Field." *Phys. Rev.* **126** (1636-1653) *91; 230*

Blount, E. I. (1971) "The Jahn-Teller Theorem." *J. Math. Phys.* **12** (1890-1896) *414*

Born, M. and Huang, K. (1954) *Dynamical Theory of Crystal Lattices.* Oxford *328; 347; 351*

Bouckaert, L. P., Smoluchowski, R. and Wigner, E. P. (1936) "Theory of Brillouin Zones and Symmetry Properties of Wave Functions in Crystals." *Phys. Rev.* **50** (58-68) *23; 367; 416*

Bowers, K. D. and Owen, J. (1955) "Paramagnetic Resonance II." *Rep. Prog. Phys.* **18** (304-373) *98*

Bradley, C. J. (1966) "Space Groups and Selection Rules." *J. Math. Phys.* **7** (1145-1152) *273*

Brinkman, H. C. (1956) *Applications of Spinor Invariants in Atomic Physics.* North Holland-Interscience#

Brockhouse, B. N. and Iyengar, P. K. (1958) "Normal Modes of Germanium by Neutron Spectroscopy." *Phys. Rev.* **111** (747-754) *325; 326; 370*

Brueckner, K. A. (1959) *Theory of Nuclear Structure in the Many Body Problem.* 1958 Les Houches Lectures, Université de Grenoble, Wiley especially Pp. 65 ff *362*

Buerger, M. J. (1956) *Elementary Crystallography; an Introduction to the Fundamental Geometric Features of Crystals.* Wiley *238*

Burnside, W. (1955) *Theory of Groups of Finite Order.* Dover#

Burstein, E. (1955) see Lax, M., *258; 259*

Callaway, Joseph (1964) *Energy Band Theory.* Academic (357) *254*

Callen, H. B. (1948) "The Application of Onsager's Reciprocal Relations to Thermoelectric, Thermomagnetic, and Galvanomagnetic Effects." *Phys. Rev.* **73** (1349-1358) *287*

Callen, H. B. and Welton T. (1951) "Irreversibility and Generalized Noise." *Phys. Rev.* **83** (34-40) *466*

Callen, H. B., Barasch, M. and Jackson, J. (1952) "Statistical Mechanics of Irreversibility." *Phys. Rev.* **88** (1382-1386) *466*

Callen, H. B. (1959) see Bernard, W., *466*

Callen, H. B. (1960) *Thermodynamics.* Wiley *287*

Casella, R. C. (1959) "Symmetry of Wurtzite." *Phys. Rev.* **114** (1514-1518) *416*

Casher, A. (1969) see Zak, W. and Gluck, M., *246*

Casimir, H. B. G. (1931) "On the Construction of a Differential Equation Belonging to the Half Integral Irreducible Representation of a Continuous Group." *Kon. Akad. Science Amsterdam* **6** (344-846) *32*

Casimir, H. B. G. (1945) "On Onsager's Principle of Microscopic Reversibility." *Rev. Mod. Phys.* **17** (343-350) *287*

Cochran, W. (1959) "Theory of the Lattice Vibrations of Germanium." *Phys. Rev. Letters* **2** (495-497) *370; 388*

Cochran, W. (1963) "Lattice Vibrations." *Reports on Prog in Phys.* **26** (1-45) *388*

Cochran, W. (1965) "Interpretation of Phonon Dispersion Curves." Pp. 75-84 *Proceedings of International Conference on Lattice Dynamics, Copenhagen, 1963* Pergamon *388*

Condon, E. U., and Shortley, G. H. (1967) *The Theory of Atomic Spectra.* Cambridge. *39; 109; 320*

Cosserat, G. and Cosserat, F. (1909) *"Théorie des Corps Deformable."* Pp. 953-1173 of O. D. Chwolson, *Traité de Physique,* (transl. E. Davaux), 2nd ed. Paris, *126*

Cotton, F. A. (1963) *Chemical Application of Group Theory.* Wiley, p. 115 *403; 415*

Crasemann, B. (1961) see Powell, J. L., *333*

Cross, P. C. (1955) see Wilson Jr., F. B. and Cross, P.C., *152; 153*

Darwin, C. G. (1928) "The Wave Equations of the Electron." *Proc. Roy. Soc.* **118A** (654-680) *91*

Daudel, R., Lefebvre, R. and Mose, C. (1960) *Quantum Chemistry: Methods and Applications.* Interscience *391*

Daudel, R. (1966) *Electronic Structure of Molecules.* Pergamon *391*

Daudel, R. (1968) *Fundamentals of Theoretical Chemistry.* Pergamon *391*

Decius, J. C. (1955) see Wilson Jr., F. B., *152; 153*

De Groot, S. R. (1951) *Thermodynamics of Irreversible Processes.* Interscience *289*

De Groot, S. R. and Mazur, P. (1962) *Non-equilibrium Thermodynamics.* North Holland -Interscience. *287; 289*

Des Cloizeaux, J. (1963) "Orthogonal Orbitals and Generalized Wannier Functions." *Phys. Rev.* **129** (554-566) *219*

Des Cloizeaux, J. (1964a) "Energy Bands and Projection Operators in a Crystal: Analytic and Asymptotic Properties." *Phys. Rev.* **135** (A685-A697) *219*

Des Cloizeaux, J. (1964b) "Analytical Properties of n-Dimensional Energy Bands and Wannier Functions." *Phys. Rev.* **135** (A698-A707) *219*

Dick, B. G., and Overhauser, A. W. (1958) "Theory of the Dielectric Constants of the Alkali Halide Crystals." *Phys. Rev.* **112** (90-103) *388*

Dimmock, J. O., and Wheeler, R. G. (1962) "Irreducible Representations of Magnetic Groups." *J. Phys. Chem. Solids* **23** (729-741) *292*

Dimmock, J. O., (1963) see Koster, G. F., Wheeler, R. G. and Statz, H., *416; 418*

Dirac, P. A. M. (1958) *Principles of Quantum Mechanics, 4th ed.* Oxford *38; 289; 313*

Doring, W. (1959) "Ray Representations of the Crystallographic Groups." *Z. Naturforsch.* **14a** (343-350) *243; 246; 452.*

Edmonds, A. P. (1957) *Angular Momentum in Quantum Mechanics.* Princeton. *38*

Elliott, R. J. and Loudon, R. (1960) "Group Theory of Scattering Processes in Crystals." *J. Phys. Chem. Solids.* **75** (146-151) *247; 273*

Erdos, Paul (1964) "The Determination of the Components of a Tensor Characterizing a Crystal." *Helvetica Physica Acta* **37** (493-509) *120*

Eringen, A. C. (1962) *Non-linear Theory of Continuous Media.* McGraw-Hill *124*

Eringen, A. C. (1967) *Mechanics of Continua.* Wiley *124*
Eyring, H., Walter, J. and Kimball, G. E. (1944) *Quantum Chemistry.* Wiley *399; 400; 404; 406; 407; 408*
Fackler, J. P. Jr. (1971) *Symmetry in Coordination Chemistry.* Academic#
Faddeev, D. K. (1964) *Tables of the Principal Unitary Representations of the Fedorov Groups.* Macmillan *246*
Fano, U. and Racah, G. (1959) *Irreducible Tensorial Sets.* Wiley *41; 80; 421*
Fano, U. (1960) "Real Representations of Coordinate Rotations." *J. Math. Phys.* **1** (417-423) *278*
Fawcett, W. and McLean, T. P. (1965) "A Generalized Perturbation Procedure." *Proc. Phys. Soc.* **85** (1315-1317) *362*
Fedorov, E. S. (1971) *Symmetry of Crystals.* American Crystallographic Association. *238*
Feenberg, E. and Pake, G. E. (1953) *Notes on the Quantum Theory of Angular Momentum.* Addison-Wesley *38*
Feher, G. (1958) "Application of ENDOR Technique to Donors in Si." *J. Phys. Radium* **19** (830-833) *227*
Fletcher, G. C. (1971) *The Electron Band Theory of Solids.* North Holland-American Elsevier#
Flood, W. F. (1959) see Haynes, J. R. and Lax, M., *370*
Foldy, L. I. and Wouthuysen, S. A. (1950) "On the Dirac Theory of Spin Half Particles and Its Nonrelativistic Limit." *Phys. Rev.* **78** (29-36) *91*
Franz, W. (1958) "Influence of an Electric Field on an Optical Absorption Edge." *Z. Naturforsch* **13a** (484-489) *314*
Fredkin, D. R. (1962) see Wannier, G. H., *230*
Fredkin, D. R. Private communication *104*
Frobenius, G. and Schur, I. (1906) "On the Real Representations of Finite Groups." *Sitzungsberichte Preuss. Akad. Wiss., Berlin* (186-208) *150; 298*
Fumi, F. G. (1952) "Physical Properties of Crystals: The Direct Inspection Method." *Acta Cryst.* **5** (44-48) *120*
Gazis, D. C. (1965) see Toupin, R. A., *127*
Glasser, M. L. (1959) "Symmetry Properties of the Wurtzite Structure." *J. Phys. Chem. Solids.* **10** (229-241) *255; 416*
Gluck, M. (1969) see Zak, J. and Casher, A., *246*
Gold, A. (1964) see Knox, R. S.,#
Goldstein, H. (1950) *Classical Mechanics.* Addison-Wesley *3; 37*
Göttlicher, Von S. and Wölfel, E., (1959) "X-ray Determination of Electron Distribution in Crystals," *Z. Elecktrochem.* **63** (891-901) 205
Goudsmit, S. (1925) see Uhlenbeck, G. E., *34*
Goudsmit, S. (1926) see Uhlenbeck, G. E., *34*
Griffith, J. S. (1961) *The Theory of Transition Metal Ions.* Cambridge *98; 417*
Griffith, J. S. (1962) *The Irreducible Tensor Method For Molecular Symmetry Groups.* Prentice-Hall#
Gurney, G. W. (1948) see Mott, N. F., *106*
Hagedorn, R. (1959) "Note on Symmetry Operations in Quantum Mechanics." *Nuovo Cimento Supplemento* **12** (73-86) *461*
Hagedorn, R. (1961) "On Symmetry Operations in Quantum Mechanics." in *Lectures on Field Theory and the Many Body Problem,* E. R. Caianiello ed., Academic, *461*
Hall, B. G. (1951) "The Molecular Orbital Theory of Chemical Valency VIII. A Method of Calculating Ionization Potentials." *Proc. Roy. Soc.* **A205** (541-552) *395*
Hamermesh, M. (1962) *Group Theory and Its Application to Physical Problems.* Addison-Wesley *25; 49; 80*

Harrison, W. A. (1960) "Band Structure of Aluminum." *Phys. Rev.* **118** (1182-1189) *201*

Harrison, W. A. (1966) *Pseudopotentials in the Theory of Metals.* Benjamin *201*

Harrison, W. A. (1970) *Solid State Theory.* McGraw-Hill *118*

Harrison, W. A. (1973) "Bond-Orbital Model and the Properties of Tetrahedrally Coordinated Solids." *Phys. Rev.* **B8** (4487-4498) *396*

Haug, A. (1972) *Theoretical Solid State Physics Vols. I, II* Pergamon#

Haynes, J. R., Lax, M. and F W. F. (1959) "Analysis of Intrinsic Recombination Radiation." *J. Phys. Chem. Solids.* **8** (392-396) *370*

Hearmon, R. F. S. (1961) *An Introduction to Applied Anisotropic Elasticity.* Oxford *130*

Hedin, L. T. (1960) "A Microscopic Derivation of the Born-Huang Relations Between Atomic Force Constants." *Ark. Fys. (Sweden)* **18** (369-378) *349*

Heine, V. (1960) *Group Theory in Quantum Mechanics.* Pergamon *50; 67*

Heitler, W. and London, F. (1927) "Reciprocal Action of Neutral Atoms and Homopolar Combinations According to Quantum Mechanics." *Zeits. f. Physik* **44** (455-472) *404*

Henry, N. F. M. and Lonsdale, K. (1952) *International Tables For X-Ray Crystallography.* Kynoch *174; 235*

Herman, Frank (1959) "Lattice Vibrational Spectrum of Germanium." *J. Phys. Chem. Solids* **8** (405-418) (421-422) *364, 366, 370, 387,*

Herring, C. (1937a) "Effect of Time Reversal Symmetry on Energy Bands of Crystals." *Phys. Rev.* **52** (361-365) *305*

Herring, C. (1937b) "Accidental Degeneracy in the Energy Bands of Crystals." *Phys. Rev.* **52** (365-373) *18*

Herring, C. (1942) "Character Tables for Two Space Groups." *J. Franklin Inst.* **233** (525-543) *247; 257; 416; 429.*

Herring, C. (1966) *Magnetism. Vol. IV* Rado, G. T. and Suhl, H., eds., Academic,#

Hopfield, J. J. (1959) "A Theory of Edge Emission Phenomena in CdS, ZnS, and ZnO." *J. Phys. Chem. Solids* **10** (110-119) *416*

Hopfield, J. J. (1960) "Fine Structure in the Optical Absorption Edge of Anisotropic Crystals." *J. Phys. Chem. Solids* **15** (97-107) *104*

Hopfield, J. J. (1960) see Thomas, D. G., *315*

Hopfield, J. J. (1961) see Lax, M., *227; 273; 370*

Huang, K. (1954) see Born, M., *328; 347; 351*

Hückel, E. (1931) "Quantum Treatment of the Benzene Problem: Part I. Electronic Configuration of Benzene and Some Related Compounds." *Zeits. f. Physik* **70** (204-286) *391*

Hückel, E. (1932) "Quantum Treatment of the Benzene Problem. Part II Induced Polarities." *Zeits. f. Physik* **72** (310-337) *391*

Hutson, A. R. Private communication *314*

Iyengar, P. K. (1958) see Brockhouse, B. N., *325; 326; 370*

Jackson, J. (1952) "A Note on 'Irreversibility and Generalized Noise'." *Phys. Rev.* **87** (471-472) *466*

Jackson, J. (1952) see Callen, H. B. and Barasch, M., *466*

Jahn, H. A. and Teller, E. (1937) "Stability of Polyatomic Molecules in Degenerate Electronic States. Part I - Orbital Degeneracy." *Proc. Roy. Soc.* **A161** (220-235) *414*

Jahn, H. A. (1938) "Stability of Polyatomic Molecules in Degenerate Electronic States. Part II Spin Degeneracy." *Proc. Roy. Soc.* **A164** (117-131) *414; 415*

Janak, J. F. (1972) see Williams, A. R., and Moruzzi, V. C., *229*

Jones, H. (1960) *The Theory of Brillouin Zones in Crystals.* North Holland-Interscience *205; 254*

Judd, B. R. (1963) *Operator Techniques in Atomic Spectroscopy.* McGraw-Hill *98*

Kane, E. O. (1961) "Theory of Tunneling." *J. Appl. Phys.* **32** (83-91) *182*

Kane, E. O. (1966) "The $\mathbf{k} \cdot \mathbf{p}$ Method." in *Semiconductors and Semimetals, Vol. II Physics of III-V Compounds* Academic, Willardson, R. K., and Beer, A. C., eds., *211*

Kaplan, T. (1956) "Relationship Between the Reciprocity Theorems of Onsager and of Callen-Greene." *Phys. Rev.* **102** (1447-1450) *466*

Kaus, P. E., and Watson, W. K. R. (1960) "Dispersion Relations for Bloch Functions." *Phys. Rev.* **120** (44-48) *182*

Keldysh, L. V. (1958) "The Effect of a Strong Electric Field on the Optical Properties of Insulating Crystals." *Soviet Physics JETP.* **34** (788-790) *314*

Kimball, G. E. (1944) see Eyring, H. and Walter, J., *399; 400; 404; 406; 407; 408.*

Kittel, C. (1963) *Quantum Theory of Solids.* Wiley#

Kittel, C. (1971) *Introduction to Solid State Physics 4th Ed.* Wiley *118*

Knox, R. S. and Gold, A. (1964) *Symmetry in the Solid State* Benjamin#

Kohn, W. (1959a) "Bloch Electrons in a Magnetic Field: The Effective Hamiltonian." *Phys. Rev.* **115** (1460-1478) *198; 230*

Kohn, W. (1959b) "Analytic Properties of Bloch Waves and Wannier Functions." *Phys. Rev.* **115** (809-821) *219*

Kohn, W., and Onffroy, J. R. (1973) "Wannier Functions in a Simple Nonperiodic System." *Phys. Rev.* **B8** (2485-2495) *219*

Koster, G. F. (1953) "Localized Functions in Molecules and Crystals." *Phys. Rev.* **89** (67-77) *219*

Koster, G. F. (1958) "Space Groups and Their Representation." *Solid State Physics* **5** Academic, F. Seitz and D. Turnbull eds., *98*

Koster, G. F., Dimmock, J. O., Wheeler, R. G. and Statz, H. (1963) *Properties of the Thirty-Two Point Groups.* M.I.T. Press *416; 418*

Kovalev, O. V., and Lyubarskii, G. Ya., (1958) "On Contact of Energy Bands in Crystals," *Sov. Phys. Tech. Phys.* **3** (1071-1077) *246*

Kovalev, O. V. (1961) *Irreducible Representations of the Space Groups.* (transl. from Russian in 1965) Gordon and Breach *246*

Kramers, H. A. (1930) "General Theory of Paramagnetic Rotation in Crystals," *Proc. Acad. Amsterdam* **33** (959-972) *285*

Kramers, H. A. (1935) "Eigenvalues in One-Dimensional Periodic Fields of Force." *Physica* **2** (483-490) *198*

Kubo, R. (1956) "A General Expression for the Conductivity Tensor." *Can. J. Phys.* **34** (1274-1277) *466*

Kubo, R. (1957) "Statistical Mechanical Theory of Irreversible Processes I. General Theory and Simple Applications to Magnetic and Conduction Problems." *Phys. Soc. Japan* **12** (570-586) *287; 466*

Kubo, R. and Nagamiya. T. (1969) *Solid State Physics.* McGraw-Hill#

Kuvshinskii, E. V. (1960) see Aero, E. L., *126*

Landau, L. D., and Lifschitz, E. M. (1958) *Quantum Mechanics, Non-Relativistic Theory.* Addison-Wesley *225; 417*

Landsberg, P. T. ed., (1969) *Solid State Theory.* Wiley-Interscience#

Laval, M. J. (1951) "Elasticity of Crystals." *C. R. Acad. Sci. (Paris)* **232** (1947-1948) *126; 128*

Laval, J. (1954) "The Atomic Theory of Elasticity Avoiding (the Assumption of) Central Forces." *C. R. Acad. Sci. (Paris)* **238** (1773-1775) *126; 128*

Lax, M. (1950) "Removal of Degeneracy in Any Order." *Phys. Rev.* **79** (200) *99; 355*

Lax, M. (1955) "Generalized Theory of Mobility." *Phys. Rev.* **100** (1808) *466*

Lax, M. and Burstein, E. (1955) "Infrared Lattice Absorption in Ionic and Homopolar Crystals." *Phys. Rev.* **97** (39-52) 258; 259

Lax, M. (1958) see Levitas, A., *167*

Lax, M. (1958) see Rosenberg, R., *224*

Lax, M. (1958a) "Quadrupole Interactions in the Vibration Spectra of Diamond Type Crystals." *Phys. Rev. Letters* **1** (133-134) *260; 387; 466*

Lax, M. (1958b) "Generalized Mobility Theory." *Phys. Rev.* **109** (1921-1926) *287; 466*

Lax, M (1959) see Haynes, J. R. and Flood, W. F., *370*

Lax, M. (1960) "Fluctuations from the Nonequilibrium Steady State." *Rev. Mod. Phys.* **32** (25-64) *287; 289*

Lax, M. and Hopfield, J. J. (1961) "Selection Rules Connecting Different Points in the Brillouin Zone." *Phys. Rev.* **124** (115-123) *227; 273; 370*

Lax, M. (1962) "Influence of Time Reversal on Selection Rules Connecting Different Points in the Brillouin Zone." Pp. 395-402 of *Proceedings of International Conference on Physics of Semiconductors,* Institute of Physics and Physical Society, London *273; 319*

Lax, M. (1964) "Quantum Relaxation, The Shape of Lattice Absorption and Inelastic Neutron Scattering Lines." *J. Phys. Chem. Solids* **25** (487-503) *466*

Lax, M. (1965a) "The Relation Between Microscopic and Macroscopic Theories of Elasticity." Pp. 583-596 of *Proceedings of 1963 International Conference on Lattice Dynamics, Copenhagen.* Pergamon *126; 336; 346; 348; 349; 351.*

Lax, M. (1965b) "Comments on the Shell Model for Lattice Vibrations." *Proceedings of 1963 International Conference on Lattice Dynamics, Copenhagen.* Pergamon *260; 364; 370; 387; 388*

Lax, M. (1965c) "Subgroup Techniques in Crystal and Molecular Physics." *Phys. Rev.* **138** (A793-A802) *120; 157; 168; 242; 258; 273; 316; 319*

Lax, M. (1966) see Birman, J. L. and Loudon, R., *273*

Lax, M. and Nelson, D. F. (1971) "Linear and Nonlinear Electrodynamics in Elastic Anisotropic Dielectrics." *Phys. Rev. B* **4** (3694-3731) *126*

Lax, M. and Birman, J. L. (1972) "Intervalley Scattering Selection Rules for Si and Ge." *Physica Status Solidi* (b) **49** (K153-K154) *120; 273; 388;*

Lax, M. (1972) see Nelson, D. F. and Lazay, P. D., *126*

Lax, M. and Nelson, D. F. (1973) "Crystal Electrodynamics." *Atomic Structure and Properties of Solids,* Vol. 52, Enrico Fermi Summer School Varenna, Academic Press (48-118) *126*

Lazay, P. D. (1972) see Nelson, D. F. and Lax, M., *126*

Ledermann, W. (1944) "Asymptotic Formulae Relating to the Physical Theory of Crystals." *Proc. Roy. Soc.* A **182** (362-377) *328*

Lefebvre, R. (1960) see Daudel, R. and Mose, C., *391*

Leibfried, G. (1955) "Lattice Theory of the Mechanical and Thermal Properties of Crystals." Pp. 104-324 of *Handbuch der Physik Vol. VII/1,* Springer *336*

Leibfried, G. and Ludwig, W. (1960) "Equilibrium Relations in the Theory of Lattices." *Z. Phys.* **160** (80-92) *349*

Leigh, D. C. (1968) *Nonlinear Continuum Mechanics.* McGraw-Hill *124*

Lennard-Jones, J. (1949) "The Molecular Theory of Chemical Valency." *Proc. Roy. Soc.* **A198** (1-26) *395*

Levitas, A. and Lax, M. (1958) "Statistics of the Ising Ferromagnet." *Phys. Rev.* **110** (1016-1027) *167*

Lie, S. (1893) "Lectures on Continuous Groups with Geometric and Other Applications." Teubner, Leipzig *25*

Liehr, A. D. (1963) "Topological Aspects of the Conformational Stability Problem. Part I: Degenerate Electronic States. Part II: Nondegenerate Electronic States." *J. Phys. Chem.* **67** (389-471) (471-494) *414*

Lifschitz, E. M. (1958) see Landau, L. D., *225; 417*

Littlewood, D. E. (1958) *The Theory of Group Characters.* Oxford#

Lomont, J. S. (1959) *Applications of Finite Groups.* Academic *237; 246*

London, F. (1927) see Heitler, W., *404*

Longuet-Higgins, H. C., Opik, U., Pryce, M. H. L., and Sack, R. A. (1958) "Studies of the Jahn-Teller Effect. II. The Dynamical Problem." *Proc. Roy. Soc. A* **244** (1-16) *414*

Lonsdale, K. (1952) see Henry, N. F. M., *174; 235*

Loudon, R. (1960) see Elliott, R. L., *247; 273*

Loudon, R. (1966) see Birman, J. L. and Lax, M., *273*

Love, W. H. (1967) see Miller, S. C., *246; 443.*

Löwdin, P. O. (1951) "A Note on the Quantum-Mechanical Perturbation Theory." *J. Chem. Phys.* **19** (1396-1401) *99*

Löwdin, P. O. (1956a) "Quantum Theory of Many Particle Systems. I. Physical Interpretation by Means of Density Matrices, Natural Spin Orbitals, and Convergence Problems in the Method of Configuration Interaction." *Phys. Rev.* **97** (1474-1489) *412*

Löwdin, P. O. (1956b) "Quantum Theory of Many Particle Systems II Study of the Ordinary Hartree-Fock Approximation." *Phys. Rev.* **97** (1490-1520) *412*

Ludwig, W. (1960) see Leibfried, G., *349*

Ludwig, W. (1967) *Recent Developments in Lattice Theory.* Springer Tracts in Modern Physics, Vol. 43 *336*

Luttinger, J. M. (1951) "Effect of a Magnetic Field on Electrons in a Periodic Potential." *Phys. Rev.* **84** (814-817) *230*

Luttinger, J. M. (1956) "Quantum Theory of Cyclotron Resonance in Semiconductors: General Theory." *Phys. Rev.* **102** (1030-1041) *230*

Lyubarskii, G. Ya., (1957) see Kovalev, O. V., *246*

Lyubarskii, G. Ya. (1960) *Application of Group Theory in Physics.* Pergamon *25; 172; 243*

Mackey, G. (1951) "On Induced Representations of Groups." *Amer. J. of Math.* **73** (576-592) *246*

McLean, T. P. (1965) see Fawcett, W., *99; 362*

Maradudin, A. A. and Vosko, S. H. (1968) "Symmetry Properties of the Normal Vibration of Crystals." *Rev. Mod. Phys.* **40** (1-37) *336; 364*

Mariot, L. (1962) *Group Theory and Solid State Physics.* Prentice-Hall#

Mase, S. (1958) "Electronic Structure of Bismuth Type Crystals." *J. Phys. Soc. Japan* **13** (434-445) *255*

Mase, S. (1959) "Algebraic Method to Obtain Irreducible Representations of Space Groups with an Application to White Tin." *J. Phys. Soc. Japan* **14** (1538-1550) *255*

Mazur, P. (1962) see De Groot, S. R., *287; 289*

Meijer, P. H. E. and Bauer, E. (1962) *Group Theory, The Application to Quantum Mechanics.* North Holland-Interscience#

Messiah, A. (1962) *Quantum Mechanics. Vol. II* North Holland, *91; 313; 409*

Miller, S. C., and Love, W. H. (1967) *Irreducible Representation of Space Groups.* Pruett Press *246; 443.*

Mindlin, R. D. and Tiersten, H. F. (1962) "Effects of Couple Stresses in Linear Elasticity" *Arch. Rational Mech. Analysis* **11** (415-418) *126*

Mirsky, L. (1955) *An Introduction to Linear Algebra.* Oxford p. 206 *40*

Moffitt, W. and Thorsen, W. (1957) "Vibronic States of Octahedral Complexes." *Phys. Rev.* **108** (1251-1255) *414*

Morgan, Jane Van W., (1972) see Williams, A. R., *229*

Mori, H. (1956) "A Quantum Statistical Theory of Transport Processes." *J. Phys. Soc. Japan* **11** (1029-1044) *466*

Moruzzi, V. C. (1972) see Williams, A. R., and Janak, J. F., *229*

Mose, C. (1960) see Daudel, R. and Lefebvre, R., *391*

Mott, N. F. and Gurney, G. W. (1948) *Electronic Processes in Ionic Crystals.* 2nd ed. Oxford, *106*

Motzkin, T. S. (1948) "Relations Between Hypersurface Crossratios, and a Combinatorial Formula for Partitions of a Polygon, for Permanent Preponderance, and for Non-associative Products." *Bull. Amer. Math. Soc.* **54** (352-360) *407*

Mulliken, R. S. (1933) "Electronic Structures of Polyatomic Molecules and Valence. IV. Electronic States, Quantum Theory of the Double Bond." *Phys. Rev.* **43** (279-302) *23; 416*
Murnaghan, F. D. (1951) *Finite Deformation of an Elastic Solid.* Wiley *124*
Nagamiya, T. (1969) see Kubo, R.,#
Nakano, H. (1956) "A Method of Calculation of Electrical Conductivity." *Progr. Theor. Phys.* **15** (77-79) *466*
Nelson, D. F. (1971) see Lax, M., *126*
Nelson, D. F., Lazay, P. D. and Lax. M. (1972) "Brillouin Scattering in Anisotropic Media; Calcite." *Phys. Rev. B* **6** (3109-3120) *126*
Nelson, D. F. (1973) see Lax, M., *126*
Nierenberg, W. A. (1957) "Spin and Moments of Radioactive Nuclei." *Ann. Rev. Nucl. Sci.* **7** (349-406) *98*
Nussbaum, A. (1971) *Applied Group Theory.* Prentice-Hall#
Nye, J. F. (1957) *Physical Properties of Crystals.* Oxford *118*
Onffroy, J. R. (1973) see Kohn, W., *219*
Onsager, L. (1931a) "Reciprocal Relations in Irreversible Processes, Part I." *Phys. Rev.* **37** (405-426) *287*
Onsager, L. (1931b) "Reciprocal Relations in Irreversible Processes, Part II." *Phys. Rev.* **38** (2265-2279) *287*
Opechowski, W. (1940) "The Double Crystallographic Groups." *Physica* **7** (552-562) *62; 63*
Opik, U. (1958) see Longuet-Higgins, H. C., Pryce, M. H. L., and Sack, R. A., *414*
Orbach, R. (1961) "On the Theory of Spin Lattice Relaxation in Paramagnetic Salts." *Proc. Phys. Soc.* **77** (821-826) *286*
Overhauser, A. W. (1958) see Dick, B. G., *388*
Owen, J. (1955) see Bowers, K. D., *98*
Pake, G. E. (1953) see Feenberg, E., *38*
Patterson, J. D. (1971) *Introduction to the Theory of Solid State Physics.* Addison-Wesley#
Pauli, W. (1927) "The Quantum Mechanics of Magnetic Electrons." *Z. f. Physik* **43** (601-623) *89*
Pauling, L. (1960) *Nature of The Chemical Bond.* Cornell, Chapter 4 *395*
Pearson, C. E. (1959) *Theoretical Elasticity.* Harvard *124*
Peierls, R. E. (1954) "Note on the Vibration Spectrum of a Crystal." *Proc. Nat. Inst. Sci. India* **20** (121-126) *328*
Peierls, R. E. (1955) *Quantum Theory of Solids.* Oxford p. 128 *189*
Phillips, J. C. (1958) "Energy Band Interpolation Scheme Based On a Pseudopotential." *Phys. Rev.* **112** (685-695) *271*
Phillips, J. C. (1973) *Bonds and Bands in Semiconductors,* Academic *396*
Pikus, G. Ye., (1972) see Bir, G. I.,#
Powell, J. L. and Crasemann, B. (1961) *Quantum Mechanics.* Addison-Wesley *333*
Prigogine, I. (1967) *Introduction to Thermodynamics of Irreversible Processes.* 3rd ed. Wiley-Interscience, *289*
Pryce, M. H. L. (1950) "A Modified Perturbation Procedure for a Problem in Paramagetism." *Proc. Phys. Soc.* **63A** (25-29) *98; 99*
Pryce, M. H. L. (1951) see Abragam, A., *98*
Pryce, M. H. L., (1958) see Longuet-Higgins, H. C., Opik, U., and Sack, R. A. *414*
Racah, G. (1959) see Fano, U., *41; 80; 421*
Rashba, E. I. (1959) "Symmetry of Energy Bands in Crystals of Wurtzite Type. I. Symmetry of Bands Disregarding Spin-Orbit Interaction." *Soviet Phys. Solid State Phys.* **1** (368-380) *416*
Roothaan, C. C. J. (1951) "New Developments in Molecular Orbital Theory." *Rev. Mod. Phys.* **23** (69-89) *391*

Rose, M. E. (1955) *Multipole Fields.* Wiley *312*

Rose, M. E. (1957) *Elementary Theory of Angular Momentum.* Wiley *38*

Rosenberg, R. and Lax, M. (1958) "Free Carrier Absorption in n-Type Ge," *Phys. Rev.* **112** (843-852) *224*

Roth, L. H. (1962) "Theory of Bloch Electrons in a Magnetic Field." *J. Phys. Chem. Solids* 23 (433-446) *230*

Roussy, Georges (1973) "An Approximate Block Diagonalization for Hermitian Matrices as an Improvement on Successive Van Vleck Transformations." *Molecular Physics* **26** (1085-1092) *362*

Ruch, E. and Schoenhofer, A., (1965) "Proof of the Jahn-Teller Theorem, with the Help of a Theorem on the Induction of Representations of Finite Groups." *Theoret. Chem. Acta* **3 (4)** (291-304) *414*

Rumer, G. (1932) "Theory of Spin Valence." *Nachr. Ges. Wiss. Göttingen Math-Phys. Klasse* (337-341) *406*

Sack, R. A. (1958) see Longuet-Higgins, H. C., Opik, U., and Pryce, M. H. L., *414*

Salpeter, E. E. (1957) see Bethe, H., *409*

Schiff, L. (1968) *Quantum Mechanics* 3rd ed. McGraw-Hill *38; 79; 416*

Schoenhofer, A. (1965) see Ruch, E, *414*

Schur, I. (1906) see Frobenius, G., *150; 298*

Seitz, F. (1934) "Matrix Algebraic Development of Crystallographic Groups." *Z. Krist.* **88** (433-459) *238*

Seitz, F. (1935a) "Matrix Algebraic Development of Crystallographic Groups, Part II." *Z. Krist.* **90** (289-313) *238*

Seitz, F. (1935b) "Matrix Algebraic Development of Crystallographic Groups, Part III." *Z. Krist.* **91** (336-366) *238*

Seitz, F. (1936) "Matrix Algebraic Development of Crystallographic Groups, Part IV." *Z. Krist.* **94** (100-130) *238*

Seitz, F. (1940) *Modern Theory of Solids.* McGraw-Hill *413*

Sharp, R. T. (1960) "Simple Derivation of the Clebsch-Gordon Coefficients." *Amer. J. Phys.* **28** (116-118) *422*

Sherwood, P. M. A. (1972) *Vibrational Spectroscopy of Solids.* Cambridge#

Shockley, W. (1950) *Electrons and Holes in Semiconductors.* Van Nostrand#

Shockley, W. (1950) see Bardeen, J., *271*

Shortley, G. H. (1967) see Condon, E. U., *39; 109; 320*

Simpson, William T. (1962) *Theories of Electrons in Molecules.* Prentice-Hall *406; 407; 408.*

Slater, J. C. (1928) "The Structure of Atoms." *Phys. Rev.* **32** (339-348) *404*

Slater, J. C. (1929) "The Theory of Complex Spectra." *Phys. Rev.* **34** (1293-1322) *404*

Slater, J. C. (1949) "Electrons in Perturbed Periodic Lattices" *Phys. Rev.* **76** (1592-1601) *219*

Slater, J. C. (1960) *Quantum Theory of Atomic Structure.* McGraw-Hill *413*

Slater, J. C. (1963) *Quantum Theory of Molecules and Solids,* Vol. I. *Electronic Structure of Molecules.* McGraw-Hill#

Slater, J. C. (1965) *Quantum Theory of Molecules and Solids,* Vol. II. *Symmetry and Energy Bands in Crystals.* McGraw-Hill#

Slater, J. C. (1967) *Quantum Theory of Molecules and Solids,* Vol. III. *Insulators, Semiconductors and Metals.* McGraw-Hill#

Smith, H. (1948) "The Theory of the Vibrations and the Raman Spectrum of the Diamond Lattice." *Phil. Trans. Roy. Soc. A* **241** (105-145) *264; 364*

Smith, R. A. (1961) *Wave Mechanics of Crystalline Solids.* Wiley#

Smoluchowski, R. (1936) see Bouckaert, L. P. and Wigner, E. P., *23; 367; 416*

Soliverez, C. E. (1969) "An Effective Hamiltonian and Time-Independent Perturbation Theory." *J. Phys. C (Solid State Phys.)* ser. 2. **2** (2161-2174) *362*

Statz, H. (1963) see Koster, G. F., Dimmock, J. O., and Wheeler, F. G., *416; 418*
Stevens, K. W. H. (1953) see Bleaney, B., *98*
Stevens, K. W. H. (1963) *Magnetism. Vol. I.* Academic, Rado, G. T., and Suhl, H., eds. *98*
Streitwolf, H. W. (1970) "Intervalley Scattering Selection Rules for Si and Ge." *Physica Status Solidi* **37** (K47-K49) *388*
Takahasi, H. (1952) "Generalized Theory of Thermal Fluctuations." *J. Phys. Soc. Japan* **7** (439-446) *287*
Teller, E. (1937) see Jahn, H. A., *414*
ter Haar, D. (1961) "Theory and Applications of the Density Matrix." *Rep. Prog. Phys.* **24** (304-362) *289*
Thomas, D. G. and Hopfield, J. J. (1960) "Excitation Spectrum of CdS." *Phys. Rev.* **116** (573-582) *315*
Thorsen, W. (1957) see Moffitt, W., *414*
Thurston, R. N. (1974) "Waves in Solids." Vol. VIa, *Handbuch der Physik* C. Truesdell, ed. Springer-Verlag *124*
Tiersten, H. F. (1962) see Mindlin, R. D., *126*
Tinkham, M. (1964) *Group Theory and Quantum Mechanics.* McGraw-Hill *404*
Titchmarsh, E. C. (1937, 1st ed.; 1948, 2nd ed.) *Introduction to Theory of Fourier Integrals.* Oxford *188*
Toupin, R. A. (1956) "The Elastic Dielectric." *J. Ratl. Mech. Anal.* **5** (849-915) *125*
Toupin, R. A. (1960) see Truesdell, C., *124; 126*
Toupin, R. A. (1962) "Elastic Materials with Couple-Stresses." *Arch. Ratl. Mech. Anal.* **11** (385-414) *126*
Toupin, R. A. (1963) "A Dynamical Theory of Elastic Dielectrics." *Inter. J. Eng. Sci.* **1** (101-126) *126*
Toupin, R. A. (1964) "Theories of Elasticity with Couple Stress." *Arch. for Ratl. Mech. Anal.* **17** (85-112) *126*
Toupin, R. A. and Gazis, D. C. (1965) "Surface Effects and Initial Stress in Continuum and Lattice Models of Elastic Crystals." Pp. 597-605 of *Proceedings of the 1963 International Conference on Lattice Dynamics, Copenhagen* Pergamon *127*
Truesdell, C. (1952) "The Mechanical Foundations of Elasticity and Fluid Dynamics." *J. Ratl. Mech. Anal.* **1** (125-300) *124; 125; 128.*
Truesdell, C. and Toupin, R. A. (1960) "The Classical Field Theories." Pp. 226-793 of *Handbuch der Physik,* Vol. III/I S. Flugge, ed. Springer. *124; 126.*
Uhlenbeck, G. E. and Goudsmit, S. (1925) "Replacement of the Hypothesis of Nonmechanical Force Through a Requirement Concerning the Inner Behavior of Each Individual Electron." *Naturwiss.* **13** (953-954) *89*
Uhlenbeck, G. E. and Goudsmit, S. (1926) "Spinning Electrons and the Structure of Spectra." *Nature* **117** (264-265) *89*
Uhlhorn, U. (1963) "Representation of Symmetry Transformations in Quantum Mechanics." *Ark. Fys. (Sweden)* **23** (307-340) *461*
Van Hove, L. (1953) "The Occurrence of Singularities in the Elastic Frequency Distribution of a Crystal." *Phys. Rev.* **89** (1189-1193) *223*
Van Vleck, J. H. (1932) *Theory of Electric and Magnetic Susceptibilities.* Oxford Section 34, *99*
Van Vleck, J. H. (1940) "Paramagnetic Relaxation Times for Titanium and Chrome Alum." *Phys. Rev.* **57** (426-447) *286*
Von der Lage, F. C., and Bethe, H. A. (1947) " A Method for Obtaining Electronic Eigenfunctions and Eigenvalues in Solids with An Application to Sodium." *Phys. Rev.* **71** (612-622) *101; 110; 432.*
Vosko, S. H., (1968) see Maradudin, A. A., *336; 364*

Walter, J. (1944) see Eyring, H. and Kimball, G. E., *399; 400; 404; 406; 407; 408;*

Wannier, G. H. (1937) "The Structure of Electronic Excitation Levels in Insulating Crystals." *Phys. Rev.* **52** (191-197) *217; 219*

Wannier, G. H. (1959) *Elements of Solid State Theory.* Cambridge *192*

Wannier, G. H. (1962) "Dynamics of Band Electrons in Electric and Magnetic Fields." *Rev. Mod. Phys.* **34** (645-655) *230*

Wannier, G. H. and Fredkin, D. R. (1962) "Decoupling of Bloch Bands in the Presence of Homogeneous Fields." *Phys. Rev.* **125** (1910-1915) *230*

Warren, J. L. (1968) "Further Considerations on the Symmetry Properties of the Normal Vibrations of a Crystal." *Rev. Mod. Phys.* **40** (38-76) *336*

Watson, W. K. R. (1960) see Kaus, P. E., *182*

Weinreich, G. (1965) *Solids: Elementary Theory for Advanced Students.* Wiley#

Welton, T. (1951) see Callen, H. B., *466*

Weyl, H. (1930) *Theory of Groups and Quantum Mechanics.* Dutton. Reprinted Dover, 1950. *321*

Weyl, H. (1946) *Classical Groups; Their Invariants and Representations.* Princeton Pp. 31, 53, 275. *21; 80; 105; 121*

Wheeler, R. G. (1962) see Dimmock, J. O., *292*

Wheeler, R. G. (1963) see Koster, G. F., Dimmock, J. O., and Statz, H., *416; 418*

Wigner, E. P. (1932) "The Time Reversal Operation in Quantum Mechanics." *Nachr. Ges. Wiss. Göttingen Math-Phys. Klasse* (546-559) (see also Chap. 26 of Wigner 1959) *275; 461*

Wigner, E. P. (1936) see Bouckaert, L. P. and Smoluchowski, R., *23; 367; 416*

Wigner, E. P. (1954) "Derivations of Onsager's Reciprocal Relations." *J. Chem. Phys.* **22** (1912-1915) *287*

Wigner, E. P. (1955) *The Application of Group Theory to the Special Functions of Mathematical Physics.* Princeton notes *236*

Wigner, E. P. (1959) *Group Theory and Its Application to Quantum Mechanics of Atomic Spectra.* Academic *32; 40; 79; 275; 292; 420; 421; 461;*

Williams, R. (1960) "Electric Field Induced Light Absorption in CdS." *Phys. Rev.* **117** (1487-1490) *314*

Williams, A. R., Janak, J. F. and Moruzzi, V. C. (1972) "Exact Korringa-Kohn-Rostoker Energy Band Method with the Speed of Empirical Pseudopotential Methods." *Phys.Rev.* **B6** (4509-4517) *229*

Williams, A. R. and Morgan, Jane Van W. (1972) "Multiple Scattering by Non Muffin-Tin Potentials." *J. Phys. C* **5** (L293-L298) *229*

Wilson, A. H. (1953) *Theory of Metals.* 2nd ed., Cambridge, p. 23 *198*

Wilson Jr., F. B., Decius, J. C., and Cross, P. C. (1955) *Molecular Vibrations* McGraw-Hill *152; 153*

Wölfel, E., (1959) see Göttlicher, Von S., *205*

Wouthuysen, S. A. (1950) see Foldy, L. I., *91*

Wyckoff, R. W. G. (1963) *Crystal Structures, Vol. I. 2nd ed.* Wiley-Interscience, *255*

Zak, J. (1960) "Character Tables of Nonsymmorphic Space Groups." *J. Math. Phys.* **1** (165-171) *246*

Zak, J. (1962) "Selection Rules for Integrals of Bloch Functions." *J. Math. Phys.* **3** (1278-1279) *273*

Zak, J., Casher, A., and Gluck, M. (1969) *The Irreducible Representations of Space Groups.* Benjamin *246*

Ziman, J. M. (1960) *Electrons and Phonons.* Oxford#

Zwanzwig, R. (1961) "Memory Effects in Irreversible Thermodynamics." *Phys. Rev.* **124** (983-992) *289*

SUBJECT INDEX

SYMBOL INDEX

Symbols used in several chapters are listed here with the numbers of the pages on which they are defined or discussed.

a	cube edge, or lattice spacing, 256
a,b	Cayley-Klein parameters, 45
a,b	one electron spatial wave functions, 404
a_B	Bohr radius, 226
a_i	Lie group parameter, 25
a_j	expansion coefficients, 391
$a,a^\dagger$	destruction, creation operators, 331
A	A-centered Bravais lattice, 172
A	reflection plane operation σ_x in group $\mathcal{C}_{3v}$, 140
A	average overlap integral, 84
A	antilinear operator, 292
A,B ...	superscripts or subscripts, denoting material reference frame components, 124
A,B,E ...	chemical notation for representations, 23; 416
A^i_{ab}	potential energy matrix in the symmetry coordinates, 148
A^{-1}	inverse of element A, 5
$\mathbf{a}_1,\mathbf{a}_2,\mathbf{a}_3$	primitive basis vectors, 171
$\mathbf{a},\mathbf{b},\mathbf{c}$	basis vectors, 172
$\mathbf{A}$	vector potential, 88
b	parameter which characterizes the type of representation of a group, 296
B	B-centered Bravais lattice, 172
B	matrix which relates equivalent time reversed representation to the original representation, 295
B	bond strength, 396
$\mathbf{b}$	phonon polarization, 325
$\mathbf{b}$	translation of origin, 231
$\mathbf{b}_i$	reciprocal lattice basis vectors, 184
$\mathbf{b}^{ia}_\mu$	symmetry vector, 146
$\mathbf{B}$	magnetic field, 223
BZ	"Brillouin Zone", 187
$\mathbf{B}^n$	effective charge matrix for atom n, 168
c	number of independent parameters among matrix elements, 312
$c(\mathbf{K})$	Fourier coefficient of wave function, 201
C	C-centered Bravais lattice, 172
C	class of a group, 10
$C_\mathbf{t}$	Fourier coefficients, 187
C_{LM}	space metric of deformed system in the material reference frame, 125
$\hat{C}_{\mu a \nu b}$	sound wave stiffness constants, 347
C^{MANB}	elasticity tensor, 127
C^{ia}_t	normalized eigenvector of potential energy matrix, 149
$C_n \equiv n$	proper counterclockwise rotation of $2\pi/n$ about some

axis, 49
$\mathcal{C}_n$ or n proper rotation group with a single n-fold axis, 49
C_i inversion group, 58
$\mathcal{C}_{nh}$ groups, $C_n \times [E, \sigma_h]$, 58
$\mathcal{C}_{nv}$ groups $C_n \times [E, \sigma_v]$, 58
$d_{\mathbf{k}}$ separation between X-ray diffraction planes, 190
ds^2 infinitesimal length squared, 125
D $2\pi/3$ counterclockwise rotation in $\mathcal{C}_{3v}$, 140
$D^j_{\mu\nu}(R)$ matrix representation of group element R; j names the representation, $\mu\nu$ names the matrix component, 14
$D^{j\pm}$ irreducible even and odd parity representations of a group containing the inversion, 55
D_n proper rotation group with two fold axes, but only one n-fold axis, n > 2, 49
D_{nd} improper dihedral group obtained from D_n by adding σ_d, 59
D_{nh} improper point group obtained from D_n by adding σ_h, 58
$\mathbf{d}$ arbitrary rigid displacement, 126
$\mathbf{D}(R)$ matrix representation of a group operation R, 15
D defined tensor quantity, 385
e electronic charge $-|e|$, 90
$e_{\mu\nu}$ infinitesimal strain tensor, 124
E abstract identity element, 5
$\overline{E}$ $\equiv R(2\pi,\mathbf{n})$, double group element whose eigenvalue is ± 1, 46
E_b first order energy at zone boundary point, 203
E_B binding energy, 226
E_{LM} finite strain tensor, 125
$E_n(\mathbf{k})$ energy of band n, 196
E^i energy eigenvalue of ith eigenfunction, 18
$\mathbf{e}$ symmetry vector, 370
e_{ijk} Levi-Civita tensor density, 38
$\mathbf{e}_i$ eigenvectors of α, 232
$\mathbf{E}$ identity dyadic or identity matrix, 42
$\mathbf{E}$ electric field intensity, 114
f scattering amplitude, 325
$f(\mathbf{k})$ momentum space wave function, 225
$f_{n'n\mu}$ oscillator strength, 224
F face centered Bravais lattice, 172
F flip operation, 120
F tight binding coefficients, 208
$F(\mathbf{r},\mathbf{r}')$ shell sum, 122
$F(\mathbf{r},\mathbf{r}',s,s')$ shell sum with spin, 123
$F(\mathbf{R})$ space wave function, 225
$F(\mathbf{x})$ invariant function, 116
$F^{ja}(\mathbf{r})$ function invariant under operations of a group, 104
$\mathbf{F}$ external force, 221
g order of a group, 6
g isomorphism on number field, 284
g structure factor, 326
G abstract group, 5
$G_{\mathbf{k}}$ group of the wave vector, 241
G_p group of the atom p, 163
G_p isomorphic proper group, 55
G_{pq} group of the bond (pq), 164
G/H factor group, 239
h order of subgroup H, 240
$\hbar$ Planck's constant $/2\pi$, 37
h_1, h_2, h_3 Miller indices, 190

H subgroup, 239
H Hamiltonian, 2
H_{spin} spin hamiltonian, 90
H Hermitian matrix, 342
H magnetic field intensity, 99
i $\sqrt{-1}$
i,j,k as superscripts, give the name of the representation, or the Cartesian component in the spatial reference frame; they are also used as (integer) cell indices, 18; 124; 325
I body centered Bravais lattice, 172
I inversion operator, 4
I_j infinitesimal generator of a Lie group, 25
I_z rotation group generator, 37
i inversion operation, 256
i,j... lattice vectors, 330
I nuclear spin vector, 99
j quantum number for total angular momentum, 39
J Fourier transform of periodic factors of a matrix element, 188
J Jacobian of a transformation, 28
J_i ith component of total angular momentum, 39
$J_{\pm}$ raising and lowering operators for total angular momentum, 39
J_{pp} numerical factor which selects only the elements of the group of the atom, 163
J_{pq} numerical factor which selects only the ordinary elements in the group of the bond, 164
J'_{pq} numerical factor which selects only the reversal elements of the bond, 164
j current, 223
J current density, 114
J total angular momentum operator, the generator of the rotation group, 39
k Boltzmann constant, 287
K generalized Wigner's time reversal operator in presence of spin, 278
K force constant in spring-mass system, 4
K_0 Wigner's time-reversal operator, in absence of spin, 276
$\mathbf{K}^{mn}$ force constant matrix, 154
$\mathbf{K}^{\alpha i,\beta j}$ crystal force constants, 261
k electron wave vector, 183
$\mathbf{k}_0$ zone boundary point, 203
l orbital angular momentum quantum number, 3
l_i dimension of matrix representation i, 20
L point of special symmetry in the Brillouin zone, 363, 443 ff
L_{ij} Onsager coefficients, 287
L orbital angular momentum operator, 3
L symmetric part of the direct product of Γ_1^- with itself, 129
L a linear operator, 290
l L/$\hbar$, angular momentum operator in units of $\hbar$, 37
l constant vector translation, 218
m quantum number of the z-component of angular momentum, 3
m free electron mass, 199
m_x,m_y,m_z basis functions which transform like a pseudo-

	vector Γ_1^+, 113; 122
m^*	effective mass (isotropic case), 226
M	mass in spring-mass system, 4
M	matrix element between Bloch functions, 188
$\mathbf{m}$	magnetic vector, transforms as Γ_1^+, 71; 113
$\mathbf{M}$	a permanent total moment, 112
$\mathbf{M}$	projectivity or mapping, 284
$\mathbf{M}$	effective mass tensor, 221
$\mathbf{M}$	total electric moment, 258
M	an arbitrary tensor, 112
$\mathbf{M}(a)$	matrix determining Lie group commutator algebra, 26
$\mathsf{M}^{\alpha i,\beta j}$	local electric moment coefficients, 264
$\mathbf{M}_h$	mass tensor of a hole, 223
n	used as subscript for band index, 194
n	number of cells per unit volume in the space lattice, 187
n	rotation by angle $2\pi/n$, 49
n_c	number of elements in class C, 10
n_t	number of phonons in mode t, 334
N	normalization constant, 287
N	number operator, 332
N	number of lattice points, 183
N	the number of independent constants in a tensor, 113
$N(R)$	number of atoms left fixed by the operation R, 158
$\|N>$	many phonon state, 334
$\mathbf{n}$	normal to the zone boundary, 204
$\mathbf{n}$	axis of rotation, 37
O	symmetry operator, 282 or semilinear transformation, 284
O	octahedral group, 50
$O(R)$	generalized notation for elements in a multiplier group, 12
$O(\alpha)$	element of space group algebra, 243
O_h	group possessing full symmetry of the cube, 60
O_h^7	space group of diamond structure, 235
O_R	rotation operator, 36
p	projective as in p-equivalent, 245
p	prototype atom, 163
p	order of a group element, 6
$\mathbf{p}$	the representation Γ_1^- that transforms like a polar vector, 71; 113
P	primitive Bravais lattice, 172
P	momenta conjugate to normal coordinates, 330
$P(\mathbf{k})$	point group of space group, 244
P_1, P_2, P_3	reciprocal lattice in units of $2\pi/a$, 205
P_m^l	Legendre polynomial, 3
$P_{\mu\nu}^i$	transfer operator, 100
P_μ	polarization component, 115
$P_{\mu\mu}^i$	projection operator, 100
$\mathbf{p}$	linear momentum operator, 3
$\mathbf{P}$	crystal momentum operator, 334
$\mathbf{P}$	invariant vector, 116
q_μ^{ia}	symmetry coordinate, 148
$q_{ti\mu}$	normal coordinates, 149
Q	reversal element of the bond, 165
Q	some element such that QK commutes with the

Hamiltonian, 294
Q normal coordinates, 328
$Q_{\pm}$ positive and negative frequency parts of normal coordinate, 330
$\mathbf{q}$ phonon wave vector, 325
$\mathbf{Q}$ quadrupole moment tensor, 260
r radial coordinate, 3
r^{+},r^{-},r^{0} basis functions of Γ_1^-, 421
R arbitrary element of a group, 7
R rhombohedral centered Bravais lattice, 172
$R(\phi,\mathbf{n})$ rotation operator, 45
$\bar{R}(\phi,\mathbf{n})$ double group element $\bar{E}R(\phi,\mathbf{n})$, 46
R_m^l Laguerre polynomial, 3
R_x,R_y,R_z basis functions, 21; 418
$R(3)$ the full rotation-reflection group, 52
$R^{+}(3)$ proper rotation group in three dimensions, 35
$\mathbf{r}$ position vector, 3
$\mathbf{R}$ 3×3 matrix representation of proper rotation group, 43
$\mathbf{R}$ position variable or operator, 224
$\tilde{\mathbf{R}}$ transpose of matrix $\mathbf{R}$, 43
$\mathbf{R}$ dynamical matrix, 327
$\mathbf{R}^{\alpha}$ atom positions within a cell, 205
$s,p,d...$ spectroscopic notation for $l = 0,1,2,\cdots$; 23, 398
S entropy, 287
S quantum number of spin, 93
$[S,A]$ commutator of two elements or operators, 9
$S,P,D...$ spectroscopic notation for many electron atoms, where total $l = 0,1,2,\cdots$; 79
S_n or $\tilde{n}$ roto-reflection operation, 52
S_x,S_y,S_z components of the spin vector or basis functions, 418
s position vector, 231
S rotational part of space group operation, 258
S spin angular momentum vector operator, 90
$\mathbf{S}$ symmetric matrix, 342
S_{2n} rotary reflection groups obtained from $\mathcal{C}_n$ by adding the inversion, 58
t time, 220
t as subscript, a mode label, 334
t_i integer multiples of lattice basis vectors, 171
t^{ij} Cauchy stress tensor, 128
T the Bravais translation group of the lattice, 233
T kinetic energy, 149
T tetrahedral group, 50
T_d the improper point group possesssing the full symmetry of the tetrahedron, 59
T_h improper point group obtained from the tetrahedral group T by adding the inversion, 60
$T_{\mathbf{k}}$ invariant subgroup of the translation group T, 241
T^{LM} Piola stress tensor, 128
T^{iM} Kirchhoff stress tensor, 128
$\mathbf{t}$ lattice translation, 171
$\mathbf{T}$ arbitrary tensor of second-rank, 116
u time variable, 290
u displacement from equilibrium, 4
$u(\mathbf{r})$ periodic part of Bloch function, 183
$u^{\mu}{}_{,\nu}$ displacement gradient, 124
u, v basis functions, 431

u_μ^α	internal shift, 350
U	unitary operator, 278
$U(2)$	group of unitary and unimodular two dimensional matrices, 45
$U(\mathbf{r})$	product of periodic factors, 188
$U(\mathbf{R})$	potential, 226
$\mathbf{u}$	displacement field, 124
$\mathbf{u}$	axis of rotation, 47
$V(\mathbf{r})$	potential energy function, 90
V_c	Coulomb part of the potential, 195
V_{ex}	exchange part of the potential, 195
$V_{\mu\nu\lambda}$	matrix element, 88
$\mathbf{v}$	group velocity, 220
$\mathbf{v}$	velocity operator, 212
$\mathbf{v}(\alpha)$	some proper fraction of a lattice translation, 112
$\mathbf{v}^{mn}$	velocity matrix elements, 212
$\mathbf{V}$	a vector which transforms as the antisymmetric part of a second-rank tensor, 116
V	unitary transformation, 343
$w(x,x')$	electron interaction, 409
W	potential energy, 115
W	distribution function, 287
x	space and spin coordinates, 409
x,y,z	the three Cartesian components of position; if used as subscripts, they may represent components of a vector or tensor or they may denote particular space group operations in the Herring notation, 3; 257
X	a special Brillouin zone point, 248
X,Y	π orbital pair, 401
X_j	thermodynamic forces, 287
$\mathbf{x}$	particle position in spatial reference frame, 124
$\mathbf{X}$	particle name, position of the particle in the material reference frame, 124
$y_{ij}(t)$	Fourier transform of admittance, 291
$Y_{ij}(\omega)$	frequency dependent admittance, 291
Y_m^l	spherical harmonic, 3
Z	partition function, 289
α	angle of rotation of a physical system, 35
α	complex number, 284
α	Dirac matrices, 211
α,β	spin functions, 315
α,β	as superscripts specify the site within a cell, 258
α,β	Hamiltonian matrix elements, 392
α,β,γ	cosine angles of basis vectors, 171
α,β,γ	multiplicative factors in multiplier representations, 246
α,β,θ	force constants, 264; 265
α_i	parameters of a physical system, 287
$\dot{\alpha}_i$	fluxes, 287
$(\alpha\|\mathbf{a})$	space group operation; α is a 3×3, rotation matrix and $\mathbf{a}$ is a translation, 34
β	inverse temperature, i.e. 1/kT, 289
β	rotational part of space group element, 232
γ	phase, 282
Γ	point $\mathbf{k} = 0$ in the Brillouin zone, 363, 443
Γ_j	name of representation, 416 ff
$\Gamma_j^{\pm}$	irreducible even and odd

parity representations of the full rotation-reflection group, 56
γ rotation matrix, 250
δ_n or n counterclockwise rotation by angle $2\pi/n$ about axis $\mathbf{n}$, 257
$\delta_{ij}, \delta(i,j)$ Kronecker symbol, 1 if $i = j$, 0 otherwise, 20
Δ representation generated by dynamical matrix, 344
Δ line of symmetry in the Brillouin zone, 366
Δ infinitesimal difference or change, 26
Δ_i irreducible representations of a subgroup, 74
$\Delta(\mathbf{k}-\mathbf{k}')$ Kronecker delta function in discrete case, 186
Dirac delta function in continuous case, 188
Δr^{ij} bond stretches, 152
$\Delta^{\pm}$ representations generated by the symmetric and antisymmetric parts of the force constant matrix, 161
Δ line of special symmetry in Brillouin zone, 363; 443 ff
ϵ a root of unity, 151, 425, 426, 428
ϵ_i time inversion factor, 288
$\epsilon_{\mu\nu}$ static dielectric tensor, 115
ϵ identity element, 165
θ angle of incidence or angle of reflection, 191
θ azimuthal angle, 3
θ angle of rotation, 47
θ commuting operator, 281
θ_m phase of quantity λ, 217
ι inversion, 309
λ wave length, 191
λ perturbation parameter, 73
$\lambda(R,S)$ multiplicative constant defining group multiplication for multiplier groups, 12
λ phase factor, 217
λ_i eigenvalues of α, 232
Λ line of special symmetry in the Brillouin zone, 363; 382; 443 ff
μ multiplier factor, 249
μ,ν,λ,τ force constants, 264, 265
$\mu,\nu...$ as subscripts, they designate the partner or member of a basis function set, or the components of a matrix representation, 14; 18
$\nu(\alpha)$ one dimensional relevant representation, 245
$\mu^{\alpha i}$ local electric moment, 264
ν wave vector, 272
ξ,η complex variables that transform as fundamental spinors, 419
ξ_{nm} vector quantities introduced in $\mathbf{k}\cdot\mathbf{p}$ method, 213
ρ density matrix, 289
ρ function space, 14
ρ mass density, 127
$\rho(\mathbf{r},\mathbf{r}')$ transition charge density, 123
$\rho(x;x')$ split charge density, 409; 411
$\rho(a)$ density of points in Lie group parameter space, 28
$\rho_{\mathbf{n}}$ reflection through plane perpendicular to $\mathbf{n}$, 257
σ time-symmetric space, 290
$\bar{\sigma}$ space which spans the time reversed states, 291
σ_d diagonal mirror plane, 59
σ_h horizontal mirror plane, 58
$\sigma_x,\sigma_y,\sigma_z$ Pauli spin matrices, 44
σ_0 the one arbitrary Hall coefficient for an isotropic

medium, 118

$\sigma_{\mu\nu\lambda}$ linear Hall effect tensor, 118

$\sigma,\pi,\delta\ldots$ one electron bond orbitals used in molecular physics, 398

Σ line of special symmetry in the Brillouin zone, 363; 377; 443 ff

Σ invariant polynomial, 121

Σ potential energy per unit undeformed volume, 126

$\Sigma,\Pi,\Delta\ldots$ molecular configurations for more than one electron orbitals, analogous to S, P, D... 398

$\sum$ summation symbol, 7

σ group operation denoting reflection through a plane, 8

σ Pauli spin operator, 44

σ conductivity tensor, 114

σ_h horizontal reflection plane, 52

σ_v vertical reflection plane, 52

τ very small time, 287

τ arguments (e.g., position and spin coordinates) of a function, 15

τ nearest neighbor translation vector in diamond structure, 204

ϕ electrostatic potential, 224

ϕ linear combination of basis functions, 14

ϕ polar spherical angle, 3

ϕ atomic orbital, 208

ϕ polar angle in cylindrical coordinate system, 17

ϕ angle of rotation, 46

$\langle\phi\rangle$ expectation value of a quantity ϕ, 35

ϕ_i functions defining group multiplication in a Lie group, 25

ϕ_n arbitrary phase, 215

ϕ^j_α basis functions for irreducible representations of the group associated with perturbed Hamiltonian, 75

ϕ^j character of the representation Δ^j, 73

(ϕ,ψ) scalar product of two functions, 15

Φ wave function, 279

Φ scalar potential, 280

Φ^i_ν basis function or base %vector, 84

χ character of a possibly reducible representation, 20

$\chi^{i\times j}$ character of outer or inner product representation, 77

$\chi^{j\pm}$ character of even or odd parity irreducible representation j, 23

ψ basis function, 14

ψ angle of rotation, 47

$\psi(\mathbf{r})$ spinor or two component wave function, 90

$\psi(\mathbf{r}_1,\mathbf{r}_2,\ldots,\mathbf{r}_n)$ wave function for many spinless particles, 38

$\psi_{ab'cd}$ structure function, 405

ψ^l_m wave function for spherically symmetric potential, 3

$\psi_n(\mathbf{k},\mathbf{r})$ the set of Bloch waves in a given band, 194

$\psi(\phi)$ wave function of a physical system described by variable ϕ, 35

ψ^i_μ eigenfunctions of H_0 belonging to representation i, 75

ψ^i eigenfunctions of a Hamiltonian, 18

Ψ wave function, 279

Ψ^j_m eigenfunction of J^2 operator, 39

$\Psi^{i\times j}_{\mu\nu}$ basis for irreducible

representation of outer or inner product group, 77

$\Psi^{\pm}_{\mu\nu}$ symmetric and antisymmetric parts of product basis function, 81

ω frequency, 291

ω infinitesimal rotation angle, 155

ω one of the Nth roots of unity, 186

ω^2_{ti} eigenvalue of potential %energy matrix, 149

Ω total volume of crystal, 187

Ω_0 volume of primitive cell, 184

Ω_c Dirac character of the class C, 10

$\boldsymbol{\Omega}$ curl $\xi_{mm'}$, 216

Ω unitary matrix, 341

∇ del or nabla, 3

∇^2 Laplacean, 3

$\times$ direct product of two groups, 56

$\times$ vector cross product,

$\cdot$ vector scalar product,

$\dagger$ hermitian adjoint,

$\in$ exists in, 7

$\sim$ transforms as, equivalence, 17

$\oplus$ direct sum 14

$|0\rangle$ vacuum state, 334

$\langle\ \rangle$ ensemble average, 287

Tr trace operation, 289

$\doteq$ equal to or equivalent to, 241

$\triangle$ three-fold rotation axis, 8

Additional Problem, Chapter 7 (see page 230):

7.8.9. Show that Eq. (7.8.21) is equivalent to the requirement that

$$\frac{\partial}{\partial k_i}\frac{\partial}{\partial k_j} = \frac{\partial}{\partial k_j}\frac{\partial}{\partial k_i}$$

or that the curl of a gradient must vanish.

[*added 2001*]